IDENTIFICATION AND ANALYSIS OF PLASTICS

IDENTIFICATION AND ANALYSIS OF PLASTICS

J. HASLAM, D.SC., F.R.I.C.
Formerly Chief Analyst
Imperial Chemical Industries Ltd
Plastics Division

H. A. WILLIS, B.SC., I.C.I. Research Associate
Leader of the Infra-red and Nuclear Magnetic
Resonance Section, Physics Division, Research Department,
Imperial Chemical Industries Ltd, Plastics Division

D. C. M. SQUIRRELL, B.SC., F.R.I.C.
Leader of the Research Section, Analytical Division,
Research Department, Imperial Chemical Industries Ltd
Plastics Division

Distributed in the United States by
CRANE, RUSSAK & COMPANY, INC.
347 Madison Avenue
New York, New York 10017

THE BUTTERWORTH GROUP

ENGLAND
Butterworth & Co (Publishers) Ltd
London: 88 Kingsway WC2B 6AB

AUSTRALIA
Butterworth & Co (Australia) Ltd
Sydney: 586 Pacific Highway, Chatswood, NSW 2067
Melbourne: 343 Little Collins Street, 3000
Brisbane: 240 Queen Street, 4000

CANADA
Butterworth & Co (Canada) Ltd
Toronto: 14 Curity Avenue, 374

NEW ZEALAND
Butterworth & Co (New Zealand) Ltd
Wellington: 26–28 Waring Taylor Street, 1

SOUTH AFRICA
Butterworth & Co (South Africa) (Pty) Ltd
Durban: 152–154 Gale Street

First published in 1965 by
Iliffe Books, an imprint of the Butterworth Group

Second edition 1972
© J. Haslam, H. A. Willis and D. C. M. Squirrell, 1972

ISBN 0592 05442 X

Filmset by Photoprint Plates Ltd, Rayleigh, Essex

Printed in England by
Hazell, Watson & Viney Ltd, Aylesbury, Bucks

IDENTIFICATION AND ANALYSIS OF PLASTICS

Preface to the Second Edition

In the first edition of this book great importance was attached to the interdependence of chemical and spectroscopic methods of examination of plastics, and particular attention was paid to infra-red methods. Combined chemical and infra-red examination of plastics frequently proved successful when the individual approach by chemical or physical tests alone was quite unrewarding.

In the intervening years, however, great changes have taken place in the types of method that are available to the analyst in the plastics industry. Our second edition takes account of these changes and this has involved a re-casting of the whole book.

The introductory chapter in the new edition deals, in a general way, with instrumental methods and their use in the examination of plastics. The chapter seeks to give the basic principles of each method and to illustrate these principles by appropriate examples of their application in the analysis of plastics. A thorough understanding of this preliminary chapter should enable the analyst in the industry to make satisfactory use of the methods in the later parts of the book.

The next chapter on qualitative analysis follows closely along the lines of the chapter on qualitative analysis in the first edition, viz. it is concerned with the qualitative detection of elements, burning, heating and solubility tests, specific tests for particular polymers, and the interpretation of the infra-red spectra of polymers and plasticisers.

This contribution is followed by a series of chapters dealing with individual polymers or polymer types, such as vinyl resins, polyesters and ester polymers, nylon and related polymers, polyolefines, fluorocarbon polymers, rubbers, thermosetting resins, and natural resins. Our observations on plasticisers, fillers and solvents are brought together in a final chapter. In this series of chapters, whenever the opportunity arises attention is drawn to methods that are particularly appropriate in certain cases. Where appropriate, S.I. equivalents of quantities expressed in other units are given throughout this volume. *The conversions are practical rather than mathematical equivalents.*

In preparing this second edition we have not sought to bring together the work of a number of other authors in a fruitless attempt to provide an unequivocal means of identifying all the plastics and plastics materials with which the analyst may be confronted. On the contrary, we have tackled the problem from the point of view of three investigators who, over a period of approximately 20 years, have worked together in close collaboration, seeking to weld together both chemical and physical methods of examination of plastics. Furthermore, it must be emphasised that the bias in this book is towards the examination of the products and the solution of the problems of the company with which we have been concerned. It is believed, however, that the practical methods of test that we have worked out will help other investigators in rather different fields of plastics analysis in solving their own problems. Most of the tests we have described are given with full working details and should require little adaptation by other workers.

We should like to acknowledge our indebtedness to the Directors of Imperial Chemical Industries Ltd, Plastics Division for permission to publish this work and to make use of the information gained as a result of our service with the company. Furthermore, we wish to thank all those members of the Plastics Division who have helped and inspired us over the years. The problems they have submitted and assisted in solving have contributed in no small measure to this effort on our part.

In addition, we are grateful to the authors of other standard textbooks in the field, notably the following: D. O. Hummel, *Kunststoff-, Lack- und Gummi-Analyse*, Carl Hanser, Munich (1958); D. O. Hummel, *Infra-red Analysis of Polymers, Resins and Additives, An Atlas*, Volume 1, Part II, Wiley Interscience, New York and London (1969); G. M. Kline, *The Analytical Chemistry of Polymers*, Interscience, New York and London, Part 1 (1959), Part 2 (1962) and Part 3 (1962); and W. C. Wake, *Analysis of Rubbers and Rubber-like Polymers*, 2nd edn, Maclaren, London (1969).
1971

J. H.
H.A.W.
D.C.M.S.

CONTENTS

1

INSTRUMENTAL METHODS

In this chapter we have sought to describe the important principles of each of the instrumental methods which we have found to be most useful in the analysis of plastics materials. We have pointed out, where appropriate, the preferred methods of sample preparation, standardisation and calibration, and have given specific examples to demonstrate the applications of the methods in plastics analysis.

The assistance given by Mr. L. H. Ruddle in the preparation of the first three sections of this chapter is gratefully acknowledged.

VISIBLE AND ULTRA-VIOLET ABSORPTION SPECTROPHOTOMETRY

Very useful general books on the theory, instrumentation and practice of visible and ultra-violet spectrophotometry are available[1-4] and a recent volume *'Ultra-Violet Spectra of Elastomers and Rubber Chemicals'*[5] translated from the Russian is of value, being an atlas of a large number of additives commonly encountered in the plastics and rubber industry.

The use of visible spectrophotometry in the analysis of plastics has in our experience been decreasing over the last few years. This is due to the fact that a large number of the routine analyses (particularly for trace metals in plastics materials) which used to rely on colorimetric methods, are now based on the use of modern instrumental methods such as atomic absorption, flame photometry and X-ray Fluorescence. This does not mean, however, that a spectrophotometer covering the visible range 350 nm to 800 nm is no longer an essential piece of equipment in the plastics analytical laboratory. It is required to deal with non-routine analysis for which the other instrumental methods may not be calibrated or applicable; for the application of referee or 'standard' methods which are normally based on relatively simple methods which can be applied in most laboratories with relatively inexpensive equipment, and for developments involving colorimetric measurements of any kind.

The instruments most generally useful are dispersive instruments which permit measurements to be made at specific wavelengths, although filter instruments can also be used successfully for many applications. For ultra-violet spectrophotometric methods a double-beam recording instrument covering the range 200 nm to 400 nm is to be preferred to a manual instrument, particularly for the methods described for the determination of antioxidants and U.V. absorbers (see Chapter 6). The complete spectrum over the wavelength range of interest can be obtained quickly, overlapping absorption bands are revealed and complex spectra more easily interpreted. Since complete bands are recorded rather than absorbances at fixed wavelengths, base-line methods of calibration can be conveniently used.

Two main properties are considered in the choice of a solvent system when spectrophotometric methods are applied to the determination of additives in plastics materials. Firstly the extraction efficiency of the solvent for the particular additive to be determined must be such as to give complete and rapid extraction of the additive and little or no solution of the polymer. Secondly the solvent should have a high U.V. transmission over the wavelength range of interest and particularly between 250 nm and 400 nm. The most useful single extraction solvent used is chloroform, otherwise ethanol, cyclohexane and methyl cyclohexane are used. For solution-precipitation methods of extraction a toluene-ethanol combination has been found most effective. When mixed solvents are used it is important that the blank solvent system placed in the comparison cell should be accurately prepared in the same ratio as is used for the sample. In the case of the toluene-ethanol system which has a relatively high U.V. absorption it is sometimes necessary to evaporate the solvent and redissolve the residue in a spectroscopically suitable solvent for examination.

A base-line method of calibration is normally used, the base line being drawn as shown in Figure 1.1(a) tangentially between the minima occurring at each side of the analytical absorption peak. When one of these minima occurs at a wavelength at which the instrument lacks sensitivity, for example because of strong absorption by the solvent, an alternative reference point (e.g. 255 nm for chloroform) may be used (Figure 1.1(b)).

Results are normally referred to calibration curves relating absorbance in a specified path length cell at a particular wavelength to mg additive per 100 ml extraction solvent. Alternatively $E_{10\,mm}^{1\%}$ values may be calculated expressing the absorbance at a particular wavelength of a 1% solution of the additive contained in a 10 mm cell.

The value of U.V. absorption measurements to the analyst in the plastics industry is seen at all stages of manufacture from raw material to finished product. Impurities are detected in monomers, solvents and diluents by examination of their U.V. spectra, and transmission curves are frequently required on clear sheet made to particular specifications. By far the most

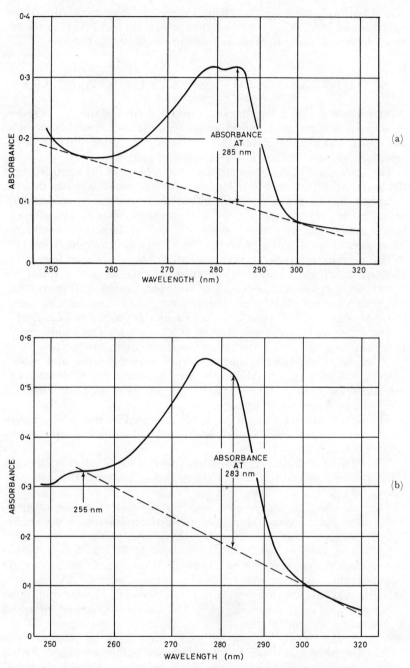

Fig. 1.1.   *Measurement of spectrophotometric absorbance values*

value, however, arises in the identification and determination of anti-oxidants and U.V. absorbers in acrylic, polyolefine and P.V.C. products, and reference is made to these methods in the following chapters.

ATOMIC ABSORPTION SPECTROPHOTOMETRY

Atomic vapours absorb light passing through them and the fact that they do so selectively has been known by astronomers for a considerable time, but it was not until 1955 that Walsh[6] indicated the possibilities of developing this phenomenon into an analytical method.

The basic principle of the method is that a beam of light emitted by a discharge lamp containing the element to be determined is passed through a flame containing atoms of this element. Some light is absorbed by atoms in the ground state and the decrease in intensity is related to the concentration of atoms in the flame. In normal flame photometry the measurement is made of light emitted by excited atoms in the flame.

The primary light source must emit the sharp resonance line of the element under examination with sufficient intensity and stability for accurate absorption measurements to be made. Commercial instruments use hollow-cathode discharge lamps, usually one element per lamp, although some multipurpose lamps are made e.g. Ca/Mg and Cu/Zn/Pb.

The flame is 50 to 100 mm long to give an adequate absorbing path: the gases used vary with the temperature required, air-propane being used for low temperatures, air-acetylene for normal work and nitrous oxide-acetylene in a special burner for high-temperature flames. The solution being examined is introduced into the flame as a very fine mist produced by a concentric ring atomiser.

The light of the wavelength required is selected by means of a mono-chromator, detected with a photomultiplier and measured after appropriate electronic amplification.

The method is calibrated by measurement of the absorption obtained when a series of solutions of known concentration are atomised sequentially into the flame.

The procedure has been used both for the determination of metal additives in plastics materials and for the determination of much smaller concentrations present as contaminants or catalyst residues.

Cadmium, calcium, lead and zinc compounds are used as stabilisers in polyvinyl chloride and the metal content is determined in the solution obtained after wet digestion to remove organic material. Aluminium, titanium and iron are determined in polyolefines after ashing and fusion of the ash with potassium bisulphate. The fusion products are dissolved in dilute acid for examination.

Atomic absorption spectrophotometry may also be applied with advantage to the quantitative analysis of non-routine samples of ferrous and non-ferrous metals. An example would be tests to establish the origin

4

of metallic particles found as impurity in a plastic moulding compound. A multi-component quantitative analysis is required in such cases, and this would be a lengthy task either by chemical methods or other spectroscopic procedures.

The procedure detailed below for the determination of aluminium in polyolefines may, by appropriate variation of the lamp, lamp current, wavelength, fuel and flame conditions, be used for the determination of other metals by atomic absorption spectrophotometry. In order to achieve a general method, either the sample is ashed and the ash fused with potassium bisulphate and dissolved in sulphuric acid, or the sample is wet oxidised with sulphuric acid, the acid diluted to about N and potassium bisulphate added. Thus one set of standard solutions will suffice for both methods of sample preparation.

A Unicam SP90 spectrophotometer fitted with a SP91 multiple turret attachment and SP94 nitrous oxide burner has been used in the development of methods for the examination of plastics materials. There is, of course, no reason why other equipment of similar sensitivity should not be equally suitable. It was also shown that better reproducibility was obtained by automatically changing the sample input to the instrument, and the SP92 automatic sample changer is thus incorporated into the equipment for most routine applications. A 10 mV recorder is used to give a permanent record of the output from the instrument.

THE DETERMINATION OF ALUMINIUM IN POLYOLEFINES

Reagents

Potassium bisulphate.
Fuse potassium bisulphate (Analar) in a large platinum dish until the melt is quiescent. Allow the melt to cool and finally grind the solid to a powder in a porcelain mortar. Store in a well-stoppered bottle to prevent absorption of atmospheric moisture.
Sulphuric acid 1 N.
Dilute 30 ml of concentrated sulphuric acid (Analar) by pouring into water. Cool and dilute the solution to 1 l.
20% Potassium bisulphate in N sulphuric acid.
Dissolve 200 g of potassium bisulphate (Analar) in water, add 30 ml concentrated sulphuric acid (Analar), cool, and dilute to 1 l with water.
Stock aluminium solution.
Dissolve 1·000 g of high purity aluminium foil in the minimum quantity of hydrochloric acid and dilute the solution to 1 l with water. 1 ml = 1 mg Al.
Dilute aluminium solution.
Dilute 10 ml of stock aluminium solution to 100 ml with water. 1 ml = 100 μg Al.

Calibration solutions.

Add 0, 1, 3, 7 and 10 ml of dilute aluminium solution and 3, 7 and 10 ml respectively of stock aluminium solution to a series of 100 ml volumetric flasks, add 20 ml of potassium bisulphate-sulphuric acid solution to each flask and dilute each solution to volume with water. Mix well and transfer to polythene bottles in which the solutions may be stored for at least 3 months. These solutions contain 0, 1, 3, 7, 10, 30, 70 and 100 μg Al per ml.

Procedure

Accurately weigh 10 g of sample into a clean platinum dish. Place the dish over a bunsen flame adjusted to give even volatilisation of the polymer without causing it to inflame. When all the organic matter has been volatilised place the crucible in a muffle furnace at 800°C for 15 min. Remove the dish from the muffle, add 2 g of the fused and ground potassium bisulphate and heat on a bunsen flame to fuse the bisulphate. Continue heating (with occasional swirling making sure that the melt comes into contact with the whole of the inside of the platinum dish) until all the residue has dissolved in the melt. Cover the dish with a clock glass and allow to cool. When cold add 10 ml of N sulphuric acid and warm the dish until a clear solution is obtained. Transfer this solution to a 50 ml volumetric flask, dilute the solution to volume with water and mix well.

Prepare a blank by fusing 2 g of potassium bisulphate and dissolving the cooled melt in 10 ml N sulphuric acid exactly as for the samples. Set up the instrument conditions as in Table 1.1.

Table 1.1. Instrument conditions for spectrophotometric determination of aluminium in polyolefines

Aluminium hollow-cathode lamp: current	10 mA
Monochromator wavelength	309·3 nm
Slit-width	0·1 mm
Burner	50 mm, nitrous oxide
Burner height	10 mm
Fuel: Acetylene pressure	10 lbf/in² (70 kN/m²)
Acetylene flow	4·2 l/min
Nitrous oxide pressure	40 lbf/in² (280 kN/m²)
Nitrous oxide flow	5·0 l/min

Adjust the wavelength control at about 309 nm to give maximum deflection and adjust the zero, coarse and fine gain controls to give 0 and 100% readings on the chart of the recorder when water is aspirated into the burner.

To a series of the polythene cups of the automatic sample changer add 3 to 4 ml of the following solutions: water, the calibration solutions

0, 1, 3, 7, 10, 30, 70 and 100 μg Al/ml, water, blank, sample solutions and finally water again.

Switch on the automatic sample changer and so aspirate all the solutions. After every five solutions examine the burner slit and remove any red-hot coke which has been deposited by scraping with a screw-driver. Also adjust the fine gain control, if necessary, to keep the base line at 100% transmission. Repeat the measurement of all the solutions. Flush the burner well by atomising water for 5 min, then extinguish the flame by following the procedure laid down by the manufacturers.

Calculation

Convert the % transmission readings from the chart to absorbances by reference to a table, and correct the absorbances of the standards and samples by subtraction of the absorbance of the appropriate blank.

Plot the corrected absorbances of the calibration solutions against μg Al/ml and draw a smooth curve through the points.

Obtain the aluminium content of the sample solution by reference to this calibration graph.

If $\varUpsilon$ is concentration of aluminium in μg/ml

$$\text{then aluminium content of the sample} = \frac{\varUpsilon \times 50}{10} \text{ p.p.m.}$$
$$= 5\varUpsilon \text{ p.p.m.}$$

EMISSION SPECTROGRAPHY

Emission spectrography has been a valuable method of elemental analysis for many years and there are a number of useful reference text-books on spectrochemical analysis and flame photometry[7-9]

In our laboratories the characteristic line emission of atoms excited in a d.c. arc is used only to identify the metallic elements present in a polymer composition or its ash. A qualitative spectrum can be obtained and interpreted on a single sample in less than 45 min (12 samples can be completed in 3 h), and the method is thus used when the interest is in all the metals that can be detected in an unknown sample. The short analysis time for a 'complete' spectrum contrasts with the comparatively long scan time (1 h 45 min) for a corresponding X-ray Fluorescence spectrum, although the latter method is quicker if there are only one or two elements of interest, since only the relevant part of the spectrum need be run, taking perhaps 5 min for each element.

It should be noted that elements such as phosphorus, lead and cadmium when present in polyvinyl chloride, and silicon when present as a silicone often tend to be lost in an ashing procedure. It is frequently our practice, therefore, to examine the plastics material directly or after only a partial low-temperature ashing procedure. Cyanogen bands do not in our experience seriously interfere in the interpretations of the spectra because, although some lines may be obscured, in most cases others are available for identification.

In our general procedure the sample, supported in suitable electrodes, is decomposed in a 220 V d.c. arc at a current between 2 and 8 A. The light emitted from the arc is focussed on to the slit of a large quartz spectrograph and the resulting spectrum recorded on a photographic plate. A spectrum of iron, produced by arcing pure iron rod electrodes, is photographed beside the spectra of any series of unknown samples. The spectral lines are identified by projection at a × 20 magnification on to comparison charts on which the lines of the elements are mapped against the spectrum of iron[10]. A large quartz spectrograph is required to give sufficient dispersion so that mixtures of elements which give very complex spectra may be more easily resolved.

Most of the spectral lines used lie within the wavelength range 2500 to 3500 Å but for certain elements, for example zinc, potassium and sodium, it is necessary to use higher wavelengths. Three ranges are normally used: 2470 to 2260 Å, 3250 to 2400 Å, and 4680 to 3180 Å. The detection and identification of an element is based on the presence of 'reliable' analytical lines and not necessarily the strongest lines for the element. A line or

Table 1.2. Principal lines used for identification of elements

			All figures are in angstroms		
Calcium	3158·8	3179·3	3181·3		
Zinc	3282·3	3302·6	3345·0	3345·5	
Vanadium	3185·4	3184·0	3183·4		
Molybdenum	3158·2	3132·6			
Tin	3034·1	2839·9			
Antimony	2877·9	2598·1			
Chromium	2849·7	2843·3	2835·6		
Magnesium	2783·0	2781·4	2779·8	2778·3	2776·7 (Magnesium group)
Lead	2802·0	2614·2	2613·7		
Aluminium	2660·0	2652·5	2675·1	2568·0	
Titanium	2646·6	2644·3	2641·1		
Phosphorus	2554·9	2553·3	2535·6	2534·0	
Silicon	2528·5	2519·2	2516·1	2514·3	2506·9 (Silicon group)
Boron	2497·7	2499·8	(Boron pair)		
Potassium	4047·2	4044·2	(Potassium pair)		
Nickel	3472·6	3483·8	3493·0		
Manganese	4034·5	4033·1	4030·8		
Barium	2335·3				
Cadmium	2288·0	2265·0			

group of lines free from interference from other elements present must be sought particularly when elements such as titanium, chromium or iron, which have many spectral lines which 'confuse' the total spectrum, are present. A list of the lines used for the detection and identification of the elements of most interest in the analysis of plastics materials is given in Table 1.2.

It may be noted that the four lines quoted for aluminium do not include the two strongest aluminium lines occurring at 3092·7 Å and 3082·1 Å. This is because these two lines often occur in the electrode spectrum, and their presence alone does not therefore confirm aluminium in the sample. The four lines quoted in the table above must also be present for a positive identification to be made.

The limit of detection by the above spectographic method varies with the element sought. In the case of phosphorus amounts below 30 p.p.m. can be easily missed, but boron on the other hand can be reliably detected at the 0·5 p.p.m. level.

The use made of Flame Photometry in the analysis of plastics materials is relatively limited and conventional procedures are used. Applications include the determination of small amounts of sodium, potassium, or lithium in polymeric materials. The solution for examination in the case of finely powdered samples can often be obtained by simple acid extraction. Other samples require wet oxidation or ashing treatments.

X-RAY FLUORESCENCE SPECTROMETRY

The measurement of X-ray Fluorescence is by no means new but developments in instrumentation over the past 15 years have produced equipment which enables the analyst to carry out the rapid determination of elements of atomic numbers greater than 10 in both solid and liquid samples, often with little or no sample preparation and without destruction of the sample. In the analysis of plastics materials X-ray Fluorescence methods are now replacing many of the conventional chemical and wet methods of qualitative and quantitative elemental analysis. These methods are therefore frequently referred to in the appropriate chapters throughout the book and include e.g. the detection and determination of catalyst residues (titanium, chlorine) and contamination (iron) in polyolefines; the characteristic elements of stabilisers (lead, calcium, phosphorus, tin, sulphur, barium, cadmium, zinc) in polyvinyl chloride compositions; the characteristic elements of fire retardants (chlorine, phosphorus) in acrylic compounds; and fillers and pigments (titanium, silicon, calcium) in all kinds of polymeric formulations.

The theory and instrumentation of X-ray spectrometry are well known and the subject is adequately described in several excellent books[11-13]. Only brief details of the principles of the method will therefore be given here.

A line diagram showing the essential arrangement of the plane-crystal X-ray spectrograph is shown in Figure 1.2.

A beam of high-energy X-rays is generated from a suitable X-ray tube having a tungsten, chromium, gold, silver or molybdenum target. This primary beam of 'white' radiation impinges on the sample contained in the specimen holder, and excites the secondary or fluorescent X-rays

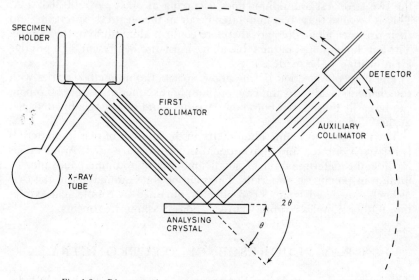

Fig. 1.2. *Diagrammatic arrangement of the plane-crystal X-ray spectrograph*

characteristic of the elements present in the sample. This secondary radiation is always of longer wavelength than the primary radiation. The wavelengths at which the characteristic lines occur for each element are dependant on the atomic number of the element, and hence qualitatively the accurate measurement of the wavelengths enables the elements present in the sample to be defined. This is carried out by diffracting the collimated beam of polychromatic fluorescent radiation at an analysing crystal, for each setting of which radiation of wavelength satisfying the Bragg relationship

$$n\lambda = 2d\sin\theta$$

is strongly reflected through the auxiliary collimator to the detector.

Here n is the order of reflection and is a small integer

λ is the wavelength in Å

d is the interplanar spacing of the analysing crystal in Å

θ is the angle between the incident radiation and the crystal surface.

10

The reflected radiation emerges at an angle of 2θ with respect to the incident beam and the 2θ value for a given value of n at which maximum intensity is detected enables the wavelength of that particular characteristic radiation, and hence the element of origin to be determined. To record a spectrum of an unknown sample the crystal is rotated at a constant speed through a desired range of θ values whilst the detector is rotated at twice the angular velocity of the crystal. The pulses received by the detector are continuously fed, after suitable amplification and electronic processing, through a rate-meter to the recorder. The 2θ angles at which strong reflections occur can thus be read directly from the recorder chart and by means of tables, appropriate to the crystal in use, the wavelength and characteristic radiation identified.

For the most accurate quantitative analysis the goniometer is set at a particular 2θ angle corresponding to the element sought and the intensity measured over an extended period of time (10 to 200 s). Provided the sample is above a critical thickness, the intensity of the characteristic fluorescent radiation from an element is proportional to the concentration in the sample. The intensity, however, not only depends on concentration but also on the mass absorption coefficient of the total matrix of the sample. The particle size of powder samples and specific inter-element effects caused by the excitation of one element in the matrix by the fluorescent radiation from another also affect the measured fluorescent intensity. When this last effect occurs the radiation from the heavier element is depressed and that of the lighter element enhanced.

It is clear, therefore, that some knowledge of the composition of the sample is necessary before a quantitative analysis can be carried out, and calibration data from one matrix cannot be indiscriminately applied to another. Inter-element effects are greatly reduced by dilution of the sample and in this respect solid solution or liquid solution methods are very useful since particle size effects are also avoided and the preparation of calibration curves is much simplified.

The accuracy of X-ray Fluorescence analysis depends on the analytical element, the matrix in which it is present, the peak/background intensity ratio, and statistically on the number of counts which can be accumulated within a reasonable time (10 to 200 s). Important, too, is the availability of standards closely matching the samples to be analysed, and when these are available in many cases it is possible to achieve a coefficient of variation better than 1% and a limit of detection in the region of 1 to 5 p.p.m.

INSTRUMENTATION AND TERMINOLOGY

It is appropriate here to define more fully some of the terminology and instrumental conditions used in describing an X-ray Fluorescence method, and their particular importance in plastics analysis.

11

X-ray tube-anodes

A tungsten anode tube is a good general-purpose tube suitable for heavy-element determinations and for light-element work at high concentrations. Since much of the work in the analysis of plastics materials concerns light elements in small concentrations however, a chromium anode giving increased sensitivity is in very common use. A molybdenum target tube is used for special determinations when tungsten or chromium lines would interfere or for elements like bromine and mercury for which a molybdenum target gives greater sensitivity. Gold and silver anode tubes are also available.

X-ray tube voltage and current

The voltage supplied to the X-ray tube must be highly stabilised and be above the minimum excitation potential required for the analytical spectral line of the element under consideration.

The current flowing in the X-ray tube must also be highly stabilised and is adjusted in conjunction with the voltage, within the specified rating of the tube, to give optimum counting rates. For much of our trace-element and light-element work we tend to work near the maximum power rating of the tube.

X-ray path

This defines the path traversed by the primary beam to the sample and of the fluorescent beam from the sample to the analysing crystal and detector. The characteristic X-rays of elements of atomic number lower than titanium are strongly absorbed by air, and hence for non-volatile samples a full vacuum is used. For volatile samples such as a solution of a polymer composition in tetrahydrofuran a total helium path is recommended[14, 15]. In our experience this is much to be preferred to the 'partial-vacuum' path in which the major part of the path of the secondary rays is isolated and evacuated, whilst the sample is maintained at atmospheric pressure with air or helium. The disadvantage of the partial-vacuum system is that an additional isolating window is required, which is itself an absorbing medium for long-wavelength radiation and is troublesome to fit if the flow of work necessitates several instrument changes to be made during the day.

Analysing crystals

A variety of analysing crystals is available, but we find that the normal requirements in the analysis of plastics materials can be met with a lithium fluoride crystal ($2d = 4 \cdot 028$ Å) for analysis in the 0·4 to 3·5 Å region and a pentaerythritol (PE) crystal ($2d = 8 \cdot 74$ Å) for analysis in the 3·5 to 8·5 Å range. The pentaerythritol crystal gives much greater sensitivity in the determination of the light elements than the alternative ethylene diamine-d-tartrate (EDDT). An ammonium dihydrogen phosphate (ADP) crystal ($2d = 10 \cdot 65$ Å) is required if magnesium determinations are to be made. Rubidium hydrogen phthalate ($2d = 26 \cdot 12$ Å) is suitable for both magnesium and sodium.

Pulse height analyser

This is used to eliminate X-rays of wavelengths different from those being measured which arrive at the detector at the same 2θ angle, usually from a higher-order line, or hard scattered radiation in the light-element region.

Detector type

A scintillation counter employing a scintillator crystal of thallium activated sodium iodide with a photomultiplier is used to detect X-rays in the 0·1 to 2 Å region, and a flow proportional counter with a very thin window is used in the 2 to 12 Å region. The flow counter is assembled inside the evacuable crystal chamber.

Lines used

Normally the most intense line available is used, but other lines are used when positional interferences occur or to obtain better resolution.

Background angles

The background intensity at the peak angle is normally calculated from the mean intensity at $2\theta \pm 1$ or $2°$.

Calibration and sample counting strategy

Intensity measurements at a fixed angle may be made either by accumulating counts over a fixed period of time or by measuring the time in which a fixed number of counts is registered. Preference for one system or the other is influenced by the type of instrument available. In our case with an instrument accepting four samples at a time we prefer to make short fixed-time counts first on an instrument reference standard

13

and then on three samples (or calibration standards) in order. The calibration graph is then normally plotted as % w/w of element against the ratio of the net intensity from the calibration standard to that from the instrument reference standard. The reference disk containing the elements to be determined in any problem is measured with each series of three samples and, since it is used only to compensate for instrumental variations, is often of entirely different composition from the sample.

SAMPLE AND STANDARD PREPARATION

Where possible calibration standards and samples are given identical preparative treatment. The closer the standards and samples are in both chemical and physical composition the more accurate is the determination likely to be. The procedures can be classified as direct and indirect methods.

Direct methods

Plastics materials may be received for analysis in several forms viz. homogeneous liquids, latices, powders, granular solids, crepes, fibres, films, irregular mouldings and flat sheets. Since for quantitative analysis the sample should be in the form of a liquid, stable latex, powder, or flat disk or pellet of suitable thickness, the following sample preparation equipment is essential.

1. A die to cut out suitable sample disks from flat sheet or film. Brittle thermoplastics can be suitably heat softened before die-cutting.
2. A press giving 20 tonf (200 kN) ram pressure and operating to 300°C, with provision for rapid cooling by air and water feeds to

Table 1.3. Suitable moulding temperatures in °C for some plastics

Polypropylene	180–200
Polythene	140–160
Polyvinyl chloride	180
Plasticised polyvinyl chloride	150
Polymethyl methacrylate	160
Urea-formaldehyde	140
Polyformaldehyde	180
Nylon	250–275
Polyethylene terephthalate	250–275

the platens, to mould thermoplastic samples to suitable disks for X-ray analysis. Suitable moulding temperatures are given in Table 1.3 and a 5 min preheat prior to application of pressure is normally given.

With 10 in (250 mm) platens and a $\frac{1}{8}$ in (3 mm) spacer die detailed in Figure 1.3, 16 samples can be moulded in one cycle with

14

no fear of contamination by flash-over from one sample to another. A polytetrafluoroethylene coating on both spacer and glazing plates facilitates removal of samples and cleaning of the mould. Care must be taken that the additive containing the element sought does not migrate to the mould surface during pressing (see Chapter 3).

3. A small lathe is very useful to prepare the analytical face of moulded

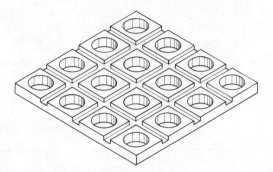

Fig. 1.3. Multiple sample disk mould for X-ray fluorescence spectrometry (made of polytetra-fluoroethylene-coated stainless steel, 10 in (250 mm) square, by $\frac{1}{8}$ in (3 mm) thick, with $1\frac{3}{16}$ in (30 mm) diameter holes and $\frac{3}{8} \times \frac{1}{16}$ in (10 × 1·5 mm) wide milled grooves)

or die-cut disks, the disk being held in the lathe, on a metal rod of the same diameter by double-sided adhesive tape.

4. A powder pellet press is required for cold compression of powder pellets and to densify fibre or film samples prior to hot compression moulding.

5. A medium-capacity mixer mill for grinding or mixing samples with suitable diluent powder.

Example

An example of a qualitative spectrum of a $\frac{1}{8}$ in sample disk from a polyvinyl chloride composition is given in Figure 1.4. The spectrum clearly shows the characteristic lines of tin and sulphur due to the presence of 0·85% of a tin thioglycollate stabiliser, and of chlorine due to the polymer. The quantitative determination of the tin and sulphur is made by a fixed-time count of the $Sn_{K\alpha}$ (second order) and $S_{K\alpha}$ intensities, in each case calculating the ratio of the net intensity to that of an instrument standard and referring the answer to the appropriate calibration graph as detailed on p. 161.

Indirect methods

In some cases it is not possible to make a quantitative determination directly on the polymer preparation. The analytical element may be unevenly dispersed throughout the sample or may require concentration

15

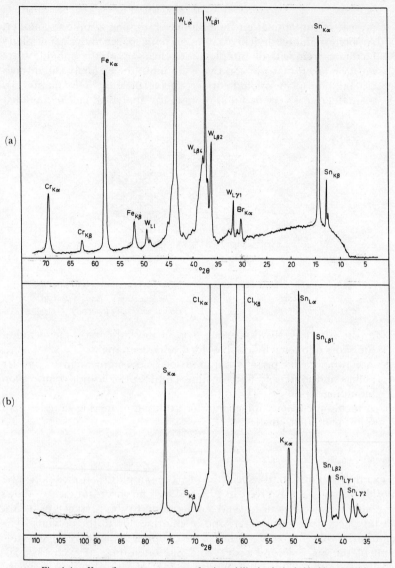

Fig. 1.4. X-ray fluorescence spectrum of a tin-stabilised polyvinyl chloride composition

	(a) 'Heavy' element scan	(b) 'Light' element scan
W tube	50 kV, 32 mA	50 kV, 32 mA
Analysing crystal	Lithium fluoride	Pentaerythritol
Counter	Flow + Scintillation	Flow
Collimator	160 μm	480 μm
Path	Vacuum	Vacuum
Rate-meter	2000 counts/s full scale	1000 counts/s full scale
Scan speed	1° 2θ/min	1° 2θ/min

to enable the required accuracy or limit of detection to be obtained. In many cases the elements of interest may be concentrated without loss by ashing a large weight of sample. The problem then becomes one of processing the ash so that it can be examined as a small flat disk in the sample holder of the spectrometer, preferably without the use of a retaining window which would reduce sensitivity to light elements.

The conventional method of dealing with oxide powders, which we use most successfully for the preparation of permanent instrument standards, is to fuse the oxides with borax and cast the melt in the form of a glass bead. This method is, however, unsuitable for small amounts of ash. This is because manipulative difficulties in the transfer of the melt to the ring mould necessitate the use of a relatively large amount of borax which so dilutes the ash that no significant concentration of the particular element from the sample is effected. A powder dilution method is thus used and for this purpose a light matrix is required which is free from contamination by the elements sought and which will also grind and mix easily with the ash. Other desirable properties of the powder diluent are that it should be non-hygroscopic, insoluble in water or methanol, have a reasonably high melting point and should bind together to form a robust polished surfaced disk when pressed in a powder pellet press. Some of the more conventional powder diluents, for example sugar, cellulose powder, sodium carbonate and stearic acid, do not fully satisfy all these requirements. We are thus indebted to Mr. I. G. Bell of I.C.I. Fibres Ltd for suggesting that we use terephthalic acid as detailed in the following method, in which the aluminium from a 5 to 10 g sample is concentrated in an analytical disk weighing only 0·5 g. We have found this method useful for the determination of aluminium in polyolefine products: the principle is of course applicable to other elements and we have used it in the determination of small amounts of titanium and silicon in polymer preparations.

The sample is ashed in a tared 50 mm diameter by 40 mm deep glazed porcelain crucible in the presence of 1 ml of a 4% w/v magnesium nitrate solution to act as carrier. The total ash is weighed and sufficient terephthalic acid added to the crucible to bring the total weight of the contents to 0·500 g. The terephthalic acid is mixed with the ash by stirring with a small spatula and transferred as completely as possible to a small agate mortar. The crucible is rinsed with a small quantity of methanol sufficient to give a creamy slurry when added to the mortar. The methanol wet mixture is ground in the mortar until the methanol has evaporated leaving a powdery residue, which is finally dried by placing the mortar in an oven at 80°C for 5 min. The bulk of the powder is now poured back into the crucible and the mixing and wet grinding procedure repeated. The final dried powder mix is transferred to a 0·5 in (13 mm) diameter powder pellet press and pressed to a disk at 10 tonf (100 kN) ram pressure.

X-ray Fluorescence measurements are made with the disk mounted on the specially designed holder shown in Figure 1.5. This design, which is used for all small-disk work, permits only the area of the disk to be at the analysing position in the spectrometer, and thus less scattered radiation background is obtained than if a conventional mask for the small disk were used. Intensity measurements are taken at the $Al_{K\alpha}$ 2θ angle,

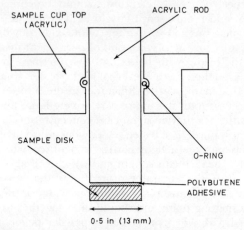

Fig. 1.5. Support for small sample disks in X-ray fluorescence measurements

excitation is with a chromium anode X-ray tube working near maximum power, and a pentaerythritol analysing crystal is used. The pulse height analyser is of course used and fixed-time counts for 100 s are taken. Calibration is carried out with a series of disks to which known amounts of aluminium oxide have been added in the range 0 to 500 μg Al. The 500 μg standard is also used as the instrument reference disk and the 'zero' standard used as a measure of the background. The method allows the accurate determination of aluminium down to 5 p.p.m. in a 10 g sample.

INFRA-RED SPECTROSCOPY

Organic substances, and many inorganic substances, exhibit absorption spectra in the infra-red region of the electromagnetic spectrum. These are widely used in both qualitative and quantitative analysis. They are particularly valuable in the qualitative analysis of polymers and compositions containing polymers, since the characterisation of these materials is often difficult by the more usual chemical and physical methods.

QUALITATIVE ANALYSIS FROM THE INFRA-RED SPECTRUM

For qualitative analysis it is useful to consider the infra-red spectrum to be made up of absorption bands which arise from the presence, in the substance examined, of pairs or small groups of atoms.

These atomic groupings have been found, empirically, to cause the appearance of absorption bands the wavelengths of which are, to a large

18

extent, independent of the nature of the rest of the molecule. The bands of some groups, such as $-CH_2-$ and $-C\equiv N$, are virtually unchanged in any compound in which they occur; the bands of other groups, such as $C=O$ and $C-O$, remain in a restricted wavelength range as structural changes are made, and in these cases not only may the group be identified but also some information is given, by the position to which the band has moved, of the nature of adjacent groups in the molecule. These bands are called 'Characteristic Bands'. Tables of such bands, known as 'Correlation Tables' have been published[16, 17] and, if used with appropriate caution, are exceedingly valuable in qualitative work[18, 19].

A number of atomic groups do not give characteristic bands. For example $S=S$, $O-O$, and $N=N$ give absorption bands too weak to be practically useful, while the absorption bands of groups such as $C-Cl$ are so variable in position as to be useless for structural diagnosis. In spite of this the infra-red spectrum will invariably give some information, whether negative or positive, on the structure of an unknown substance.

Qualitative analysis tables giving particular emphasis to those structures likely to be encountered in polymers and associated substances, together with some observations on qualitative analysis from the infra-red spectrum by the 'Molecular Fingerprint' method, appear in Chapter 2.

SPECTROMETER FOR QUALITATIVE ANALYSIS

For simple qualitative work, a double-beam spectrometer, operating between 2·5 μm and 15 μm is adequate[4, 20]. The region between the visible and 2·5 μm is of little use for qualitative work, while beyond 15 μm very few characteristic bands have been established, although occasional reference is made in the text to bands at greater wavelength, usually within the range of a potassium bromide prism instrument (15 μm to 25 μm).

PRESENTATION OF INFRA-RED SPECTRA

There are several methods of presenting infra-red spectral data. In the spectra presented here, we have plotted linear wavelength on the horizontal axis versus percentage transmission on the vertical axis, wavelength increasing to the right and percentage transmission increasing upwards. This method is chosen because the major collections of spectra to which reference will be made are plotted in this way. The wavelength unit is the micron:

$$1 \text{ micron } (1 \ \mu\text{m}) = 0\cdot001 \text{ mm}$$

An alternative method is to plot linear frequency instead of linear wavelength, the unit being the wavenumber (cm^{-1}):

$$\text{Frequency in wavenumber } (cm^{-1}) = \frac{10,000}{\text{wavelength in microns}}$$

19

Conventionally wavenumber decreases to the right. Unfortunately the appearance of the spectrum is rather different in these two methods of presentation. The vertical axis is occasionally plotted in reverse, with zero transmission at the top of the diagram; the effect is to produce the spectrum as a mirror image. Occasionally 'Fractional Transmission' or 'Fractional Transmittance' is quoted as the ordinate scale. This is the percentage transmission divided by 100, and the presentation is identical. The vertical axis is sometimes marked 'Percentage Absorption', but it has then been (incorrectly) assumed that:

$$\text{Percentage absorption} = 100 - \text{percentage transmission}$$

The effect is simply to reverse the direction of the vertical scale, and the presentation is not affected. Precalibrated charts are sometimes used on which the vertical axis is marked 'Absorbance':

$$\text{Absorbance} = \log_{10}\frac{100}{\text{percentage transmission}}$$

but the absorbance scale is non-linear, and the vertical axis is linear in percentage transmission so the result is not affected. Occasionally a linear absorbance scale is used. This modifies the vertical scale to some extent, but since the horizontal scale is not affected the spectrum is easily recognised qualitatively.

REPRODUCIBILITY OF WAVELENGTH OR FREQUENCY

In comparing spectra obtained in the laboratory with those in the literature and collections, the band maximum of the same substance, recorded in the same physical state, may be expected to agree within ± 0.1 μm. If plotted in frequency, maxima should agree within ± 5 cm^{-1} between 650 cm^{-1} and 2000 cm^{-1}. Above 2000 cm^{-1} the error may rise to ± 50 cm^{-1}.

The calibration of one's own spectrometer should be checked frequently using the standard tables[21] and if the error exceeds ± 0.03 μm (or ± 2 cm^{-1} between 650 cm^{-1} and 2000 cm^{-1} and 20 cm^{-1} above 2000 cm^{-1}) adjustment is required.

Data in old publications should be treated with suspicion.

REPRODUCIBILITY OF PERCENTAGE TRANSMISSION

The numerical values of percentage transmission in qualitative spectra are of no great consequence, since they depend upon sample thickness and upon various spectroscopic dodges which improve the apparent 'contrast' of the spectrum. Nevertheless the overall 'shape' of the spectrum is not affected, and bands should maintain approximately the correct relative intensities unless the physical state of the specimen is changed, when considerable differences may appear[4].

EXPERIMENTAL PROCEDURE FOR QUALITATIVE ANALYSIS

With modern double-beam spectrometers the procedure is simple. Place the sample in the sample beam, then run through the spectrum rapidly without recording it. Attenuate the reference beam as necessary to bring the highest transmission point to about 95% transmission. Then record the spectrum.

For qualitative work, a sample of about 0·001 in (0·03 mm) thickness is often satisfactory, but an examination of the spectrum so obtained will indicate whether some other thickness would be more suitable. For a general qualitative examination it is best to aim at a spectrum in which two or three of the stronger bands reach zero transmission.

Purity of the sample

The spectrum should be obtained on the purest sample readily available, and it is pointless to record the spectrum of a sample known to be a mixture if some ready means is available for separating it. However, polymer compositions may be intractable, insoluble solids, or mixtures about which little is known. In such cases the complete sample must be examined; the resulting spectrum will usually contain all the absorption bands of each constituent. Consideration of this spectrum will often give sufficient information to establish the general nature of at least some of the constituents, and may suggest a method of separation.

SAMPLE PREPARATION

Methods of sampling are well described by many authors[4, 17, 19, 20, 22]. In the case of monomeric substances there is little difficulty, but resins, both natural and synthetic, often call for considerable ingenuity in sample preparation.

The method of sample preparation to be used will depend upon the physical nature of the substance, and in the descriptions which follow the kind of sample for which each method is appropriate is indicated. In the case of resins which are partially crystalline (e.g. nylon, polyethylene terephthalate and polypropylene) the spectrum will be influenced, usually in respect of the relative intensities of the absorption bands, by the conformation of the molecules in the sample as placed in the spectrometer. If the object of the examination is primarily qualitative or semi-quantitative identification it will be better, where possible, to endeavour to destroy the original physical state of the sample by a preparation method involving solution or substantial heat treatment (e.g. hot pressing), whereby all samples of a particular resin, irrespective of their history (which is usually unknown), are reduced to a common physical state. In the standard spectra reproduced in this book, the sample preparation method is stated in each case, and it is hoped that, if the same method is followed, the spectrum obtained will be substantially identical with that given here. Where different preparation methods are known to lead to substantially different spectra, we have endeavoured to include the spectra obtained by the different methods.

On the other hand, where infra-red evidence is required on the physical state of the original sample[23] (which is largely outside the scope of this book), a preparation method such as the alkali halide disk, or sectioning in the microtome is essential.

Casting from solvents and latices

Casting is often preferred for latices and soluble resins, as it requires no special apparatus and, with practice, thin samples are easily prepared. Furthermore, if the resin can be taken into solution, fillers can be removed by filtering or centrifuging before casting the film.

There is little point in trying a wide range of solvents for casting films for infra-red work. It is better to try a few solvents known to be effective for a wide range of resins and which do not introduce practical difficulties. In this respect tetrahydrofuran is particularly troublesome, leaving high-boiling residues. Very volatile solvents (e.g. ether, acetone, methylene chloride) are best avoided, as they evaporate so rapidly that atmospheric moisture condenses on the sample surface, and a very poor film results.

If nothing is known of the nature of a resin, or its behaviour to solvents, we try the following solvents in the order stated.

1. Ethylene dichloride (1 : 2-dichloroethane). This dissolves a wide range of thermoplastic resins, including the majority of vinyls and acrylics.
2. Toluene. This is particularly useful for polythene and α-olefin polymers and copolymers.
3. Methyl ethyl ketone. This solvent is more difficult to remove than those mentioned above, but appears to be particularly suitable for butadiene copolymers.
4. Water. A number of resins rich in hydroxyl, acid, or amine groups are water soluble.
5. Formic acid. This is a most satisfactory solvent for polyamides and linear polyurethanes. It should be tried after the solvents previously mentioned as it reacts chemically with some resins (e.g. polyvinyl alcohol). Films cast from formic acid must be well washed with water to remove impurities (possibly sodium formate).

6. Dimethyl formamide. This is suitable for polyacrylonitrile, polyvinyl fluoride, polyvinylidene chloride and some other resins which are almost insoluble in the other solvents suggested. After drying, the film must be well washed with water to remove residual solvent.

If the resin proves insoluble after these solvents have been tried, or separates from the solvent on cooling, this method is abandoned unless there is some good reason for doing further work, e.g. in the case of heavily filled samples from which satisfactory spectra cannot be measured by any other method. For rubbers, in particular, Barnes et al.[24] suggest a prior extraction with an acetone-chloroform mixture, followed by refluxing with a p-cymene-xylene mixture to dissolve the polymer. This method is found to be particularly useful with carbon-black filled compositions.

Dinsmore and Smith[25] recommend a number of solvent mixtures and suggest that shaking with cold solvent is more effective than refluxing. This is in agreement with our experience, particularly with butadiene based rubbers.

If the rubber does not prove to be completely soluble, sufficient material may be dissolved to cast a film suitable for examination. Such a process however may preferentially remove extraneous matter (plasticisers etc.) and the resulting spectrum may be that of a very impure specimen. Nevertheless solution methods are particularly valuable with heavily filled samples.

Procedure for preparing films by solvent casting

We attempt to make a solution of about 0·05 g of resin in about 1 ml of solvent (these quantities may be decreased proportionately if only small amounts of sample are available). The solvents are tried in the order given above, first warming gently, then boiling if solution is slow. Test tubes of appropriate diameter are used according to the quantity of solution, and electric heating mantles are particularly recommended as the risk of burning the resin, or igniting the solvent vapour, is thereby much reduced.

A quantity of the solution is poured on to a small glass plate (e.g. a microscope slide) which is then placed under an infra-red lamp for about 10 min. The plate is allowed to cool, and the film stripped from the plate, easing it away with a razor blade. A little water run over the film may assist stripping. A uniform portion of film of about 0·001 in (0·03 mm) thickness is cut out and mounted in a card mount.

If the film proves too brittle to strip, the solution is poured on to a rock salt or potassium bromide plate and dried off in the same way. The plate is then mounted in the spectrometer beam. This method is not generally recommended, as the film so produced is usually of poor quality. For aqueous solutions, silver chloride foil or a thallium bromo-iodide (KRS-5) plate is used instead of rock salt or potassium bromide.

Casting on liquid surfaces, aluminium foil, etc. is often suggested, but in our opinion these procedures are of no particular value for general work.

The film-casting methods suggested above are well suited to the direct examination of samples submitted as film-forming latices or dispersions in liquid media, the sample being treated in the same way as a solution. Such samples usually contain surface active agents. The majority of these may be removed from the cast film by immersing the film, with its card mount, in methanol, then washing with water, then immersing in ether, and finally drying under the infra-red lamp. Because the sample is so thin, the action of these solvents is quite rapid, and a few minutes in each solvent are usually sufficient. The same procedure is often successful for obtaining the spectrum of the resin in a plasticised composition, such as plasticised polyvinyl chloride, and in some cases lengthy extraction procedures may be avoided if only qualitative identification is required.

Hot pressing

Hot pressing is a most convenient and quick method of preparing samples from thermoplastic resins, although it is not suitable for producing films less than 0·001 in (0·03 mm) thick. It is the preferred method for the examination of polythene and α-olefin resins such as polypropylene, and is generally worth trying for any resin which does not dissolve readily. With a minor modification, as described below, it may be used to prepare samples from some rubbers.

A small hydraulic press, 20 tonf (200 kN) ram pressure and 6 in (150 mm) square platens, is convenient. Provision is necessary both for heating the platens to 270°C and for cooling them rapidly with water, as some resins decompose if held for extended periods at high temperature. The sample is pressed between polished stainless-steel plates, $\frac{1}{8}$ in (3 mm) thick, cut to the same size as the platens. To assist in stripping the pressed film, the plates are coated with polytetrafluoroethylene. Details of the coating process are given in Reference 26. Samples pressed between these coated plates have a matt surface. This destroys the specular reflection from the film, and consequently interference fringes, which may otherwise obscure some features of the spectrum, do not appear with samples prepared in this way.

Feeler gauge blades are used as spacers for films over 0·003 in (0·08 mm) thick, but for thinner samples no spacer is used, the sample thickness being adjusted by altering the amount of material used, or by changing the pressing temperature. In many cases it is advantageous to preheat the sample in the press before applying pressure. The best temperature for a particular sample may be found by trial and error, but generally temperatures between 150°C and 200°C are suitable. Samples of rubbers may contract after pressing. A useful method is to press a small specimen

24

between two thin sheets (0·002 to 0·003 in or 0·05 to 0·08 mm thick) of silver chloride. The complete sandwich is placed in the spectrometer without removing the silver chloride, although subsequently these supporting sheets can be removed and used again.

Cold sintered disks

The alkali halide disk method, which has been frequently described in the literature[4, 20, 22], is probably the most widely used procedure for preparing samples from simple (non-polymeric) solids. Unfortunately the majority of resins are very difficult to disperse in alkali halides, and the spectra obtained by this method are inferior, both qualitatively and quantitatively, to those measured on solid films. Hence, in polymer work, the halide disk method is applied only to those samples, particularly insoluble resins and rubbers, which do not respond to the various methods of film preparation.

Brittle resins respond moderately well to the normal disk-making procedure, although an extended grinding period is usually necessary. In some cases it is advantageous to pre-grind the resin before adding the halide powder. Suitable samples may be powdered by rubbing with a diamond-faced spatula. The addition of a solvent to swell an insoluble resin or rubber is often useful, as a swollen gel is broken down more readily; the solvent is removed by heating the powder after grinding and before pressing. Alternative suspending media, such as silver chloride powder or polythene powder, appear to offer no particular advantages in normal work.

It is worth noting that some polymers in powder form, particularly polytetrafluoroethylene, and some linear polythenes, will sinter into disks without the addition of alkali halide powder, by the usual disk-making procedure.

Grinding in the frozen state

A modification of the disk method is often successful with rubbers. We are indebted to the Natural Rubber Producers' Research Association for details of their method, which depends upon cooling the rubber below its brittle point and grinding it, while in this condition, with potassium bromide. The normal type of potassium bromide ball mill may be used but while grinding is in progress the grinding capsule is suspended in a container (e.g. a vacuum flask) filled with liquid nitrogen. The capsule is loaded in the usual way with substance, potassium bromide powder, and grinding balls. Prolonged grinding is usually necessary (up to 1 h), after which a disk is pressed from the powder. During the grinding the powder may become wet owing to moisture condensation. If the effect of this is sufficiently pronounced to interfere with the interpretation of the spectrum, the disk may often be dried by heating in the oven at 140°C for about 10 min. The dried disk is placed (whole) in the disk press and pressed again. This usually reduces the water content of the disk substantially. We have found this method satisfactory, particularly for urethane rubbers, but it is very slow, and the A.T.R. method (see p. 29) is likely to be preferred for such samples.

Thin film preparation with the microtome

When the original sample is presented as a moulding which is stiff and tough, rather than rubbery or brittle, thin films may be prepared by sectioning with a robust microtome of the base-sledge type. A much wider range of samples can be treated in this way if the temperature of the specimen is changed, by warming brittle samples and cooling rubbers with solid carbon dioxide or liquid nitrogen. By trial and error it will frequently be found that there is a narrow temperature range within which good sections may be cut. Some success may be obtained with hard resins by swelling them with a solvent. The microtoming method has been advocated by Corish[27] for use with filled rubbers. He obtains the transmission spectra of extremely thin layers ($\sim$0·0001 in thick) improving the spectrum by the use of a scale-expansion accessory. Corish[28] has extended this work to the identification of a number of common inorganic fillers for polymers again using thin microtomed sections. Unfortunately these processes are tedious and rarely worth the effort involved, unless other methods of thin-film preparation have failed.

Even when sectioning can be done at room temperature, the method must be applied with caution, as the spectrum obtained from a microtomed section of a resin may well differ from that measured on a solvent-cast or hot-pressed specimen, if the resin can exist in a partially crystalline state, e.g. like polyethylene terephthalate or polyvinylidene chloride.

INFRA-RED EXAMINATION OF PYROLYSATES

For many years the pyrolysis method suggested by Harms[29] was the only means available for the infra-red examination of many insoluble resins and rubbers. The more recently discovered A.T.R. method (see p. 29) has greatly facilitated the examination of a number of these intractable materials. Nevertheless, there are many heavily filled insoluble samples from which it is not possible to obtain satisfactory spectra directly. Furthermore, there are many cases of copolymers and polymer blends where the information on the nature of minor polymeric constituents available from the infra-red spectrum of the total composition does not admit of their unambiguous identification, and the low concentration of the constituent sought, or the insolubility of the sample, precludes n.m.r. examination.

It is to these samples that the pyrolysis method is particularly appropriate. The original method consists in sealing off a small sample ($\sim 0 \cdot 1$ g) of the unknown substance under vacuum in a U-shaped glass tube with limbs 2 to 3 in (50 to 80 mm) long. One limb of the tube is heated while the other limb is maintained in a cold bath (e.g. solid CO_2-methanol); decomposition products slowly accumulate in the cold limb. When pyrolysis is complete the tube is cut through and the condensate (usually a liquid or paste) is removed for examination, usually as a capillary film between transparent plates.

It has been claimed[29] that many resins can be pyrolysed, under standard conditions of heating, to yield products with reproducible spectra which may be used empirically to characterise the original sample. In our experience, however, we have found, with a few notable exceptions, that the spectra of pyrolysates tend to be indefinite and characterless as would be expected for complex mixtures, and not reproducible, except in their major features, if a pyrolysis is repeated. To some extent the lack of reproducibility arises from poor heat transfer during pyrolysis, but is also difficult to ensure that a representative sample of the pyrolysate is taken since this is frequently not homogeneous, or the lower molecular-weight species present, which are usually responsible for the more prominent spectral features, tend to evaporate when the sample is transferred from the pyrolysis tube. In some cases pyrolysis leads to considerable evolution of water which must be removed before a useful spectrum can be obtained. Some resins, notably polyvinyl chloride compositions, yield products with a high vapour pressure, and it is wise to open the pyrolysis tube behind a safety screen.

Nevertheless, in particular cases, as for example in the identification of cross-linked and heavily filled rubbers, infra-red examination of the total pyrolysate can be a very valuable test, and the method has been strongly advocated by some workers[30, 31].

Harms[29] and Hummel[32] give the spectra of a number of pyrolysates of resins, including several rubber compositions. Hummel modifies the method by collecting separately a liquid and gaseous fraction from the pyrolysis, and examining each of these. Yet another procedure is to pyrolyse a sample under vacuum within a specially constructed heated infra-red gas cell. The cell contains a pyrolysis filament, fitted in such a position that it does not obstruct the radiation beam. Any volatile products are retained in the gas phase and their spectra may be measured. This apparatus is available from Techmation Ltd, Edgware, London. These methods seem to be advantageous as the gaseous fraction is often a moderately pure component, which can be identified unambiguously. Working on this general principle, however, we have found it preferable to concentrate on the chemical information available from the pyrolysate spectrum, rather than to attempt to use it to 'fingerprint' the resin.

27

For example the presence of nitrile groups in the pyrolysate will suggest acrylonitrile or methacrylonitrile in the resin; strong bands between 10 μm and 11 μm will suggest butadiene or polythene, while a mono-substituted benzene pattern in the 14 μm region of the spectrum indicates polystyrene. Any deductions of this nature may be confirmed by the pyrolysis of known samples under conditions as comparable as possible with those used for the unknown sample. It must be remembered that some components of a mixture or copolymer may not decompose, or may not yield recognisable products, and a complete identification of an unknown is therefore rarely possible by this method. In our opinion the method is best confined to a preliminary examination which may give sufficient information to enable the unknown to be decomposed by some more satisfactory method to yield products which can be properly characterised.

Present experience suggests that pyrolysis and infra-red spectroscopy can be more effectively combined in a different way. If it is the intention to produce a 'fingerprint' for comparison against standards as a method of characterisation, a more meaningful 'fingerprint' is likely to be the pyrogram obtained by direct pyrolysis of a small sample under controlled conditions into a G.L.C. column (see p. 57). The major experimental problems referred to above are then overcome. A much smaller sample is used, hence heat transfer difficulties are minimised. The total pyrolysate enters the column for examination, and the presence of water is not an embarrassment. However, the pyrogram of an unknown may be impossible to interpret, or it may be known from experience that the interpretation of certain peaks is not unambiguous. In such cases a most satisfactory approach will be the G.L.C./infra-red procedure (see p. 69). Sufficient sample will be required to produce up to 10 mg of pyrolysate. Examination, preferably directly in the gas phase, of at least the major peaks, should enable these peaks to be identified unambiguously from their spectra. It is then a matter of experience and deduction to ascertain the nature of the original sample from the nature of the pyrolysis products which have been identified.

The limitation of the examination of pyrolysates by a method involving G.L.C. is that any constituent too involatile to pass through the column will be missed. This would apply, for example, in the case of *o*-phthalates, where *o*-phthalic acid is generated on pyrolysis. Useful information has been obtained in such cases by a method similar to that described by Lindley[33]. The pyrolysis head is that used for the normal pyrolysis/gas chromatography method (p. 58). This is adapted for the present method by enclosing the element within a short tube with a B7 joint (see Figure 1.6). Up to 10 mg of material are placed in the heating element and the tube is immersed in an ice bath. The apparatus is evacuated through the side arm, and the sample is pyrolysed at 750°C for 30 s. After releasing the vacuum, the products are washed from the tube with the minimum amount of acetone. The solution is evaporated on to finely powdered potassium bromide. After thorough mixing, the powder is pressed into a disk, and the infra-red spectrum measured. This procedure has been found useful in the preliminary identification of polyesters, epoxy and urethane resins.

In conjunction with other methods of examination, the pyrolysis method is sometimes advantageous in that the pyrolysis residue may be a useful sample of the filler present in the original composition. If, by some

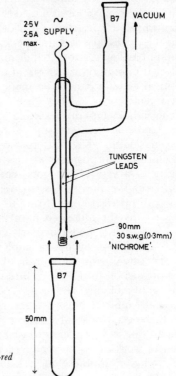

Fig.1.6. Modified pyrolysis head for infra-red examination of involatile pyrolysates

other means, the spectrum of the total composition has been obtained, comparison of this with the spectrum of the recovered filler will enable the true absorption bands of the organic constituents present to be differentiated in the spectrum of the total sample. The method is to be preferred to examination of the ash from the sample prepared in the usual way, since the character of the spectra of inorganic substances is often dependent upon the crystalline state of the filler, which may well be modified by drastic heating.

ATTENUATED TOTAL REFLECTION

The A.T.R. (Attenuated Total Reflection) method, reported originally by Fahrenfort[34] and Harrick[35], enables a reflection spectrum to be obtained which is superficially very similar to an absorption spectrum. The radiation beam in the instrument is passed through a small prism of a material of high refractive index (e.g. silver chloride or thallium bromo-iodide) in such a way that the beam suffers total internal reflection at one face of the prism. The substance whose spectrum is to be measured is pressed into contact with the prism face at which the internal reflection occurs; the sample attenuates the reflection from this surface in such a manner that a spectrum is obtained which shows a remarkable similarity to a transmission spectrum, provided that good contact is obtained between the prism face and the specimen. There is however some distortion, and in particular strong bands may be shifted to greater wavelength. This effect is minimised if a prism of high refractive index is used. Where good optical contact cannot be made, the interface between prism face and specimen may be filled with a high refractive index liquid such as methylene iodide, but the resulting spectrum will be inferior, and will have the bands of the liquid superimposed on those of the sample. Thallium bromo-iodide (KRS–5) appears to be the most suitable prism material for general purposes as it is tough and will withstand the firm pressure necessary to obtain good contact between the sample and the prism surface. We find it best to use a multi-reflection unit (~ 25 reflections) as imperfect contact due to roughness of the sample surface is counteracted by the increased number of reflections. With rubbers and soft samples which make particularly good contact, the area of the sample may be reduced if the spectrum is too intense. Furthermore, recognisable spectra may often be obtained from fibres (wound around the prism) and powders pressed on to the prism surface.

An attractive feature of the method is that with samples over about 0·001 in (0·02 mm) thick the spectrum obtained is independent of the sample thickness, thus laborious sample preparation procedures are avoided. In our experience, however, the method is of doubtful value in quantitative work except in some special applications, as the band intensities, in relation to those in a transmission spectrum, increase with increasing wavelength, and are subject to variation due to inadequate contact with the prism surface. A particularly important application of the A.T.R. method is in the examination of surface coatings and laminates (see p. 125), and some further observations on the method appear in that section.

QUANTITATIVE ANALYSIS FROM THE INFRA-RED SPECTRUM

Infra-red spectroscopy is being used increasingly in quantitative analysis in the field of synthetic resins, and examples of these analyses are included in the text. In many cases these methods have been devised by Mrs. V. Zichy and we are happy to acknowledge her assistance in this respect.

If an absorption band in the infra-red spectrum arises from a particular component in the sample the concentration of the component is related to the percentage transmission at the band maximum by Beer's Law:

$$\log_{10}\frac{100}{\text{Percentage transmission}} = Kcl$$

where c is the concentration of the absorbing component
  l is the thickness of the sample
  K is a constant.

The value of K is determined by measuring the spectra of samples containing a known concentration of the compound sought.

Quantitative analysis by infra-red spectroscopy has been described in detail by many authors[4, 19] and we will consider only the particular problems which arise in the plastics field.

QUANTITATIVE INFRA-RED ANALYSIS WITH SAMPLES PREPARED AS SOLID FILMS

Conventionally, infra-red quantitative analysis is performed with samples in the gaseous or liquid state. Solids are normally examined in solution.

However, if the sample for analysis is a polymer or polymer composition, it is more convenient to prepare it in film form, for example by hot pressing. This is much quicker than making a solution and since, to be handled, polymer solutions must be dilute, it may be difficult or impossible to find a solvent which does not obscure the spectral region of interest for the quantitative determination.

Nevertheless, there are particular difficulties associated with the use of pressed film samples. These must have flat parallel sides, otherwise their infra-red absorption properties do not follow Beer's Law. Furthermore, in partially crystalline polymers or copolymers (e.g. vinylidene chloride/methyl acrylate copolymers) the intensity of many of the absorption bands will depend upon the degree of crystallinity of the sample examined, and this in turn may vary according to the sample preparation conditions. In such cases the extent of any changes observed in the spectrum with, for example, variation in the pressing procedure, particularly the rate of cooling, must be determined. In this way it may be possible to decide upon some reproducible sampling method. Again, it is not wise to assume that the small sample of film used for infra-red examination is representative of the bulk of the material; blends of one polymer with another, or polymers containing additives, are frequently inhomogeneous and mill blending may be necessary.

There are, in addition, optical effects to be considered. The more important is probably the suppression of interference fringes by pressing between coated plates (see p. 24), but the problem of light scattering from solid samples must also be considered. While a first order correction for scattering is achieved by the 'base-line density' method of measurement[4], it is important that as far as possible all samples and standards should be similar in scattering properties. It has been noted, for example, that pressings from insufficiently dried samples, and heterogeneous polymer blends, may vary considerably in opacity, and under these conditions the correction for scattered light may well be inadequate.

Almost all the difficulties noted above disappear if a solution method can be used and we would suggest that, where suitable solvents are available, and speed is not important, solution methods are much to be preferred, and in many cases will be the only reliable methods of analysis. A further point in favour of solution methods is that where a strongly absorbing constituent is present (e.g. the ester constituent of an ester/olefin copolymer) it may be impossible to produce films thin enough to permit the bands of this constituent to be measured, whereas by a solution method this is allowed for merely by decreasing the concentration of the solution until the bands are reduced to a convenient intensity for measurement.

It must not be inferred, from this catalogue of problems, that solid-film methods cannot be used for the analysis of copolymers, polymer blends, or additives in polymers. A number of quantitative methods employing solid polymer films which have been found completely reliable are described in this book. However, as it is our intention to describe methods which other workers may wish to adapt to similar but not identical systems, it is important to indicate those problems which are likely to arise.

If solid-film methods are applied to the direct determination of additives in polymers, it may be important to consider also the physical properties of the additive. It must not decompose at the temperature at which the film is prepared, nor must it volatilise or 'bleed out' of the composition. It is also important that the additive is well dispersed in the resin, and not present as agglomerates. In routine determinations, particularly when large numbers of samples are to be examined, we have found that a small film extruder will produce in appropriate cases film samples of reproducible thickness, quality, and composition, and will add greatly to the speed and reproducibility of the analysis.

When there are indications that the direct determination of additives in a resin is likely to be unreliable, it may be possible to extract the additive with, for example, ether or alcohol, to evaporate the solution to dryness, and to dissolve the residue in a known volume of a solvent suitable for the infra-red measurement. Alternatively, the entire composition may be dissolved, and the polymer precipitated with a further solvent. The solution is evaporated to dryness and re-dissolved as before. This second method is usually quicker, but the speed may be outweighed by the extra manipulation required. In particularly fortunate cases, when the sample is presented in a finely divided form, it may be sufficient to shake a known weight of polymer in the cold with a suitable solvent, filter, make up the solution to known volume, and proceed with the infra-red examination.

SPECTROMETER FOR QUANTITATIVE INFRA-RED MEASUREMENTS

It is pointless to attempt to use a cheap spectrometer for quantitative work. A minimum specification should be a double-beam instrument with a repeatability of the transmission scale of 0·2% and with an absolute accuracy of 1%, both of full scale deflection. A grating instrument, having a wide range of traverse speeds and manual and programmed slits, is preferred. The spectral range should be from 1 μm to 25 μm. For some tests it will be advantageous to operate in true single-beam mode. An example would be in cases where samples submitted for analysis are different in their transmission properties because scattering fillers are present[4].

A wide chart (e.g. 10 in; 250 mm) is most desirable, and for many purposes direct absorbance recording is a convenience, as is ordinate scale expansion, particularly in conjunction with direct absorbance recording.

EXAMPLES OF QUANTITATIVE INFRA-RED ANALYSIS

Many examples of quantitative infra-red analysis are included in this book. Among the applications to copolymers is the analysis of vinyl chloride/vinyl acetate copolymer for both free and combined vinyl acetate (p. 184). The method is also well suited to blends of hydrocarbon resins, for example the determination of the polyisobutene content of polythene/polyisobutene blends (p. 328). In both these examples solid films are used but, in the case of polyisobutene blends, if there is reason to suppose that the sample is inhomogeneous, an alternative method, based on separation of the polyisobutene, is given (p. 329).

Some examples are given of the direct determination of additives in polypropylene film (p. 358). Alternative methods depend upon hot ether extraction in a Soxhlet extraction apparatus and, for polypropylene powder, direct extraction into carbon tetrachloride.

N.M.R. SPECTROSCOPY

During the last 20 years, nuclear magnetic resonance spectroscopy has been developed as a powerful method of qualitative and quantitative analysis. More recently it has been shown to be of particular interest in the examination of polymers, and we are indebted to Mr. M. E. A. Cudby for much of the work on this subject reported here. N.M.R. spectra can be observed from a number of atomic nuclei but for the organic chemist spectra of the H^1 nucleus, proton resonance spectra, are of the greatest practical importance. The reason is that the majority of compounds of interest in organic chemistry contain hydrogen, and a great deal can be deduced about the structure of an organic compound if the relative positions and the chemical environment of the hydrogen atoms in the molecule can be established. This information may be obtained from proton resonance spectra. These spectra are comparatively simple to measure, as H^1 is a naturally abundant isotope, and the proton gives a very intense signal. In principle with the majority of organic compounds carbon nuclear magnetic resonance spectra would give complementary information. However, the naturally abundant isotope C^{12} is inactive, and although the C^{13} nucleus is magnetically active, it has low natural abundance and gives only weak resonance signals. Thus nuclear magnetic resonance spectroscopy of organic compounds is confined mainly to proton spectra. These are now well understood, and are of proven value in analysis. Fluorine (F^{19}) nuclear magnetic resonance is of some interest to the polymer chemist, and has yielded valuable information on the structure of some fluorinated polymers[36].

GENERAL PRINCIPLES

The magnetically active nucleus is considered to behave as a small spinning bar magnet. When placed in a strong magnetic field it will precess about the direction of the field axis with a frequency f, the Larmor frequency

$$f = \gamma \frac{H_0}{2\pi}$$

where H_0 is the strength of the magnetic field in G
  γ is the magnetogyric ratio; this is a constant having a unique value for each active nucleus.

The nucleus is irradiated simultaneously with an alternating magnetic field which has its axis at right angles to that of the principal field H_0. When the frequency of this alternating field becomes equal to the Larmor frequency for the nucleus, resonance is established and energy transfer occurs. This energy transfer is measured with an appropriate detector.

The atomic nuclei in a molecule are subjected to shielding effects from the surrounding electrons which modify the field strengths in the vicinity of the nuclei. Furthermore, the extent of this shielding varies according to the chemical environment of the nucleus. Thus, at constant irradiation frequency, when the strength of the main field is changed, nuclei of the same chemical species in different chemical environments may be brought successively into resonance. In this way the n.m.r. spectrum for a particular nuclear species (e.g. the proton) may be measured, with a series of resonances occurring at different values of the main field strength. In the case of protons, a change of field strength of 10 parts in 10^6 (10 p.p.m.) covers the majority of the resonances of interest. The experiment above is described as a 'field sweep', but it is also possible to hold the field constant and to slowly change the irradiating frequency (frequency sweep). The two methods produce results which are in practice indistinguishable.

EQUIPMENT

The n.m.r. spectrometer consists of five important pieces of equipment.

1. The magnet

Either a permanent magnet or an electromagnet may be used to produce a strong and homogeneous magnetic field. It is difficult to maintain homogeneity over a period of time due, for example, to small changes in the magnet temperature. Since field inhomogeneity is the principle cause of line broadening, 'shim' coils are provided by means of which the field may be 'tuned' to obtain the best resolution in the spectrum. The most commonly used magnets have the following characteristics:

Magnetic field strength (G)	Proton resonance frequency (MHz)	F^{19} nuclear resonance frequency (MHz)
14 092	60	56·5
23 490	100	94·1

Higher field strength instruments are preferred, as the dispersion of the spectrum is increased linearly with the field strength, and there is an improvement in signal to noise ratio with increasing field strength. In commercial instruments the pole pieces of the magnet are arranged to generate the magnetic field with its axis in a horizontal plane to facilitate sample handling.

2. Sampling system

The sample, usually a liquid or a solid in solution, is contained in a precision glass tube of external diameter approximately 5 mm for proton studies. About 20 mm length of the tube is filled with liquid. The sample tube is suspended with its axis vertical so that the portion filled with liquid is between the magnet pole-pieces. The tube is rotated about its axis with a pneumatic spinner. This has the effect of averaging the field in relation to the sample, thus improving the field homogeneity. The unit which supports the sample tube is known as the 'probe'. In many spectrometers provision is made for heating or cooling the sample, often by means of a stream of nitrogen. For successful examination of polymers, a variable-temperature 'probe' is essential. Polymers, even if they are sufficiently soluble to give a solution strong enough to produce good signals, will show, in most cases, very broad and diffuse spectra at room temperature. Increasing the temperature to 100°C or 150°C will usually effect a remarkable improvement in the spectrum by sharpening the resonances. This effect has been discussed by Willis and Cudby[37].

3. The transmitter (or oscillator)

This is the source of energy, and provides the alternating field by energising a transmitter coil contained within the probe. The axis of the coil is at right angles to the axis of the principal magnetic field, as described earlier.

4. Sweep mechanism

Two methods are available for 'sweeping' through the spectrum. 'Field sweep' is generated by 'sweep coils' wound on the magnet pole pieces. Current flowing in the sweep coils is varied so as to change the main magnetic field strength through a small range. In the case of protons a field change of ten parts in 10^6, or 10 p.p.m. will bring almost all protons into resonance at constant oscillator frequency.

Alternatively, 'frequency sweep' may be used with constant magnetic field strength, in this case changing the oscillator frequency over the same range (10 p.p.m.) to sweep the spectrum.

5. The detector

The active part of the detector is a coil wound inside the probe, although in some designs the same coil is used in the transmitter and detector circuit. Current is induced in the detector coil at resonance, and the magnitude of this current after appropriate amplification is presented as the vertical axis of the recorded spectrum.

APPLICATIONS OF THE N.M.R. SPECTRUM

THE CHEMICAL SHIFT

We have stated above that nuclei of the same chemical species may show resonances at different values of magnetic field strength at constant frequency (or different values of frequency at constant field). This effect, due to the screening of the nucleus by the surrounding electron cloud, is known as the 'Chemical Shift'. The value of the chemical shift gives direct information on the chemical environment of the nucleus.

The effect is relatively small, and to enable shifts to be measured with the necessary precision (1 part in 10^8) they are in practice measured as shifts relative to the position of the resonance of a standard substance. This substance is usually added to the sample or solution and is then an 'internal reference'. External references are occasionally needed (e.g. if the standard is insoluble in the sample or solution); the procedure is then to immerse in the sample a capillary tube containing the reference substance. If an external standard is used, a magnetic susceptibility correction should be made[38].

The position of a resonance is defined in terms of the 'chemical shift' parameter δ, expressed in p.p.m.

$$\delta = \frac{H_{sample} - H_{standard}}{H_{standard}} \times 10^6$$

where H_{sample} is the field strength at which the resonance of the nucleus of interest occurs

$H_{standard}$ is the field strength at which resonance of the nucleus occurs in the reference standard.

By this definition, if the same reference standard is used, the 'chemical shift' of a nucleus is the same measured on any spectrometer, since δ is independent of the operating frequency.

For practical purposes, δ may also be defined as

$$\delta = \frac{v_{standard} - v_{sample}}{v_{standard}} \times 10^6$$

where $v_{standard}$ and v_{sample} are the frequencies at which the respective resonances appear at constant field.

The accepted standard for proton resonance work is tetramethyl silane (TMS). This substance shows a strong single resonance, as all protons in the molecule are equivalent. It is particularly useful as the resonance appears to high field of (i.e. at a higher field strength than) any proton signal encountered in most work in organic chemistry. Thus most chemical shift values for protons have the same sign and, by the above definitions, because at constant frequency the resonance of almost all protons occurs at a field strength lower than that of TMS, the chemical shift δ is in most cases negative. (It is usually said that such resonances occur 'to low field' of TMS.) Thus strictly almost all δ values are negative, although the negative sign is often omitted. To overcome this difficulty, the τ scale has been introduced where

$$\tau = 10 + \delta$$

where δ is the chemical shift of a resonance relative to TMS. Thus almost all τ values are positive, and further the τ value increases with increased shielding of the nucleus and increasing field strength. Tables of δ values and τ values for all commonly occurring proton-containing groups appear in the literature[39, 40].

AREAS OF RESONANCES

In the resonance spectrum of nuclei of a particular chemical species the appearance of resonances at different chemical shifts (or τ values) is due to shielding of the nuclei, and there is no effect on the signal intensity. Thus the area of a resonance found in a particular spectrum is proportional to the number of nuclei giving rise to the resonance. This has the important application that the relative numbers of nuclei in different chemical environments can be measured directly simply by measuring the relative areas of the resonances.

SPIN-SPIN COUPLING

The n.m.r. spectrum of a substance is usually much more complex than would be expected solely from the chemical shift effect. The resonances are often split into a number of components and may appear as doublets, triplets, quartets, or more complex patterns. The effect arises from spin-spin coupling between adjacent nuclei. This may be heteronuclear coupling, e.g. between protons and adjacent fluorine nuclei, but the most important effect is homonuclear coupling, especially that between protons which are in different chemical environments and are very close together in the molecule.

The effects of spin-spin coupling can be illustrated by the n.m.r. spectrum of ethanol (Figure 1.7). Three resonances are observed near $8\cdot8\ \tau$, $6\cdot3\ \tau$ and $7\cdot4\ \tau$ corresponding to methyl, methylene and hydroxyl protons respectively. The methyl resonance at highest field is a triplet, and the methylene resonance at lowest field is a quartet. The separation between adjacent lines of the triplet is the same, and equal to the separation between the lines of the quartet. This separation, quoted in Hz (cycles per second) is known as the 'Coupling Constant' $\mathcal{J}$ between the nuclei which interact.

Provided that the difference in frequency between the resonances of the interacting protons is large compared with $\mathcal{J}$,

$$\text{Number of lines into which a resonance is split} = n+1$$

where $n = $ number of identical nuclei in the interacting group.

Following this rule, the CH_3 group resonance is split into three because two identical nuclei (i.e. the two methylene protons) are adjacent to it (i.e. on the next carbon atom). Similarly the methylene resonance is

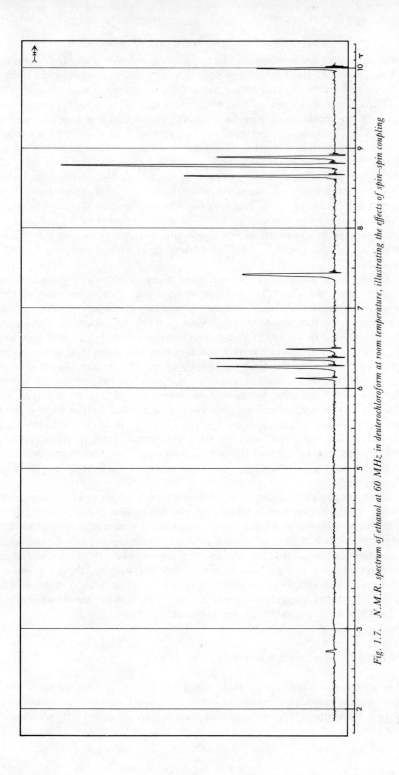

Fig. 1.7. N.M.R. spectrum of ethanol at 60 MHz in deuterochloroform at room temperature, illustrating the effects of spin–spin coupling

split into four by interaction with the three identical methyl protons. (The proton of the OH group does not partake in spin-spin coupling, otherwise further splitting of the methylene resonance would occur.)

Thus if in a spectrum we find resonances split with the same frequency spacing, the resonances arise from nuclei which lie close together in the molecule. Furthermore, the number of components into which a particular resonance is split is related to the number of identical nuclei in the interacting group. The magnitude of the coupling constant J is indicative of the relative positions of the interacting nuclei. Tables of coupling constants for commonly encountered groups appear in a number of publications (e.g. Reference 40). Examples of the application of coupling constants in structural analysis include the differentiation of *o*-, *m*-, and *p*-arrangement of hydrogen atoms about the benzene ring, and the determination of the relative positions of hydrogen atoms about the $C{=}C$ bond.

Doublets, triplets, quartets, etc. may sometimes be recognised, even though they are partially overlapped by other bands, because of the relative intensities of the component peaks. These are equal to the coefficients of the binomial expansion $(x+1)^n$, i.e. doublets consist of bands of equal intensity, triplets have intensity ratios $1:2:1$ and quartets $1:3:3:1$. The frequency separation between components of a resonance remains constant as the field strength is changed, whereas the frequency separation between resonances (which is due to chemical shift) increases linearly with increasing frequency. Hence, when plotted on a p.p.m. or τ scale, a spin-spin coupling pattern is compressed in the higher field strength spectrum. Thus spectra recorded at, say, 60 MHz frequency and 100 MHz frequency will usually show differences, and in principle the differences may be used to distinguish spin-spin coupling from chemical shift. In practice, however, it is often more convenient to apply spin decoupling[41].

Such decoupling is most easily performed at constant field by frequency sweep. In the simplest case, when only two nuclei A and B are interacting, if the sample is irradiated during the frequency sweep with a second source at the resonance frequency of nucleus A, the coupling pattern of nucleus B collapses into a singlet. In the more general case, where nucleus B is interacting with nucleus A and with other nuclei also, the coupling of nucleus B will simplify, because the coupling to nucleus A is removed, but that to other nuclei is retained. A series of these experiments will reveal which nuclei in the molecule are coupled.

PRESENTATION OF THE N.M.R. SPECTRUM

The n.m.r. spectrum is presented as a plot of p.p.m. or τ on the horizontal axis against resonance energy as a vertical axis, usually on precalibrated paper. Field strength conventionally increases to the right. The horizontal axis is usually calibrated in p.p.m. of shift, τ value, or both, and also in Hz.

Unless otherwise indicated, it is presumed that a spectrum is recorded with the resonance of TMS coincident with the zero of the p.p.m. scale, which is on the right hand edge of the calibrated portion of the recorder chart. This corresponds also with zero on the Hz scale, and 10 τ. The Hz scale differs in relation to the other horizontal scales according to the operating frequency of the instrument, e.g. 600 Hz for a 60 MHz instrument and 1000 Hz for a 100 MHz instrument are each equivalent to 10 p.p.m. shift, or 10 τ. It is usual to read p.p.m. or τ values to the nearest 0·01 p.p.m. or 0·01 τ. Coupling constants may be read approximately from the Hz scale, although to improve reading accuracy the horizontal scale may be expanded by simple multiples.

RESONANCE AREA MEASUREMENT

Most spectrometers have provision for integration of areas under the resonances. The integral is presented on an arbitrary vertical scale, on the same chart as the normal spectrum, and substantially in alignment with it along the horizontal axis.

PREPARATION OF SAMPLES FOR N.M.R. MEASUREMENTS

Monomeric substances

If full advantage is to be taken of spin-spin coupling patterns and area measurements, the resonances in the n.m.r. spectrum must be sharp and well defined. This is achieved by ensuring that chemically equivalent nuclei in each molecule in the sample experience effectively the same magnetic environment. For this reason samples are examined in the liquid state and preferably in solution. Each molecule is then so mobile that it can assume all possible orientations with respect to the magnetic field. Thus each nucleus experiences the same average magnetic environment, and the resonances are sharp.

For n.m.r. studies of monomeric substances a wide range of proton-free solvents is available, including carbon tetrachloride, deuterochloroform, perdeutero-benzene and many others. Interactions between solvent and solute are often significant, hence the solvent is usually noted on the spectrum. A blank run on the solvent is advisable as it may have proton-containing impurities. With field-locked instruments sufficient internal standard (e.g. TMS) must be added to provide a strong locking signal ($\sim 5\%$). This is likely to distort the spectrum in the region of the TMS signal (10 τ). When this is important, it is possible to use a proton-containing solvent with a simple spectrum (e.g. benzene or chloroform)

41

and lock on to the solvent signal, adding just sufficient TMS to observe its resonance and adjust it to 10 τ.

Polymers

The rapid overall motion of molecules of monomeric substances in solution is not achieved with polymers, but in practice if rapid segmental motion of the polymer chain can be achieved comparatively sharp resonances may be observed. In most cases this occurs only at elevated temperatures. Thus most polymers examined in solution show a substantial reduction in the width of the resonances as the temperature is raised, and the best temperature for observation of the spectrum is often 100°C to 150°C. However, even under these conditions the fine detail characteristic of the spectra of monomeric substances is rarely achieved; resonances are comparatively broad, and it is often impossible to utilise spin-spin coupling patterns for structural diagnosis when analysing the spectra of polymers.

The purpose of the solvent is to break up inter- or intra-molecular forces which restrict segmental motion; powerful solvents are needed when there is strong interaction between polar groups in the polymer. Thus, for example, nylon will give good spectra only in solvents such as tri-chloroacetic acid.

Where inter- and intra-molecular forces are comparatively weak e.g. in hydrocarbon polymers, satisfactory spectra may be measured in the melt rather than in solution. In practice this method of investigation is of limited value as the sample in this state is liable to thermal oxidation and decomposition.

Most of the common proton-free solvents either will not dissolve polymers, or are too volatile for use at elevated temperatures. Alternatives which have been suggested include arsenic and antimony trichlorides. Although these are often good solvents for polymers, we do not feel that their use can be recommended for general work because of their toxicity, and we prefer to use proton-containing solvents chosen so that their resonances do not interfere with the spectrum to be measured, even though more than one solvent may be necessary to record the complete spectrum of the sample. This occasionally causes difficulties because solvent—solute interactions occur and the different portions of the spectrum measured in the different solvents are not strictly comparable.

Some solvents which we have used for particular resins are listed in Table 1.4. Although solvents such as trichloro- and trifluoro-acetic acid are useful because the resonance of the single proton in the solvent is at such low field ($\sim 0\ \tau$) that it rarely interferes with the spectrum, many polymers are decomposed by the strongly acid medium; an indication of decomposition is a darkening in the colour of the solution. Furthermore, the resonance of the carboxylic acid proton is temperature sensitive, and if this resonance is used as a lock signal the spectrum may drift when operating at elevated temperatures. The strength of the polymer solution needed is not usually critical, and about 10% w/v is often satisfactory. The solution in the sample tube should be free of air bubbles.

TMS is too volatile to provide a reference signal for high-temperature

Table 1.4. Selected n.m.r. solvents for polymers

Polymer	Solvent	Temperature (°C)
Polyacrylonitrile	Dimethyl sulphoxide-d_6	~100
Polyethylene	α-Chloronaphthalene, pentachloroethane	
	o/p-dichlorobenzene	100–160
Polyester resins	Chloroform-d_1, acetone-d_6	ambient
Polymethylmethacrylate	Chloroform-d_1, benzene, o/p-dichlorobenzene,	ambient → 160
	α-chloronaphthalene	
Polypropylene	o/p-Dichlorobenzene, α-chloronaphthalene,	
	pentachloroethane	120–170
Polyvinylacetate	Benzene, o/p-dichlorobenzene	ambient → 160
Polyvinyl alcohol	D_2O	~80
Polyurethane	o/p-Dichlorobenzene	120–170
Polyvinylchloride	o/p-Dichlorobenzene	~140
Polyethylene terephthalate	Trichloro-acetic acid	100–140
Polyoxymethylene	o/p-Dichlorobenzene, p-chlorophenol	~170
Poly(diphenylene ether) sulphones	Dimethylsulphoxide, chloroform-d_1	~100
Vinylchloride/vinyl isobutyl ether copolymer	o/p-Dichlorobenzene	140–160
Ethylene/propylene copolymer	o/p-Dichlorobenzene, diphenyl ether	160–200
Ethylene/vinyl acetate copolymer	Benzene, o/p-dichlorobenzene	ambient or 160

Note: o/p-dichlorobenzene is a saturated solution of p-dichlorobenzene in o-dichlorobenzene at room temperature.

work, and hexamethyl disiloxane (HMDS) is preferred. The resonance appears at approximately 9·95 τ, but this varies slightly with different solvents. The resonance position may be found by adding small amounts of both HMDS and TMS to the solvent in question and measuring the value of the HMDS resonance relative to that of TMS at 10 τ. Since polymer solutions are always dilute, the further effect of adding the polymer to the system is usually neglected.

EXAMPLES OF THE APPLICATION OF N.M.R. SPECTROSCOPY

Many examples of the application of n.m.r. spectroscopy in the polymer field are given in this book. An example of the identification of a stabiliser is given on p. 137. The qualitative identification of some vinyl and acrylic polymers appear on p. 204, and the quantitative analysis of a copolymer is considered on p. 211. The application of n.m.r. spectroscopy to the detection of isotactic and syndiotactic sequences in polymers is illustrated by reference to polymethyl methacrylate on p. 272.

GAS CHROMATOGRAPHY

The contribution of gas chromatographic methods to more rapid and improved analysis in many fields has been seen, no less, in the analysis of plastics materials. For those without experience of the method there are many textbooks detailing the theory, instrumentation and procedures in gas chromatography[42-45]. Although primarily a method of separation of mixtures in the vapour state, involving partition differences between the components (transported through the column by a carrier gas) and either a liquid stationary phase or a solid absorbent, the high efficiency of the columns, detectors, amplifiers, recorders and other ancillary equipment available allows quantitative data to be obtained on the eluate gases coming from the column.

A mass of commercially produced equipment is available, some of it highly specialised in its application, and so many variations in essential components such as columns and detectors have been described that it would be difficult to generalize further than stating that the choice of column (length, diameter and packing), detector and to a lesser extent the type of ancillary equipment will depend on the analysis which has to be carried out. Equipments incorporating dual columns and detectors, pre-columns, column switches, back-flushing facilities, temperature programmers and pyrolysis attachments etc. all have their value in the analysis of plastics materials, and we are grateful to Mr. A. R. Jeffs for his help in preparing the following comments and suggestions made as a result of his considerable experience of the application of gas chromatography in the plastics industry.

COLUMNS*

It has been our experience that packed columns in the internal diameter range $\frac{3}{16}$ to $\frac{1}{16}$ in (4·5 to 1·5 mm) are the most valuable for general use, particularly when the analyst may be called upon to carry out quality-control analysis on the monomers used by his company. Such analysis, although outside the detailed scope of this book, includes the detection and determination of often less than 5 p.p.m. quantities of objectionable impurities in monomers such as ethylene, vinyl chloride and tetra-fluoroethylene. For this type of work, although open capillary columns

---

*Tube couplings for gas chromatographic columns are not generally available in metric sizes. Outside diameters of tubing are therefore also quoted in inches and these are the accurate dimensions.

may provide excellent separations, their very small capacity precludes their efficient use. Narrow-bore packed columns on which relatively large sample loadings can be used are much to be preferred, particularly when used in conjunction with sensitive flame ionisation detectors.

With regard to column packings, a few simple rules may be of value. For the solid support, C22 firebrick (Chromosorb P, Gas Chrom R,

Table 1.5. Stationary liquid phases for gas liquid chromatgraphy

Stationary phase material	Maximum temperature of operation (°C)	Comments and uses in plastics analysis
Silver nitrate in ethylene glycol solution	40	These stationary phases are
Silver nitrate in benzyl cyanide solution	40	normally used at 0°C, alone or in
$\beta\beta$-Oxydipropionitrile	60	combination to give specific
Propylene carbonate	40	separation of C_2–C_6 paraffins and
Tetraethyleneglycol dimethyl ether	60	olefines and dienes.
Glycerol	70	Used normally to retard water
Diglycerol	110	and lower alcohols to give a specific separation.
Squalane	120	Non-polar
Dinonyl phthalate	130	Moderately polar
Di-n-decyl phthalate	150	General Moderately polar
Polyethylene glycol 400	100	Purpose Strongly polar
Carbowax 1500	120	Polar
Tricresyl phosphate	130	Moderately Polar
Bentone 34	150	Phenols and aromatics generally
Silicone oil (Embaphase)	200	Separates methyl methacrylate and ethyl acrylate at 60°C.
Ucon 50 HB 2000	200	See example on p. 55.
Nitrile Silicone XE60	200	
Carbowax 20 M	250	
Nonyl-phenoxy-polyethylene-oxyethanol	200	Plasticiser alcohols, and gives reasonable separation of hydrocarbon fractions for fingerprinting in solvent analysis.
Versamid 900	300	Polar: separation of monomers and dimers in latex odour analysis.
Silicone gum rubber SE 30 (or E 301)	300	Non-polar
Porapak	225	Glycols

*Note: for operation of a flame ionisation detector at the highest sensitivity the maximum temperature of operation should be 30° to 50°C lower than that quoted.

Anakrom P), when sieved to a mesh size between 60 and 80 mesh and washed or blown free from fines, makes for an excellent, easily packed column, but should only be used with between 20% and 40% w/w stationary phase and when non-polar or relatively non-polar substances such as saturated and unsaturated hydrocarbons are to be separated (e.g.

in the determination of paraffins, olefines and acetylenes in 1 : 3-butadiene). Polar substances exhibit tailing due to adsorption on the support. Celite (Chromosorb W, Gas Chrom. CL, Anakrom U, Embacel) can be used with 1 to 30% w/w stationary phase, but is slightly more difficult to pack. Strongly polar substances will however still show some tailing. Silanised grades of both the above types of support are also available. This treatment does in fact reduce the adsorption properties of the support but does not in all instances eliminate it completely. For separations involving acids, obtained for example by the alkaline hydrolysis of peroxides used as polymerisation initiators, the Celite should be acid washed and an acid incorporated with the stationary phase. 1 or 2% w/w phosphoric acid, sebacic acid and stearic acid have all been used with good effect. For separations involving basic substances such as amines, the Celite should be acid and alkali washed, and 5% w/w of KOH evaporated and dried on to the support.

Powdered polytetrafluoroethylene (Chromosorb T, Teflon 6, Fluon CD 4) is recommended as a support for the separation of inorganic polar substances, e.g. water, ammonia and hydrogen chloride. Because of the lower surface area of the support 10% w/w is the maximum amount of stationary phase which should be added. Coating the particles is more difficult because the polymer holds a static charge and packing a column uniformly is also not easy[46].

A relatively new support sold under the name of 'Porapak' consists of cross-linked polystyrene beads. This material can be coated with stationary phase in the usual manner but will perform many separations on its own account. The beads may be used up to 225°C. 'Porapak' columns have been used in our own laboratory to give excellent separations of ethylene, propylene and butylene glycols derived from the hydrolysis of polyester resins.

Literally hundreds of organic substances have been reported in the literature as being used as stationary liquid phases for gas liquid chromatography. Table 1.5 lists about 20 of these which in our experience are very valuable.

DETECTORS

Katharometers and flame ionisation detectors have so far proved adequate for the gas chromatographic problems arising in the plastics and polymer field. Both detectors are suitably described in the literature and it is sufficient here to indicate their chief characteristics and when one detector is used in preference to the other.

THE FLAME IONISATION DETECTOR

This detector is extremely sensitive to almost all organic substances; it has a wide linear range and is reasonably insensitive to changes in temperature and flow rate. This detector is insensitive to all permanent and inorganic gases, water and carbon disulphide. Unlike the argon ionisation detector, the flame ionisation detector is not desensitised by the passage of water through the detector. One disadvantage of this detector is of course that it is destructive, but it is normally used in our laboratory for most organic analysis unless permanent gases and water are also to be determined or fraction collecting or direct infra-red analysis of the eluate is required.

THE KATHAROMETER

Although the katharometer lacks the high sensitivity of the flame ionisation detector and has a narrower range over which the response is linear, it is extremely valuable particularly for non-destructive work and in analysis in which response to water and the permanent gases is required. It is temperature and pressure sensitive and must therefore be adequately housed. It has the advantage that only a relatively simple Wheatstone bridge ancillary circuit is required together with an attenuator to reduce the signal as needed.

The flame ionisation detector can be used in series with the katharometer, but whether in single or series mode both detectors require calibration for the components being determined since their response to different functional groups varies.

RECORDING

The recording device commonly used is the potentiometric recorder. A good recorder for chromatographic use should have a span of not more than 2·5 mV and a full scale response time of 1 s or less. The recorder should be kept well adjusted to the correct gain so that the pen exhibits no dead band and does not oscillate. If changes in signal attenuation are to be made during the development of the chromatogram, it is important that the pen be zeroed to the electrical zero, so that the baseline does not shift with each attenuation step. Chart speeds tend to vary with the application. For instance if peak areas are to be measured and the chromatogram shows components eluted at an early stage, a relatively fast chart speed is required to give a more accurate measurement of the peak width. If peak heights only are compared, then a slower speed is more usual. Chart speeds between 6 in/h and 60 in/h (120 mm/h and 30 mm/min) are commonly employed.

TEMPERATURE PROGRAMMING

The elution time of a component from a chromatographic column and the shape of the peak obtained depend on the temperature of the column; hence in those cases where mixtures containing both high- and low-boiling

47

constituents are to be separated, and a reasonable chromatogram is to be obtained within a practical time limit, temperature programming is essential. The limitations of temperature programming, particularly as applied to single columns, should however be appreciated. Programmers can be obtained to give a temperature rise in the chromatographic oven of from 1°C/min to 40°C/min. These permit excellent separations to be effected but often the repeatability of the analysis, particularly at high programming rates, is such that retention times measured after approximately 30 min with a chart speed of 15 in/h (600 mm/h) can differ by times corresponding to as much as 5 mm on the chart on consecutive applications of the same sample, even when the cool-down time and re-stabilisation time at the starting temperature are well standardised.

With a single column and detector only the lower sensitivity ranges of the ionisation detector amplifier can be used since at the higher sensitivities the base line drift and background noise due to uncompensated temperature rise and 'bleeding' of the stationary phase, respectively, give increasing interference. In general therefore this application is normally confined to the separation of percentage components.

In the plastics industry, however, much attention must often be paid to the determination of small amounts of monomer or undesirable low polymers, e.g. dimers in high-polymer samples, since such components may impart an objectionable odour to the plastic, be present in concentrations of only a few p.p.m. and also have only a low vapour pressure. In such cases we find compensated programming facilities to be essential.

Two identical columns are packed and mounted in the same oven, and twin detectors are employed. The sample is injected into one of the columns and the oven programmed to the desired temperature. Since both columns are subjected to the same treatment and the output of the 'blank' detector is offset against the output of the analysis detector, a difference chromatogram is obtained.

THE USE OF PRE-COLUMNS

The term pre-column denotes a short length of column which can readily be detached and replaced at regular intervals. Its purpose is to retain non-volatile or comparatively non-volatile material, preventing contamination of the main gas chromatographic column.

A similar short length of column attached externally to the injection port of the chromatograph can be used for the determination of toxic organic substances in the atmosphere[47]. In this case the short column or absorption tube is sleeved with a heater and a by-pass valve fitted. The tube containing a normal gas liquid chromatographic packing or solid absorbent is taken to the plant or laboratory to be tested. A given volume of the atmosphere is drawn slowly through the packing at the ambient temperature, by means of a small hand pump. The tube is brought back to the chromatograph, connected to the inlet and heated. At this stage the tube is by-passed by the carrier gas and after a suitable heating period (2 to 3 min) the carrier gas is diverted through the tube, carrying the volatile components on to the column where the chromatogram is developed in the normal manner. This method, for example, is the only convenient way that we have found that enables methyl methacrylate and ethyl acrylate monomers to be separately determined in a plant atmosphere when both are present at the same time; this is important particularly as the toxic limit for the acrylate monomer is one tenth that of methyl methacrylate.

Precisely the same principle can be used to separate and determine the volatile components present in a polymer. The sample is weighed into a short empty tube, which is heated in a closed loop prior to chromatography of the released volatiles. Moisture, blowing agents (in the case of foamed polymers), residual polymerisation solvents, dimers and in some cases residual monomers can be determined in this manner. Some monomers will, of course, undergo further polymerisation during the preliminary heating period, and each system must be carefully evaluated for quantitative work. To facilitate this, a rather more elaborate apparatus has been developed[48] specifically for the determination of volatile components in polymers and polymer preparations (see p. 52).

BACK FLUSHING AND COLUMN SWITCHING

When only the early peaks in a chromatogram are of interest the time between analyses can be shortened by employing back flushing. In this procedure the flow of carrier gas through the column is reversed after the components of interest are eluted. It is important however that the injection port is included in the back flushing system.

Column switching and cutting can also be useful procedures. This facility is often adopted when no one column can be found which separates all the components of the mixture and two or more components remain unresolved. That portion of the eluate corresponding to the unresolved part of the chromatogram can be diverted to a second column which separates these particular components. Under these circumstances the pressure-balance system of Deans[49, 50] can be used as an alternative to the switch valve.

STANDARDISATION AND CALCULATION IN QUANTITATIVE WORK

In calculating the sample composition from the chromatogram, three methods are in common use.

1. Internal standardisation

This method is generally used when one or two components present in the mixture in the range 0·1 to 50% are to be determined accurately. A known weight (or volume) of an organic substance not present in the original mixture is added to a known weight (or volume) of the mixture. The internal standard must be resolved from all other components and should have a similar concentration to the components being determined. Ideally it should be chemically similar and elute close to the peaks of interest. Calibration mixtures are prepared containing the same concentration of the chosen internal standard and their chromatograms obtained. The peak heights (or areas) of the particular components and the internal standard are measured and a calibration graph constructed relating the ratio

$$\frac{\text{Peak Height (or area) of component}}{\text{Peak Height (or area) of internal standard}}$$

to the known % of component. The internal standard procedure compensates for handling errors (i.e. a precise volume need not be injected each time) and for small vagaries of the instrument (temperature and flow fluctuations). This method is used in particular in the quantitative analysis of liquid monomers and solvent mixtures etc.

2. Normalisation

Normalisation is often the preferred method for calculating the composition of a multi-component mixture, when most of the components are present in the percent range, for example hydrocarbon fractions. The area of each peak is calculated (the preferred manual method being Peak Height × Peak Width at half height), and for the most accurate work each area is corrected for the response factor for the corresponding component. All the areas are summed and the % of component x is then given by the formula:

$$\%(x) = \frac{A_x \times 100}{\sum\limits_{n} A_n}$$

where A_x is the corrected area of component x and n is the number of components.

A limitation of this method of calculation is that the whole of the applied sample must be completely eluted from the column and registered by the detector. In many cases this assumption is not correct, however, as anyone examining the inlet port of a gas chromatograph after a few weeks use can see from the tarry residues there deposited. Furthermore, the gradual falling-off of column performance (or in a few cases an increase of column performance) is often due to modification of the stationary liquid phase with non-volatile material from a large number of sample injections.

Ideally, the normalisation method of calculation is best performed with the aid of an integrator reading directly from the electrical output of the chromatographic instrument. The more expensive types of integrator are obviously to be preferred; they display and then print out each peak area together with its retention time in seconds, and will compensate for base-line drift, account for partially resolved peaks and integrate off-scale peaks. On demand at the completion of the chromatogram, the total area is printed.

3. Equal-volume injection method

When the sample under examination is reasonably pure and is being examined for impurities in the range 5 p.p.m. up to 1%, the equal-volume injection method is often used. A precise volume of the sample under examination is injected under standard conditions by means of a Hamilton syringe (or gas-sampling valve). The peak heights (or areas) of the impurity components are directly compared with those obtained by injecting the same volume of standard samples under the same con-ditions. Dilute standard solutions can be prepared conveniently by adding the impurity to a pure sample by a similar syringe.

EXAMPLES OF THE APPLICATION OF GAS CHROMATOGRAPHY

Examples of gas chromatographic analysis in the plastics industry are described throughout this book and include the examination of products derived from the chemical attack or depolymerisation of acrylic polymers (p. 230), the determination of free acrylonitrile, methyl methacrylate and styrene in butadiene copolymer latices (p. 433), the determination of free furfural and the nature and amount of free phenols and cresols in phenol-formaldehyde resins (p. 473), the examination of solvents (p. 592) and the separation of alcohols obtained by hydrolysis of ester plasticisers (p. 568).

A particular example demonstrating the points mentioned in this preliminary description of the method is given in some detail below and concerns the determination of water in polyolefine products.

APPARATUS

The apparatus as designed by Jeffs[48] for general use in the determination of water and other volatiles in polymeric materials consists of a sample tube or chamber forming an external loop connected to a gas chromato-graph. This loop can be isolated and the sample heated to the required

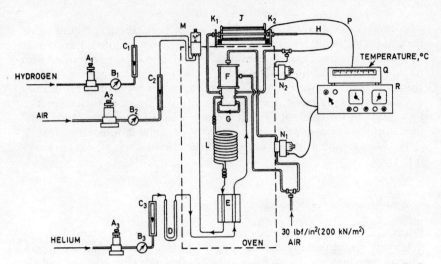

Fig. 1.8. *Details of general-purpose instrument for the determination of water and other volatiles in polymeric materials (after 'Analyst')*

A_1, A_2 *and* A_3 = *Edwards VPC 1 pressure controller*

B_1, B_2 *and* B_3 = *Pressure gauges 0 to 30 lbf/in²* *(0 to 200 kN/m²)*

C_1, C_2 *and* C_3 = *Rotameter-type flow gauges*

D = *Clear plasticised PVC tubing (1·5 m long × 6 mm bore), packed with copper sulphate crystals,*

E = *Katharometer*
$$CuSO_4 \cdot 5H_2O, > 44 \ mesh$$

F = *Pneumatic sample valve (Pye Cat. No. 12900), fitted with P 9904 change-over block*

G = *Internal loop*

H = *External loop made in part of 18-gauge stainless-steel capillary tubing (0·048 in or 1·22 mm o.d.)*

J = *Split sample heater*

K_1 *and* K_2 = *Straight reducing couplings, captive-seal type for ⅜ to ¼ in o.d. tubing (Drallim, Cat. No. L 50/D/B)*

L = *Chromatographic column*

M = *Flame ionisation detector*

N_1 *and* N_2 = *Electrically actuated three-port pilot valves (Martinair, Type 557C/IZ)*

P = *Nickel-chromium/nickel-aluminium thermocouple embedded in the split heater*

Q = *Electronic temperature controller, proportional type*

R = *Sample valve time-delay unit, containing three synchronous timers and a semiconductor proportional energy controller for the sample heater* temperature. After an initial heating period, the volatiles liberated from the sample are 'flushed' on to a gas chromatographic column by a flow of carrier gas through or over the sample, and the required components separated and determined quantitatively. The essential parts of the apparatus (Figure 1.8) are the pneumatic sample valve and the split heater mounted on horizontal travel in a plane at right angles to the sample tube. To prevent oxidation, provision is made for the sample to be purged free from air in the cold, and the excess pressure of inert gas to be released to atmosphere before the sample is heated. This operation can proceed automatically with the aid of sequential timers. The sample heater is then moved over the sample and the sample is heated for a pre-set period of time (controlled by a third timer) after which the volatiles are flushed on to the chromatographic column.

The components from which the apparatus is assembled can obviously be varied according to availability. Almost any commercially available katharometer is satisfactory. The oven must be large enough to accommodate the column katharometer and the pneumatic valve, since condensation must be prevented.

The sample heater (Figure 1.9) consists of a cylindrical aluminium block with a 9 mm hole through the centre. The block is split axially and the two halves hinged. Each half of the block contains two cartridge

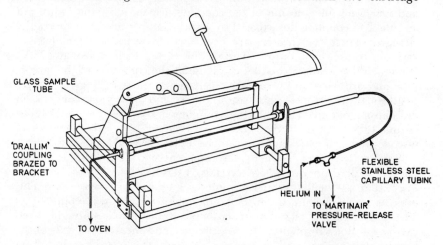

GLASS SAMPLE TUBE

'DRALLIM' COUPLING BRAZED TO BRACKET

FLEXIBLE STAINLESS STEEL CAPILLARY TUBING

HELIUM IN

TO 'MARTINAIR' PRESSURE-RELEASE VALVE

TO OVEN

Fig. 1.9. Split sample heater (see J in Fig. 1.8), shown in open position advancing over sample tube (After 'Analyst')

heater elements $5\frac{1}{2}$ in (140 mm) long and $\frac{3}{8}$ in (9 mm) o.d. (210 W). One half of the block contains a thermocouple pocket accommodating a Ni-Cr/Ni-Al thermocouple. The heater is supplied by a semiconductor energy controller which in turn is controlled by a galvanometer two-position temperature controller.

53

The heater is mounted on a horizontal travel in a plane at right angles to the sample tube, and the jig for mounting the sample tube is also part of the heater assembly. One $\frac{3}{8}$ in coupling is brazed to the mounting bracket and another $\frac{3}{8}$ in coupling is brazed to the flexible carrier-gas inlet tube and rests loosely in the second mounting bracket.

The disposable sample tubes are made by cutting 9 mm o.d. Pyrex glass tubing into 200 mm (± 2 mm) lengths. The cut ends of the tubing are fire polished in a flame.

The pneumatic sample valve may be operated manually or the whole sequence of the analysis governed automatically by a control unit incorporating three synchronous timers and associated circuitry. Tube connections between the column, detectors and sample heater are in 18-gauge stainless-steel capillary tubing (0·048 in; 1·22 mm o.d.) and non-flexible couplings between the switch valve and other components are in $\frac{1}{8}$ in (3 mm) o.d. copper tubing.

Although for the sake of clarity this is not shown in Figure 1.9, the inlet and exit tubes to the sample-tube coupling are wrapped with an electric-heating tape or element to maintain the temperature of the whole assembly at approximately 100°C. The inlet tube is wrapped over a length of 4 to 5 in (100 to 130 mm) and the 'L'-bend of the exit tube is wrapped to a point 3 in (75 mm) inside the oven. Both tubes are wrapped up to, and including, the $\frac{1}{4}$ in nut of the coupling, leaving only the centre nut and the $\frac{3}{8}$ in coupling nut exposed so that the sample tube may be readily changed. These two heaters are connected in series and run from a variable transformer, output 0 to 270 V (2 A), maintained at 170 V. Without these heaters, volatiles condense at the cold ends of the sample tube and are swept away comparatively slowly on switching the carrier gas through the sample. This leads to diffuse and unsymmetrical peaks in the chromatogram. The reducing couplings for connection of the glass sample tube must be of the captive-seal type and the seals should be of silicone rubber to withstand continuous use at elevated temperatures.

A 1·5 m length of 6 mm bore transparent polyvinyl chloride tubing packed with copper sulphate crystals, $CuSO_4 \cdot 5H_2O$, (>44 mesh) is inserted on the helium carrier gas inlet line to 'wet' the carrier gas. This imparts to the helium a constant amount of water (≈ 3 p.p.m. w/v). The determination of water in a sample is unaffected since the 'wet' carrier gas flows through both the reference and the analysis cells of the katharometer. The 'wet' carrier gas prevents the entire gas line from 'drying out' completely. Dry pipework tends to absorb moisture which can then be desorbed leading to spurious results.

THE DETERMINATION OF WATER IN POLYPROPYLENE

The following gas chromatographic conditions have been found suitable for the determination of water in polypropylene powder and granules.

The method covers the range 0·01 to 0·08% water when a 2 g sample is used.

The column consists of 4 ft (1·2 m) of $\frac{1}{4}$ in (6 mm) o.d. stainless-steel (or copper) tubing packed with 10% w/w Ucon 50 HB 2000 fluid on polytetrafluoroethylene powder prepared as described below.

Weigh out sufficient polytetrafluoroethylene powder into a shallow evaporating dish, weigh the appropriate amount of Ucon 50 HB 2000 fluid (ex Perkin-Elmer) into a beaker and dissolve in methylene chloride. Then pour the methylene chloride solution over the polytetrafluoroethylene powder and rinse the beaker with methylene chloride. Add this methylene chloride to the powder. Continue the rinsing until the powder is thoroughly wetted. Stir the mixture gently and drive off the solvent with a gentle stream of compressed air. *(Note: Carry out this operation in a fume hood: the operator should not smoke at any time while polytetrafluoroethylene powder is being handled.)* Continue the evaporation until all the methylene chloride has been removed. Cool the resulting mixture in a refrigerator for a short time before packing the column (polytetrafluoroethylene powder is handled and packed into a column more easily below its phase-transition point of 19°C). Pack the column lightly without excessive tapping.

The temperatures of the oven, column and katharometer switch valve should be 100°C.

The carrier gas is helium, at an inlet pressure of 5 lbf/in^2 (35 kN/m^2), or adjusted to give an indicated flow rate, at the inlet, of 110 ml/min.

The sample heater temperature is 260°C.

Procedure

Fit the glass sample tubes each with a small quartz-wool plug approximately 1 in (25 mm) from one end, and store in an oven until just before they are required. Transfer several tubes from the oven to a desiccator containing phosphorus pentoxide and allow to cool. (The desiccant should be examined at weekly intervals and renewed when necessary.) Use the tubes from the desiccator as required. Weigh a sample tube and pour the sample (powder or granules) into the tube. Reweigh and then insert a second quartz-wool plug with tweezers. These plugs are pre-prepared, oven dried and stored in the desiccator. Place the sample tube in the heater jig between the captive seal couplings and tighten the couplings. By manual operation or by automatic control by means of the sequential timers arrange for the following sequence of operations to be carried out.

1. At room temperature operate pneumatic switch valve to allow helium carrier gas to purge sample tube free of air for 15 s.
2. Operate second exhaust valve to allow excess helium pressure in closed sample loop to escape to the atmosphere.

3. Open split heater and move it forward on its travel to close over sample tube.
4. Allow 5 min heating period at 260°C.
5. Operate penumatic switch valve to allow carrier gas to flow through and/or over the sample, flushing the volatiles released during the heating period on to the gas chromatographic column.

Two chromatograms are obtained. The first results from the 15 s purge of helium through the cold tube, which shows a peak for air followed by a small peak of water. This is developed while the sample is being heated for 5 min. The second chromatogram, after the heating period, shows the liberated water followed by a peak for any residual hydrocarbon diluent in the sample. Carry out a blank determination on an 'empty' tube

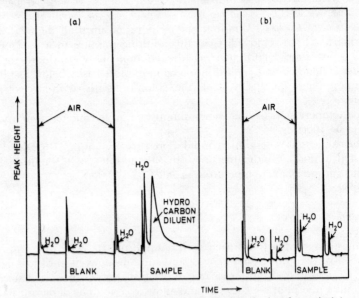

Fig. 1.10. *Typical chromatograms obtained in the determination of water in (a) polypropylene powder and (b) high-pressure polyethylene (After 'Analyst')*

containing two quartz-wool plugs only, in the same manner. Sum the two water peaks resulting from each determination and correct the sample figure for the blank value. Figure 1.10 shows typical chromatograms obtained on a blank tube and on a sample of polypropylene powder and high-pressure polyethylene.

Calibration

Calibration is best carried out with barium chloride crystals, $BaCl_2 \cdot 2H_2O$, weighed into empty sample tubes. Barium chloride loses its water of crystallisation, 14·75% w/w, at 115°C. Chromatograms obtained in this manner are similar to those obtained on samples, i.e. a small water peak, due to the atmospheric moisture, is obtained when the tube is purged with helium for 15 s, followed by a larger water peak after the heating period when the water is released from the crystals.

If calibration is carried out by direct injection of water on to a quartz-wool plug in an 'empty' tube using a Hamilton syringe, 75% to 90% of this water is purged on to the column during the 15 s period. Chromatograms are therefore obtained which are quite unlike those obtained for a sample, although the total amount of water should be unchanged.

PYROLYSIS/GAS CHROMATOGRAPHY

The examination of the products of thermal degradation (pyrolysis) *in vacuo* has for a long time been one of the methods used in the characterisation of polymers and copolymers. Over the years it has been in the methods of examination of the pyrolysate rather than in the methods of pyrolysis that advances have been made through the application of instrumental methods as these became available. These instrumental methods, infra-red spectroscopy and gas liquid chromatography, have the great advantage over the chemical methods originally used in that it is not always necessary to completely identify the components of the pyrolysate, since identification of the original polymer may be made by comparison of the spectrum or chromatogram of the pyrolysate with spectra obtained from known polymers pyrolysed under the same conditions, e.g. under vacuum in a sealed tube at a given temperature.

From the gas chromatographic examination of pyrolysates it is a logical step to carry out the pyrolysis in the carrier gas stream so that the products are carried directly on to the column. This requires a rapid method of pyrolysis in order that the products may enter the column as a 'plug', or else the separation will be impaired.

Of the several ways of carrying out a pyrolysis we prefer that of Barlow, Lehrle and Robb[51] in which the sample is pyrolysed on an electrically heated filament mounted in the carrier-gas stream at the column inlet. These authors used a katharometer detector but far superior results are obtained with a flame ionisation detector. Pyrolysis directly coupled with the gas chromatograph requires only a small amount of sample and no isolation or transfer of the pyrolysate. Furthermore, the method is rapid and the separation on the column combined with the rapid removal of pyrolysate from the hot zone reduces the possibility of secondary reactions taking place between the products.

The description of the equipment and procedures which follow and applications of the method given throughout this book result largely from the work of our colleague Mr. P. E. Arnold, to whom grateful acknowledgement is made.

The essential features of a suitable general-purpose instrument are detailed below.

Coil and head—Figure 1.11

A 90 mm length of 30 S.W.G. (0·304 mm) Nichrome wire is wound on a former to produce a cylinder 2 mm i.d. closed at the lower end by a single turn about 1 mm diameter. The coil is spot welded to tungsten leads with nickel intermediates and the whole made to fit into the inlet of the gas chromatographic column. These coils have been found to give long service (hundreds of pyrolyses) and no detrimental effects due to the presence of residues have been detected.

Before use the coils are calibrated by placing substances of known melting points in the coil, slowly increasing the current and recording the current passed when the substance melts. This is carried out with the head fitted in the column inlet and the carrier gas flowing. A typical calibration

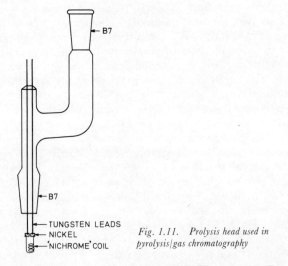

TUNGSTEN LEADS
NICKEL
'NICHROME' COIL

Fig. 1.11. Prolysis head used in pyrolysis/gas chromatography

graph showing the substances used is illustrated in Figure 1.12. From such a calibration graph the current required to produce a particular temperature may be deduced. All the substances used in the calibration are water soluble and may be removed by immersing the coil in warm water.

Controller

The temperature of pyrolysis may be varied by a variable resistance connected in series with the filament and the time controlled by reference to a stopwatch, but more reproducible pyrograms are obtained if these variables are controlled mechanically or electronically. The mechanical controller described below, based on the original design of Barlow, Lehrle and Robb[51], has given many years of trouble-free service. It consists of two essential parts. The first part controls the pyrolysis temperature and contains a constant-voltage transformer which may be set to produce different pyrolysis temperatures rising in 100°C steps between 150°C and 950°C, the required temperature being obtained by means of a selector switch. The controller may be operated manually or automatically. There is also a 'burn-off' switch which passes a large current through the coil raising it to about 1000°C, at which temperature any carbonaceous residues will be burnt off in air. The second part of the controller contains

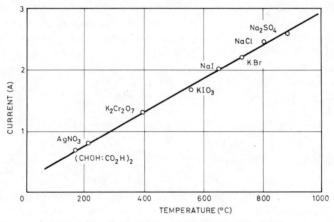

Fig. 1.12. Pyrolysis filament calibration (see Fig. 1.11)

a mechanical timer based on a synchronous motor driving a plate at 1 rev/min. Studs on this plate activate micro-switches which in turn operate the relays to switch the current to the coil on and off, once the circuit has been primed. A boost heater is incorporated which switches in for the first seconds of a heating cycle to raise the temperature to 250°C very rapidly. The timer is calibrated against a stopwatch and may be varied to give pyrolysis times between 5 and 30 s.

GAS CHROMATOGRAPH

Any standard gas chromatograph fitted with a flame ionisation detector and an inlet system modified to accept the pyrolysis head may be used. The column we prefer for general work is a $5\,ft \times \frac{1}{4}\,in$ ($1\cdot5\,m \times 6\,mm$) diameter column packed with 30% w/w Embaphase silicone oil on 'acid washed' Celite 545 (30–80 mesh). It is necessary to wash the Celite by soaking it overnight in concentrated hydrochloric acid and to wash with water in order to produce a support from which acidic pyrolysates such as acetic acid can be eluted.

Connection between the pyrolysis coil and the column is made by a tube (length $6\frac{1}{2}$ in or 165 mm overall) with a B7 socket at the top to form the pyrolysis chamber and a B7 cone at the bottom to fit into the column. The tube contains a 2 in (50 mm) length of 30% w/w silicone oil on Celite packing, which acts as a trap to retain tarry materials produced in the pyrolysis and to prevent these contaminating the column. This tube is repacked periodically.

PROCEDURE

Gas chromatographic conditions

In order to obtain a set of standard pyrograms for the rapid 'fingerprint' method of polymer identification it is necessary to have one set of conditions under which all the standards and samples are chromatographed. Suitable conditions are as follows:

Column temperature	90°C
Nitrogen carrier gas flow rate	28 ml/min (inlet)
Hydrogen flow rate	50 ml/min
Air flow rate	9000 ml/min
Recorder chart speed	12 in/h (300 mm/h)

SAMPLE PREPARATION

Normally a suitable weight of sample in the form of a single piece is placed in the coil on the small bottom turn. Powder samples are hydraulically pressed to form a film or disk from which a suitable piece is cut.

When pressing is not feasible either due to the nature or to the amount of sample it has been found possible to mix the powder with a drop of a dilute aqueous solution of Polycel*, which when dry acts as a binder. The small amount of Polycel present causes little interference in the pyrogram obtained. For very small samples (< 100 μg) pyrolysis in a glass capillary as described by Stanford[52] may be used with the filament temperature increased 100°C above normal.

In the case of soluble polymers it is also possible to place drops of a solution of the polymer in a suitably volatile solvent on the coil and evaporate the solvent prior to pyrolysis, and this method is used in the quantitative analysis of certain copolymers as described in Chapter 4.

PYROLYSIS

Two methods of pyrolysis have been used:

1. Complete pyrolysis at one temperature.
2. Stepwise pyrolysis over a range of temperatures below that at which complete pyrolysis occurs. In this method the residue from the

* Polycel is a proprietary adhesive.

pyrolysis at one temperature is available for pyrolysis at the next higher temperature to give a series of pyrograms.

Normally pyrograms obtained by method (1) are sufficiently characteristic in the relative retention times of the peaks and the ratios of the peak heights for identifications to be made. In certain instances method (2) may be used to distinguish between polymers or copolymers and between copolymers and blends.

Pyrolysis time

15 s has been found to be sufficient to completely pyrolyse 1 mg of polymer at the maximum temperature permissible before excessive amounts of low molecular weight hydrocarbons are produced. 15 s pyrolysis over a range of lower temperatures also gives a reproducible series of pyrograms provided that the weights of the samples are not excessively different.

Complete pyrolysis at one temperature

With the coil removed from the column, the coil is brought to red heat to burn off any organic matter and then allowed to cool. 0·5 mg to 1 mg of the sample is placed in the pyrolysis coil and the head inserted into the connecting tube. The sample is heated at 250°C for 15 s to remove any volatile solvents and the chromatogram allowed to develop. If volatiles are detected this process is repeated until all traces have been removed. The pyrolysis proper is now carried out for 15 s at 550°C and the pyrogram allowed to develop (30 to 60 min). If the sample does not appear to pyrolyse completely, indicating a polymer of high thermal stability, e.g. polytetrafluoroethylene or silicone rubber, the procedure is repeated at a pyrolysis temperature of 750°C.

Stepwise pyrolysis over a range of temperatures

The pyrolysis is carried out for 15 s at 100°C intervals over a specific range. The amplifier sensitivity is reduced at each temperature to keep all the chromatogram peaks within the full scale deflection of the recorder.

IDENTIFICATION FROM THE PYROGRAM

Identifications are made by comparing the pyrogram obtained from the sample with those obtained from known polymers under the same conditions, no effort being made to isolate or identify the products.

The retention times of the peaks are first measured relative to styrene, the retention time of which is measured daily by placing a drop of a 0·2% w/v solution of polystyrene in chloroform on the coil, evaporating the

61

chloroform, pyrolysing the polystyrene and measuring the retention of styrene (the main peak) which under the chromatographic conditions given should be about 55 mm on the chart (10 to 12 min). In this way allowance can be made for slight variations in chromatographic conditions.

To aid the direct measurement of relative retention times a ruler has

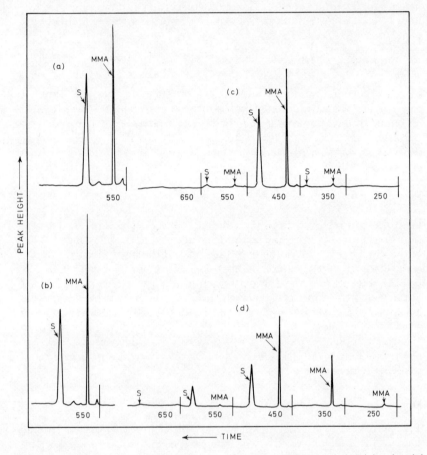

Fig. 1.13. Pyrograms of (a and c) methyl methacrylate/styrene copolymers and (b and d) polymethyl methacrylate/polystyrene blends of similar composition; (a and b) on direct pyrolysis and (c and d) on stepwise pyrolysis. (Figures are pyrolysis temperatures in °C; short vertical lines indicate pyrolysis starts)

been made from clear plastic with a series of scales marked on it, the basic units of these scales varying from 52 to 59 mm.

A library of standard pyrograms from samples of known origin is needed for comparison purposes and these are arranged so that those which have a similar relative retention time (R_T) for the main peak are grouped together. It is then only necessary to search through one particular section when attempting to match sample pyrograms, the relative height and shape of the minor peaks being compared to find the exact match.

EXAMPLES OF THE APPLICATION OF PYROLYSIS/GAS CHROMATOGRAPHY

The method is used with great success in the identification of many of the polymers described in the chapters which follow, including acrylics, rubbers, polyesters, polyolefines, vinyls and fluorocarbon polymers. It is of particular value in the examination of insoluble materials containing inorganic filler, and for small samples, e.g. fibres from mixed blends.

The method can also be used quantitatively when the more conventional approaches fail or give rise to difficulty, e.g. the analysis of methyl methacrylate/2-ethyl hexyl acrylate copolymers (see p. 224).

In certain cases it is possible to distinguish between copolymers and blends of similar composition by means of the stepwise pyrolysis procedure. This is illustrated by a methyl methacrylate/styrene copolymer and blend. Although the pyrograms obtained by complete pyrolysis are similar (Figures 1.13(a) and (b)), the stepwise procedure pyrograms obtained by pyrolysis of 1 mg for 15 s at each of the temperatures indicated reveal differences in the proportions of methyl methacrylate (MMA) and styrene (S) produced at the various temperatures. With the copolymer (Figure 1.13(c)), the first products are detected at 350°C, while after 450°C very little sample remains to be pyrolysed. At each temperature the proportion of methyl methacrylate to styrene is constant. However, with the blend (Figure 1.13(d)), only methyl methacrylate is detected at 250°C and 350°C, styrene first being detected at 450°C, while at 550°C it is essentially styrene that is produced. Although only the monomers are obtained from the copolymer and blends, in the blend the polymethyl methacrylate, having the lower heat stability, is decomposed preferentially, permitting a differentiation to be made.

GAS CHROMATOGRAPHY/INFRA-RED ANALYSIS

In the examination of unknown mixtures by G.L.C., it is not satisfactory to establish the identity of a fraction from retention time data alone. The infra-red spectrum of the fraction will however frequently give enough information to complete the identification.

In principle any G.L.C. column may be used for this work, if the appropriate apparatus is available to transfer the sample to the spectrometer. However, in practice it is necessary to collect about 0·05 mg for infra-red examination, and, even if the method used permits this quantity to be accumulated from successive runs, it is preferable to use a large ($\frac{1}{4}$ in or 6 mm diameter) column, as repeated collection is tedious and usually wasteful. Furthermore, a column with a non-destructive detector (e.g. hot-wire detector) is preferred as no modifications are necessary; if a

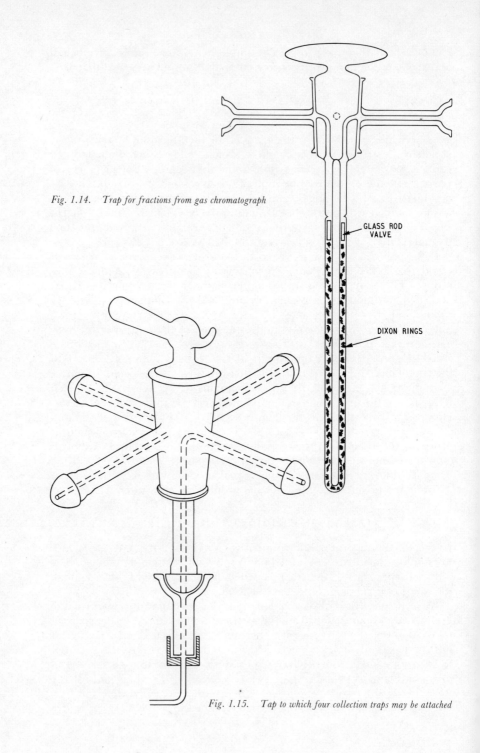

Fig. 1.14.   *Trap for fractions from gas chromatograph*

GLASS ROD VALVE

DIXON RINGS

Fig. 1.15.   *Tap to which four collection traps may be attached*

destructive detector (e.g. flame ionisation detector) is used, the effluent stream must be split into a fine capillary to feed the detector and a coarse capillary to feed the collection system.

Whatever sample collection procedure is used, it is very important that any valves or pipework between the chromatographic oven and the collecting apparatus are held at elevated temperatures, otherwise condensation may occur, with loss of efficiency of collection and cross-contamination of the fractions.

Various collection methods are possible and each may have advantages for particular applications. The simplest method is direct collection in a packed glass trap (Figure 1.14). A number of such traps, suspended in a liquid nitrogen bath, may be fed from a multiway tap (Figure 1.15). The course of the separation is followed on the column recorder. The effluent is passed to a particular trap, but with the parallel tap open to by-pass the cooled portion of the trap. Immediately a fraction shows on the column recorder, the by-pass is closed, and collection commences. When collection is complete, the effluent is diverted to the next trap, and the procedure is repeated.

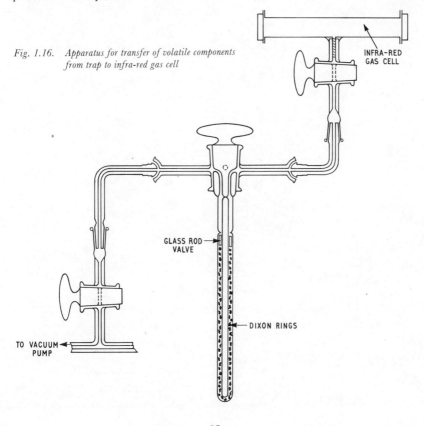

*Fig. 1.16.   Apparatus for transfer of volatile components
from trap to infra-red gas cell*

INFRA-RED
GAS CELL

GLASS ROD
VALVE

DIXON RINGS

TO VACUUM
PUMP

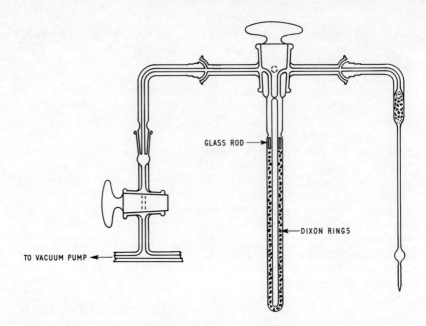

GLASS ROD

DIXON RINGS

TO VACUUM PUMP

Fig. 1.17.   *Apparatus for transfer of liquid components from trap to tip of side arm*

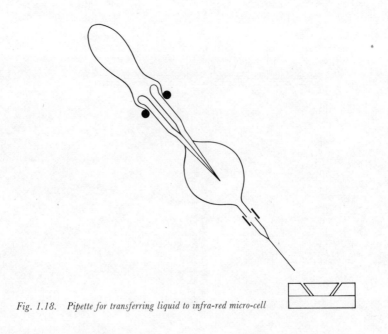

Fig. 1.18.   *Pipette for transferring liquid to infra-red micro-cell*

Any fraction boiling below about 100°C is best examined in the gas phase; suitable transfer apparatus is shown in Figure 1.16. With the trap still in the nitrogen bath, the apparatus is pumped down to low pressure. The pump is isolated, and the trap allowed to attain room temperature. The apparatus is then opened to air on the side remote from the gas cell; the entering air-stream draws the residue of the fraction from the trap into the gas cell, which is then closed off and removed to the spectrometer for examination. The construction of a suitable gas cell of small volume is described in the original paper[53].

Higher boiling components are distilled from the trap into a capillary tube (Figure 1.17). Again the apparatus is pumped down with the trap in the nitrogen bath; after isolating the pump the nitrogen bath is transferred to the capillary, thus causing the liquid to distil from the trap. When distillation is complete, the liquid is encouraged to collect in the capillary tip by gradually lowering the nitrogen bath until only the tip is submerged. The apparatus is opened to the air and the tip cut off. The liquid is transferred with a pipette (Figure 1.18) into a micro-cell, shown in Figure 1.19. The method of constructing this cell is given in Reference 53.

Considerable skill is required to transfer fractions to such a cell, otherwise transfer losses are substantial. For this reason direct collection in a cell which may be submerged in a solid carbon dioxide-methanol cooling bath may be preferred. Suitable cells (made from silver chloride sheet) are available from the Research and Industrial Instrument Company Ltd, London, and Beckmann Instruments Inc., Fullerton, California.

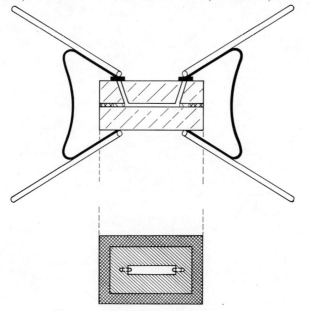

Fig. 1.19. Infra-red micro-cell for liquid samples

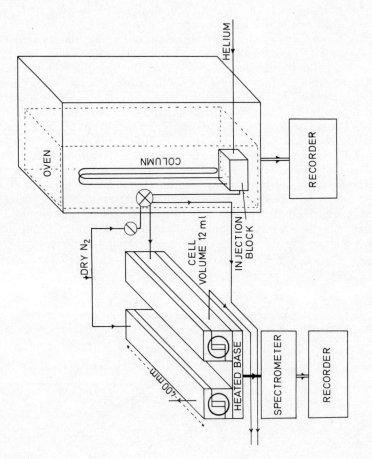

Fig. 1.20.   *Vapour phase chromatography/infra-red analyser*

The methods so far described are particularly useful if the fractions have to be stored to await spectroscopic examination. Furthermore, successive collections can be made if necessary to produce sufficient material. Nevertheless, a fair degree of manipulative skill is required, and the whole procedure is time-consuming and laborious for routine work.

We have therefore turned our attention to a method which is experimentally much simpler. The fractions are collected directly in a heated gas cell mounted in the spectrometer. The apparatus used is illustrated in Figure 1.20. The spectrometer source optics are drawn back on an extended base plate from their normal position so that the focus which previously was in the slit plane is now formed in the entrance window of the gas cell. The cell acts as a light pipe to transfer the focal position to the exit window of the cell; this is placed as close as possible to the entrance slit of the spectrometer. To preserve optical balance a similar cell placed in the reference beam is continuously flushed with nitrogen. The cells are of brass, made in two pieces, as shown, to facilitate gold-plating the interior surfaces. Good quality plating is important, as the surface is required to be both highly reflective for good light-pipe action, and to be chemically inert, otherwise there is serious risk of the interior of the cell becoming contaminated with decomposition products. The cells rest on a hot plate and are permanently held at 150°C.

The effluent pipe from the chromatograph is connected to the cell through a Loenco valve (L–206–6V six port, Viton A O-rings) mounted in the column oven, or otherwise heated; the connections to the valve are shown in Figure 1.20. With the valve in the open position, as shown, the column is operated in the normal way with effluent going to waste. As soon as a fraction is indicated on the column recorder, the valve is pushed through to the closed position, thus diverting the effluent through the absorption cell. When the column recorder reaches the band peak, the valve is returned to the open position, and the spectrum of the fraction is recorded. Immediately the recording is complete, the nitrogen flush valve is opened for a few seconds to displace the fraction from the cell. During this period the spectrometer is returned to the start position. This sequence of operations is repeated for subsequent fractions. Thus the manipulation is reduced to operating two valves, a process well suited to simple automatic operation, actuated from the column recorder.

It is evident that while a spectrum is being recorded, any fraction emerging from the column is lost. Thus the preferred spectrometer is that which records a recognisable spectrum in the shortest possible time; we have successfully used the Perkin-Elmer 157 with 1 min recording speed and 3 s reset time. A continuous paper recorder is most convenient, as the time taken to change the chart paper is significant in this procedure. Even under the best conditions it is almost inevitable that one or more peaks will be lost, and it will be necessary to repeat the separation to collect these.

If the column recorder is to be used to indicate when the fraction is in the gas cell, the volume of pipework and valves must be kept to a minimum, otherwise there will be a significant time delay before the fraction reaches the cell, and this time delay will vary with the flow rate. Since the gas cell contains the fraction and the carrier gas as diluent, the column conditions must be chosen to obtain the highest possible concentration of the fraction in the carrier gas. This is indicated by the sharpness of the peaks on the column recorder. In this respect temperature programming of the chromatograph oven is most desirable, as by this means the peaks of higher boiling constituents can be sharpened without eluting low boilers too rapidly.

With this apparatus, materials boiling up to 250°C can be measured, the practical limit being the temperature at which the switch valve may be safely operated.

The whole operation is extremely simple and, because the transfer efficiency to the spectrometer is good, the method may be made very sensitive. Furthermore, if some other identification method, such as n.m.r. or mass-spectrometer examination is required, the sample exhausted from the gas cell is easily transferred to another apparatus. The principal disadvantage is that there is no simple means of increasing the sensitivity by multiple collection, as each fraction is blown to waste after examination. Furthermore, samples are examined as gases rather than liquids, and very few spectra have been recorded for reference standards in the gaseous state. In practice this does not prove to be a great handicap; the usual procedure for qualitative analysis from the infra-red spectrum can be applied, and the knowledge thus gained of the chemical nature of the fraction, together with an estimate of its boiling point from the retention time, will generally reduce the number of possibilities sufficiently to make it practicable to examine these pure substances under the same conditions as those used for the unknown sample.

The value of the method is illustrated by the following practical example.

IDENTIFICATION OF THE CONSTITUENTS OF A PRINTER'S INK SOLVENT BY G.L.C./INFRA-RED SPECTROSCOPY (DIRECT GAS PHASE METHOD)

The apparatus for this examination is that shown in Figure 1.20. The gas cell, of our own construction (described above), is used together with the following equipment.

Chromatograph. Perkin-Elmer 452, fitted with thermistor detector. A Loenco valve (described above) is fitted inside the column oven.

Spectrometer. Perkin-Elmer 157, modified with extended base plate to accommodate the gas cells.

For solvent mixtures, we have found the following column suitable:
Column. $\frac{1}{4}$ in (6 mm) diameter, 2 m long.

Packing. Chromosorb W AW DMCS from John Manville (Great Britain) Ltd, London.

Stationary Phase. 8% Antarox CO-990 from Fine Dyestuffs and Chemicals Ltd, Manchester, England.

The sample as presented was a viscous pigmented liquid. Vacuum distillation yielded a colourless mobile liquid, of which approx. 0·5 μl was available for test. The column was loaded with approx. 0·1 μl of

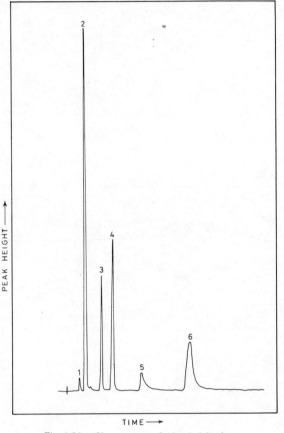

Fig. 1.21.   Chromatogram of printer's ink solvent

liquid. After a trial run with the column recorder in use, but without fraction collection, the following conditions were established:

Column temperature. 140°C isothermal.

Gas flow. 25 ml helium per min.

This gave the chromatogram shown in Figure 1.21. Six major peaks were evident. Components 1 to 5 were closely spaced and it was decided

71

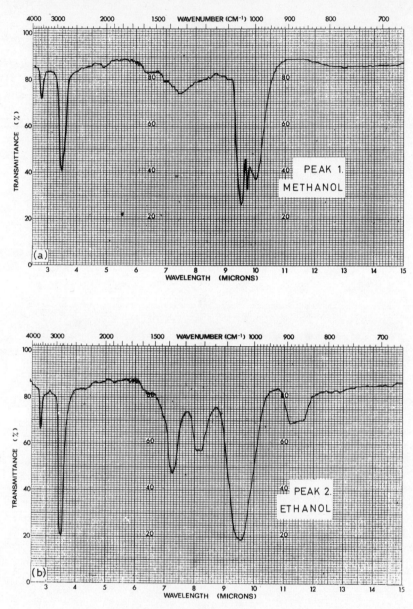

Fig. 1.22.   Infra-red spectra of chromatograph fractions of printer's ink solvent: (a) peak 1 (methanol) and (b) peak 2 (ethanol)

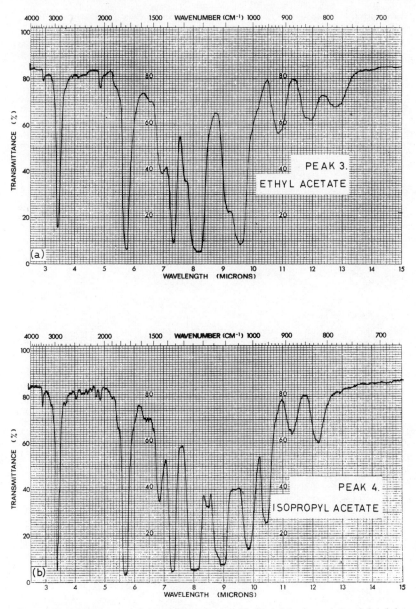

Fig. 1.23. Infra-red spectra of chromatograph fractions of printers ink solvent: (a) peak 3 (ethyl acetate) and (b) peak 4 (isopropyl acetate)

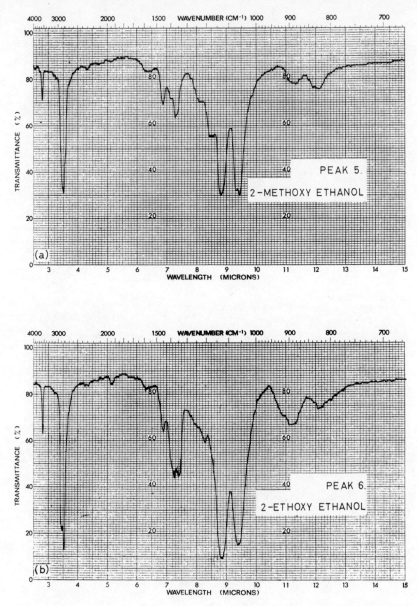

Fig. 1.24. *Infra-red spectra of chromatograph fractions of printer's ink solvent: (a) peak 5 (2-methoxy ethanol) and (b) peak 6 (2-ethoxy ethanol)*

to trap 1, 3, 5 and 6 on one run, and 2 and 4 on a repeat run. The general procedure for trapping was that described above. Spectra of all six fractions were obtained and all were of good quality (Figures 1.22 to 1.24) with a column load of 0·1 μl for each run. The spectra were identified as follows.

Fractions 1 and 2 (Figures 1.22(a) and (b)). Both spectra have sharp bands at 2·8 μm and a broader strong band near 9·5 μm. This strongly suggests primary alcohols. Comparison with standard vapour spectra[54] shows that Fraction 1 is methanol, and Fraction 2 is ethanol.

Fractions 3 and 4 (Figures 1.23(a) and (b)). These are evidently chemically similar. The prominent band at 5·8 μm and the strong band at 8·0 μm strongly suggest acetate (see Table 4.4). Comparison with spectra of simple acetates in the liquid phase[55] suggests that the materials are probably ethyl acetate and isopropyl acetate respectively. This conclusion was confirmed by injecting each of these substances directly into the gas cell.

Fractions 5 and 6 (Figures 1.24(a) and (b)). These again are evidently similar chemically as they have very similar spectra, and have the characteristics of primary alcohols (2·8 μm and 9·5 μm) and ethers (8·8 μm). A search of the spectra of simple substances of this type shows that the spectra are very similar to those of 2-methoxy ethanol and 2-ethoxy ethanol. Again these conclusions were confirmed by injecting each of these substances directly into the gas cell.

Thus from 0·5 μl of sample, including the quantity used for trial runs to establish the correct conditions for examination of the sample, we were able to identify each of six constituents of the solvent mixture.

AUTOMATIC TITRATION

Automatic titrators are now commonplace in many laboratories and their value to the analyst is well documented in the literature[56, 57]. Two forms of titrator are however of particular value in the analysis of plastics materials and are thus worthy of some brief mention in this Chapter.

AUTOMATIC RECORDING TITRATORS

These assemblies contain four essential component instruments.

1. A constant-speed syringe burette driven by a synchronous motor, preferably through a gear box to give a selection of titrant delivery rates for each size of syringe.
2. A titration and stirrer assembly with provision for supporting the titration cell and sensing electrodes as required.

3. A metering instrument with recorder output to monitor the changes occurring during the titration. Depending on the type of titration being performed, this meter may be a pH meter, millivolt meter, micro or milliammeter, spectrophotometer, spectrofluorimeter, conductivity meter or thermometric assembly.
4. A recorder, preferably with at least an 8 in (200 mm) chart and a choice of chart speeds.

The above combination of units provides a simple and versatile titrator assembled from normal laboratory instruments with little or no modification. Composite instruments are of course available commercially and these may incorporate such refinements as facilities for recording the first, second and even third differential of the normal titration curve as well as variable-speed titrant injection. In this case the injection rate is automatically slowed down as the end-point is approached, with synchronised slowing of the recorder chart drive so that the recorded titration curve remains true-scale. Whatever the system used, the advantage of automatically recorded titration is that the whole course of the titration is followed and, when two or more components reactive with the titrant are present, quantitative differentiation is often possible.

Extensive use is made of this facility in the differentiating titration of mixed amine and amide additives extracted from polyolefines; the titration of chloride and mixed halides ionised by hydrolysis or by combustion in oxygen, from plastics materials; the differentiating titration of mixtures of metals (e.g. Ca-Mg or Cu-Ni) by means of complexones[58]; the titration of polyethylene oxide type emulsifying agents with sodium tetraphenyl boron; and in the many cases when acid mixtures present in, or extracted from, a polymer composition require resolution.

The titration medium may be aqueous, partially aqueous, or non-aqueous, and is chosen to bring out the sharpest end-points and clearest differentiation of the components of interest.

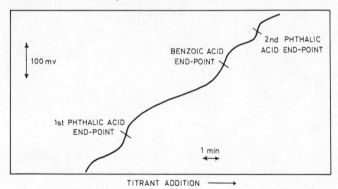

100 mv

BENZOIC ACID
END-POINT

2nd PHTHALIC
ACID END-POINT

1st PHTHALIC ACID
END-POINT

1 min

TITRANT ADDITION ⟶

Fig. 1.25. Differentiating titration of mixed benzoic and o-phthalic acids in t-butyl alcohol (titrant: 0·1 N tetrabutylammonium hydroxide)

The example given in Figure 1.25 shows the differentiating titration of o-phthalic and benzoic acids obtained from the hydrolysis products of a mixed plasticiser extracted from a polyvinyl chloride composition. The resolution of the acids was of great value in the final analysis of the plasticiser. In the example given, 0·15 to 0·20 g of the total acids were dissolved in 50 ml t-butyl alcohol and the solution titrated with 0·1 N tetrabutylammonium hydroxide titrant delivered at a rate of approximately 1 ml per minute with a recorder sensitivity of 500 mV full-scale. The electrode pair used was a glass indicator electrode and a modified calomel (saturated solution of potassium chloride in methanol electrolyte solution) reference electrode. The first end-point corresponds to the titration of the first carboxyl group of the phthalic acid, the second to the titration of this group plus the benzoic acid and the third end-point to the titration of the total acidity.

In our experience a convenient method for the determination of the end-points on conventional S-shaped curves is by the method of circles[59]. In this method the largest circle of best fit for each side of the curve is found by means of a plastic cursor, with a small hole at the common centre of several concentric circles so that this can be marked on the chart with a pencil. The point at which the line joining the two centres cuts the titration curve is taken as the end-point. For the type of record obtained, for example in conductometric or spectrophotometric titrations, the end-point is taken as the intersection of the straight portions of the recorded trace.

PRE-SET END-POINT TITRATORS

In the pre-set end-point titrator the sensing and meter unit incorporates relay controls which stop the flow of titrant from the burette at a pre-set end-point. It is often necessary that the final one or two ml of titrant before the end-point are added slowly to allow equilibration of the solution and avoid overshoot of the end-point. This facility is normally achieved by an anticipation control which, through a relay system, automatically alters the flow of titrant from 'fast' to 'slow' at a predetermined point in the titration.

It is thus a prerequisite of the use of this type of titrator that the nature and shape of the full titration curve under the exact conditions of test be known and furthermore that they remain constant. These conditions are often satisfied, however, particularly in routine pH and potentiometric titrations. Examples of value in the analysis of plastics materials are the back titration of the excess acid remaining after the steam distillation of ammonia in the determination of nitrogen in polymers by the Kjeldahl method[60] and the potentiometric titration of chloride ionised by sodium peroxide fusion[60] or Oxygen Flask Combustion of chlorine containing polymers.

Another advantage of the pre-set end-point titrator is that it will automatically switch the slow flow of titrant on and off to maintain the pre-set end-point condition in the titration solution. This facility is of great value in dealing with titrations involving relatively slow reactions, for example the iodometric determination of peroxides by the method of Mathews and Patchan[61], and in the determination of water in gases by the Karl Fischer method[57]. The gas, for example a monomer or volatiles from a polymeric sample, may be metered through the titration cell and the 'dry state' corresponding to the pre-set end-point will be maintained by automatic addition of reagents by the titrator.

REFERENCES

1. EDISBURY, J., *Practical Hints on Absorption Spectroscopy (200–800 nm)* Hilger & Watts, London (1966)
2. GILLAM, A. E., and STERN, E. S., *An Introduction to Electronic Absorption Spectroscopy in Organic Chemistry*, Edward Arnold, London (1960)
3. RAO, C. R., *Ultra-Violet and Visible Spectroscopy*, Butterworths, London (1961)
4. BEAVEN, G. H., and JOHNSON, E. A.; WILLIS, H. A., and MILLER, R. G. J., Molecular Spectroscopy, Heywood, London (1961)
5. FIKHTENGOL'TS, V. S., ZOLOTAREVA, R. V., and L'VOV, YU. A., (Transl. STUBBS, A. E.), *Ultra-Violet Spectra of Elastomers and Rubber Chemicals,* Plenum Press, Data Division, New York (1966)
6. WALSH, A., *Spectrochim. Acta,* **7,** 108 (1955)
7. AHRENS, L. H., and TAYLOR, S. R., *Spectrochemical Analysis,* Pergamon, Oxford (1961)
8. *Methods for Emission Spectrochemical Analysis,* A.S.T.M. Committee E2 (1964)
9. HERRMANN, R., and ALKEMADE, C. T. J., (Trans. GILBERT, P. T., JR), *Flame Photometry,* Interscience, New York (1963)
10. NICKOLLS, L. C., *J. Soc. chem. Ind., London.,* **62,** 31 (1943)
11. BIRKS, L. S., *X-Ray Spectrochemical Analysis,* 2nd edn, Interscience, New York (1969)
12. LIESHAFSKI, H. A., PFEIFFER, H. G., WINSLOW, E. H., and ZEMANY, P. D., *X-Ray Absorption and Emission in Analytical Chemistry,* John Wiley, New York and London (1960)
13. JENKINS, R., and DE VRIES, J. L., *Practical X-Ray Spectrometry,* Philips Technical Library (1967)
14. SQUIRRELL, D. C. M., (Ed. SHALLIS, P. W.), in *Proceedings of the S.A.C. Conference, Nottingham 1965,* W. Heffer, Cambridge, 102 (1965)
15. SQUIRRELL, D. C. M., and WARREN, P. L., in *Proceedings of the 6th Conference on X-Ray Analytical Methods, Southampton 1968,* N. V. Philips Gloeilampenfabricken, Scientific Equipment Department, Eindhoven, The Netherlands, 72 (1968)
16. COLTHUP, N. B., DALY, L. H., and WIBERLY, S. E., *Introduction to Infra-Red and Raman Spectroscopy,* Academic Press, New York (1964)
17. WEST, W. (Ed.), *Chemical Applications of Spectroscopy,* Interscience, New York (1956)
18. BELLAMY, L. J., *Infra-Red Spectra of Complex Molecules,* Methuen, London (1958)
19. BRUGEL, W., (Transl. KATRITZKY, A. R., and KATRITZKY, A. J. D.), *An Introduction to Infra-Red Spectroscopy,* Methuen, London (1962)
20. CROSS, A. D., *Introduction to Practical Infra-Red Spectroscopy,* Butterworths, London (1960)
21. *Tables of Wavenumbers for the Calibration of Infra-Red Spectrometers,* I.U.P.A.C., Butterworths, London (1961)
22. MILLER, R. G. J., *Laboratory Methods in Infra-Red Spectroscopy,* Heyden, London (1965)
23. ELLIOTT, A., *Infra-Red Spectra and Structure of Organic Long-Chain Polymers,* Edward Arnold, London (1969)
24. BARNES, R. B., WILLIAMS, V. Z., DAVIS, A. R., and GEISECKE, P., *Ind. Engng Chem. analyt. Edn,* **16,** 9 (1944)

25. DINSMORE, H. L., and SMITH, D. C., *Analyt. Chem.,* **20,** 11 (1948)
26. *Surface-Coating with Aqueous Dispersions,* Imperial Chemical Industries, Plastics Division, Technical Service Note F.7, Welwyn Garden City, England (1966)
27. CORISH, P. J., *J. appl. Polym. Sci.,* **4,** 86 (1960)
28. CORISH, P. J., *J. appl. Polym. Sci,* **5,** 53 (1961)
29. HARMS, D. L., *Analyt. Chem.,* **25,** 1140 (1953)
30. *Methods for Identification of Rubbers,* BS 4181 (1967)
31. JASPER, B. T., *Jl* I.R.I., **3,** No. 2, 72 (1969)
32. HUMMEL, D., *Kautschuk Gummi,* **11,** W.T. 185 (1958)
33. LINDLEY, G., *Lab. Pract.,* **14,** 826 (1965)
34. FAHRENFORT, J., *Spectrochim. Acta,* **17,** 698 (1961)
35. HARRICK, N. J., *Internal Reflection Spectroscopy,* Interscience, New York (1967)
36. FERGUSON, R. C., *J. Am. chem. Soc.,* **82,** 2416 (1960)
37. WILLIS, H. A., and CUDBY, M. E. A., (Ed. BRAME, E. G.), *Applied Spectroscopy Reviews,* Marcel Dekker, New York, **1,** 237 (1968)
38. EMSLEY, J. W., FEENEY, J., and SUTCLIFFE, L. H., *High Resolution Nuclear Magnetic Resonance Spectroscopy,* Pergamon, Oxford, **1,** 65 (1965)
39. JACKMAN, L. M., *Applications of Nuclear Magnetic Resonance Spectroscopy in Organic Chemistry,* Pergamon, Oxford (1959)
40. BIBLE, R. H., *Interpretation of n.m.r. Spectra. An Empirical Approach,* Plenum Press, New York (1965)
41. EMSLEY, J. W., FEENEY, J., and SUTCLIFFE, L. H., *High Resolution Nuclear Magnetic Resonance Spectroscopy,* Pergamon, Oxford, **1,** 240 (1965)
42. KEULEMANS, A. I. M., and VERVER, G., *Gas Chromatography,* Reinhold, New York (1957)
43. ETTRE, L. S., and ZLATKIS, A., *The Practice of Gas Chromatography,* Interscience, New York (1967)
44. ETTRE, L. S., *Open Tubular Columns in Gas Chromatography,* Plenum Press, New York (1965)
45. JEFFERY, P. G., and KIPPING, P. J., *Gas Analysis by Gas Chromatography,* Pergamon, Oxford (1964)
46. KIRKLAND, J. J., *Analyt. Chem.,* **35,** 2003 (1963)
47. CROPPER, F. R., and KAMINSKY, S., *Analyt. Chem.,* **35,** 735 (1963)
48. JEFFS, A. R., *Analyst, Lond.,* **94,** 249 (1969)
49. DEANS, D. R., *J. Chromat.,* **18,** 477 (1965)
50. DEANS, D. R., *Nature, Lond.,* **209,** 1296 (1966)
51. BARLOW, A., LEHRLE, R. S., and ROBB, J. C., *Polymer,* **2,** 27 (1961)
52. STANFORD, F. G., *Analyst, Lond.,* **90,** 266 (1965)
53. HASLAM, J., JEFFS, A. R., and WILLIS, H. A., *Analyst, Lond.,* **86,** 44 (1961)
54. WELTI, D., *Infra-Red Vapour Spectra,* Heyden, London (1970)
55. *Sadtler Standard Spectra,* The Sadtler Research Laboratories, Philadelphia 2
56. PHILLIPS, J. P., *Automatic Titrators,* Academic Press, New York and London (1959)
57. SQUIRRELL, D. C. M., *Automatic Methods in Volumetric Analysis,* Hilger & Watts, London (1964)
58. REILLEY, C. N., and SHELDON, M. V., *Talanta,* **1,** 127 (1958)
59. SQUIRRELL, D. C. M., *Automatic Methods in Volumetric Analysis,* Hilger & Watts, London, 43 (1964)
60. HASLAM, J., and SQUIRRELL, D. C. M., *Analyst, Lond.,* **82,** 511 (1957)
61. MATHEWS, J. S., and PATCHAN, J. F., *Analyt. Chem.,* **31,** 1003 (1959)

2

QUALITATIVE ANALYSIS

The first part of this chapter is concerned with the chemical detection and semi-quantitative determination of elements such as nitrogen, sulphur, phosphorus, chlorine and fluorine in plastics materials. Information provided by spectrographic examination is described. Solubility tests and procedures involving burning and heating are dealt with and specific chemical tests for certain polymers, e.g. melamine, urea-formaldehyde, polyethylene terephthalate and polyvinyl chloride are described.

The second part of this chapter deals with the interpretation of infrared spectra and, furthermore, gives an example of the interpretation of the n.m.r. spectrum of an unknown substance; this interpretation is rendered much more satisfactory if the nature and rough concentration of the elements present are already known.

Information is then provided which has particular reference to the qualitative examination of film laminates, fibre blends and surface active agents.

In the first place it is important to realise the particular point of view from which this chapter is written.

Essentially we are concerned with the preliminary qualitative examination of polymers, polymer preparations, plasticisers and the like, which may be submitted for test and which may have their origin in all parts of the globe. The samples may be submitted in innumerable forms, i.e. as polymer powders, as sheet, as rod, tube or film and as finished articles. Moreover, specific questions may be asked about the samples, i.e. questions that do not always involve the complete identification of all the constituents of e.g. a polymer preparation. That complete identification might be very prolonged and involve the isolation and identification of a polymer, or copolymer, or blend of polymers, the isolation of a mixed plasticiser and the resolution of this mixed plasticiser into its individual components, the isolation and identification of a stabiliser, filler, antioxidant or surface active agent and so on.

80

Moreover, a great deal of the preliminary work in the identification of plastics materials is merely concerned with separation procedures, that serve to isolate components for subsequent infra-red or other final identification.

In some respects, one of the most useful papers on the subject of identification of synthetic resins and plastics is that of T. Gladstone Shaw[1]. Shaw seeks to develop systematic procedures for the separation of resins from solvents, plasticisers etc., for the separation of mixtures of resins into individual resins, for the classification of individual resins into groups, and for the identification of resins within a group.

In our experience, however, strict application of this procedure would fail to answer a lot of questions submitted to the analytical chemist in the plastics industry, e.g. is a nylon type polymer nylon 6 : 6 or nylon 6 : 10 or a copolymer or terpolymer? Is a given chlorine containing polymer polyvinyl chloride or is it a copolymer of vinyl chloride with a very small proportion of vinyl acetate?

It is for reasons such as this that we regard our initial qualitative work as being preliminary work that can provide helpful information about a polymer or polymer preparation. Often enough it will provide useful clues as to the subsequent quantitative tests that may have to be carried out, and in many cases, if supplemented by infra-red evidence on the right preparations, will enable a whole host of questions to be answered with great speed.

Moreover, it is emphasised how important it is that the analyst in this field should assemble a collection of known polymers, copolymers, plasticisers etc. for comparative purposes. These known substances will prove to be very valuable when making observations on solubility, burning and heating tests on unknown preparations.

TESTS FOR 'ADDITIONAL ELEMENTS'

By far the most important test is that for 'additional elements' in a plastics material, i.e. elements other than carbon, hydrogen and oxygen such as sulphur, fluorine, phosphorus, chlorine and nitrogen, whether or not our examination is carried out directly on a polymer composition, or on a separated polymer, or plasticiser, or other constituent of a polymer composition.

For many years this test for additional elements was carried out by means of the sodium fusion test.

Experience has shown however, that this test as ordinarily carried out is often quite inadequate. One good reason is that the analytical chemist in the plastics industry is not, as a rule, concerned solely with the differentiation of simple polymers such as polyvinyl chloride, polyethylene terephthalate, polytetrafluoroethylene, polymethyl methacrylate and the like. He may, for instance, meet a preparation that, by the sodium fusion test, may obviously contain a high proportion of chlorine and yield no evidence of nitrogen, e.g. a blend of a high proportion of polyvinyl chloride and a proportion, small but not insignificant, of a terpolymer of butadiene, styrene and acrylonitrile containing of the order of 0·3% nitrogen overall. In such a case the failure to detect nitrogen would be serious because it would mean that an appreciable amount of acrylonitrile, e.g. equivalent to 0·3% nitrogen, had been missed.

It is likely that the sodium fusion test does not proceed as effectively in the presence of polyvinyl chloride as in its absence. We are assured by experts in the field that nitrogen would not be missed on application of the sodium fusion test to raw rubber, which also contains of the order of 0·3% nitrogen.

Then again, in the qualitative examination of certain polymers it may be quite important not to miss quite small proportions of sulphur; the detection of this element may give a clue to the presence of a surface active agent or other additive.

Moreover, by its very nature, the sodium fusion test cannot be a very efficient or quantitative process.

In our laboratories it was thought desirable to approach the whole problem of detection of additional elements in a plastics material from quite a new angle, and to provide tests that, at the same time, would furnish, broadly speaking, semi-quantitative information. The method is known as the Oxygen Flask Combustion Method[2, 3]. In this method, the plastics material, wrapped in filter paper, is ignited electrically and burnt in a closed glass flask filled with oxygen in the presence of sodium hydroxide solution as absorbent. Appropriate tests for the additional elements are then made on the sodium hydroxide solution.

A full description of the method is given below.

OXYGEN FLASK COMBUSTION METHOD

Apparatus

The complete combustion flask and firing adapter together with the firing spark source are shown in Figure 2.1 and a more detailed diagram of the firing adapter together with the necessary constructional details are shown in Figure 2.2.

The H.F. tester (model T.1) for producing the spark used to initiate the combustion is manufactured by Edwards High Vacuum Company Ltd, Crawley, Sussex. It is mains operated and relatively inexpensive.

The leads from the H.F. tester to the firing terminals may be of any convenient length, allowing remote control firing from behind a suitable safety screen. The position of the platinum gauze combustion basket in

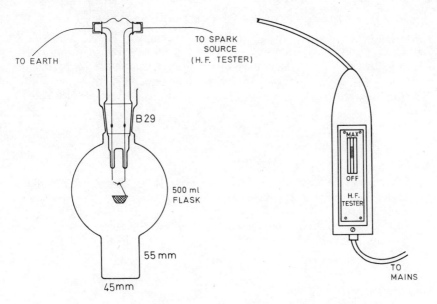

Fig. 2.1.   Oxygen Flask Combustion unit and spark source (flask and adapter made of Pyrex glass)

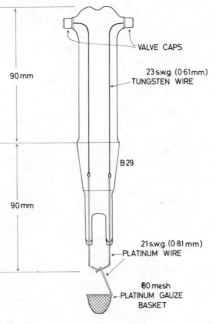

Fig. 2.2.   Firing adapter for Oxygen Flask Combustion unit

relation to the ancillary firing electrode is as shown in Figure 2.2, so that the flame from the burning sample does not play on the platinum wire, which, by a cooling effect, may on occasion lead to the formation of unburnt carbon in the flask. It is important that the joint between the firing adapter and the combustion flask should show no signs of leakage.

General reagents

Filter paper. Whatman No. 541 filter paper. When not in use, this filter paper is stored in a sealed container out of contact with the laboratory atmosphere.
Cotton wool B.P.C. Super Quality.
Sodium hydroxide N.

Procedure

Weigh out approximately 20 mg of sample and transfer it to the centre of a small piece of the Whatman No. 541 filter paper weighing approximately 0·1 g. Fold the filter paper so that the sample is completely enclosed, and, before making the final fold, insert a small wick of cotton wool weighing about 6 mg.

Prepare the combustion vessel by washing the flask and adapter and heating the end of the platinum wire and the basket to red heat in a bunsen flame. Place 5 ml of N sodium hydroxide solution in the bottom of the flask to act as absorbent. Fill the flask with oxygen and insert the stopper. Place the wrapped sample in the platinum basket and twist the cotton wool wick so that it lies between the basket and the ancillary firing electrode. Remove the stopper from the flask and replace it with the adapter containing the sample. Seal the joint with a suitable amount of distilled water, and attach the retaining springs. Attach the electrical and earth leads to the adapter and place the safety screen in front of the combustion vessel, or, alternatively, arrange the whole apparatus inside a suitable safety box.

Switch on the H.F. tester to ignite the cotton wool and sample and allow to burn to completion.

Set the flask aside for 15 min, shaking occasionally to ensure good absorption of the combustion products, then open up the flask and add 20 ml of distilled water and mix the contents thoroughly.

The test solution, prepared as above, has a volume of approximately 25 ml. Aliquots of this solution are taken for the detection of the individual elements by the colorimetric and turbidimetric methods detailed below. For comparison purposes in the tests, prepare a blank test solution by carrying out the combustion procedure on the filter paper and cotton wool only.*

CHLORINE

Reagents

Ammonium ferric sulphate solution.
Dissolve 12 g of Analar ammonium ferric sulphate in water and add 40 ml of Analar nitric acid. Dilute to 100 ml and filter.
Mercuric thiocyanate solution.
Dissolve 0·4 g of mercuric thiocyanate that has been recrystallised from ethanol in 100 ml of absolute ethanol.

Procedure

Transfer 5 ml of the test solution to a 50 ml beaker and add 1 ml of ammonium ferric sulphate solution. Mix the solution and add 1·5 ml of mercuric thiocyanate solution. When chlorine is present in the sample an orange red colour will be developed in the test solution. If a semi-quantitative estimation of chlorine is required, set the solution aside for 10 min, and measure the optical density of the test solution against the blank solution at 460 nm in 20 mm cells.

Useful calibration figures for the examination of plastics materials are given below.

Chlorine in Plastics Material %	Optical Density $D^{20\,mm}_{460\,nm}$ measured against blank
1	0·4
2	0·75

SULPHUR

Reagents

Hydrochloric acid N.
Precipitating reagent solution A.

* The data given in the test procedure are for a flask of the type shown in Figure 2.1. It may, on occasion, be more convenient to use a lipped conical combustion flask, provided that appropriate calibration is made.

A further point concerns the combustion process. If the platinum gauze has fused and developed a hole during previous combustions, this hole should be covered up with a fresh piece of gauze for the next test. When the gauze can no longer be used it is returned to the makers for recovery. The cost of an individual test is negligible.

Dissolve 0·2 g of peptone in 50 ml of 1% w/v barium chloride ($BaCl_2 \cdot 2H_2O$) solution. Buffer to a pH of 5·0 with 0·02 N hydrochloric acid, add 10 g of sodium chloride (Analar) and dilute to 100 ml. Heat on a water bath for 10 min and add a few drops of chloroform. Filter if necessary.

Precipitating reagent solution B.

Dissolve 0·4 g of gum ghatti in 200 ml of distilled water by warming slightly. When solution is complete, add 2·0 g of barium chloride ($BaCl_2 \cdot 2H_2O$). Filter if necessary.

Store solutions A and B separately, and prepare the final reagent just before use by diluting 10 ml of solution A to 100 ml with solution B.

Hydrogen peroxide. 100 volume. Analar.

Procedure

Transfer 5 ml of the test solution to a 6×1 in (150×24 mm) test tube, and add 2 drops of 100 volume hydrogen peroxide and then 1·2 ml of N hydrochloric acid. Mix well and add 2·0 ml of precipitating reagent with continued shaking. A distinct turbidity will be produced in the mixed solution if sulphur is present in the sample; the blank test under the same conditions will be perfectly clear. If a semi-quantitative estimation of the sulphur content is required, add 5 ml of distilled water to both blank and test solutions, mix and set aside for 30 min. Mix the solutions and measure the optical density of the test solution in a 40 mm cell at 700 nm with the blank solution in the comparison cell.

Useful calibration data for the examination of plastics materials are given below.

Sulphur in Plastics Material %	Optical Density $D_{700\,nm}^{40\,mm}$ measured against blank
1	0·2
2	0·4

NITROGEN

Reagents

Resorcinol. Analar.
Acetic acid glacial.
Ammonium ferrous sulphate. Analar.

Procedure

Weigh 0·1 g of resorcinol into a clean dry 50 ml beaker and dissolve in 0·5 ml of glacial acetic acid, add 5 ml of the test solution, and after mixing, add 0·1 g of ammonium ferrous sulphate. Carry out the same test on the blank test solution. The development of a green colour in the sample test solution, compared with a pale yellow in the blank, indicates the presence of nitrogen in the sample.

If a semi-quantitative estimation of the nitrogen content of the sample is required, set both sample and blank solutions aside for 20 min. Add 10 ml of distilled water to each, mix and measure the optical density of the sample solution against the blank at 690 nm in a 40 mm cell. In our experience, the blank in this test is low, giving an optical density of 0·07 measured against water.

Useful calibration data for the examination of plastics materials are given below.

Nitrogen in Plastics Material %	Optical Density $D_{690\,nm}^{40\,mm}$ measured against blank
1	0·6
2	1·0

PHOSPHORUS

Reagents

Ammonium molybdate solution.
Dissolve 10 g of Analar ammonium molybdate [$(NH_4)_6Mo_7O_{24}\cdot4H_2O$] in about 70 ml of water, and dilute to 100 ml. Add this solution with stirring to a cooled mixture of 150 ml of sulphuric acid and 150 ml of water.
Ascorbic acid. B.D.H. laboratory reagent grade.

Procedure

Transfer 2 ml of the test solution to a 100 ml beaker. Add 40 ml of distilled water and 4 ml of ammonium molybdate solution. Mix thoroughly, then add 0·1 g of ascorbic acid, and boil the solution for 1 min. Cool in running water for 10 min and dilute to 50 ml with distilled water. Treat the blank solution in a similar manner. When phosphorus is present in the sample, a blue colour will be developed in the test solution as compared with a pale yellow in the blank. If a semi-quantitative estimation of the phosphorus is required, measure the optical density of the test solution against the blank solution at 820 nm in 20 mm cells.

Useful calibration data for the examination of plastics materials are given below.

Phosphorus in Plastics Material %	Optical Density $D_{820 nm}^{20 mm}$ measured against blank
1	0·45
2	0·95

FLUORINE

Reagents

Buffered alizarin complexan solution.
Weigh 40·1 mg of 3-aminomethylalizarin-NN-diacetic acid (Hopkins and Williams Ltd) into a beaker and add 1 drop of N sodium hydroxide and approximately 20 ml of distilled water. Warm the solution to dissolve the reagent, cool and dilute to 208 ml. Weigh into another beaker 4·4 g of sodium acetate $(CH_3 \cdot COONa \cdot 2H_2O)$ and dissolve in water. Add 4·2 ml of glacial acetic acid and dilute to 42 ml. Pour this sodium acetate solution into the alizarin complexan solution and mix to give the final buffered alizarin complexan solution.
Cerous nitrate 0·0005 M.
Dissolve 54·3 mg of cerous nitrate $[Ce(NO_3)_3 \cdot 6H_2O]$ in water and dilute to 250 ml.

Procedure

Transfer 20 ml of distilled water and 2·4 ml of buffered alizarin complexan solution to a 50 ml beaker. Add 1 ml of test solution and mix by swirling the solution. Finally add 2 ml of cerous nitrate solution and mix again. Treat the blank solution in a similar manner. When fluorine is present in the sample a mauve colour will be developed in the test solution compared with the pink coloured blank solution. If a semi-quantitative estimation of fluorine is required, set the solution aside for 10 min and measure the optical density of the test solution against the blank solution at 600 nm in 10 mm cells.

Useful calibration data for the examination of plastics materials are given below.

Fluorine in Plastics Material %	Optical Density $D_{600 nm}^{10 mm}$ measured against blank
1	0·07
2	0·135

For most purposes, application of the procedure as described provides sufficient information about the sample under test, but, if required, the sensitivity of any of the individual tests may be significantly increased.

This may be accomplished by repeating the combustion, absorbing the products in 5 ml of 0·2 N sodium hydroxide and making a direct test for the particular element on this solution without dilution and with appropriate calibration.

Moreover, it should be realised that a lot of time may be saved in the actual tests by using the solid reagents in the form of solid pellets, of known weight, as sold by Messrs. Ridsdale & Co. Ltd, Newham Hall, Middlesbrough, and dispensing the liquid reagents from dispensers e.g. Zippette dispensers as manufactured by Jencons Ltd, Hemel Hempstead, Hertfordshire, England or Quickfit dispensers made by Quickfit and Quartz Ltd, Stone, Staffordshire, England.

Note: Although bromine and iodine are not normally found in plastics materials, the original paper on the Oxygen Flask Combustion procedure[3] includes information about the fluorescein test given by both iodides and bromides and the starch test for iodine. Moreover, it should be realised that iodides and bromides, if present in the Oxygen Flask Combustion products, will give positive results in the mercury thiocyanate test for chloride.

SPECTROGRAPHIC EXAMINATION

Spectrographic examination often provides useful information, particularly about inorganic additives.

The emission spectrum, determined directly on a sample or its ash, may be obtained quickly. It is a sensitive method and detects the majority of metallic elements. Chlorine, bromine, iodine and sulphur are not detected and small amounts of phosphorus may be missed.

The X-ray Fluorescence spectrum may be determined directly on a sample. Although it takes longer to obtain than the emission spectrum, it is a non-destructive test. Bromine, iodine, sulphur and phosphorus, even when present in very small proportions, are readily detected. Elements of atomic number less than 11 are not detected, except by the use of specialised equipment.

GENERAL EXAMINATION OF POLYMERS AND POLYMER PREPARATIONS

A useful book to have available in laboratories concerned with the preliminary examination of plastics is that written by A. Krause and A. Lange[4]. This text provides a wealth of information on the physical properties such as density, solubility, melting point and refractive index of polymers, plasticisers etc. In addition, the colour reactions of, and the results of quantitative tests on, individual polymers are given in considerable detail.

Furthermore, a considerable amount of analytical information may be derived from the tests adopted by the American Society for Testing and Materials, and the British Standards Institution (see Reference 5).

Amongst the more important specifications that contain analytical information and that have been published by the British Standards Institution in recent years are the following:

BS 903. Methods of Testing Vulcanised Rubbers.

BS 1673. Methods of Testing Raw Rubbers and Unvulcanised Compounded Rubber.

BS 3397. Methods of Testing Synthetic Rubber Latices.

BS 2782. Methods of Testing Plastics.

Our observations on the general tests used in the qualitative examination of polymers and polymer preparations are given below.

SOLUBILITY

The analytical chemist in the plastics industry can often provide useful information about a test preparation by obtaining knowledge of its behaviour towards solvents, or solutions that attack the substance, e.g. polytetrafluoroethylene may be characterised by its inertness towards almost all chemical solvents.

Very useful information about the solubility or otherwise of a whole host of polymers etc. in various solvents is provided in text books such as those of Collins[6] and Marsden[7]. In practice, however, it is probably much more realistic to provide information about those solvents or solutions that, over the years, have proved to be useful in actual determinations on polymers or polymer preparations. They can all be adapted by the competent analyst to preliminary qualitative examination of miscellaneous preparations, provided he takes care to work at room temperature, and at temperatures up to boiling temperature of the solvent or attacking solution.

This information is summarised in Table 2.1.

Sound information about the correct solvents to employ in particular cases has been provided by our experience in casting films for infra-red examination, and information on this topic is given on p. 22.

From the point of view of solubility, rubber, modified rubber and rubber-like synthetic elastomers, silicone resins, protein resins such as casein and soya bean, and natural resins such as colophony, copal and shellac are regarded as special cases and are not treated here.

The analytical chemist is interested in the solution of practical problems. Again and again he is called upon to identify coatings on fabrics, wire etc., and paper and particularly the last. The real problem involves the

Table 2.1.

Polymer or Polymer Preparation	Solvent or solution used	Purpose
Polyvinyl chloride	Ethylene dichloride	Determination of solution viscosities and K values
Polyvinyl chloride compositions	Tetrahydrofuran	Recovery of polymer or copolymer free from impurities
Polyvinyl chloride compositions	Ethylene dichloride	Determination of lead stabilisers
Vinyl chloride copolymer	Methyl ethyl ketone	Determination of acidity
Vinyl chloride/vinylidene chloride copolymer compositions	Tetrahydrofuran	Recovery of polymer or copolymer free from impurities
Vinyl chloride/vinyl acetate/ maleic acid terpolymer	Pyridine	Determination of acid groups
Vinyl chloride/methacrylic acid copolymer	Pyridine	Determination of acid groups
Vinyl chloride/acrylamide copolymer	Acetic anhydride	Titration of acrylamide
Polyvinyl chloride/chlorinated polythene blends	Carbon tetrachloride or chloroform	Solution and separation of chlorinated polythene
Polymethyl methacrylate	Acetone	Plasticiser determination
	Acetone	Chemical determination of U.V. absorber
	Chloroform	U.V. determination of U.V. absorber,
	Chloroform	Residual monomer determination
	Acetone	Mercaptan determination
	Chloroform	Relative viscosity determination
	Acetone	Butyl lactate determination also dibutyl citric acid determination
	Chloroform or isopropyl alcohol	Determination of residual benzoyl peroxide

Table 2.1—*continued*

Polymer or Polymer Preparation	Solvent or solution used	Purpose
Polymethyl methacrylate (emulsion polymerised)	Acetone	Stearic acid determination
Polymethyl methacrylate moulding powder containing lead pyrophosphate	Ethylene dichloride	Determination of lead pyro-phosphate
Polymethyl methacrylate and related polymer preparations	Phenol and Hydriodic acid	Determination of alkoxyl groups
Methyl methacrylate/ethyl acrylate copolymer containing opacifying agent (?)	Acetone	Isolation of butadiene/styrene copolymer used as opacifying agent (?)
Methyl methacrylate/ methacrylic acid copolymer	Acetone Isopropyl alcohol	Potentiometric titration of acid groups
Methyl methacrylate/styrene copolymer	Methylene dichloride	Residual benzoyl peroxide determination
Polymethyl α-chloroacrylate	Methylene dichloride	Determination of polymer
Polyethylene terephthalate	o-chlorophenol Mixtures of phenol and cresols	Determination of relative viscosity
	o-cresol and chloroform	Determination of end groups
	Phenol and tetrachloro-ethane	U.V. determination of U.V. absorbers
	Monoethanolamine	Breakdown to yield crystal-line derivative for identifi-cation
Polycarbonates	Monoethanolamine	Breakdown to yield con-stituent phenol
Nylon and related polymers	90% v/v Formic acid	Determination of relative viscosity
	90% v/v Formic acid	Direct chromatography
	Hydrochloric acid (1 : 1) either in sealed tube or boiled under reflux	Chemical or chromato-graphic examination of breakdown products

Table 2.1—*continued*

Polymer or Polymer Preparation	Solvent or solution used	Purpose
Nylon and related polymers continued	Phosphoric acid with distillation	Determination of acetyl end groups
Nylon (methoxy methyl modified)	Alcohol	Direct determination of free formaldehyde
Polythene	Tetrahydronaphthalene	Determination of relative viscosity
	Toluene	Determination of anti-oxidants, additives and contaminants e.g. oxidised polythene
	Methyl cyclohexane	Determination of anti-oxidants and additives
Polythene/polyisobutene blends	Toluene	Determination of polyiso-butene
Polystyrene	Chloroform	U.V. Spectrophotometric examination
Polystyrene	Acetone	Determination of dibutyl citric acid
Styrene/acrylonitrile copolymer	Chloroform	U.V. Spectrophotometric examination
Acrylonitrile/styrene copolymer Acrylonitrile/butadiene copolymer blends	Acetone	Determination of acrylo-nitrile/styrene copolymer by taking advantage of its solubility
Phenol-formaldehyde Phenol/cresol-formaldehyde resins	Sodium hydroxide solution	Determination of free phenols
Phenol-formaldehyde moulding powder	Acetone	Determination of resin
Phenol-formaldehyde mouldings	2-Naphthol in sealed tube	Determination of filler
Urea-formaldehyde and melamine-formaldehyde resins	Acetic acid	Characterisation of urea and melamine after breakdown

Coated Material	Solvent	Polymer etc. isolated
Urea-formaldehyde and melamine-formaldehyde resins and compositions	0·1 N Hydrochloric acid	Determination of melamine by U.V. spectrophotometry
Urea-formaldehyde compositions	1% v/v Hydrochloric acid	Detection of urea in break-down products by urease
Urea-formaldehyde resins	1 : 1 Phosphoric acid	Determination of formal-dehyde
	Benzylamine	Determination of urea
Thiourea-formaldehyde resins	Acetic acid	Detection of thiourea
Epoxy resins	Conc. sulphuric acid	Application of Foucry's tests for epoxy resins
Polyvinyl alcohol	Water	Cloud point determination

isolation of the coating material, but repeatedly we have found that methyl ethyl ketone is an excellent solvent in this type of work. Table 2.2 will give some idea of the kind of problem that is encountered and solvents that have been successfully used in particular cases to isolate polymers as a preliminary to identification.

All the extracted polymers and copolymers were readily identified by direct infra-red examination.

An alternative method of identification of resins presented as coatings on substrates is to measure the spectrum of the coating in situ by 'attenuated total reflection' as described in Chapter 1, but this is not always successful.

BURNING TESTS

Few tests have created more controversy than burning and heating tests. Nevertheless, such tests can often provide useful information and especially so to a good organic chemist with a good sense of smell who has previously obtained sound experience on known polymers and copolymers and who retains a stock of such materials for comparative tests.

In the burning test the substance under test is held on a spatula in the flame of a bunsen burner. The spatula is removed from the flame at frequent intervals in order to find out whether or not burning is taking place.

Table 2.2

Coated Material	Solvent	Polymer etc. isolated
Coated leather	Tetrahydrofuran	Polyethyl acrylate
Collar with stiffening agent	Acetone	Cellulose acetate
Coated wire	Tetrachloroethane	Polyethylene terephthalate
Laminated film	Hot toluene then tetrachloroethane	Polythene Polyethylene terephthalate
Coated cloth	Ether then tetrahydrofuran	Plasticiser Polyvinyl chloride
Coated paper	Hot toluene	Polythene *
Plastic coated cord	Formic acid Tetrachloroethane	Polyvinyl acetate
Coated hessian	Acetone	Cellulose acetate
Coated cotton	Toluene	Polythene
Coated nylon	Acetone	Cellulose nitrate
Coated hessian	Tetrahydrofuran	Polyvinyl chloride
Laminate (subsequently shown to consist of polyethylene terephthalate, polymethyl methacrylate)	Chloroform	Polymethyl methacrylate
Non-woven material	Methyl ethyl ketone	Butadiene/acrylonitrile copolymer
Coated paper	Methyl ethyl ketone	Butadiene/methyl methacrylate copolymer
Coated paper	Methyl ethyl ketone	Butadiene/styrene copolymer
Coated paper	Methyl ethyl ketone	Polyvinylidene chloride/ ester copolymer
Coated paper	Methyl ethyl ketone	Polyvinyl acetate/butyl acrylate copolymer
Coated paper	Methyl ethyl ketone	Polyvinyl acetate
Coated paper	Methyl ethyl ketone	Vinyl acetate/ethyl acrylate or methyl acrylate copolymer
Coated paper	Methyl ethyl ketone	Butadiene/styrene copolymer

1. If the material does not burn and retains its shape, then the presence of thermosetting resins such as: phenol-formaldehyde, urea-formaldehyde and melamine-formaldehyde may be suspected.
2. If the substance burns in the flame but this flame is readily extinguished on removal of the spatula and substance from the flame then the following types of substance may be suspected.

   Polyvinyl chloride and related polymers i.e. copolymers of vinyl chloride and vinyl acetate, and copolymers of vinyl chloride and vinylidene chloride.

   Casein and cellulose acetate behave similarly.
3. If the substance burns in the flame and continues to burn after removal from the flame, then many substances may be suspected and attention should be paid to the colour of the flame.

   Nylon and the polyacrylates burn with a blue flame.

   Polystyrene, cellulose and cellulose acetobutyrate burn with a yellowish white flame, but polystyrene, because of its high carbon content, yields a very smoky flame.

   With ethyl cellulose and cellulose acetate the flame is tinged yellowish green.

   Cellulose nitrate burns extremely rapidly with an intense white flame.

HEATING TEST

This test is carried out by heating the substance under test in a small dry test tube, at first over a small flame and finally to ignition. Particular attention is paid to the odour and the acidity or alkalinity of the vapours that are given off.

1. With phenol-formaldehyde a phenolic odour is evident along with an odour of formaldehyde.
2. Urea-formaldehyde and melamine-formaldehyde give off the odour of formaldehyde, and in the case of melamine-formaldehyde a characteristic fish-like odour is noted.
3. Substances such as polyvinyl chloride and related polymers give off a strong acrid odour. By placing the end of a glass rod carrying a drop of silver nitrate solution at the mouth of the test tube the presence of chloride in the volatile products can be shown.

   Polyvinylidene chloride leaves a heavy black ash. With vinyl chloride/vinyl acetate copolymers containing the equivalent of more than 10% polyvinyl acetate, the background odour of acetic acid can also be detected.
4. Polymethyl methacrylate and related compounds, as well as polystyrene, give off a sweet odour of the monomer, which is fairly characteristic in each case.

5. Cellulose gives off a smell resembling that of burnt paper. Cellulose acetate gives off an odour resembling that of acetic acid whilst a smell of rancid butter suggests the presence of cellulose aceto-butyrate.
6. Nylon and related compounds generally give off an odour of burning vegetation.
7. Polyacrylonitrile smells of cyanide.

Roff[8] has devised useful tests based on the behaviour, on heating, of such substances as nylon and polyethylene terephthalate.

SPECIAL TESTS

Having found out a certain amount of information about a test sample, it is desirable to supplement this preliminary information by special tests, some of which are rather specific in character.

These tests are described in such a way as to bring them into line with the tests for elements

NITROGEN PRESENT. NYLON AND RELATED COMPOUNDS

If a nylon type polymer is suspected, then a melting point should be determined on a few scrapings of the sample contained in a thin-walled capillary tube. Nylon homopolymers have sharp and characteristic melting points whereas nylon copolymers have indefinite melting points. In simple cases the infra-red spectrum of the sample together with knowledge of the melting characteristics will lead to immediate identification. In doubtful cases it may be desirable to hydrolyse the sample with hydrochloric acid solution and to examine the hydrolysis products (see p. 292).

NITROGEN PRESENT. UREA-FORMALDEHYDE AND MELAMINE-FORMALDEHYDE TYPE RESINS

Here, use may be made of a test developed by Widmer[9] for the identification of urea-formaldehyde and melamine-formaldehyde in wet strength paper.

TEST FOR UREA

0·05 g of the sample are refluxed for $\frac{1}{2}$ h with 25 ml of 20% v/v acetic acid solution, cooled and filtered. To 10 ml of the filtrate are added 0·5 to 1·0 ml of 1% w/v xanthydrol solution in methanol. The mixture is then evaporated to dryness on a water bath. On warming the solution on the water bath a white bulky precipitate is formed. The residue from the evaporation is transferred to a small test tube and dissolved by warming gently with a few drops of pyridine.

If urea is present crystals of urea xanthydrol are obtained on cooling. Their microscopic appearance is characteristic, i.e. small needles split or tapered at the ends.

Alternatively, the test may be carried out on a further portion of 10 ml of the hydrolysis filtrate as follows. 0·5 to 1·0 ml of 1% xanthydrol in methanol are added. The mixture is heated to boiling and boiled for 1 to 2 min when a white bulky precipitate separates out in the presence of urea. The precipitate is centrifuged, the mother liquor is decanted off, the precipitate then washed with 8 to 10 ml water and the centrifuging process repeated. The moist precipitate is dissolved by immersing the tube in a boiling water bath and adding pyridine until solution is complete. On cooling, characteristic crystals of urea xanthydrol separate out.

TEST FOR MELAMINE

0·05 g sample are refluxed for 30 min with 25 ml of 80% v/v acetic acid solution, cooled and filtered. 10 ml of the filtrate are evaporated to dryness in a small evaporating basin on a boiling water bath. The residue is transferred to an ignition tube ($7·5 \times 150$ mm approx.). The tube is evacuated to a pressure of about 5 torr ($0·7 \text{ kN/m}^2$), sealed and placed in a Woods' metal bath that is then heated at 300°C for about 10 min. In the presence of melamine a white ring of sublimate forms on the tube just above the level of the bath. The tube is allowed to cool, opened, and the sublimate transferred to a small test tube. 3 to 4 drops of water are added to the tube and the sublimate is dissolved by warming gently. One drop of this solution is allowed to evaporate slowly on a microscope slide to yield, in the presence of melamine, characteristic rhombic melamine crystals.

To the remaining solution are added a few drops of 1% w/v picric acid solution. In the presence of melamine an immediate precipitate is obtained.

This is dissolved by warming gently and allowed to crystallise at room temperature when characteristic hair-like needles of melamine picrate are deposited.

A useful test for melamine in this type of preparation is based on the work of Hirt, King and Schmitt[10]. 0·1 g of the resin is boiled with 100 ml of 0·1 N hydrochloric acid solution for 1 h, filtered, and the filtrate divided into two parts.

A 25 ml aliquot is diluted to 50 ml with water. Another 25 ml aliquot is treated with 3 ml of N sodium hydroxide solution, then diluted to 50 ml with water. The U.V. spectra of the two solutions, on acid and alkaline sides, are determined. On the acid side melamine shows a peak absorption band at 235 nm which becomes a slight shoulder in neutral or slightly alkaline solution, whereas urea does not exhibit this behaviour.

TEST FOR FORMALDEHYDE

This test gives good results with resins and moulding powders, but not so with mouldings, particularly phenol-formaldehyde mouldings.

A small amount (about 5 to 10 mg of the sample) is mixed with 2 ml of sulphuric acid solution (prepared by mixing 150 ml of sulphuric acid with 100 ml of water), and a few crystals of the sodium salt of chromotropic acid in a $6 \times \frac{3}{8}$ in $(150 \times 12 \text{ mm})$ test tube. The tube is heated for 10 min in a water bath at 65 to 75°C and the development of a violet colour is indicative of the presence of formaldehyde.

Alternatively the test for formaldehyde may be made on the distillate obtained by hydrolysing the sample with 50% v/v phosphoric acid solution.

NITROGEN PRESENT. CELLULOSE NITRATE

The Molisch test for cellulose, described on p. 103 does not give a very satisfactory answer with cellulose nitrate unless a preliminary removal of the nitrate is first effected by reduction with Devarda's alloy in alkaline solution. The substance, however, does give a satisfactory anthrone test (see p. 104).

The nitrate in cellulose nitrate may be detected by dissolving 20 mg (approx.) of diphenylamine in 1 ml of conc. sulphuric acid and pouring two or three drops of this solution on to a small piece of the sample placed on a spotting plate. The immediate development of an intense blue colour indicates the presence of nitrate.

The resorcinol test is useful. The sample, freed from any film of gelatin by treatment with hot water, is dissolved in conc. sulphuric acid. On addition of a small amount of resorcinol, a purplish blue colouration is produced.

Another test to which cellulose nitrate responds is carried out by first grinding with sodium thiosulphate in a mortar. The mixture is then placed in a porcelain crucible and heated on a sand bath until the material begins to burn. On cooling, the product is boiled with water, acidified with hydrochloric acid to a pH of 2 to 3 and again boiled for a short time to remove any sulphur that has separated out. On adding ferric chloride to the filtrate, a deep red colouration is obtained.

NITROGEN PRESENT. POLYACRYLONITRILE

Polyacrylonitrile may be suggested by the presence of nitrogen and the odour of cyanide on heating. When polyacrylonitrile is depolymerised in a vacuum depolymerisation tube a pale straw coloured liquid is obtained in the receiver. The liquid gradually darkens to a dark yellow and finally to a dark brown liquid on standing, i.e. without opening the tube.

On releasing the vacuum in the apparatus a distinct smell of cyanide is observed, and, on standing for some time, the liquid in the receiver gives off a strong smell of ammonia.

The presence of cyanide may be confirmed as follows. 2 ml of 10% w/v sodium hydroxide solution are added to the receiver tube and two to three drops of ferrous sulphate solution are added. The solution is then boiled, cooled and acidified with 10% v/v hydrochloric acid solution, when a blue colour or precipitate of ferric ferrocyanide is obtained.

CHLORINE

If polyvinyl chloride or a related polymer is suspected it is desirable to carry out tests as follows:

1. For polyvinyl chloride (method of Whitnack and Gantz[11])

About 50 mg of the resin are placed in a test tube and dissolved in pyridine by heating on a water bath. The addition of a few drops of a methanol solution of sodium hydroxide or potassium hydroxide produces a brown colour when polyvinyl chloride is present. This test detects as little as one drop of a 0·01% w/v solution of polyvinyl chloride in pyridine. Aromatic chlorides do not interfere but aliphatic chlorides give a faint positive test. The product of the reaction is a cyanine dye that precipitates from the pyridine solution as a brown powder.

2. For polyvinylidene chloride (Verdcourt's test[12])

About 50 mg of sample are dissolved in about 10 ml of morpholine by heating in a water bath. The presence of polyvinylidene chloride is shown by the appearance of a brownish black colour, the depth of which depends on the amount of polyvinylidene chloride present. Polyvinyl chloride itself does not affect the colour of the morpholine solution.

With vinyl chloride copolymers it may be desirable to have available a chemical test for acetate and this is given on p. 104.

FLUORINE ONLY

Here it is probable that the polymer concerned is polytetrafluoroethylene or polyvinyl fluoride. Other fluorine containing resins are mentioned in Chapter 7.

FLUORINE AND CHLORINE

Here it is probable that the polymer concerned is polytrifluorochloro-ethylene.

NITROGEN AND SULPHUR

In the event of nitrogen and sulphur being detected it will probably be useful to make a test for a thiourea-formaldehyde resin with Fearon's pentacyanoammonio ferrate III reagent[13]. This is prepared by reaction of sodium nitroprusside with 0·880 ammonia solution for 48 h at 0°C. The precipitate of pentacyanoammonio ferrates II and III is filtered off, and ethyl alcohol is added to the mother liquor to precipitate II which is filtered off, washed with ethyl alcohol and dried. II is dissolved in water and stood in sunlight for 24 h to yield a solution of III.

In making the test on the resin 0·1 g of the sample is boiled with 2 ml N sulphuric acid solution, 2 ml N sodium hydroxide solution are added followed by 1 drop of N hydrochloric acid solution. The solution is filtered and 5 drops of the pentacyanoferrate III reagent are added to the filtrate.

Thiourea resins, under these conditions, give a bright blue colouration that remains bright blue on addition of 1 ml of 30% v/v acetic acid solution.

In general, the blue pentacyanoferrate test gives positive results with compounds containing the $=C=S$ group but not with the $-SH$ group.

NITROGEN, SULPHUR AND PHOSPHORUS PRESENT

Casein may be suspected and the following tests for protein may be carried out.

Xanthoproteic reaction

5 to 10 mg of the sample are heated in a boiling water bath with 0·1 ml of conc. nitric acid for 5 to 10 min after which most proteins will have dissolved. The solution is then diluted with 0·3 ml of water. In the presence of protein a white precipitate is formed that redissolves on boiling to yield a yellow solution. On making alkaline an orange solution is produced.

Biuret reaction

10 to 20 mg of the sample are heated in a boiling water bath with 0·5 ml of conc. hydrochloric acid for 5 to 10 min. The solution is diluted to

2 ml with 10% w/v sodium hydroxide solution. The solution is filtered, if necessary, and to 1 ml of the filtrate 0·3 ml of 40% w/v sodium hydroxide solution are added followed by 1 drop of a dilute solution of copper sulphate. A violet colouration indicates the presence of a protein.

ONLY CARBON AND HYDROGEN PRESENT
OR ONLY CARBON, HYDROGEN AND OXYGEN PRESENT

POLYMETHYL METHACRYLATE AND RELATED COMPOUNDS

Most of these materials, unless cross linked, are soluble in chloroform. It is useful to depolymerise this type of substance in vacuo and to examine the monomers produced or to carry out pyrolysis/G.L.C. examination.

POLYSTYRENE

This may be depolymerised as for polymethyl methacrylate, although it may be desirable to carry out the depolymerisation for a much longer time, i.e. 4 to 5 h.

The styrene in the depolymerisation products may be identified as follows:

8 to 10 drops of the monomer that is obtained are measured into a clean test tube $(5 \times \frac{3}{8}$ in; 100×12 mm) and liquid bromine is added dropwise until a definite excess is indicated by the colour of the mixture. The excess of bromine is removed by heating on a water bath and the mixture is then cooled.

The solid residue that is obtained is broken up with a glass rod and then dissolved by heating with 5 ml of 80% v/v ethyl alcohol.

The filtered solution is cooled and the crystalline deposit filtered off and recrystallised from 3 ml of 80% v/v ethyl alcohol. The recrystallised product is dried in air and the m.p. determined. Styrene dibromide melts at 73°C.

POLYTHENE, POLYPROPYLENE AND α-OLEFINE POLYMERS

These materials are not readily identified by chemical means because of their inert nature. Many of them have melting points which can assist in characterising them. Polythene may be dissolved in hot toluene from which it separates in characteristic form on cooling. Determination of the carbon and hydrogen contents of these polymers may provide useful information but the figures obtained do not distinguish e.g. between polythene and

102

polypropylene. Infra-red examination and pyrolysis/gas chromatography are often employed in the characterisation of this type of polymer.

PHENOL-FORMALDEHYDE TYPE RESINS

Here we are concerned with the reaction products of phenols and cresols with formaldehyde. The test for formaldehyde, previously described under urea-formaldehyde, gives positive results with resins and moulding powders, but not, as a rule, with mouldings.

TEST FOR PHENOLS, CRESOLS ETC.

A small quantity of the powdered sample (20 to 50 mg) is mixed with a similar bulk of soda lime and the mixture heated carefully over a small flame until decomposition starts to take place. After cooling, the products are mixed with 2 ml of water and 1 ml of the mixture is withdrawn and centrifuged. The clear liquor is made just acid with hydrochloric acid and any precipitate obtained is removed by centrifuging. 2 drops of the solution are mixed with 2 drops of buffer solution (23·4 g borax and 3·2 g of sodium hydroxide dissolved in water and the solution diluted to 1 l) and 1 drop of a fresh solution which contains 0·1 g of 2:6-dibromo-quinone chloroimide in 25 ml of ethyl alcohol is added. The appearance of a deep blue colour indicates the presence of phenols.

CELLULOSIC TYPE RESINS

The most common types are cellulose, cellulose acetate, cellulose nitrate and ethyl and methyl cellulose. All these substances respond to the Molisch test, except cellulose nitrate, which requires pretreatment.

This Molisch test is carried out as follows:

25 to 50 mg of the sample are allowed to stand overnight with 3 to 5 ml of 50% v/v sulphuric acid solution. 2 to 3 drops of the resulting solution are diluted with 1 ml of water and 2 drops of a freshly prepared 10% w/v solution of α-naphthol in chloroform are added; 1 ml of conc. sulphuric acid is allowed to run in slowly down the sides of the test tube in such a way that the two liquids do not mix. A violet ring at the junction of the two liquids indicates the presence of a cellulosic compound.

If a yellow coloured mixture is produced on addition of the conc. sulphuric acid, cellulose nitrate may be suspected.

In that case a small piece of the sample is digested with 3 ml of N sodium hydroxide solution for $\frac{1}{2}$ h. The warm sodium hydroxide solution is decanted off and to it a small amount of Devarda's alloy is added. The mixture is allowed to stand until no further gas is evolved and is then filtered through glass wool.

The resultant solution is halved. To one half are added 2 drops of the α-naphthol solution and 1 ml of conc. sulphuric acid is poured carefully down the side of the test tube used. To the second half of the solution 1 ml of dilute sulphuric acid solution is added prior to the addition of α-naphthol and conc. sulphuric acid. A violet ring at the junction of the two liquids in either test indicates the presence of cellulose.

TEST FOR ACETATE ESTERS

Cellulose acetate gives a positive result in the following test for acetate. Copolymers of vinyl chloride and vinyl acetate also respond to the test. When applied to the latter type of polymer the test is carried out as follows:

0·5 g of the sample is refluxed for 1 h with 5 ml of 2 N ethylene glycol potash. After cooling and acidifying with 1·5 ml of 50% v/v phosphoric acid solution the mixture is distilled in steam, 25 ml of steam distillate being collected. 5 ml of this distillate are titrated with 0·1 N sodium hydroxide solution using phenolphthalein as indicator. The remaining 20 ml of distillate are then neutralised without indicator addition and the neutral solution boiled down to 0·5 ml. 2 drops of the concentrated neutral solution, 2 drops of 5% w/v lanthanum nitrate solution, 2 drops of 0·01 N iodine solution and 2 drops of N ammonia solution are added successively, without mixing and in that order. Acetates give a blue colour.

The presence of acetate may be confirmed by the ferric chloride test.

Acetates such as cellulose acetate and copolymers of vinyl chloride and vinyl acetate give a positive result in Roff's test for polyethylene terephthalate[8], which is carried out as follows:

Approximately 0·1 g sample is placed at the bottom of a $3 \times \frac{3}{8}$ in (75 × 10 mm) test tube and covered with 0·3 g calcium oxide. A small plug of glass wool is placed in the mouth of the tube which is then covered by a disk of filter paper ($\frac{1}{4}$ in or 6 mm diameter) saturated with a freshly prepared saturated solution of o-nitrobenzaldehyde in 2 N sodium hydroxide. The paper is held in position by means of a small filter funnel inserted over the tube. The tube is heated over a small bunsen flame until the sample is decomposed. After a while the paper is removed and washed with hydrochloric acid solution. The appearance of an indigo blue colour or ring on the test paper indicates the presence of acetate in the sample. 0·5% vinyl acetate in a vinyl chloride/vinyl acetate copolymer gives a positive reaction in the test.

Cellulose, cellulose acetate, cellulose nitrate and ethyl and methyl cellulose all give positive results in the anthrone test which is carried out as follows:

1 ml of water is placed in a test tube containing about 1 mg of the sample and 2 ml of a freshly prepared 0·2% w/v solution of anthrone in

conc. sulphuric acid. The solution is mixed, the heat produced on dilution being a necessary part of the test.

Cellulose, cellulose acetate and ethyl and methyl cellulose yield blue coloured solutions in this test. With cellulose nitrate a green colour is first produced and this changes rapidly to give a dark blue solution.

As anthrone itself gives a faint positive reaction a blank test must always be carried out alongside.

EPOXY RESINS

From time to time various tests have been proposed for the detection of epoxy resins, including spot tests of a solution of the resin in conc. sulphuric acid on filter paper, reaction with glucose sulphuric acid or with paraformaldehyde sulphuric acid, but Foucry's tests[14] have often proved most useful. They are carried out as follows.

Special reagents

Ferric Sulphate Solution (10% w/v).
10 g of ferric sulphate are dissolved in water and the solution is filtered.
Denigé's Sulphomercuric Reagent.
5 g of yellow mercuric oxide are dissolved in a mixture of 20 ml conc. sulphuric acid and 100 ml of water and the solution is filtered.

Procedure

1 g of the test resin is dissolved without heating in 100 ml of 95% v/v sulphuric acid solution. Each of the following tests is carried out on 1 ml of the test solution.

(a) 1 ml of conc. nitric acid is added, the mixture shaken and allowed to stand for 5 min. The mixture is then poured into 100 ml of N sodium hydroxide solution, when an orange red colouration will be produced in the presence of epoxy resin.

(b) 3 drops of 40% w/v formaldehyde solution are added. In a positive test a red brown colour is produced, that, on dilution with water, changes to green.

(c) 5 ml of the Denigé's sulphomercuric reagent are added. In a positive test an orange precipitate is produced immediately.

(d) 5 ml of the ferric sulphate reagent solution are added. In a positive test a dark violet colouration is produced.

105

Although these colour tests are very useful, the results should be accepted with caution in the case of suspected epoxy resins and more positive evidence should be obtained from infra-red spectra.

POLYETHYLENE TEREPHTHALATE

A useful test for polyethylene terephthalate is that given by Haslam and Squirrell[15] that depends on the quantitative hydrolysis of the polyethylene terephthalate with monoethanolamine and the simultaneous production of a crystalline derivative of the terephthalic acid.

This hydrolysis with monoethanolamine is of value in the examination of polycarbonates; the hydrolysis is quickly performed with the liberation of the constituent phenolic body.

PLASTICISERS

Although this preliminary part of the book is concerned, in the main, with the qualitative and semi-quantitative examination of polymers and polymer compositions, many of the tests will be found useful in the preliminary examination of plasticisers.

More detailed methods of examination of plasticisers are given in Chapter 11.

THE INFRA-RED SPECTRUM IN QUALITATIVE ANALYSIS

Before considering the value of infra-red spectroscopy in the qualitative analysis of polymer compositions, it should be appreciated that, contrary to what has often been suggested, in our laboratories chemical examination has almost invariably preceded infra-red examination. This method has the advantage that the infra-red spectroscopist has a knowledge of the elements present in a sample before his examination of the spectrum; it follows a logical process of first performing the test which gives unequivocal information, rather than wasting time on tests which can give only a partial answer. For example, if one requires to know whether a compound contains sulphur, one could examine the infra-red spectrum for sulphur-oxygen groups or sulphur-hydrogen groups but since the sulphur-sulphur linkage is almost inactive in the infra-red, a firm answer on the presence of sulphur could not always be given by examining the spectrum. On the other hand this information is given by a simple chemical test already described. We cannot too strongly emphasise the necessity for putting the various tests, both chemical and physical, in their true perspective as being complementary rather than competitive methods of acquiring the same information.

Moreover, the preliminary chemical work may have pointed to the necessity for a separation procedure to be effected before submitting the sample to infra-red examination. The spectroscopist sees, in the spectrum, the functional groups of all constituents of a mixture superimposed, and often key features of one constituent are obscured by the features of another. Thus time is wasted, and useful information often concealed, by submitting easily separable mixtures for the infra-red test.

'MOLECULAR FINGERPRINTS' AND MECHANICAL SORTING FOR THE IDENTIFICATION OF SUBSTANCES

The infra-red spectrum may be considered to be made up of absorption bands due to particular groups of atoms, modified by the way in which the groups are built into the molecule. However, since a monomeric substance has a unique arrangement of atoms, it follows that it must have a unique absorption spectrum, which can be characterised by the position and intensity of the absorption bands. In view of this, several methods have been devised for the identification of unknown substances using the spectrum as a 'molecular fingerprint', matching the spectrum of the unknown against the previously recorded spectra of known pure substances. Although both position and intensity of bands might be used, in practice no satisfactory method has been devised for using band intensities other than in a qualitative way. The intensity of the bands in the spectrum depends upon the thickness of the sample and to some extent upon the instrument used and until all spectra are converted to an agreed intensity scale, for example, absorbance per cm thickness, it will not be possible to use intensities in a strictly quantitative fashion for sorting purposes.

Nevertheless good progress has been made with sorting methods which are based on recording the wavelengths of the absorption bands in the spectra of known compounds either on data sheets, punched cards, or magnetic tape, so that unknown spectra may be compared with those of known compounds by manual sorting, mechanical sorting, or, in the case of magnetic tape, by computer. In many cases additional parameters such as physical constants, elements present and empirical formula are included in the data recorded, and when these data are available on the unknown compound they may be used in conjunction with band positions in the matching process.

There is no doubt that much time may be saved in identification work by using these mechanical sorting methods, and we use computer methods regularly in our laboratories, although we find them much more satisfactory for monomeric substances than for polymers. This is not surprising because whereas all the molecules of a monomeric substance are chemically identical the molecules of even a homopolymer are not, and cannot therefore be expected to have the same simple fingerprints. They may differ in length—i.e. the number of consecutive monomer units in the different molecules can vary, although in practice with regular

107

linear molecules as soon as a reasonable degree of polymerisation has been achieved (say more than ten consecutive monomer units) the spectrum appears to be substantially constant, provided the substance is in a constant physical state, apart from those (usually weak) features of the spectrum associated with the terminal groups of the polymer chain which are necessarily chemically different from the repeating unit. However, polymers produced by the successive addition of the same unsaturated monomer unit to the end of a growing polymer chain do not necessarily have a regular structure. One example of a type of irregularity is that which occurs in the polymerisation of diene monomers, where successive units can add in different ways to produce chemically different polymer structures, each of which has distinct spectral features (see Chapter 8). Again the molecules of a monomer with only one unsaturated group may add to the end of the growing polymer chain with the substituents on one or the other side of the backbone; this produces chemically distinct species which are usually described as isotactic or syndiotactic sequences (see discussion in Chapter 6). These species have different spectra which can be simply due to the difference in chemical structure, but may arise because the spatial arrangement of substituents along the polymer chain causes the molecule to take up a particular shape, such as a fully extended or coiled form. These different physical forms may have quite distinct spectral features.

Different physical forms may also appear without any change of chemical composition in polymers such as nylon or polyethylene terephthalate where the molecules can pack together and crystallise. The transition from the crystalline to the amorphous state is a well known cause of differences in the spectrum of a monomeric substance, but whereas monomeric substances usually have a sharp melting point, at which the spectrum changes abruptly from that of the crystalline to that of the amorphous substance, polymeric materials change slowly in their degree of crystallinity as the temperature is changed below the crystalline melting point, so that both crystalline and amorphous forms occur simultaneously. Furthermore the crystalline/amorphous ratio of a sample depends upon previous heat treatment, and may be changed, at one temperature, by physical means such as stretching the sample and may be influenced also by the presence of additives. Thus with a polymer which is able to crystallise, the spectrum recorded at normal temperatures will contain the superimposed spectra of the crystalline and amorphous phases, in varying amounts according to the past thermal or physical treatment which the sample has received.

Examples can also be found of monomeric substances which can exist in different crystalline forms with different spectra—the same phenomenon is observed in the spectra of some synthetic polymers (see particularly nylons, Chapter 5).

When considering the spectra of copolymers it will be seen that further difficulties arise because changes in the distribution of the different monomer units along the polymer chain again produce molecules which are chemically different, and can therefore have significantly different spectra either due directly to these chemical differences, or to indirect effects on molecular shape or molecular packing.

It will be evident therefore that the interpretation of the spectra of synthetic polymers is much more complex than that of monomeric substances, and an approach in terms of 'molecular fingerprints' is, except in the simplest cases, inadequate to deal with the effects found in practice. On the other hand, irrespective of these effects, the interpretation of the spectrum in terms of bands due to chemical groups is valid, and it is clear that this is a more satisfactory and rewarding approach.

QUALITATIVE ANALYSIS FROM THE SPECTRUM

It is desirable when studying the spectrum of a compound to have available the result of the qualitative elements test. There is considerable overlapping of the wavelength ranges in which different groups absorb, and much time is saved by immediately rejecting groups excluded by the elements test.

The stronger bands in the spectrum are first considered, as these are likely to be due to those groups present in greatest concentration.

It is not proposed here to discuss qualitative analysis from the spectrum in general, since this subject is adequately covered elsewhere[16-19]. However, attention will be drawn to those groups more frequently encountered in polymers and associated substances. These observations are arranged according to elements detected, beginning with compounds of carbon and hydrogen, and carbon, hydrogen and oxygen. These two classes are treated together as they are not usually separated in the simple elements test. They are in most cases distinguished without difficulty from the spectrum.

In examining the spectra of materials containing additional elements such as chlorine and sulphur, characteristic bands for hydrocarbon and various oxygen containing groups must be considered as well as those arising directly from groups containing the additional elements. In some cases the presence of additional elements modifies the positions of characteristic bands of other groups. This is noted in the tables.

NO ELEMENTS DETECTED

It is first necessary to establish whether the unknown contains oxygen. With the exception of the simple peroxide group,

$$\diagdown C-O-O-C \diagup$$

which has no useful absorption, oxygen containing groups give rise to bands of at least medium intensity (see Table 2.3).

109

The occurrence of one or more bands of moderate or high intensity in the O—H and C=O regions can be taken to indicate the presence of oxygen in a compound.

Table 2.3. Absorption bands of oxygen containing groups in C, H, O compounds

Group	Wavelength range (μm)	Frequency range (cm⁻¹)	Band Character
—OH	2·7—2·8 (free) 2·8—3·2 (bonded)	3700—3550 3550—3150	Weak, sharp Strong, broad
>C=O	5·5—6·5	1825—1550	Strong, medium width
>C—O—	7·5—11·0	1350—900	Strong, variable width often multiple

Note. Some compounds containing both C = O and O—H show little apparent —OH absorption. However, the presence of oxygen is apparent from the C=O absorption band.

C—O bands are usually very intense, and if a band falling within the 7·5 μm to 11·0 μm region exceeds in intensity the C—H bands at 3·45 μm or near 7 μm, this also indicates the presence of oxygen.

COMPOUNDS CONTAINING CARBON AND HYDROGEN ONLY

The principle bands to be expected are summarised in Table 2.4.

Saturated hydrocarbons

All hydrocarbons (with the exception of one or two simple gaseous compounds) exhibit a band, usually multiple in character, near 3·4 μm and another between 6·8 μm and 7·0 μm. Most hydrocarbons also contain methyl groups (characteristic band near 7·3 μm).

Hydrocarbons are classified as saturated, unsaturated or aromatic mainly by the absorption bands beyond 10 μm. The least characteristic class is the alicyclics for which no useful general rules can be given except that the region 7 μm to 13 μm often contains several rather sharp bands of medium intensity.

Unsaturated groups

These hydrocarbons produce highly characteristic bands between 10 μm and 15 μm and all unsaturated structures other than

may be identified. (Table 2.5). These bands represent the only simple

Table 2.4. Absorption bands of saturated hydrocarbon groups

Structure	Type of vibration	Approximate band position		Remarks
		Wavelength (μm)	Frequency (cm^{-1})	
$-CH_2-$	C—H stretch	$\begin{cases} 3 \cdot 42 \\ 3 \cdot 51 \end{cases}$	$\left.\begin{matrix} 2925 \\ 2850 \end{matrix}\right\}$	Reliable, strong bands
$-CH_3$	C—H stretch	$\begin{cases} 3 \cdot 38 \\ 3 \cdot 48 \end{cases}$	$\left.\begin{matrix} 2960 \\ 2870 \end{matrix}\right\}$	Difficult to observe unless methyl concentration is high
$\begin{matrix} R_1 \\ R_2 \\ R_3 \end{matrix}\!\!>\!C-H$	C—H stretch	3·46	2890	Difficult to observe
$R-\underset{\underset{O}{\parallel}}{C}-H$	C—H stretch	3·68	2720	Useful band of aldehydes
$-CH_2-$	C—H bend	6·82	1465	Reliable band. Shifts to greater wavelength with adjacent chlorine or oxygen
$-CH_3$	C—H bend (asymm.)	6·90	1450	Difficult to distinguish from $-CH_2-$ band
$\begin{matrix} R_1 \\ R_2 \\ R_3 \end{matrix}\!\!>\!C-CH_3$	C—H bend (symm.)	7·28	1375	
$\begin{matrix} R_1 \\ \\ R_2 \end{matrix}\!\!>\!C\!\!<\!\begin{matrix} CH_3 \\ \\ CH_3 \end{matrix}$	C—H bend (symm.)	$\begin{cases} 7 \cdot 24 \\ 7 \cdot 33 \end{cases}$	$\begin{matrix} 1380 \\ 1365 \end{matrix}$	A most useful series of bands. Very reliable for methyl groups in complex structures
$R-C\!\!<\!\!\begin{matrix} CH_3 \\ CH_3 \\ CH_3 \end{matrix}$	C—H bend (symm.)	$\begin{cases} 7 \cdot 19 \\ 7 \cdot 33 \end{cases}$	$\begin{matrix} 1390 \\ 1365 \end{matrix}$	
$\begin{matrix} R_1 \\ R_2 \\ R_3 \end{matrix}\!\!>\!C-H$	C—H bend	7·58	1320	Occasionally useful

Table 2.4—*continued*

Structure	Type of vibration	Approximate band position		Remarks
		Wave-length (μm)	Fre-quency (cm^{-1})	
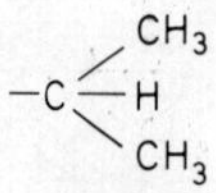 $-C\begin{smallmatrix} CH_3 \\ H \\ CH_3 \end{smallmatrix}$	$\left\{\begin{array}{l} \text{C—C stretch} \\ \text{C—H bend} \end{array}\right\}$	8·55 8·66	1170 1155	
$-C\begin{smallmatrix} CH_3 \\ CH_3 \\ CH_3 \end{smallmatrix}$	$\left\{\begin{array}{l} \text{C—C stretch} \\ \text{C—H bend} \end{array}\right\}$	8·0	1225	Useful bands for hydrocarbon systems, but less useful when other elements are present
$R_1-\overset{\overset{\displaystyle CH_3}{\mid}}{\underset{\underset{\displaystyle CH_3}{\mid}}{C}}-R_2$	$\left\{\begin{array}{l} \text{C—C stretch} \\ \text{C—H bend} \end{array}\right\}$	8·3	1205	
$R_1-\overset{\overset{\displaystyle H}{\mid}}{\underset{\underset{\displaystyle CH_3}{\mid}}{C}}-R_2$	$\left\{\begin{array}{l} \text{C—C stretch} \\ \text{C—H bend} \end{array}\right\}$	8·65	1155	
$R_1-\overset{\overset{\displaystyle H}{\mid}}{\underset{\underset{\displaystyle CH_3}{\mid}}{C}}-\overset{\overset{\displaystyle H}{\mid}}{\underset{\underset{\displaystyle CH_3}{\mid}}{C}}-R_2$	$\left\{\begin{array}{l} \text{C—C stretch} \\ \text{C—H bend} \end{array}\right\}$	8·64 8·90 9·32	1157 1125 1072	
$>\!\!C-CH_3$	CH_3 rock	10·3	971	
$>\!\!C-CH_2-CH_3$	CH_3 rock	10·8	926	Principal application to chain branching in polythene
$>\!\!C-CH_2-CH_2-CH_3$	CH_3 rock	11·0	909	
$>\!\!C-(CH_2)_n-CH_3$	CH_3 rock	11·24	890*	

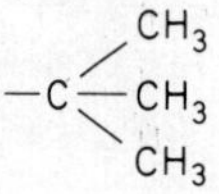

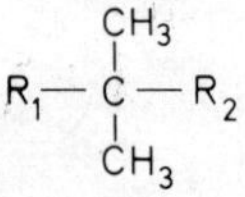

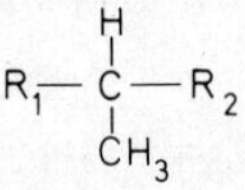

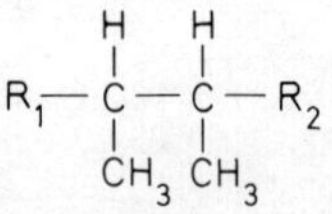

*$n>3$

Table 2.4—*continued*

Structure	Type of vibration	Approximate band position		Remarks
		Wave-length (μm)	Fre-quency (cm^{-1})	
$>$CH—CH$_2$—CH$_3$	CH$_2$ rock	13·00	770	
$>$CH—(CH$_2$)$_2$—CH$_3$	CH$_2$ rock	13·60	740	Applicable to higher α-olefins and chain branching in poly-thene
$>$CH—(CH$_2$)$_3$—CH$_3$	CH$_2$ rock	13·79	725	
$>$CH—(CH$_2$)$_n$—CH$_3$	CH$_2$ rock	13·86	721	
$=$C—CH$_2$—C$\lessgtr$	CH$_2$ rock	~ 13	~ 770	Useful band in vinyl polymers, but sensitive to nature of sub-stituent on adjacent carbon atom
$=$C—(CH$_2$)$_2$—C$\lessgtr$	CH$_2$ rock	13·3	752	
$=$C—(CH$_2$)$_3$—C$\lessgtr$	CH$_2$ rock	13·64	733	
$=$C—(CH$_2$)$_4$—C$\lessgtr$	CH$_2$ rock	13·70	730	
$=$C—(CH$_2$)$_5$—C$\lessgtr$	CH$_2$ rock	13·75	727	Very useful in examination of olefin polymers and copolymers
$=$C—(CH$_2$)$_n$—C$\lessgtr$	CH$_2$ rock	13·88	720*	
$=$C—(CH$_2$)$_n$—C$\lessgtr$	CH$_2$ rock	13·70 13·88	730† 720	

*Amorphous $n > 5$ †Crystalline $n > 5$

Table 2.5. Absorption bands of unsaturated hydrocarbon groups

Structure	C—H stretch Wavelength μm	Frequency cm⁻¹	C—C stretch Wavelength μm	Frequency cm⁻¹	In plane C—H bend Wavelength μm	Frequency cm⁻¹	Out of plane C—H bend Wavelength μm	Frequency cm⁻¹
R_1H C=C HH	{ 3·25 / 3·30	3085 / 3030	6·08	1643	7·10 / 7·68	1410 / 1300	10·10 / 11·00	990 / 910 }
R_1,R_2 C=C HH	3·24	3080	6·06	1650	7·10	1410	11·27	888
R_1H C=C HR_2	3·30	3030	6·02	1660	7·70	1300	10·35	965
$R_1$$R_2$ C=C HH	3·30	3030	6·05	1653	7·10	1410	~14·5*	~690*
$R_1$$R_3$ C=C R_2H	3·30	3030	6·00	1675	7·40	1350	12·05	830
C=C—C=C	{ 3·25 / 3·33	3085 / 3000	6·25	1600	7·30 / 7·64	1370 / 1310	9·90 / 11·00	1010 / 910 }
R—C≡C—H	3·30	3030	~4·45	~2200	—	—	~15·5	~650
Benzene							14·50	690
mono sub.							{ 13·35 / 14·50	750 / 690 }
ortho-disub.	~3·30	~3030	6·2—6·5†	1600—1550†	~6·75	~1490	13·35	750
meta-disub.							12·90	775
para-disub.							12·35	810
1:2:3 trisub.							12·82	780

*Position very variable   †Several weak sharp bands.

method of determining the structure of polythene and of diene polymers. Unfortunately they are of limited value outside the hydrocarbon field, being strongly modified by adjacent structural groups[20].

Aromatic structures

These produce relatively intense absorption bands in the 12 μm to 15 μm region characteristic of the type of substitution about the benzene ring, (Table 2.5). These so-called substitution patterns are reliable for hydrocarbons, chlorinated aromatics and phenols; in some compounds, however, particularly where the benzene nucleus is conjugated with the carbonyl group, the pattern may be modified and is less reliable as a method of ascertaining the type of substitution, although an aromatic compound is indicated.

Most aromatic compounds have one or more sharp bands of weak intensity between 6 μm and 7 μm wavelength.

COMPOUNDS CONTAINING CARBON, HYDROGEN AND OXYGEN

The prominent absorption bands of the common oxygen containing groups (Table 2.3), form a useful method of classifying these compounds.

Compounds found to contain carbonyl groups

While the precise wavelength of the carbonyl group absorption band is sometimes sufficient to distinguish the main types of carbonyl containing compound, additional evidence from other spectral regions, or independent tests, may be necessary. Spectral data on the common carbonyl compounds are presented in Table 2.6. Other groups are considered in References 16 and 19. Aliphatic esters, aldehydes and ketones are difficult to differentiate. The difference in the carbonyl band position is small, and many aldehydes and ketones have bands of quite high intensity between 7·5 μm and 11·0 μm which can be confused with the —C—O— absorption bands of aliphatic esters. Most aldehydes have an additional C—H band at 3·68 μm which is useful for confirming the presence of this group. Fortunately aliphatic aldehydes and ketones are not commonly encountered in polymer work.

The carbonyl group may bond so strongly to the hydroxyl group that the band of the latter near 3·0 μm is difficult, or impossible, to observe. In carboxylic acids it is evident only as a broadening of the C—H band at 3·45 μm, while in hydroxy-benzophenones of the type

and

115

Table 2.6. Absorption bands of carbonyl containing compounds

Compound	Structure	Carbonyl band approximate position		Confirmatory bands
		Wave-length μm	Fre-quency cm^{-1}	
Aliphatic acid anhydride	$R_1 \cdot CO$ — O — $R_2 \cdot CO$	$\left\{\begin{array}{l}5 \cdot 48 \\ 5 \cdot 72\end{array}\right.$	$\left.\begin{array}{l}1825 \\ 1750\end{array}\right\}$	C—O bands $1150 - 1050$ cm^{-1} ($8 \cdot 7 - 9 \cdot 5$ μm)
Aliphatic acyl peroxide	$R_1 \cdot CO - O$ — $R_2 \cdot CO - O$	$\left\{\begin{array}{l}5 \cdot 52 \\ 5 \cdot 60\end{array}\right.$	$\left.\begin{array}{l}1810 \\ 1785\end{array}\right\}$	No useful bands
Aromatic acyl peroxide	(phenyl)—CO — O — (phenyl)—CO	$\left\{\begin{array}{l}5 \cdot 56 \\ 5 \cdot 60\end{array}\right.$	$\left.\begin{array}{l}1800 \\ 1785\end{array}\right\}$	Aromatic structure
Aromatic acid anhydride	(phenyl)⟨CO — O — CO⟩	$\left\{\begin{array}{l}5 \cdot 57 \\ 5 \cdot 75\end{array}\right.$	$\left.\begin{array}{l}1795 \\ 1740\end{array}\right\}$	C—O bands $1150 - 1050$ cm^{-1} ($8 \cdot 7 - 9 \cdot 5$ μm)
Imide	(phenyl)⟨CO — N— — CO⟩	$\left\{\begin{array}{l}5 \cdot 62 \\ 5 \cdot 82\end{array}\right.$	$\left.\begin{array}{l}1780 \\ 1718\end{array}\right\}$	Aromatic structure
Aryl ester of aliphatic acid	$R \cdot CO \cdot O$ — (phenyl)	$5 \cdot 65$	1770	C—O bands 1220 cm^{-1} ($8 \cdot 2$ μm) 1150 cm^{-1} ($8 \cdot 7$ μm)
Aryl ester of aromatic acid	(phenyl)—$CO \cdot O$—(phenyl)	$5 \cdot 73$	1745	Several C—O bands $1300 - 1100$ cm^{-1} ($7 \cdot 7 - 9 \cdot 0$ μm)
Alkyl ester of aliphatic acid	$R_1 \cdot CO \cdot O \cdot R_2$	$5 \cdot 76$	1735	C—O bands near 1170 cm^{-1} ($8 \cdot 6$ μm)
Alkyl aldehyde	$R \cdot CO \cdot H$	$5 \cdot 78$	1730	C—H band 2720 cm^{-1} ($3 \cdot 68$ μm)
Alkyl ester of aromatic acid	(phenyl)— $CO \cdot O \cdot R$	$5 \cdot 82$	1720	C—O bands 1120 cm^{-1} ($8 \cdot 9$ μm) 1260 cm^{-1} ($7 \cdot 9$ μm)
Dialkyl ketone	$R_1 \cdot CO \cdot R_2$	$5 \cdot 83$	1715	C—O bands $1325 - 1215$ cm^{-1} $7 \cdot 55 - 8 \cdot 23$ μm)

Table 2.6—*continued*

Compound	Structure	Carbonyl band approximate position		Confirmatory bands
		Wave-length μm	Fre-quency cm^{-1}	
Alkyl carboxylic acid (dimer)	R·COOH	5·85	1710	C—O ~ 1300 cm^{-1} (7·7 μm) OH ~ 930 cm^{-1} (10·7 μm)
Aryl carboxylic acid (dimer)	—COOH	5·92	1690	Aromatic bands for aryl acid
Diaryl ketone	—CO—	6·00	1665	Aromatic structure
o-Hydroxy diaryl ketone	OH —CO—	6·08	1645	Aromatic structure
Carboxylate ion	R—C O O $^-$	6·40 7·15	1560 1400	Alkyl acid salts —CH$_2$—structure 1350—1100 cm^{-1} (7·5 μm–9 μm) Aryl acid salts— aromatic structure

no evidence of the hydroxyl stretching band is found. Carboxylic acids may be identified from other spectral features, while a large carbonyl shift to longer wavelengths occurs in the latter compounds as compared with simple benzophenones (Table 2.6).

Compounds found to contain hydroxyl groups

The hydroxyl group absorption band appears (except as noted above) within the range of 2·7 μm to 3·2 μm.

In the majority of compounds, intermolecular hydrogen bonding occurs in the solid or liquid state, and the band then appears in the 2·8 μm to 3·2 μm region, broad and intense. Such compounds, examined as dilute solutions in inert solvents, show a marked change in their spectra in this region, the O—H band becoming very sharp, weaker, and falling in the 2·7 μm to 2·8 μm range.

Certain compounds show no intermolecular hydrogen bonding, and the band is in the 'free' position, 2·7 μm to 2·8 μm, under all circumstances. Such compounds include phenols with large substituents adjacent to the hydroxyl group which prevent bonding.

Alternatively, some compounds show the O—H band in the 2·8 μm to 3·2 μm region (bonded) even in dilute solution. Such compounds remain in the hydrogen bonded condition because each molecule contains the hydroxyl group so situated that it may hydrogen bond with another polar group within the molecule.

It will be seen that a careful examination of the hydroxyl stretching region can be extremely valuable in the determination of the structure of unknown compounds, particularly phenols. The presence of a band in the 2·7 μm to 3·2 μm region is a very reliable indication of the presence of hydroxyl groups, except for the following points:

1. Water shows strong absorption in this region and for hydroxyl observation the sample must be dry. Halide disk suspensions are rarely dry, and for observation of hydroxyl groups this method is best avoided.
2. Interfering absorption arises from N—H groups, and the overtone of the 6 μm carbonyl band.

Table 2.7. Absorption bands of common hydroxyl containing compounds

| Compound | Type of vibration | Approximate band position | | Remarks |
		Wave-length μm	Fre-quency cm^{-1}	
Alcohols, phenols, acids, hydro-peroxides	—OH stretch	2·7–3·2	3700—3150	See text
Aliphatic acids	C=O stretch	5·85	1710	In dimer state. Moves to shorter wavelength in dilute solution
Aromatic acids	C=O stretch	5·92	1690	
Acids	C—O stretch	7·8	1280	Often several bands. Also present in aryl ethers
Phenols	C—O stretch	8·2	1220	
Tertiary alcohols	C—O stretch	8·7	1150	Tertiary alcohol band unreliable. For details see Ref. 16 and 18
Secondary alcohols	C—O stretch	9·1	1100	
Primary alcohols	C—O stretch	9·5	1050	
Acids	O—H bend	10·8	930	Only in dimer state

The principal bands of hydroxyl containing compounds are noted in Table 2.7.

ETHERS

Compounds showing their strongest absorption bands between 7·5 μm and 11 μm are likely to be ethers. However alkyl ethers may be confused with alcohols, while aryl ethers are similar to phenols and some aromatic esters; thus the absence of carbonyl and hydroxyl must be confirmed before concluding the unknown to be an ether.

NITROGEN DETECTED

Many of the commonly occurring nitrogen containing groups have useful diagnostic bands. These are summarised in Table 2.8. The structures

have no useful absorption, and this may cause difficulty in considering the spectra of compounds likely to be azo compounds or tertiary amines. Similarly quaternary ammonium compounds show no bands associated with nitrogen linkages.

The amide group

The two bands of the secondary amide group at 6·1 μm and 6·4 μm are about equal in intensity and dominate the spectra of these substances. The spectrum between 6 μm and 12 μm is often complex; in simple synthetic polyamides this region contains a large number of sharp bands whose position and intensity are much influenced by the crystalline state of the sample. The corresponding structure in urea-formaldehyde resins and natural proteins is usually broad and indefinite.

The 6·1 μm and 6·4 μm bands are replaced in primary amides by one band near 6·1 μm, often complex in shape, but very intense.

Polyimides

The most useful bands of the polyimide structure are those due to the carbonyl groups in the imide ring, which appear as a doublet at 5·62 μm

119

Table 2.8. Principal bands for nitrogen containing groups

Compounds	Remarks
Aliphatic amines	3 μm band split in primary amines, absent in tertiary amines, strong in aromatics, weak in aliphatics. Secondary amines broad band 13 μm to 14 μm
Aromatic amines	
Amides, primary and tertiary	3 μm bands absent in tertiary amides
Amides, secondary	Cyclic secondary amides (lactams) 6·4 μm band missing. All secondary amides have broad band 13 μm to 14 μm
Urethanes	Distinguished from secondary amides by position of 6 μm band
Imides	Longer wavelength band stronger
Amido-imides	Show two bands of imides and two bands of amides
Ammonium salts NH_4^+	
Substituted ammonium NH_2^+ and NH_3^+	Series of weak sharp bands between 3·6 μm and 5 μm
Nitriles, isocyanates	Sharp weak band in nitriles. Broad strong band in isocyanates
Sulphonamides (see also Table 2.10)	With NH_2, double at 3 μm. With NH, single at 3 μm. No NH, no band at 3 μm
Covalent nitrates	Strong bands
Ionic nitrates	Also a weak sharp band between 11·8 μm and 12·5 μm
Nitro compounds	Strong sharp bands
Triazine compounds	Strong and broad

WAVENUMBER (cm⁻¹)
5000   3000   2000   1500   1200

WAVELENGTH (μm)
2   4   6   8

TWO BANDS · DOUBLET · TWO BANDS · FOUR BANDS

and 5·82 μm. The 5·62 μm band is sharp, while the 5·82 μm is broader and stronger.

Polyamido-imides

In the spectra of polyamido-imide resins the characteristic bands of the carbonyl groups of the imide and amide structure may be seen. Thus a series of four bands of relatively high intensity may be seen in the 6 μm region which are a very reliable indication that the unknown is a polyamido-imide.

Polyurethanes

The prominent band pair noted for the secondary amide structure is present, but now occurs at about 5·9 μm and 6·5 μm. This easily recognised structure is useful both for simple polyurethanes and in polyether and polyester urethane rubbers.

Nitriles and isocyanates

Both these groups produce an absorption band near 4·4 μm, a position relatively free of other well defined absorption bands and therefore the band is easily observed. Differentiation is sometimes difficult, although the isocyanate band is far stronger (up to 100 times as intense as the nitrile band) and is often doubled or irregular in shape.

Other nitrogen containing compounds

Covalent nitrates and nitro compounds produce strong easily recognised bands (Table 2.8); the triazine ring, for example in melamine compounds, shows a useful strong band at 6·4 μm to 6·5 μm with a weaker band near 12 μm. For other nitrogen compounds consult Reference 18.

CHLORINE DETECTED

The C—Cl group produces a band of medium intensity, often broad, but too variable in position to be useful. The CCl_2 group in polyvinylidene chloride produces a useful band at 9·5 μm which, in crystalline polymers splits into a sharp and characteristic doublet (Table 2.9).

SULPHUR, PHOSPHORUS OR SILICON DETECTED

The sulphur-sulphur and sulphur-carbon linkages show no useful absorption, while that of the S—H group is extremely weak. However, the sulphur-oxygen linkage, encountered in polymer work as sulphates, diaryl sulphones, sulphonates and sulphonamides is very intense. A strong band between 8 μm and 9 μm will suggest the presence of one of these

Table 2.9. Absorption bands of chlorinated hydrocarbons

Group	Type of vibration	Approximate band position		Remarks
		Wave-length μm	Fre-quency cm^{-1}	
—CHCl—CH$_2$—CH$_2$—	CH$_2$ bend	6·95	1440	The —CH$_2$— band also occurs near 7 μm in vinyl resins
—CHCl—CH$_2$—CHCl—	CH$_2$ bend	7·00	1430	
—CCl$_2$—CH$_2$—CHCl—	CH$_2$ bend	7·05	1420	
—CCl$_2$—CH$_2$—CCl$_2$—	CH$_2$ bend	7·15	1400	
—CH$_2$—CHCl—CH$_2$—	—CH bend	7·5 / 8·0	1333 / 1254	Bending modes of single C—H group
—CHCl—CHCl—CHCl—	—CH bend	7·85	1273	
—CH$_2$—CCl$_2$—CH$_2$—	C—C stretch	9·5	1060	In polyvinylidene chloride and copolymers
—C—Cl	C—Cl stretch	12–16	830–620	Very variable in position

groups in a sulphur containing substance. If nitrogen also is detected, and an additional strong band occurs near 7·6 μm, the sulphonamide group is indicated (Table 2.10).

As with sulphur, the most useful bands of phosphorus compounds arise from phosphorus-oxygen linkages. The P—H group is an exception, showing a useful band of medium intensity at about 4·2 μm.

Phosphorus is usually encountered as phosphates in polymer systems. The P—O—C group in aryl phosphates shows a strong band at 10·3 μm, while the corresponding band in alkyl phosphates at 9·7 μm is even stronger, but rather broad.

Silicon compounds exhibit a number of useful absorption bands detailed on pp. 544–546. The band due to the Si—H group occurring at about 4·6 μm is exceedingly prominent and is easily recognised. Absorption due to the Si—O linkage occurs between 9 μm and 10 μm (a broad, complex and intense band). Sharp bands near 8 μm and 7 μm arise from

Table 2.10. Absorption bands of sulphur and phosphorus compounds

Substance	Type of vibration	Approximate band position		Remarks
		Wavelength μm	Frequency cm^{-1}	
Mercaptans	S—H stretch	3·9	2550	Weak band, difficult to find in long chain compounds
Sulphonamides (see also Table 2.8)	N—H stretch	3·0—3·1	3290—3270	
	S=O stretch	7·5—7·6	1340—1320	
	S=O stretch	8·5—8·8	1170—1140	S=O bands strongest in spectrum
Diaryl sulphone	S=O stretch	7·6—7·8	1320—1280	
	S=O stretch	8·8—9·0	1140—1110	
Primary alkyl sulphate ion R—O—SO$_2$—O$^-$	S=O stretch	7·8—8·2	1280—1220	
	S=O stretch	9·1—9·3	1100—1070	
	S=O stretch	11·9—12·1	840—820	S—O stretch in 10–12 μm region appears to be the chief distinguishing feature.
Secondary alkyl sulphate ion R$_1$ | CH—O—SO$_2$—O$^-$ | R$_2$	S=O stretch	8·0—8·2	1250—1220	8 μm band usually strongest in the spectrum
	S=O stretch	9·3—9·4	1070—1050	
	S=O stretch	10·6—10·8	940—920	
Alkyl sulphonate ion R—SO$_2$—O$^-$	S=O stretch	8·2—8·7	1220—1150	
	S=O stretch	9·3—9·6	1080—1040	Distinguished mainly by aromatic bands in aryl sulphonates
Aryl sulphonate ion Ar—SO$_2$—O$^-$	S=O stretch	8·2—8·6	1220—1160	
	S=O stretch	9·4—9·7	1070—1030	
Ionic sulphates (SO$_4^{--}$)	S=O stretch	9·0	1100	Broad band with shoulders
Phosphines, phosphonates	P—H stretch	4·15	2400	Medium intensity, fairly sharp
Phosphinates	P—H stretch	4·25	2350	
Acid phosphates, phosphites etc., containing P—OH	O—H stretch	3·9 ~5·0	2600 ~2000	Very broad bands
Phosphates, phosphonates, phosphinates	P=O stretch	~8·1	~1240	Strong and very reliable
Alkyl phosphates, phosphites, etc.	P—O stretch	9·5—10·5	1050—950	Strong, variable in position
Aryl phosphates, phosphites, etc.	P—O stretch	10·3	970	Strong, constant in position

For further information on phosphorus compounds consult Corbridge[21]

Si-methyl and Si-phenyl respectively. The Si—O—H group absorption is similar to that of the alcoholic O—H group.

COMPOUNDS CONTAINING METALS

Salts of carboxylic acids

These show a most intense absorption band in the region 6·3 μm to 6·5 μm. The position appears to depend more upon the nature of the cation than upon that of the carboxylic acid. Although very intense, the band is remarkably sharp and sometimes double; at low concentrations the 'salt' band may be confused with aromatic structure.

Covalent organo-metallic compounds

These compounds are disappointing in their spectra since absorption bands due to most metal-carbon bonds lie in the longer wave infra-red (well beyond 15 μm) and little work has been done to characterise them.

Inorganic compounds (Table 2.11)

Inorganic compounds containing oxygen in the anion often have useful spectra[22]. Silicates occasionally have well defined spectra, but more often

Table 2.11. Absorption bands of common inorganic ions

| Group | Structure | Approximate band position | | Remarks |
		Wavelength μm	Frequency cm^{-1}	
Carbonate	CO_3^{--}	6·9—7·1	1450—1410	Broad and strong
		11·3—12·0	880—830	Split in basic carbonates (6·7 μm and 7·1 μm)
Bicarbonate	HCO_3^-	7·4—7·7	1350—1300	Broad and strong
		9·8—10·2	1020—980	Broad, medium intensity
		11·5—12·2	870—820	Sharp and weak
		14·2—14·4	700—690	Broad, medium intensity
Cyanide	CN^-	4·8	2200	Sharp, medium intensity
Nitrate	NO_3^-	7·1—7·2	1410—1390	Strong, broad
		11·9—12·2	840—820	Sharp and weak
Phosphate	PO_4^{---}	9—10	1100—1000	Structure often complex with several strong bands[21]
Sulphate	SO_4^{--}	8·8—9·2	1130—1080	Broad and strong
Silicate		9·1—11·1	1100—900	Many silicates have complex band patterns useful for identification[23, 24]

Ammonium—See Table 2.8
Other inorganic ions, see Ref. 25

have broad and featureless bands which are of little use. This is also true of metallic oxides. The spectra of phosphates, while rich in strong bands[21] are not easy to understand. On the other hand sulphates, nitrates and carbonates have spectra with prominent bands sufficiently constant in position for the respective groups to be recognised, and sufficiently distinct to enable the different cations to be distinguished.

QUALITATIVE EXAMINATION OF FILM LAMINATES, FIBRE BLENDS AND SURFACE ACTIVE AGENTS

Methods for the qualitative examination of film laminates and coated surfaces, fibre blends, and surface active agents, which present particular difficulties, are dealt with in this third part of the chapter on qualitative analysis.

QUALITATIVE EXAMINATION OF FILM LAMINATES AND COATED SURFACES

Commercial packaging films are frequently laminates in which both the composition of the layers and the position of these layers in the laminate are of importance. Similar problems are encountered in the examination of adhesive tapes and the methods that are proposed may, in appropriate cases, be applied to the qualitative examination of coatings on paper, leather and cloth.

DETECTION OF 'ADDITIONAL ELEMENTS'

Oxygen Flask Combustion of samples will yield evidence of 'additional elements'. The presence of chlorine will suggest vinyl or vinylidene chloride polymers or copolymers, that of nitrogen will suggest acrylonitrile, nylon or polyurethane and so on, as discussed earlier in this chapter.

In the case of metallised and printed films the nature of the metal and the elemental constituents of the printing ink can often be determined by X-ray Fluorescence methods. Films of hydrocarbon polymers, polyesters and cellulose polymers do not contain significant amounts of elements detectable specifically by X-ray Fluorescence. Many laminates of these films, however, have at least one layer of a chlorine containing constituent such as a vinylidene chloride copolymer and, in such a case, by X-ray Fluorescence examination of each side of the sample over the Cl $K\alpha$ region, qualitative evidence of this element can be obtained even with layers of only 10^{-6} in (25 m) in thickness. The relative intensities obtained often provide useful information as to whether both or only one side of the base film has been coated. The detection of silicon by X-ray Fluorescence often points to the presence of a silicone treated film.

DIRECT EXAMINATION BY INFRA-RED SPECTROSCOPY

Where the sample is transparent and of suitable thickness direct infra-red examination is a rapid and informative test, as many of the resins used in film and tape manufacture have spectra easily recognised under these conditions, e.g. polypropylene, polyvinyl chloride, polyvinylidene chloride copolymers and polyethylene terephthalate. The results of this test, however, must be interpreted with caution, as the transmission spectrum of a laminate of two or more different resins cannot be distinguished from a blend or, in some cases, a copolymer based on the same constituent monomers. This problem will arise, for example, in the examination of a laminate of polypropylene and polyethylene, which cannot be distinguished, by the transmission spectrum, from some forms of propylene/ethylene copolymer.

EXAMINATION BY A.T.R. SPECTROSCOPY

This test can often be applied to obtain information about the surface layer of an opaque sample e.g. coated paper or leather. Many such samples have irregular surfaces which give poor contact with the A.T.R. prism face. To some extent this is overcome by using a multi-A.T.R. unit, rather than a single reflection prism, but a prior treatment of the sample by hot pressing between polished plates is often very effective in improving the surface contact. In the case of heavily filled surface layers, e.g. coated papers, the filler particles are depressed, forcing the resin binder into a flat coherent layer on the surface.

In the case of transparent laminates, A.T.R. measurements are particularly useful when considered alongside the transmission spectrum. It is advantageous to use a high refractive index prism, e.g. KRS-5, for the A.T.R. measurements, thus restricting the depth of penetration of the radiation into the laminate. The absorption of lower layers in the sample is suppressed, so that the spectrum of the surface layer in contact with the prism suffers less interference from the absorption of subsequent layers of the laminate. Again, the multi-A.T.R. unit is preferred to enhance the intensity of the spectrum.

Comparison of the A.T.R. spectra of each surface of the laminate with that obtained by transmission will often enable conclusions to be drawn on both the composition and the position of individual layers.

If there is reason to suggest that the laminate may contain intermediate layers, it will be necessary to open the laminate. A convenient method is to rub adhesive cellulose tape firmly on to both surfaces of the laminate, and to strip the tape away. If the outer layers are removed by this treatment, and an intermediate layer exposed, this will be revealed by measuring the A.T.R. spectra of the sticky side of the adhesive tape and of the newly exposed surface of the laminate. A blank A.T.R. run on the sticky side of the unused adhesive tape is necessary, as the layer of resin adhering to it may be discontinuous, or so thin that the adhesive shows through in the A.T.R. spectrum.

Example of the application of A.T.R. spectroscopy to the examination of a transparent bag used for packing coffee

The test for 'additional elements' by Oxygen Flask Combustion suggested that the bag contained a minor proportion of nitrogen but no other 'additional elements'. The infra-red transmission spectrum (Figure 2.3) had a strong peak at 5·82 μm indicating an ester (see Table 2.6). Reference to Table 4.4 showed that the band combination 5·82 μm, 7·9 μm, 9·0 μm, 9·8 μm, 11·5 μm and 13·7 μm arose from a terephthalate. Comparison with

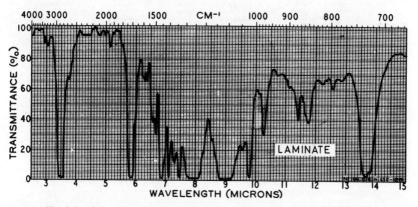

Fig. 2.3. Infra-red transmission spectrum of transparent bag used for packing coffee

reference spectra (Spectra 4.29(a) to 4.33) showed that a major constituent of the bag was polyethylene terephthalate. The band at 3·4 μm, however, was relatively too strong, indicating that other constituents must be present and, since it was probable that the bag was a laminate, A.T.R. spectra were measured on the two outer surfaces with a 45° KRS-5 A.T.R. prism (Figures 2.4 and 2.5). One spectrum (Figure 2.4) appeared to be consistent with polyethylene terephthalate, and hence it was concluded that this material formed one outer face of the laminate. The A.T.R. spectrum of the second surface (Figure 2.5), on the other hand, showed no evidence of polyethylene terephthalate and was clearly that of polythene.

127

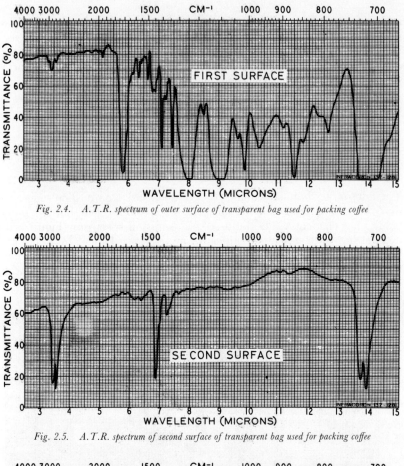

Fig. 2.4. *A.T.R. spectrum of outer surface of transparent bag used for packing coffee*

Fig. 2.5. *A.T.R. spectrum of second surface of transparent bag used for packing coffee*

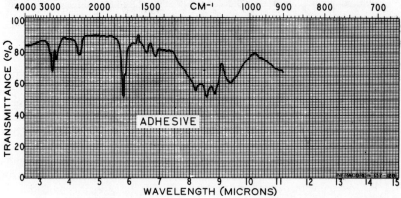

Fig. 2.6. *A.T.R. spectrum of adhesive used to laminate films of transparent bag used for packing coffee*

The question then arose as to whether there was an adhesive layer between the polythene and polyethylene terephthalate.

By patient work, proceeding from one edge of the specimen, it was found possible to strip the sample into two films. The inner surfaces thus revealed were examined separately by means of the KRS-5 multi-A.T.R. prism, but no evidence of an adhesive layer could be obtained.

On the other hand, examination with the 45° germanium multi-A.T.R. prism (Figure 2.6) revealed a new constituent attached to the polyethylene terephthalate surface.

The band at 4·35 μm and the presence of nitrogen suggested cyanate or isocyanate, and this again suggested that the unknown might be a urethane polymer, with a spectrum similar to that of Spectra 4.36.

Hence it was concluded that the sample consisted of polyethylene terephthalate film laminated to polythene film by means of a polyester urethane adhesive.

SOLUTION METHODS

Selective solvent extraction of laminates often provides a satisfactory method of examination. Cotton wool is first extracted with methyl ethyl ketone. The laminate surfaces are then swabbed off, separately, with this clean cotton wool soaked in methyl ethyl ketone. The solution is evaporated to dryness on a salt plate for the infra-red examination or, alternatively, the dry residue is examined by pyrolysis/G.L.C. Some care is required otherwise more than one layer may be removed. Plasticisers may often be removed selectively by this treatment.

After removal of surface layers by this method, further examination of the film by A.T.R. may indicate the presence of other resin layers.

The solvent-extraction method has been found particularly useful with coated papers since the coating contains such a small amount of resin that this cannot be revealed by A.T.R. measurement in the presence of large amounts of fillers, particularly clay or talc. In some cases swabbing the surface of a coated paper with methyl ethyl ketone may be satisfactory but, in general, most success has been achieved by preliminary removal of the gums and starches from the paper by Soxhlet extraction with water or petroleum ether (b.p. 40–60°C). This is followed by extraction with boiling methyl ethyl ketone for between 18 h and 4 d. The resin recovered from the methyl ethyl ketone solution is examined by infra-red spectroscopy and pyrolysis/G.L.C.

X-RAY FLUORESCENCE MEASUREMENT OF COATING THICKNESS

When a chlorine containing polymer of known composition has been used, quantitative measurements of film thickness may be made by X-ray Fluorescence. For example, the total thickness variation of a vinylidene chloride/acrylonitrile copolymer coating on polypropylene film (both sides coated) may be required over a large area of film. Comparative figures may be obtained by an absorption method in which the intensity of the potassium $K\alpha$ radiation from a borax bead containing potassium is measured after passage through the sample. Chlorine in the coating absorbs the potassium $K\alpha$ radiation very strongly and there is consequently a decrease in the measured intensity with increasing total vinylidene chloride copolymer thickness. With the provision of suitable standards, and provided also that the thickness of the polypropylene can be assumed to be reasonably constant, absolute coating thicknesses can be calculated.

When only a single side is coated or when the base film is thick enough to prevent interference from the coating on one side on examination of the other, then direct measurement of the chlorine $K\alpha$ intensity may be used. Calibration standards may be prepared by placing known microvolumes of a standard solution of a known vinylidene chloride copolymer within a masked (20 mm diameter) area of base film and examining the dried standard coatings in a revolving sample holder. The test samples are examined behind the same diameter mask aperture.

QUALITATIVE EXAMINATION OF FIBRE BLENDS

Blends of man-made and natural fibres are now used for many applications and the identification of the constituents of such blends is of some importance. Ford and Roff[26] have developed a simple system by which man-made and natural fibres are separated by microscopical examination. The natural fibres are further separated into groups by the application of morphological tests. Further morphological and chemical tests are applied in order to identify the individual fibres within each group. Man-made fibres are separated by tests for elements and by determination of the physical characteristics of the fibres. Further identification is carried out by the application of chemical tests. In our experience microscopy, coupled with staining methods, with stains such as Shirlastain, Herzberg's stain and Phloroglucinol-HCl[27], provides the most convenient method of identification of natural fibres when sufficient material is available.

In cases where only small amounts of sample are available, however, the methods developed by Longhetti and Roche, based on consideration of such properties as birefringence, melting point, refractive index, fibre diameter and solubility in solvents are of particular value[28].

The use of selective solvents for the quantitative analysis of mixed fibres has been described by Praeger[29].

QUALITATIVE EXAMINATION OF SURFACE ACTIVE AGENTS

In the examination of surface active agents the general procedure that is followed is based on a scheme described by Wurzschmitt[30]. The surfactants are classified as anion active, cation active and non-ionic depending on their mode of activity. After isolation of the surface active agent from the polymer composition or latex, it is then placed in one or other of three broad groups by application of the following series of simple tests.

Reagents

   Methylene blue 0·02% w/v aqueous solution.
   Bromophenol blue 0·02% w/v aqueous solution.
   N Hydrochloric acid solution.
   N Sodium hydroxide solution.
   Chloroform.

Procedure

Approximately 10 mg of the emulsifying agent are dissolved in 50 ml of distilled water.

Test for anion activity

In a 6 × 1 in (150 × 24 mm) stoppered test tube mix 10 ml of test solution, 2 drops of N hydrochloric acid solution and 1 ml of methylene blue solution.
   Shake the solution vigorously with 5 ml of chloroform and allow the layers to separate.

Anion Active: Chloroform layer	Distinctly blue
Cation Active: Chloroform layer	Almost colourless
Non-Ionic: Chloroform layer	Colourless or faintly blue

Test for cation activity

In a 6 × 1 in (150 × 24 mm) stoppered test tube mix 10 ml of test solution with 2 drops of N sodium hydroxide solution and 1 ml of bromophenol blue solution. Shake the solution vigorously with 5 ml of chloroform and allow the layers to separate.

Cation active: Chloroform layer	Distinctly blue
Anion active: Chloroform layer	Almost colourless
Non-ionic: Chloroform layer	Colourless or faintly blue

The additional elements nitrogen, sulphur, chlorine etc. are now detected either by the sodium fusion test or, when the substance is combustible in oxygen, by the Oxygen Flask Combustion procedure.

The classification tests outlined by Wurzschmitt[30] are now used in order to decide on the particular type of emulsifying agent that is being dealt with. The types of compound in Wurzschmitt's three broad classes are as follows.

CATIONIC SURFACE ACTIVE COMPOUNDS

(a) Ternary and quaternary compounds, e.g.
Ternary oxonium compounds
Ternary sulphonium compounds
Quaternary ammonium compounds
Quaternary phosphonium compounds.

(b) Compounds with mostly trivalent nitrogen which are present in solution as non-quaternary ammonium compounds, e.g. poly-alkylene imine addition products and the corresponding substituted polyamides.

(c) Polyalkylene oxide addition products in aqueous solution, present as polyoxonium compounds, e.g. condensation products with:
Fatty alcohols
Alkyl phenols
Alkyl naphthols
Fatty acid alkylol amine ethylene oxide addition products
Fatty acids.

NON-IONIC COMPOUNDS

Alkylol amine condensation products.
Polyalcohol condensation products.

ANIONIC SURFACE ACTIVE COMPOUNDS

In our experience, this is by far the most important in polymer work and in dealing with this class Wurzschmitt[30] divides the twenty types of compound into four groups, based purely on a knowledge of the elements present.

These groups are as follows:

1. Compounds containing neither sulphur nor nitrogen.
   (a) Soaps.
   (b) Fatty acid condensation products with oxyalkyl carboxylic acids.
   (c) Aromatic carboxylic acids.

2. Compounds containing nitrogen but not sulphur.
   (a) Fatty acid condensation products with aminoalkyl carboxylic acids.
   (b) Fatty acid-protein condensation products.
   (c) Gall soaps.

3. Compounds containing sulphur but not nitrogen.
   (a) Fatty acid condensation products with oxyalkyl sulphonic acids.

(b) Sulphonated oils.
(c) Sulpho-dicarboxylic acid esters.
(d) Fatty alcohol sulphates.
(e) Alkyl sulphonic acids.
(f) Oxy-fatty alkyl sulphonic acids.
(g) Alkyl-aryl sulphonic acids.
(h) Fatty sulphonic acids.
(i) Sulphonated alkylene oxide addition products.
(j) Sulphonated polyalcohol-fatty acid condensation products.

4. Compounds containing both sulphur and nitrogen.
(a) Fatty acid condensation products with amino alkyl sulphonic acids.
(b) Amides of sulphonated oils.
(c) Condensation products of protein decomposition products or aminocarboxylic acids and alkyl sulphochlorides.
(d) Condensation products of fatty acid amides and oxyalkyl sulphonic acids.
(e) Sulphonated condensation products of fatty acids and aromatic *o*-diamines. (Benzimidazol sulphonates.)
(f) Sulphonated alkylamino fatty acid condensation products.

Note: The compounds in these groups may be present as the metal, ammonium, amino or alkyl amino salts.

Having thus decided whether the surface active agent is anion active, cation active or non-ionic, i.e. to which of the above three classes described by Wurzschmitt the substance belongs, two further simple tests A and B are carried out, viz.

Heteropolyacid test A

This test gives a positive result with polyalkylene imines, polyalkylene imine addition products and cation active materials. All other surface active materials give a negative result.

Reagent

A 5% w/v solution of phosphotungstic acid adjusted to pH 5 with sodium hydroxide solution.

Procedure

To a few ml of the aqueous solution of the test material are added a few drops of the reagent. A dense white precipitate indicates a positive result.

Heteropolyacid test B

This test gives a positive result with polyglycols of molecular weight 194 upwards (tetraethylene glycol), polyalkylene imines, polyalkylene imine

133

addition products, cation active compounds and polyalkylene oxide addition products.

Reagents

1. A 5% w/v solution of phosphotungstic acid.
2. A 1 : 1 v/v mixture of 25% v/v hydrochloric acid solution and 10% w/v barium chloride solution.

Procedure

To a few ml of an aqueous solution of the test material is added an equal volume of reagent 2. If a precipitate forms due to precipitation of the barium salt of anionic compounds then more reagent 2 is added until precipitation is complete. After heating and cooling the solution is filtered. Reagent 1 is now added to the filtrate when a copious white precipitate indicates a positive result. If a precipitate is not formed with reagent 2 then reagent 1 may be directly added.

A negative result in Test A and a positive result in Test B indicate the presence of a polyalkylene oxide type of compound.

An infra-red spectrum of the surface active agent is now obtained and this, together with the chemical information provided by the application of the above tests, normally suffices to characterise the substance sufficiently for most purposes. On occasion, however, in the case of anion active compounds, it is necessary to prepare the barium salt of the compound by the method described by Jenkins and Kellenbach[31] for further confirmation by infra-red spectrometry. The paper by Jenkins and Kellenbach includes the spectra of 10 organic sulphates and sulphonates.

If a more detailed examination of the surface active agent is required then more of the appropriate Wurzschmitt tests are applied. The following example illustrates the kind of procedure that is employed.

Identification of the emulsifier in a polyvinyl chloride emulsion polymer

A polyvinyl chloride emulsion polymer was received for examination. Continuous extraction of 500 g of the polymer with ether yielded an ether extract that amounted to 0·68% of the original sample. This extract was shown to contain a minor proportion (15·5%) of low molecular weight polyvinyl chloride. The total residue was slightly acid, i.e. it possessed an acidity equivalent to 26 mg KOH/g and, further, had a saponification value of 129 mg KOH/g.

The residue was hydrolysed with N absolute alcoholic potash, and the precipitated potassium salts isolated by filtration. These salts were then acidified and the liberated acid extracted with ether. On treatment with alkali the acid behaved like a fatty acid, and, moreover, infra-red examination indicated that it was palmitic acid. The conclusion was therefore reached that the sample contained a small amount of free acid and a larger proportion of an ester of palmitic acid, such as ethyl palmitate.

Direct methanol extraction of 1000 g of the sample, however, yielded a residue amounting to 0·012% of the sample. A small portion of this residue, after solution in water containing 2 drops of aqueous methylene blue, was shaken with chloroform. The chloroform layer turned blue indicating the presence of an anionic surface active compound. The classification procedure devised by Wurzschmitt was then applied to the methanol extract as follows:

1. It was shown that the residue contained the elements nitrogen and sulphur as well as a small amount of chlorine, which, it was decided, was derived from low molecular weight polymer soluble in the methanol. The presence of nitrogen and sulphur indicated that the surface active agent was one or other of Group 4 in Wurzschmitt's classification of anionic surface active compounds (see p. 132).
2. Heated with triethylene glycol and solid magnesium oxide, the surface active agent yielded alkaline vapours, indicating that it contained labile nitrogen. This in its turn showed that the surface active agent was present either as the ammonium, amino or alkyl-amino salt, or alternatively it was a condensation product of a fatty acid amide with an oxyalkyl sulphonic acid.
3. After removal of all labile nitrogen by the magnesium oxide reaction, potassium hydroxide pellets were added to the boiling solution. Further alkaline vapours were evolved indicating the presence of a substance containing firmly bound nitrogen.
4. A fresh portion of the methanol extracted surface active agent was saponified with ethylene glycol potash, the product diluted with water and the solution extracted with ether.

   This solution was now acidified with hydrochloric acid, and again extracted with ether. The aqueous solution was then tested with barium chloride.

   A precipitate of barium sulphate was not obtained, indicating that no sulphate had been produced in the alkaline hydrolysis. This excluded the presence of condensation products of fatty acid amides with oxyalkyl sulphonic acids, or sulphonated alkylolamino fatty acid condensation products, in the sample.
5. The Biuret test for protein type compounds was negative, indicating the absence of condensation products of protein decomposition products in the sample.

Hence it was concluded that the substance was one or other of two types, viz. either a fatty acid condensation product with an amino alkyl sulphonic acid or a benzimidazol sulphonate. Infra-red examination yielded the following evidence:

1. The material was aliphatic in nature.
2. There was strong evidence of the presence of the sulphonate group.
3. The group

was probably present and there was little evidence of the presence of the grouping

4. The material was therefore in all probability a condensation product of a fatty acid with an amino alkyl sulphonic acid.

On the total evidence, therefore, the conclusion was reached that the analytical data were consistent with the presence, in the original sample,

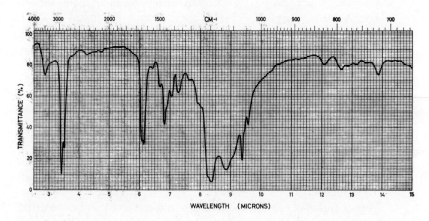

Fig. 2.7. Infra-red spectrum of methanol extract from polyvinyl chloride emulsion polymer

on the one hand of a small amount of an ester of a fatty acid such as ethyl palmitate, together with a little free acid, and, on the other an emulsifying agent of the type

$$R - C - N - (CH_2)_x - SO_3M$$

136

Where R = long hydrophobic radical
      R′ = alkyl radical
and M = ammonium amino or alkylamino salt.

The infra-red spectrum of the methanol extract is given in Figure 2.7.

QUALITATIVE ANALYSIS FROM THE N.M.R. SPECTRUM

Proton n.m.r. spectroscopy has so far found its principal chemical application in qualitative analysis, and as such is very valuable in the polymer field in the identification of polymers and associated substances such as plasticisers, stabilisers, antioxidants, and so forth. As we have emphasized also in the discussion on the application of infra-red spectroscopy, a prior knowledge of the elements present in a compound is extremely useful. Again, the most effective analysis is conducted on pure substances; this is perhaps even more important in n.m.r. than in infra-red spectroscopy.

We have referred previously (p. 36) to the chemical shift, the measurement of the areas of resonances, and spin-spin coupling, and have indicated how these may be used in qualitative analysis. Examples of the identification of polymeric substances from their n.m.r. spectra appear frequently in the text, but the following example will show how the n.m.r. spectrum may be applied in the case of the identification of a monomeric material of interest in the analysis of polymer compositions, namely a U.V. stabiliser for hydrocarbon polymers.

Example of qualitative analysis from the n.m.r. spectrum

A sample of pale yellow solid, claimed to be a U.V. stabiliser, was submitted for identification. Qualitative elemental analysis showed the presence of nitrogen and chlorine. Quantitative analysis for carbon, hydrogen, nitrogen and chlorine gave the following empirical formula (oxygen by difference).

$$C_{17}H_{18}N_3ClO$$

The mass spectrum showed that the molecular weight was 315, thus the formula above is the molecular formula. The infra-red spectrum was measured as a paraffin mull, and showed a band at 3·3 μm, indicating a strongly bonded hydroxyl group, and sharp bands between 8·0 μm and 8·2 μm suggesting that the material was phenolic in nature. The infra-red spectrum in the hydroxyl stretching region near 3 μm was measured also with the sample in dilute solution in carbon tetrachloride. The band at 3·3 μm remained almost unchanged in solution; this showed the hydroxyl group to be strongly intramolecularly bonded (see p. 118).

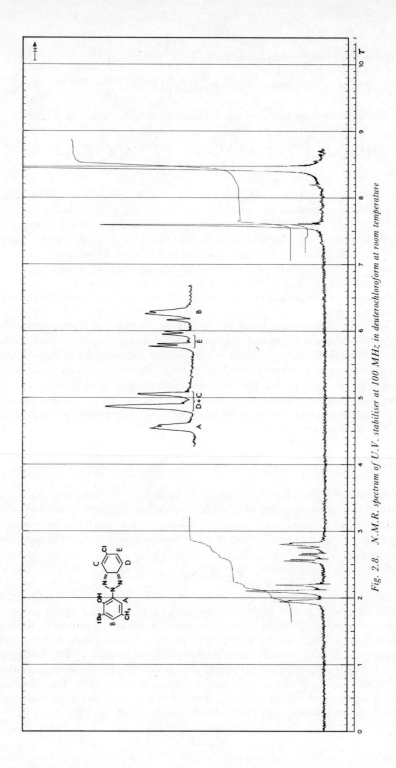

Fig. 2.8. N.M.R. spectrum of U.V. stabiliser at 100 MHz in deuterochloroform at room temperature

The n.m.r. spectrum of the sample was recorded in deuterochloroform solution at room temperature (Figure 2.8). The hydroxyl group proton resonance occurs below 0 τ.

The high-field resonances are both singlets, and arise from alkyl protons. The 8·45 τ resonance could well arise from a tertiary butyl group, and that at 7·60 τ from a methyl group, each attached to an aromatic nucleus. Considering now the areas of the resonances, if that at 8·45 τ is assumed to have an area of 9 (i.e. the tertiary butyl group contains 9 protons) that at 7·60 τ has an area of 3. The combined area of the aryl proton resonances between 1·90 τ and 2·80 τ is 5, and the area of the hydroxyl proton resonance is 1. The total of these areas is equivalent to 18 protons, which is also the number of hydrogen atoms in the molecular formula. Subtracting the 5 carbon atoms required for the tertiary butyl and the methyl groups, there are 12 carbon atoms remaining. There are 9 substituents which must be on benzene rings, i.e.

1 tertiary butyl group
1 methyl group
1 hydroxyl group
5 hydrogen atoms.

There are also

3 nitrogen atoms
1 chlorine atom.

Evidently the compound contains two benzene rings, bridged together with the nitrogen atoms, since all the other substituents are monofunctional, i.e. the skeleton is

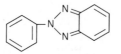

The infra-red evidence that the OH group is strongly internally bonded is good evidence that it must be placed where a six-membered ring can be formed with a nitrogen atom

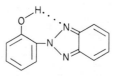

The pair of doublets near 2·55 τ and 2·65 τ have a total area equivalent to 1 proton. This structure must arise from a resonance which is both *ortho*-split ($\sim$9 Hz) and *meta*-split ($\sim$2 Hz), i.e. it is due to the 2 proton in a 1:2:4-proton substitution on a benzene ring. The resonance from proton 1 must also show a 9 Hz splitting. Half of this doublet is seen at 2·19 τ with the other half concealed under the resonance at 2·10 τ. This is consistent with the total area of the structure between 2·1 τ and 2·2 τ,

139

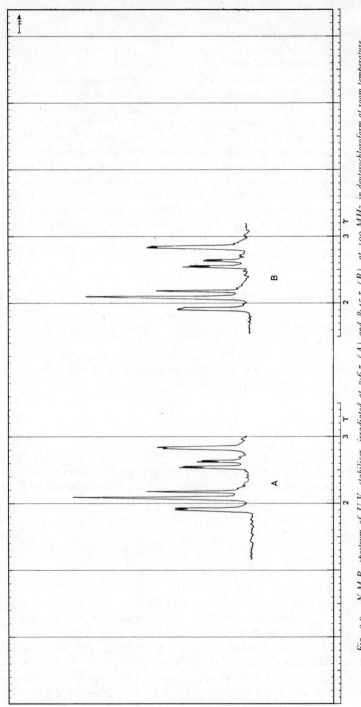

Fig. 2.9. N.M.R. spectrum of U.V. stabiliser, irradiated at 7·6 τ (A) and 8·45 τ (B), at 100 MHz in deuterochloroform at room temperature.

which is equivalent to 2 protons. Since the resonances at 2·10 τ, 1·90 τ and 2·80 τ all show some indication of splitting consistent with *meta*-coupling, it is not possible at this stage to identify the resonance of proton 4 in the 1 : 2 : 4-substitution. However there are no aryl proton resonances above 3 τ and this strongly suggests that there is no proton *ortho* to the hydroxyl group. Thus the 1:2:4-trisubstitution must be on the ring without the hydroxyl group.

Further assignment of the aryl resonance pattern is possible with spin decoupling. When the sample is irradiated at 7·60 τ (the methyl group resonance) resonances at 1·9 τ and 2·80 τ change into sharp doubtlets with splitting of 2 Hz (see A in Figure 2.9). Thus these resonances arise from two protons *meta*-coupled, which are also coupled to the protons of the methyl group. Thus we have the arrangement

and this must be on the benzene ring containing the hydroxyl group. By the previous argument there is no proton *ortho* to the hydroxyl group, hence we can write the structure

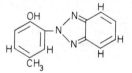

There are two more groups to be placed, the tertiary butyl group and the chlorine atom. Irradiation of the sample at 8·45 τ (the tertiary butyl group resonance) affects the shape of the resonance at 2·80 τ (see B in Figure 2.9) which is due to the proton ortho to the methyl group. Thus the tertiary butyl group must be placed in the same ring as the methyl group, and we can complete the structure

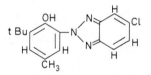

REFERENCES

1. SHAW, T. P. G., *Ind. Engng Chem. analyt. Edn*, **16**, 541 (1944)
2. HASLAM, J., HAMILTON, J. B., and SQUIRRELL, D. C. M., *Analyst, Lond.*, **85**, 556 (1960)
3. HASLAM, J., HAMILTON, J. B., and SQUIRRELL, D. C. M., *Analyst, Lond.*, **86**, 239 (1961)
4. KRAUSE, A., and LANGE, A., (Transl. ed. HASLAM, J.), *Introduction to the Chemical Analysis of Plastics*, Iliffe, London (1969)
5. CHANCELLOR, S. F., (Ed. JOLLY, S. C.), *Supplement to Official, Standardised and Recommended Methods of Analysis*, The Society for Analytical Chemistry, London (1967)

6. COLLINS, J. H., *Testing and Analysis of Plastics, Part 1, The Identification of Plastics*, 2nd edn, Plastics Institute, Monograph No. C.1, London (1955)

7. MARSDEN, C., *Solvent and Allied Substances Manual*, Cleaver-Hume, London (1954)

8. ROFF, W. J., *Fibres, Plastics and Rubbers*, Butterworths (1956)

9. WIDMER, G., *Paper Trade J.*, **128,** 19 (1949)

10. HIRT, R. C., KING, F. T., and SCHMITT, R. G., *Analyt. Chem.*, **26,** 1273 (1954)

11. WHITNACK, G. C., and GANTZ, E. ST. C., *J. Polym. Sci.*, **11,** 235 (1953)

12. VERDCOURT, B., *PATRA Jl*, **10,** 135 (1947)

13. FEARON, W. R., *Analyst, Lond.*, **71,** 562 (1946)

14. FOUCRY, M. J., *Peintures, Pigments et Vernis*, **30,** 11, 925 (1954)

15. HASLAM, J., and SQUIRRELL, D. C. M., *Analyst, Lond.*, **82,** 511 (1957)

16. WEST, W. (Ed), *Chemical Applications of Spectroscopy*, Interscience, New York and London (1956)

17. JONES, R. N., *N.R.C. Bulletin No. 5*, National Research Council, Ottawa (1959)

18. BELLAMY, L. J., *Infra-Red Spectra of Complex Molecules*, Methuen, London (1958)

19. BRUGEL, W., (Transl. KATRITZKY, A. R., and KATRITZKY, A. J. D.), *An Introduction to Infra-Red Spectroscopy*, Methuen, London (1962)

20. POTTS, W. J., and NYQUIST, R. A., *Spectrochim. Acta*, **15,** 679 (1959)

21. CORBRIDGE, D. E. C., *J. appl. Chem., Lond.*, **6,** 456 (1956)

22. MILLER, F. A., and WILKINS, C. H., *Analyt. Chem.*, **24,** 1253 (1952)

23. HUNT, J. M., WISHERD, M. P., and BONHAM, L. C., *Analyt. Chem.*, **22,** 1478 (1950)

24. *Infra-Red Spectroscopy. Its use in the Coatings Industry*, Federation of Societies for Paint Technology, Philadelphia (1969)

25. YOGANARASIMHAN, S. R., and RAO, C. N., *Chemist Analyst*, **51,** 21 (1962)

26. FORD, J. E., and ROFF, W. J., *Shirley Inst. Mem.*, **27,** 77 (1954)

27. STOVES, J. L., *Fibre Microscopy*, National Trade Press, London (1957)

28. LONGHETTI, B. A., and ROCHE, G. W., *J. forens. Sci.*, **3,** 304 (1958)

29. PRAEGER, S. S., *Mater, Res. & Stand.*, **1,** 381 (1961)

30. WURZSCHMITT, B., *Z. analyt. Chem.*, **130,** Nos. 2–3, 105 (1950)

31. JENKINS, J. W., and KELLENBACH, K. O., *Analyt. Chem.*, **31,** 1056 (1959)

3

VINYL RESINS

With vinyl resins the most interesting problems and those that demand most of the analytical chemist are probably connected with the examination of compositions such as polyvinyl chloride compositions. Some idea of the complexity of these problems may be gained from a consideration of the components of such a composition.

It may contain a polymer such as polyvinyl chloride or a copolymer such as a vinyl chloride/vinyl acetate copolymer, a primary plasticiser such as dibutyl phthalate or tritolyl phosphate, a monomeric secondary plasticiser such as butyl acetyl ricinoleate or a polymeric plasticiser such as polypropylene adipate, a plasticiser extender such as Cereclor (a chlorinated paraffin wax), or Iranolin, (a hydrocarbon component), a stabiliser such as basic lead carbonate or an organo-tin compound or diphenylurea, a lubricant which may be a metallic stearate or stearic acid or even a wax, a filler as e.g. whiting, china clay, barium sulphate, anhydrous calcium sulphate, powdered slate, or talc, and a pigment or colouring agent that may be inorganic or organic in origin.

The analytical problems that arise are obviously innumerable but, over the years, our experience is that the most useful tests and those most frequently asked for, are concerned with:

1. The determination of the proportion of plasticiser, the subsequent examination of which is carried out as given in the chapter on plasticisers (Chapter 11).
2. The isolation from the composition of the component polymer or copolymer in as pure a condition as possible.
3. The identification and determination of stabilisers and other additives present in the composition.

Many interesting analytical problems are involved in the examination of recovered polymers. For instance, with vinyl chloride polymers and co-polymers one may be concerned with granular or suspension polymers,

143

with emulsion polymers, with paste polymers, and with copolymers, whether granular or suspension, with other monomers such as vinylidene chloride and vinyl acetate.

Other problems are encountered and these are dealt with in this chapter which has been constituted on the following lines:

The first part is concerned with the method of recovery of monomeric and polymeric plasticisers from the composition and with the qualitative and quantitative examination of recovered vinyl resins. In particular, the determination of chlorine and the pyrolysis/gas chromatographic examination of such resins are discussed.

The second part of the chapter is concerned with the detection and determination of additives or constituents other than plasticisers in polyvinyl compositions. Attention is chiefly directed to:

(a) Polymeric additives such as methyl methacrylate/ethyl acrylate copolymers, chlorinated polythene, polythene, epoxy resins and copolymers and terpolymers with acrylonitrile, butadiene and styrene components.

(b) Inorganic and organo-metallic constituents, for example the determination of elements such as cadmium, zinc, barium, tin and lead in compositions by chemical and X-ray Fluorescence methods and the identification of tin stabilisers.

(c) U.V. absorbers.

(d) The important determination of water in polyvinyl chloride powders and compositions—a recurring problem in analytical laboratories concerned with plastics.

The third part deals with the separation and identification of emulsifying agents used in polyvinyl chloride preparations and finally, the fourth part of the chapter deals with the application of both infra-red spectroscopy and n.m.r. to the examination of polyvinyl chloride compositions. Guidance is given on the identification of resins such as vinyl chloride, alcohol, acetate, laurate, stearate and other vinyl esters, together with their copolymers, terpolymers and polymer blends which are encountered commercially. The descriptions given will enable compositions not mentioned to be analysed qualitatively. Other subjects discussed are the spectra of the linear and cyclic vinyl ethers and attention is drawn to the features that distinguish these resins from e.g. the cellulose ethers. Moreover, information is provided which will help to resolve difficulties encountered in differentiating copolymers from compositions containing polymeric plasticisers.

RECOVERY OF MONOMERIC AND POLYMERIC PLASTICISERS

The method of determination and isolation of monomeric plasticisers in polyvinyl chloride compositions is based on the work of Haslam and

Soppet[1] and involves extraction with ether. Originally, a method was used in which the sample was treated with acetone followed by precipitation of any resin in the acetone extract with petroleum ether, and subsequent recovery of any plasticiser from the petroleum ether solution by the evaporation of the petroleum ether.

An examination of known polyvinyl chloride-tritolyl phosphate-lead carbonate compositions, however, indicated that this method gave low results for the proportion of tritolyl phosphate plasticiser.

The method we prefer is as follows:

Weigh accurately 2 to 3 g of the finely divided sample and transfer to a Soxhlet extraction thimble (Figure 3.1). Dry, cool and weigh an extraction flask (100 ml) and connect to a hot Soxhlet extraction apparatus.

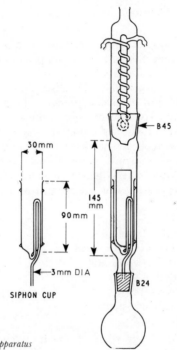

Fig. 3.1. Soxhlet extraction apparatus

Place the thimble containing the sample in the cup of the Soxhlet and add 50 to 70 ml of Analar ether and connect the condenser. After soaking overnight, heat on a boiling water bath and allow the hot extraction to continue for 6 to 7 h. Disconnect the flask and remove the ether by evaporation on the water bath. Dry the flask in an oven at 100°C for 1 h. Cool, weigh, and continue drying in the oven for 30 min periods until the difference between two consecutive weighings is not more than 2 mg.

A blank test should be carried out on the ether, but the figure obtained is usually negligible. Calculate the % ether extract, and hence the plasticiser content of the sample.

This method gives good results with monomeric plasticisers, but difficulties arise with polymeric plasticisers, as has been shown by Haslam and Squirrell[2].

These authors first took a preparation that was prepared from the following constituents:

	% by weight
P.V.C. polymer	59·17
Tritolyl phosphate	17·75
Dioctyl phthalate	17·75
White lead paste (7 : 1)	4·73
Calcium stearate	0·59

Extraction with ether of the final composition gave a plasticiser content of 34·5%, that, taking into account plasticiser losses on compounding, was regarded as satisfactory.

Moreover, when this plasticiser free composition was submitted to a tetrahydrofuran recovery process, the recovered polymer was of first class quality. Its infra-red spectrum was identical with that of the polyvinyl chloride used in the preparation of the composition and its chlorine content was 55·8% whereas that of the original polymer was 55·4%.

The whole sequence of operations was repeated on a fresh composition this time containing polypropylene adipate.

The preparation was prepared from the following constituents:

	% by weight
P.V.C. polymer	50·24
Tritolyl phosphate	15·08
Dioctyl phthalate	15·08
Polypropylene adipate	15·08
White lead paste (7 : 1)	4·02
Calcium stearate	0·50

Extraction with ether of the final composition gave plasticiser contents of 33·1, 33·0, 33·2 and 33·2%, in separate experiments. Allowing for small plasticiser losses in compounding, this yield of plasticiser was quite unsatisfactory.

Moreover, when this plasticiser free composition was submitted to a tetrahydrofuran recovery process, the recovered polymer was of very poor quality. Its infra-red spectrum gave ample evidence of the presence of polymeric ester plasticiser as well as polyvinyl chloride. Its chlorine content, though variable, was invariably low and usually of the order of only 50% as compared with the 55·4% of the original polymer.

It was therefore decided to supplement the ordinary ether extraction of the plasticiser with a subsequent hot methanol Soxhlet extraction for 16 h in an attempt to recover the polymeric plasticiser.

This was much more satisfactory as is shown by the following figures:

	% by weight	
Ether extract	33·2,	33·1
Hot methanol extract	11·0,	11·2

The figure for the total plasticiser was now very satisfactory indeed. In addition, when this plasticiser free composition was submitted to a tetrahydrofuran recovery process, with methanol precipitation of the polymer, the product was of very high quality, i.e. of correct chlorine content viz. 56·0, 55·7, 55·7% and with correct infra-red spectrum showing evidence of a negligible proportion of polymeric ester.

This procedure has another advantage. If the ether and hot methanol extractions are carried out separately and the products reserved, then the hot methanol extract invariably presents the polymeric ester in a very pure condition for infra-red identification, and this is particularly useful when unknown compositions are being examined.

So long as the polymeric ester plasticiser has reasonable solubility in hot methanol, the above procedure will yield useful information.

In recent years, however, the tendency has been to use polymeric plasticisers of high molecular weight and quite low solubility in hot methanol. In such cases alternative procedures have to be employed and the work of M. W. Robertson and R. M. Rowley[3] is of particular value.

They find that a constant boiling mixture of carbon tetrachloride and methanol is a very good solvent for polymeric plasticisers and its use is to be recommended in difficult cases.

The analysis of the recovered plasticiser or ether extract is described in Chapter 11.

RECOVERY OF THE PLASTICISER FREE POLYMER OR COPOLYMER

Having removed the plasticiser, it is now necessary to recover the polymer or copolymer from the plasticiser free composition. Over very many years an adaptation of the procedure originally devised by Haslam and Newlands[4] has proved to be of great value.

In this method 1 g of the deplasticised sample is transferred to a 50 ml centrifuge tube and treated with 35 ml tetrahydrofuran by placing the centrifuge tube in a 500 ml beaker standing on a boiling water bath until the polymer is completely dissolved. The solution or mixture is cooled, diluted to about 35 ml with tetrahydrofuran in order to compensate for any losses by evaporation and centrifuged for 30 min at 3000 rev/min. If the solution is still slightly turbid at the end of this time the centrifuging process is repeated for a further 30 min. If carbon black is present in the composition it may be necessary to use a centrifuge of higher speed (17 000 rev/min) or, if this is not available, to filter several times through a sintered glass crucible containing Celite filter aid (about 1 to 2 in or 25 to 50 mm deep), applying gentle suction.

The clear supernatant liquor is transferred to a 250 ml beaker and the polymer precipitated by the dropwise addition of 100 ml ethyl alcohol to the vigorously stirred solution. The precipitate is allowed to settle for about $\frac{1}{2}$ h, filtered off on a 1 G 3 sintered glass crucible and washed with hot ethyl alcohol using 5 portions each ~ 30 ml. The washed polymer is dried in a vacuum oven at 60°C and the main filtrate from the precipitation is reserved for the detection of modifiers.

In their original work on this subject Haslam and Newlands followed the progress of the recovery by determining the chlorine contents of original and recovered polymers and by the examination of the infra-red spectra of the same polymers. They showed, with polyvinyl chloride compositions containing dibutyl phthalate, tritolyl phosphate, dioctyl phthalate, white lead paste, calcium stearate, and pigments such as carbon black, titanium dioxide and ultramarine, that the recovered polyvinyl chloride was of first class quality.

Similarly, with compositions prepared from vinyl chloride/vinyl acetate copolymers of both low and high acetate content, dibutyl phthalate, tritolyl phosphate, white lead paste, and calcium carbonate, the recovered copolymer was again of first class quality. Moreover, they satisfied themselves that rather abnormal copolymers such as

Vinyl chloride/diethyl fumarate
Vinyl chloride/ethyl acrylate and
Vinyl chloride/methyl methacrylate
would behave satisfactorily in the test.

They indicated, however, that the position with regard to vinyl chloride/vinylidene chloride copolymers was not quite so satisfactory. Polyvinylidene chloride itself is insoluble in tetrahydrofuran even on prolonged boiling. On solution of 1 g of a vinyl chloride/vinylidene chloride copolymer (chlorine content 58·6% equivalent to 12·6% polyvinylidene chloride) in 35 ml tetrahydrofuran followed by precipitation with 80 ml ethyl alcohol, a rather lumpy precipitate was obtained. The chlorine content of the recovered polymer was 57·9% i.e. rather lower than the original. In general the application of the tetrahydrofuran procedure to either a vinyl chloride/vinylidene chloride copolymer or its compositions leads to the recovery of lumpy, rather impure polymers of slightly lower chlorine content than the original.

Nevertheless, the method outlined for polyvinyl chloride compositions is of very definite value in the examination of vinyl chloride/vinylidene chloride copolymers and compositions where the copolymer does not contain much above the equivalent of 15% vinylidene chloride; this is particularly true if the examination is supported, as it invariably is, by the application of the morpholine test for polyvinylidene chloride and by infra-red examination of the recovered polymer.

Incidentally, in favourable cases where polymeric plasticiser is absent, the tetrahydrofuran recovery process can often be applied directly to 1·5 g of the original composition without preliminary removal of plasticiser.

Having recovered a polymer or copolymer by this tetrahydrofuran recovery process, the product is now available for chemical and infra-red examination.

QUALITATIVE AND QUANTITATIVE EXAMINATION OF RECOVERED VINYL RESIN

The qualitative tests for polyvinyl chloride and polyvinylidene chloride (see Chapter 2) will help in the general classification of the recovered polymer or copolymer. Together with an infra-red spectrum, the evidence may be sufficient to identify the polymer conclusively. In some cases, however, evidence from other chemical tests, and from the application of such methods as pyrolysis/G.L.C. and n.m.r. may be required.

THE DETERMINATION OF CHLORINE

The chemical test that yields most information is that involving the determination of chlorine.

For very many years, at least on the macro scale, we have favoured sodium peroxide fusion of the polymer sample, with final potentiometric titration of the ionised chloride.

Initially, we used the method of titration of Haslam and Soppet[5]. The silver nitrate titrant was added manually. The method, as originally described, is quite valid and is still being used in some laboratories.

Nevertheless, with the advent of automatic titrimeters titrating to preset end points, the titration procedure has become exceedingly simple and accurate, and, in our case, we proceed as follows.

Apparatus

A monel metal bomb, capacity 18 ml, suitable for electrical firing, as manufactured by Hudson's Engineering Company, Latimer Road, Luton. The Electrode assembly. This is as shown in Figure 3.2. The reference electrode consists of a silver wire (diameter 1 mm) immersed in a solution containing an excess of silver ions. The reference electrode is connected to the solution to be titrated by means of a ground glass sleeve which is kept moistened with the solution contained in the electrode. The indicator electrode is again a piece of silver wire (diameter 1 mm) arranged in

149

such a way that the wire touches the glass stirrer. This avoids poisoning of the electrode by precipitated silver chloride.

An Automatic Titrimeter.

Reagents

Sodium peroxide with low chlorine content manufactured by I.C.I Ltd.

Starch catalyst.

Analar soluble starch that has been heated for 3 h at 100°C before use.

Nitric acid concentrated. Analar.

Nitric acid 2 N.

Sodium hydroxide solution N.

Standard 0·1 N silver nitrate solution.

Sodium chloride.

Analar sodium chloride that has been heated to 270°C±10°C before use.

Electrode Solution

A preparation containing 64 ml of saturated sodium sulphate solution, 6·0 ml of 2 N nitric acid, 30·0 ml of 0·1 N sodium chloride solution and 30·1 ml of 0·1 N silver nitrate solution is made.

The mixture is allowed to stand until the bulk of the silver chloride has settled out. The reference electrode is filled with the supernatant solution.

Procedure

Weigh accurately 0·25 to 0·30 g of the finely divided sample into the cup of the monel metal bomb containing 0·6 g of the catalyst and mix intimately with 12 to 14 g of sodium peroxide. Tap gently to densify the bomb charge. Assemble the bomb and fire electrically.

Allow the bomb to cool, unclamp, and place the bomb top in 40 ml of distilled water contained in a 500 ml beaker. Cover the beaker and boil for a few minutes, then remove and wash the bomb top.

Place the bomb cup in the beaker so that no water touches the peroxide melt. Cover the beaker with a watch glass and add cold water to the melt dropwise and with care from a polythene wash bottle. Continue this addition until the vigorous reaction between melt and water ceases. Immerse the cup in the water until all the peroxide is leached out. Remove the cup and wash well.

Boil the alkaline solution for 10 min, cool, and make just acid to litmus with conc. nitric acid added from a burette.

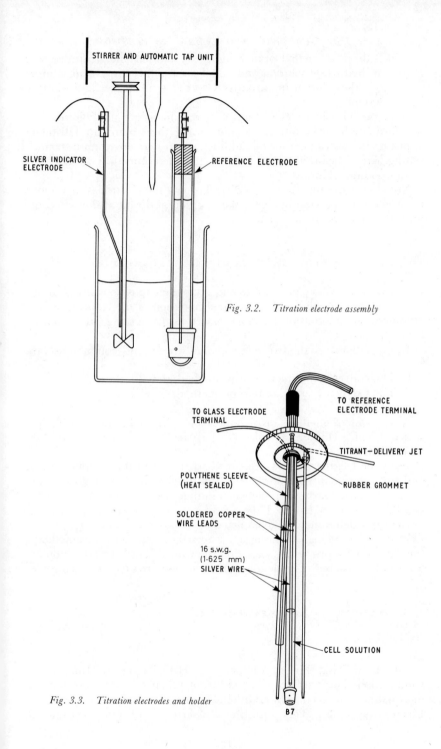

STIRRER AND AUTOMATIC TAP UNIT

SILVER INDICATOR
ELECTRODE

REFERENCE ELECTRODE

Fig. 3.2.   Titration electrode assembly

TO GLASS ELECTRODE
TERMINAL

TO REFERENCE
ELECTRODE TERMINAL

TITRANT—DELIVERY JET

POLYTHENE SLEEVE
(HEAT SEALED)

RUBBER GROMMET

SOLDERED COPPER
WIRE LEADS

16 s.w.g.
(1·625 mm)
SILVER WIRE

CELL SOLUTION

Fig. 3.3.   Titration electrodes and holder

B7

Add 3 drops of methyl orange indicator solution and neutralise with N sodium hydroxide solution and add 1·5 ml of 2 N nitric acid solution.

Transfer the solution quantitatively to a 250 ml beaker and dilute to 100 to 200 ml.

Immerse the electrodes in the solution, and titrate the chloride with standard 0·1 N silver nitrate solution, using the Automatic Titrimeter, adjusting the instrument so that addition of titrant stops at an instrument setting 60 mV before the final end point, as determined by a calibration run on sodium chloride.

Now lower the titration beaker, wash down the electrodes, reassemble the apparatus and continue with the addition of titrant at slow drop rate until the final end point is reached.

THE DETERMINATION OF CHLORINE (MICRO-METHOD)

In all work on polyvinyl chloride and related compositions it is desirable to have available methods for the determination of the chlorine content of quite small amounts of substance as well as the macro method that has been outlined.

In our laboratory the two methods that are in most common use are:

1. The conventional micro-Carius method and
2. The Oxygen Flask Combustion method.

A semi-micro method involving fusion with sodium peroxide in a sealed bomb (Haslam and Hall[6]) has also been used in particular investigations.

In method 1, the micro-Carius method, 10 mg of the sample are used and the chlorine in the sample is finally weighed as silver chloride. This method is particularly valuable when samples can be saved until 10 or 12 can be processed together. They can then be digested overnight and the gravimetric determinations completed on the following day.

When results of high accuracy are required quickly, however, method 2, the oxygen flask procedure, based on the work of Haslam, Hamilton and Squirrell[7, 8], is used. The details of this method are given below.

THE OXYGEN FLASK COMBUSTION METHOD

Apparatus

Combustion Flask, Firing Adapter and H.F. Tester for Initiation of Combustion. These are described in Chapter 2 (p. 82) in connection with the qualitative detection of additional elements in polymers etc.
Electrode Assembly. This consists of a silver wire indicator electrode and

a silver wire in contact with a dilute solution of silver ions as a reference. The electrode system is shown in Figure 3.3 and is conveniently supported in a Perspex holder designed to fit into the top of the combustion flask and be retained by springs hooked to the lugs on the rim of the flask. A hole is drilled in this holder to permit the titrant inlet tube from the titration syringe to be inserted into the flask.

Full scale automatic titrimeter. The full scale automatic titrimeter designed by ourselves and previously described[9, 10] is used. It may, of course, be replaced by any other titrimeter of comparable performance.

A 30 ml syringe is used to deliver the titrant and the recorder sensitivity is adjusted to give a full scale deflection of 400 mV registered by the pH meter. Other instrument settings are: injection motor speed, 0·1 in/min (150 mm/h), and chart speed, 1·0 in/min (30 mm/min).

Reagents

Sodium hydroxide N.
Sodium hydrogen sulphite solution 35% w/v M.A.R. grade.
Hydrogen peroxide 100 volume. Analar.
Silver nitrate 0·01 N.
Nitric acid N.
Methyl red indicator solution.
Sodium chloride.
Analar sodium chloride that has been heated for 6 h at 270°C before use.

Reference Electrode Solution

This solution is prepared by adding 0·1 ml of 0·01 N silver nitrate to a mixture of 0·15 ml of N nitric acid and 30 ml of 1% w/v potassium nitrate solution.

Procedure

Weigh accurately a suitable amount of the sample to give 5 to 6 mg of chlorine (this weight should not exceed 30 mg) and transfer it to the centre of a small piece of Whatman No. 42 filter paper weighing approximately 0·1 g. Fold the filter paper so that the sample is completely enclosed but before making the final fold insert a small wick of cotton wool.

Prepare the combustion vessel by washing the flask and adapter and heating the end of the platinum wire and the basket to red heat in a bunsen flame. Put 1 ml of N sodium hydroxide, 3 drops of 35% w/v sodium hydrogen sulphite solution and 3 ml of water in the flask, and agitate it in a horizontal plane to wet the walls of the flask with the absorption solution.

153

Fill the flask with oxygen and insert the stopper. Place the wrapped sample in the platinum basket and twist the cotton wool wick with tweezers so that it lies between the basket and the ancillary firing electrode. Remove the stopper from the flask and replace it with the adapter containing the sample, seal the joint with a suitable amount of distilled water and attach the retaining springs. Attach the electrical and earth leads to the adapter, and place the safety screen in front of the combustion vessel. Switch on the H.F. tester to ignite the cotton wool and sample and allow to burn to completion. Set aside for 10 min and then disconnect the electrical and earth leads to the adapter and remove the springs. Disconnect the adapter from the combustion vessel, wash down the platinum wires and basket into the lower compartment of the combustion vessel, and rinse the sides of the flask to give a final volume of approximately 35 ml. Add 6 drops of 100 volume hydrogen peroxide to destroy any excess of sulphite, neutralise the solution with N nitric acid to methyl red indicator, and add 0·15 ml of nitric acid in excess.

Place a $\frac{3}{4}$ in (20 mm) polythene covered magnet in the combustion vessel, fit the electrode assembly, securing it by means of the springs provided, and place the flask on a magnetic stirrer. Titrate with 0·01 N silver nitrate until the titration record shows that the end point has been passed. Carry out a blank combustion omitting only the sample and titrate this at the same instrument settings. Finally, standardise the silver nitrate solution by carrying out a blank combustion with a known weight of sodium chloride (about 10 mg) added to the absorption solution in the combustion flask before the test.

Titrate this solution under the same conditions as used for the sample and blank. Deduce the end points for all sample and standardisation titrations from the points of maximum inflexion on the titration curves. This point is conveniently found by means of a plastic cursor inscribed with a series of concentric circles. The circle of most perfect fit is found for each side of the titration curve and the intersection of the line joining the centres of these circles cuts the titration curve at the required point of maximum inflexion.

Correct each titre for the blank titre corresponding to the same end point potential difference (this blank is only of the order of 0·07 min titration time—about 0·05 ml) and hence calculate the chlorine content of the sample.

PYROLYSIS/GAS CHROMATOGRAPHIC EXAMINATION OF VINYL RESINS

Polymers and copolymers of vinyl chloride, vinyl acetate and vinyl alcohol are readily distinguished by pyrolysis/gas chromatography (p. 57). The pyrolysis/gas chromatography method has proved of great value in the identification of small amounts of modifying polymers in polyvinyl chloride preparations. The differing behaviour of polyvinyl chloride and

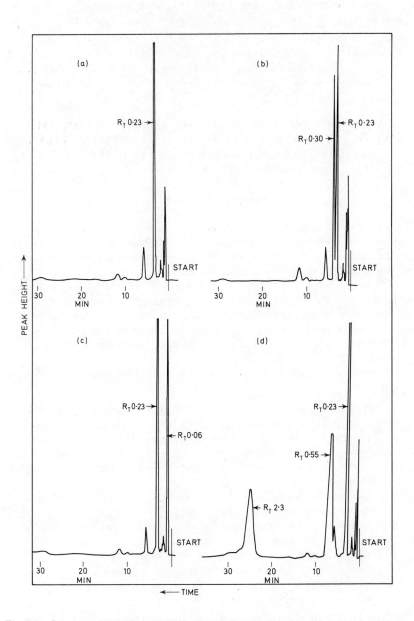

Fig. 3.4.   Pyrograms of (a) polyvinyl chloride, (b) polyvinyl chloride/polymethyl methacrylate (95:5) blend, (c) polyvinyl chloride/polymethyl acrylate (87:13) blend and (d) polyvinyl chloride/polyethylhexyl acrylate (80:20) blend

three different acrylic modified compositions in the test is shown in the pyrograms illustrated in Figure 3.4.

It should be realised that the pyrograms are obtained by application of the test to deplasticised resins; plasticisers produce additional peaks which complicate the pyrogram and render interpretation difficult.

THE DETECTION AND DETERMINATION OF ADDITIVES OR CONSTITUENTS OTHER THAN PLASTICISERS IN POLYVINYL CHLORIDE COMPOSITIONS

POLYMERIC ADDITIVES

In recent years, one of the major changes in the formulation of polyvinyl chloride compositions has been in the addition of other polymeric materials to improve the impact strength, facilitate injection moulding or improve the adhesion to metal surfaces etc.

When the tetrahydrofuran–ethanol method for the recovery of the polymer from the deplasticised sample is applied, some of the modifiers are insoluble in the tetrahydrofuran and are recovered with the 'filler'; others are soluble in the tetrahydrofuran and are precipitated along with the polymer on the addition of alcohol, whilst others are soluble in both solvents.

Observations on some of these modifiers quoted by Ruddle et al.[11] are given below.

1. Polyacrylate/polyethylhexyl acrylate and methyl methacrylate/ethyl acrylate copolymers may both be detected in the tetrahydrofuran soluble ethyl alcohol soluble matter. The infra-red spectrum of polyethylhexyl acrylate is rather indecisive for identification purposes and this polymer is best identified by pyrolysis/gas liquid chromatography.
2. Chlorinated polythene and polythene may be added to polyvinyl chloride compositions as lubricants.
3. Chlorinated polythene may be detected in the tetrahydrofuran soluble ethyl alcohol insoluble matter.
4. Polythene, however, is insoluble in tetrahydrofuran and may be recovered from the filler fraction by extracting this with hot toluene.
5. Epoxy resins, in conjunction with phenol-formaldehyde resins, are used in compositions for coating metals etc.

If the sample is in paste form the additive will be extracted with ether along with the plasticiser. Epoxy resins may be separated from the plasticiser by thin layer chromatography and phenol-formaldehyde by carbon tetrachloride precipitation. If, however, the preparation submitted for test is a finished product, the epoxy resin will be rendered insoluble by cross-linking and will be present in the tetrahydrofuran insoluble matter along with inorganic fillers. Prolonged extraction of this fraction with methyl ethyl ketone, however, will usually yield sufficient epoxy resin for infra-red identification.

ACRYLONITRILE/STYRENE AND ACRYLONITRILE/ BUTADIENE/STYRENE ETC. COPOLYMERS AND TERPOLYMERS

The solubilities of these copolymers and terpolymers vary with molecular weight and, in practice, they may be found in the tetrahydrofuran soluble ethyl alcohol soluble matter, in the precipitated polyvinyl chloride polymer and in the tetrahydrofuran insoluble filler fraction. Application of a fractional precipitation procedure to recovered polyvinyl chloride polymers containing this type of copolymer is often very rewarding.

The method of application of the test to a polyvinyl chloride polymer containing 3% of a mixture of an acrylonitrile/butadiene/styrene terpolymer and an acrylonitrile/styrene copolymer is given below.

2 g of the preparation were warmed with 50 ml of tetrahydrofuran and the mixture spun in a centrifuge at approximately 3000 rev/min. The insoluble residue was separated from the tetrahydrofuran solution, washed once with tetrahydrofuran and respun in the centrifuge. The insoluble residue was separated and reserved as Fraction 1.

The combined tetrahydrofuran solution and washings were then spun in a centrifuge at approximately 17 000 rev/min and the small amount of insoluble residue obtained, after washing with tetrahydrofuran, reserved as Fraction 1a.

The combined tetrahydrofuran solution and washings were evaporated to 20 ml, then diluted with 75 ml of methylene chloride. 10 ml of ethanol were now added slowly, with stirring, and the precipitate, Fraction 2, was isolated by centrifugation. This precipitation was repeated, two further portions each of 20 ml of ethanol being used, giving Fractions 3 and 4. 50 ml of light petroleum (boiling range 40–60°C) were added to the residual solution, and the precipitate, Fraction 5, was allowed to settle overnight, leaving the mother liquor, which was evaporated to give Fraction 6.

All the fractions were dried and examined by infra-red spectroscopy with the following results:

Fraction 1. Inorganic filler and acrylonitrile/butadiene/styrene terpolymer.
Fraction 1a. Similar to 1 but with less filler.
Fractions 2 and 3. Polyvinyl chloride.
Fraction 4. Polyvinyl chloride and a small amount of acrylonitrile/ styrene copolymer.
Fraction 5. Acrylonitrile/styrene copolymer and a small amount of polyvinyl chloride.

Fraction 6. Mixture of polyvinyl chloride, acrylonitrile/styrene and other residues.

Further separation of Fraction 5 was achieved by adding 100 ml of carbon disulphide to a solution of the separated fraction in 20 ml of tetrahydrofuran. The small amount of polyvinyl chloride present was soluble in this mixture and the precipitated copolymer could be separated in an almost pure condition.

INORGANIC AND ORGANO-METALLIC CONSTITUENTS

Metals present in vinyl compositions usually occur in such substances as fillers e.g. titanium dioxide, china clay, calcium carbonate and asbestos; in stabilisers e.g. calcium, lead, cadmium, and barium stearates, tribasic lead sulphate and basic lead carbonate; or they may be present as organo-metallic compounds, particularly the tin stabilisers such as dibutyl tin dilaurate and dibutyl tin maleate.

The identification of inorganic additives is considered in Chapter 11 and the organo-tin stabilisers are in this chapter considered on p. 162. In the case of metallic salts of organic acids, such as stearates, the metal present is identified and the amount determined by the methods given below, but a method is often required for the direct detection and the semi-quantitative determination of the stearate. The test is carried out as follows:

Approximately 1 g of the finely divided deplasticised resin is heated under reflux on a water bath with 10 ml concentrated hydrochloric acid and 50 ml ether for 8 h. The solution is cooled, filtered into a separating funnel and the aqueous layer discarded. The ether layer is washed with water and filtered into a weighed flask from which the ether is evaporated, leaving a residue of acid which is dried to constant weight at 100°C. The acid is identified from the infra-red spectrum.

The inorganic elements are determined either by chemical, atomic absorption or X-ray Fluorescence methods.

CHEMICAL METHODS

Although time consuming, chemical methods, with modification, are normally applicable to all compositions. The principles of the different methods are as follows.

Cadmium and zinc

The sample is wet oxidised and the oxidation products evaporated to dryness. The residue is dissolved in ammonia-ammonium chloride base solution in the polarographic test for cadmium and in ammonium chloride-ammonium thiocyanate-methyl red base solution in the polarographic test for zinc. Alternatively, an atomic absorption finish may be used.

Barium

This element is not normally determined in small amounts. Barium in substantial amounts, e.g. as with barium containing fillers, is determined gravimetrically as barium sulphate.

Lead

The sample is dissolved in ethylene dichloride and the lead extracted with 50% v/v hydrochloric acid solution for direct determination in the acid extract from the absorbance at 270 nm[12], correcting, if necessary, for iron.

Tin

The wet oxidation products of the sample are evaporated to dryness and the tin determined in the residue by a modification of the turbidimetric method of Karsten, Kies and Walraven[13]. The residues are dissolved in 4 ml 50% v/v sulphuric acid solution, 25 ml 30% w/v tartaric acid solution is added, followed by 10 ml of a 2% w/v solution of 3-nitro-4-hydroxyphenylarsonic acid reagent. The turbid solution is mixed, allowed to stand overnight, remixed and the turbidity measured at 450 nm.

Calcium

The calcium in the wet oxidation products of the sample is isolated as the oxalate after removal of interfering ions (for example lead as sulphide). The washed calcium oxalate is decomposed in acid solution and titrated with standard potassium permanganate solution.

Phosphorus

Phosphate in the wet oxidation products is determined by conversion to ammonium phosphomolybdate and reduction of this, by ascorbic acid

159

under appropriate conditions, to the blue coloured reduction product. The blue colour is measured at 820 nm.

X-RAY FLUORESCENCE METHODS

Determination of inorganic elements in rigid unfilled compositions

These compositions usually consist of simple combinations of polymer, stabiliser, and only small amounts of plasticiser and the samples for analysis may be received as finished compositions or as intermediate process samples in the form of powders.

The elements of interest in these samples occur usually as combinations such as lead and calcium; barium, cadmium and zinc; calcium and phosphorus; or tin and sulphur.

Standards for calibration purposes are prepared by Henschel mixing and milling known amounts of chemically analysed stabiliser additives into polyvinyl chloride polymer together with sufficient plasticiser to give easy processing. The resultant crepe is hot pressed to a $\frac{1}{8}$ in (3 mm) biscuit from which X-ray sample disks are die stamped. Random samples of the disks are chemically analysed to confirm no loss of additive during compounding and the analytical surface of each disk skimmed on a lathe. The concentration ranges covered by three typical sets, each of six standards, are as follows:

Set 1. Sn: 0–0·5%; S: 0–0·15%.
Set 2. Ba: 0–0·25%; Cd: 0–0·035%; Zn: 0–0·015%.
Set 3. Ca: 0–0·25%; P: 0–0·15%.

Suitable instrumental conditions for the individual calibrations and for the examination of samples in the form of $\frac{1}{8}$ in (3 mm) thick moulded disks are summarised in Table 3.1, and the following points should be noted:

Instrument Standards: Polyvinyl chloride is not stable to repeated exposure to X-rays and loss of chlorine occurs[14]. Borax glass instrument standards are thus preferred particularly for determinations involving light elements.

Phosphorus: The method cannot be applied in the presence of a large excess of calcium due to line interference of the $Ca_{K\beta}$ second order line on the $P_{K\alpha}$ line used in the determination of phosphorus.

Calcium: The determination of calcium present as calcium stearate in samples which are received for examination in the form of milled, moulded, or extruded pieces from which the analytical specimen can be cut after simply flattening a suitable area at 110°C presents no real problems. With samples received in powder form however, which require hot compression

Tables 3.1. X-ray Fluorescence examination of rigid and unfilled polyvinyl chloride samples

Element	Range	X-ray Tube. Anode, kV/mA	Crystal and Order	Lines Used	Background ±°2θ	Automatic Pulse Height Analyser	Instrument Standard (IS)	Counter	Calibration Graph
Lead	0·1–5·0%	W 40/16	LiF, 1st	$Pb_{Lα}$, $W_{Lβ}$	—	in	P.V.C. + 1·4% Pb	S.C.	% Pb v. ratio $\dfrac{Pb_{Lα}}{W_{Lβ}}$
	0·01–0·3%	W 60/32	LiF, 1st	$Pb_{Lα}$	+1	in	P.V.C. + 0·145% Pb	S.C.	% Pb v. ratio to I.S.
Tin	0·01–0·5%	W 40/32	LiF, 2nd	$Sn_{Kα}^{(2)}$, W_{L1}	—	in	P.V.C. + 0·5% Sn	S.C.	% Sn v. ratio $\dfrac{Sn_{Kα}^{(2)}}{W_{L1}}$
	0·005–0·1%	W 60/32	LiF, 2nd	$Sn_{Kα}^{(2)}$	−1	in	Borax Glass Disk containing 2·2% Sn 0·5% S	S.C.	% Sn v. ratio to I.S.
Sulphur	0·01–0·15%	W 60/32	PE, 1st	$S_{Kα}$	+1	in		F.C.	% S v. ratio to I.S.
Calcium	0·01–0·25%	W 60/32	PE, 1st	$Ca_{Kα}$	+1	in	Borax Glass Disk containing 6·3% $Ca(H_2PO_4)_2$	F.C.	% Ca v. ratio to I.S.
Phosphorus	0·01–0·15%	Cr 40/32	PE, 1st	$P_{Kα}$	+1	in		F.C.	% P v. ratio to I.S.
	0·01–0·15%	W 60/32	PE, 1st	$P_{Kα}$	+1	in		F.C.	% P v. ratio to I.S.
Barium	0·01–0·25%	W 60/32	LiF, 1st	$Ba_{Lα}$	+2	in	Borax Glass Disk containing 0·4% Ba	F.C.	% Ba v. ratio to I.S.
Cadmium	10–400 p.p.m.	W 60/32	LiF, 2nd	$Cd_{Kα}^{(2)}$	+2	in	0·25% Cd	S.C.	p.p.m. Cd v. ratio to I.S.
Zinc	5–150 p.p.m.	W 60/32	LiF, 1st	$Zn_{Kα}$	−0·7	in	0·10% Zn	F+S	p.p.m. Zn v. ratio to I.S.
Bromine	10–500 p.p.m.	W 60/32	LiF, 1st	$Br_{Kα}$	+1	in	Borax Glass Disk containing 0·26% Br	S.C.	p.p.m. Br v. ratio to I.S.

moulding to give a sample disk, a serious problem arises from extensive migration of the calcium stearate to the mould surface during moulding. The X-ray Fluorescence method is not therefore used for powder samples when the calcium content is expected to be greater than 100 p.p.m. For calcium levels below 100 p.p.m. the method gives a precision equal to the chemical method (p. 159) but, of course, cannot be used in the presence of lead since due to line interferences the $Ca_{K\beta}$ line must be used and absorption effects are too marked.

The analysis of plasticised and filled polyvinyl chloride compositions by X-ray Fluorescence

The number and complexity of plasticised and filled polyvinyl chloride formulations is very great and the analyst will thus meet a correspondingly large variety of inter-element and matrix effects if he attempts the direct X-ray Fluorescence analysis of such compositions. A method which can be generally applied for the determination of, for example, chlorine, sulphur, calcium, lead, titanium, aluminium and silicon in samples of unknown origin is thus not a practical proposition, unless perhaps the handling of the intensity data and matrix corrections can be made by computer, or a simple means of homogenising each sample with about ten times its weight of a compatible diluent is available to minimise the matrix effects.

X-ray Fluorescence can be very valuable, however, for the comparative analysis of production formulations of known nominal composition for which calibration standards in the same matrix can be prepared. It is often sufficient to prepare for each formulation a sample known to be correct in all respects. All that is then required is a direct comparison of the relevant line intensities of both sample and standard, which should be identical if there have been no errors in the sample compounding, such as bad dispersion or omitted additives.

An indirect method for the rapid X-ray Fluorescence analysis of plasticised polyvinyl chloride compositions has also been used[15]. This method involves some preliminary separation of the components of the composition by solvent solution and centrifuge methods.

TIN STABILISERS

Tin stabilisers may be divided into two distinct groups i.e. those containing sulphur and those from which sulphur is absent. Indeed, a useful preliminary in this work is to make a qualitative test for sulphur and, if necessary, a quantitative determination of the same element in the substance under examination. To the sulphur containing group belong compounds such as alkyl tin mercaptides, mercapto-esters, and mercaptocarboxylates. To the second group belong alkyl tin carboxylates and their esters.

All these compounds, and especially those containing dialkyl tin groups, are important because of their wide application as stabilisers for polyvinyl chloride compositions.

A very useful paper on the subject i.e. that deals with the detection and determination of such substances as dibutyl tin dichloride, dibutyl tin dilaurate, dioctyl tin dichloride, diethyl tin dichloride and diphenyl tin dichloride in polyvinyl chloride compositions is that of Chapman, Duckworth and Price[16].

Although, of course, only concerned with the recovery of tin compounds the paper is useful in giving the analyst information about the solubility and separation of these tin stabilisers. Moreover it is exceedingly informative on the colorimetric dithizone reactions of the stabilisers.

Udris[17] has investigated both types of tin compound, i.e. sulphur containing and non-sulphur containing, in considerable detail. The methods he has worked out for the examination of tin stabilisers may be applied with equal success to tin stabilisers recovered from polyvinyl chloride compositions by extraction with ether in a Soxhlet apparatus as described on p. 144.

THE EXAMINATION OF TIN STABILISERS CONTAINING SULPHUR

Tin stabilisers belonging to this group usually have characteristic sulphurous odours, but the presence of sulphur should be confirmed by a specific chemical test, or X-ray Fluorescence examination, prior to the commencement of chemical analysis. Compounds belonging to this group include dialkyl tin dialkyl thioglycollates, dialkyl tin dilauryl mercaptides,

(e.g. $R-Sn\begin{smallmatrix}SCH_2COOR'\\SCH_2COOR''\end{smallmatrix}$ or $R-Sn\begin{smallmatrix}SCH_2(CH_2)_{10}CH_3\\SCH_2(CH_2)_{10}CH_3\end{smallmatrix}$ respectively)

and stabilisers in which the dialkyl tin group is combined with both mercaptan and carboxyl groups. In the above formulae R, R' and R'' denote alkyl groups—they can be the same or different. The principles of the methods worked out by Udris for the identification of these compounds may be illustrated by reference to the identification of two stabilisers, dibutyl tin di-2-ethylhexyl thioglycollate and dibutyl tin dilauryl mercaptide.

1. DIBUTYL TIN DI-2-ETHYLHEXYL THIOGLYCOLLATE

(a) Identification of the alkyl radical derived from the thioglycollate

Approximately 1 g of the sample is dissolved in acetone and treated with an excess of silver nitrate solution. This decomposes the stabiliser and precipitates the silver mercaptide of the thioglycollate ester and leaves dibutyl tin nitrate in solution. The filtered precipitate is washed with a little water, followed by acetone and finally with ether, prior to suspension in ether and treatment by shaking vigorously with 1 ml concentrated hydrochloric acid. This precipitates all the silver as chloride and the ether solution containing the free ester is isolated. The ether is evaporated off

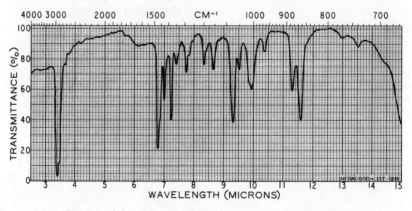

Fig. 3.5.   Infra-red spectrum of dibutyl tin oxide. Suspension in KBr disk

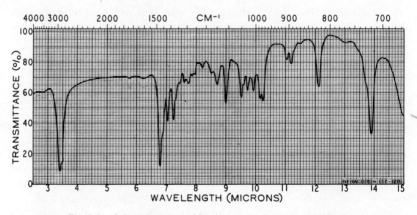

Fig. 3.6.   Infra-red spectrum of dioctyl tin oxide. Suspension in KBr disk

and the ester hydrolysed by boiling under reflux for $1\frac{1}{2}$ h with 20 ml 10% w/v sodium hydroxide solution. On completion of the hydrolysis the alcohols are isolated by distillation and/or ether extraction and identified by the gas chromatographic and infra-red methods, the principles of which have been described in Chapter 1. In the example cited above a clear identification of 2-ethyl-hexanol is obtained.

164

(b) Identification of the thioglycollic acid from the thioglycollate ester

The alkaline solution remaining after the removal of the alcohols contains
the sodium salt of the thioglycollic acid. This solution is acidified with
hydrochloric acid and the liberated thioglycollic acid extracted with
ether. The acid recovered on simple evaporation of this extract may be
contaminated with by-products from the hydrolysis treatment and a
further purification stage is recommended. A few drops of 50% w/v
aqueous silver nitrate solution are added to the ether solution, dropwise
and with shaking, and the precipitated silver mercaptide isolated, washed
with ether and dried in a desiccator, prior to identification by infra-red
examination (nujol mull) or by n.m.r. examination. In the latter case a
small amount of the precipitate is suspended in chloroform, a little
hydrochloric acid added and the n.m.r. examination carried out on the
chloroform extract.

(c) Identification of the alkyl groups attached directly to tin

The filtrate containing dibutyl tin nitrate from section 1(a) above is
treated with hydrochloric acid to remove the excess silver and the
solution refiltered. Sodium hydroxide solution is now added to yield a
precipitate of dibutyl tin oxide which is recovered by filtration and subsequently identified by infra-red examination. The infra-red spectrum of
dibutyl tin oxide is given in Figure 3.5. This compound is easily distinguished from dioctyl tin oxide (Figure 3.6). Alternatively, the dialkyl tin
oxide may be obtained by boiling a methanol solution of the original
stabiliser with sodium hydroxide solution. The precipitated oxide is
washed with distilled water and methanol prior to drying at 105°C.

The thin-layer chromatographic method described by Belpaire[18] for
the identification of dialkyl groups attached directly to tin has also been
found to be a valuable confirmatory test.

(d) Determination of tin

The sample is wet digested with sulphuric acid and hydrogen peroxide[19].
The tin in the digest is precipitated with cupferron and finally weighed
as SnO_2.

2. DIBUTYL TIN DILAURYL MERCAPTIDE

(a) Identification of the mercaptan derived from the mercaptide

The sample is dissolved in acetone and treated with an excess of silver
nitrate solution to decompose the stabiliser and precipitate the mercaptan

as silver mercaptide. The precipitate is washed with acetone followed by ether and dried in a desiccator prior to identification by infra-red and n.m.r. examination as described in section 1(b) above. If required, the free mercaptan can be isolated for examination by suspending the mercaptide in ether and shaking vigorously with a little concentrated hydrochloric acid to precipitate all the silver as silver chloride and leave the free mercaptan in the ether phase.

(b) Identification of the alkyl groups attached directly to tin

Dibutyl tin oxide is isolated and clearly identified by the procedure described in section 1 (c) above.

(c) Determination of tin

The Cupferron method as described in section 1(d) above is preferred.

TITRATION OF THIO-TIN COMPOUNDS

Udris has examined several published methods for the direct titration of thio-tin compounds[20] and has found these satisfactory. In addition he has carried out potentiometric titrations of thio-tin compounds in benzene-methanol solution in the presence of sodium acetate, with a standardised solution of silver nitrate in isopropanol. The indicator electrode is a silver wire coated with silver sulphide, and the reference electrode a copper wire immersed in a pool of mercury covered with a solution of 0·1 N sodium acetate. Contact with the test solution is made via an agar gel bridge.

THE EXAMINATION OF TIN STABILISERS THAT DO NOT CONTAIN SULPHUR

To this group belong stabilisers like dialkyl tin maleates, dialkyl tin dialkyl maleates and dialkyl tin dilaurates, e.g. respectively

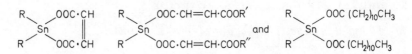

where R, R′ and R″ are alkyl groups, which can be the same or different.

166

The principles of the methods preferred for the analysis of this group of compounds may be illustrated by reference to the identification of dibutyl tin di-2-ethylhexyl maleate and dioctyl tin dilaurate.

3. DIBUTYL TIN DI-2-ETHYLHEXYL MALEATE

(a) Identification of the alkyl radical derived from the maleate

Approximately 1 g of the sample is hydrolysed by boiling with 50 ml 10% w/v sodium hydroxide solution under reflux. The alcohols that are liberated are isolated by distillation and identified by gas chromatographic and infra-red methods as previously described.

(b) Identification of the butyl groups attached directly to tin

The alkaline residues after distillation of the alcohols are extracted with ether to remove residual alcohols and unhydrolysed material, care being taken to avoid emulsification during the extraction. The dibutyl tin oxide is filtered off, washed and recovered for infra-red identification, as previously described.

(c) Identification of the maleic acid derived from the maleate

The alkaline filtrate after removal of the dialkyl tin oxide contains the sodium salt of the maleic acid. The solution is acidified with hydrochloric acid and extracted by shaking with ether to remove any acids or contaminants readily soluble in this solvent. The residual aqueous phase is now re-extracted with ether in a continuous liquid–liquid extractor for 4 h to isolate the maleic acid for identification by infra-red spectroscopy.

(d) Determination of tin

The Cupferron method previously described is used.

4. DIOCTYL TIN DILAURATE

The identification of the alkyl groups attached directly to tin and the acid component of this stabiliser proceeds along similar lines to the above example. The following minor differences are made in the procedure.

1. The treatment with alkali is much shortened and, in the absence of ester, no recovery of alcohols is required.
2. The dialkyl tin oxide is dioctyl tin oxide.
3. The acid is recovered in the initial extraction with ether and is best identified from the infra-red spectrum of its barium salt.

TITRATION OF TIN CARBOXYLATES

Udris has investigated the direct titration of several compounds of the type indicated above and has found the principle of the method worked out by Groàgova and Pribyl[21] to be very useful. Using this method RCOO— groups in these compounds may be titrated with sodium methoxide in pyridine. In the potentiometric titration, antimony and calomel electrodes are used.

THE DETECTION AND IDENTIFICATION OF TIN STABILISERS IN ETHER EXTRACTS FROM P.V.C. COMPOSITIONS

Tin stabilisers may be encountered in highly plasticised polyvinyl chloride compositions. In such cases the ether extract from the composition may contain e.g. 5 to 10% tin stabiliser and 90 to 95% of a plasticiser such as dioctyl phthalate.

In our experience, in these cases the tin stabiliser cannot be recovered as such by thin-layer chromatographic examination of the mixture. The separation is effected by column chromatography on a 15 mm diameter column packed with 10 g of a 1 : 1 w/w mixture of Celite 545 and 200–300 mesh silica gel as slurry in carbon tetrachloride. 2 to 3 g of the sample are transferred to the top of the column and the plasticiser eluted with carbon tetrachloride (~ 200 ml). The tin stabiliser is now eluted with ether and recovered for identification.

The infra-red spectra of a number of common tin stabilisers are given in Spectra 11.4, 11.41 and 11.44. However, the material recovered from the column separation may show evidence of partial decomposition and, in particular, carboxylic acid or carboxylic acid salt may be present. For this reason the test is best conducted by comparing the spectrum of the eluted material with those of known tin stabilisers eluted from the column in the same way.

Moreover, wherever practicable in identification work, advantage should always be taken of the principles enumerated in the general observations on tin stabilisers outlined above.

An alternative test, which may be applied to tin stabiliser-phthalate mixtures in order to identify the alkyl groups on tin, is to pass a methanol solution of the tin stabiliser–phthalate mixture through a column of cation exchange resin in its H-form, washing free of plasticiser, and finally eluting

168

with methanol-hydrochloric acid solution. The dialkyl tin dichloride is extracted with ether from the acid solution remaining after evaporation of the methanol under an air jet, and identified from its infra-red spectrum (see Figures 3.7 and 3.8).

In his work Udris has stressed the importance of infra-red and n.m.r. examination of separated fractions and derivatives; he has also shown how

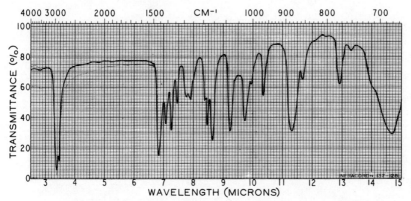

Fig. 3.7.   Infra-red spectrum of dibutyl tin dichloride. Crystallised from melt between NaCl plates

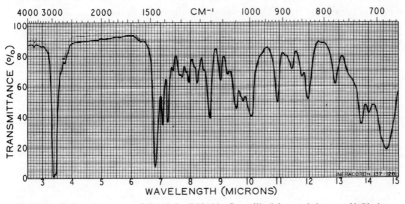

Fig. 3.8.   Infra-red spectrum of dioctyl tin dichloride. Crystallised from melt between NaCl plates

it is possible to detect additives including phenolic compounds and phthalate esters in tin stabilisers by U.V. spectrophotometric methods.

U.V. ABSORBERS

U.V. absorbers or light stabilisers, and other aromatic constituents such as α-phenyl indole, which is added to rigid polyvinyl chloride formulations

169

as a heat stabiliser, may be present in the ether extractable matter, particularly that obtained from unfilled compositions.

These materials are detected and determined by measurement of the U.V. absorption of 0·5% w/v solution of the original sample in ethylene dichloride. Some ultraviolet absorbing constituents and their wavelengths of maximum absorption are listed in Table 3.2, together with the corresponding $E_{10\,mm}^{1\%}$ values.

Table 3.2. Some ultraviolet absorbing constituents of polyvinyl chloride formulations

Constituent	Wavelength of maximum absorption, nm	$E_{10\,mm}^{1\%}$
2-Hydroxy-4-methoxy benzophenone	328	450
2 : 2'-Dihydroxy-4-octoxy benzophenone	352	430
2 : 4-Dihydroxy benzophenone	325	440
2-Hydroxy-4-n-octoxy benzophenone	325	350
2-(2'-Hydroxy-5'-methylphenyl)benzotriazole	341	770
Phenyl salicylate	312	260
α-Phenyl indole	308	1325

It should be noted that phthalate does not absorb at these wavelengths and does not interfere, whereas these materials all absorb at 275 nm and would interfere in the U.V. determination of phthalate.

SURFACE MOISTURE

With many polyvinyl chloride powders surface moisture is a particularly important determination. The gas chromatographic method (p. 51) may be used or an approximate determination may often be carried out by Karl Fischer titration of a methanol slurry of the sample.

With polyvinyl chloride compositions, however, a much more rigorous and searching treatment is carried out as follows:

Apparatus

The distillation apparatus is shown in Figure 3.9.

A is a 1 l distillation flask connected by a B 24 joint to the trap, T, which has a capacity of about 10 ml below the overflow tube, U. The trap, T, is fitted with a condenser, B, which is protected from atmospheric moisture by a silica gel tube. C is a 2 mm precision bore capillary about 560 mm long, graduated at 1 mm intervals over a 500 mm length, X–Y, and fitted with hemispherical joints. The waste trap, R, is connected to the top of the capillary tube and to a 250 ml bulb, P, which is connected in turn with the levelling bulb, L. The bulbs P and L contain water.

The capillary tube, trap and condenser are coated with a film of silicone polymer to prevent water droplets adhering to the walls. The film is applied by pouring a solution of 5 ml mixed methyl chloro-silanes in 100 ml dry toluene into the clean air-dried apparatus and allowing it to stand for 15 min before pouring off. The apparatus is then air-dried and

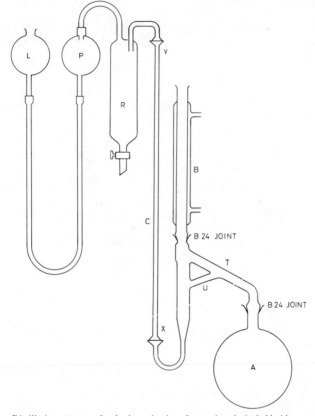

Fig. 3.9.   *Distillation apparatus for the determination of water in polyvinyl chloride compositions*

heated in an oven at 105°C overnight. Operations involving the chloro-silanes should be carried out in a fume hood since these substances are toxic, inflammable and corrosive.

Calibration of the capillary

The trap is filled to the overflow with dry toluene and a known weight of water added. The water in the trap is transferred to the capillary by means of the levelling bulb, the bottom meniscus of the water column adjusted to the zero mark and the height of the water column noted. The water is returned to the trap and the procedure repeated with successive additions of water.

Drying the toluene

The toluene is dried by passing it through a column of silica gel as shown in Figure 3.10. A is a glass tube of 25 mm diameter and about 300 mm

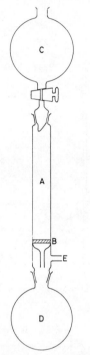

Fig. 3.10.   Apparatus for drying toluene

length fitted with a porosity 1 sintered glass disk, B, and B 24 joints at either end. The tube is filled with 10–20 mesh activated silica gel. Toluene is allowed to percolate slowly through the column from the 2 l tap funnel, C, into the receiver flask, D. The outlet tube, E, is fitted with a moisture trap. The silica gel may be re-activated by removing the toluene in a current of air and finally drying the gel in an oven at 100 to 105°C.

Procedure

80 to 100 g of the sample is weighed in a weighing bottle and transferred to a 1 l distillation flask containing 500 ml of dry toluene. The trap is

172

filled to the overflow with dry toluene and connected to the distillation flask. The flask is heated on an electric heating mantle until the liquid begins to distil, and the distillation continued at such a rate that the refluxing liquid reaches a height of between 10 and 40 mm in the condenser. After refluxing for 45 min any water droplets in the condenser are washed down with dry toluene, refluxing continued for a further 15 min and the condenser again washed down with dry toluene. By means of the levelling bulb the level in the trap is adjusted so that the water is in a continuous column. The levelling bulb is slowly lowered to draw the water column into the graduated capillary tube, the lower meniscus of the water column adjusted to the zero mark and the height of the water column read off.

The water and toluene are drawn into the waste receiver, the condenser and trap washed down with dry toluene and the toluene again drawn into the waste receiver. The apparatus is then ready for a further determination.

A blank determination is carried out on 500 ml of the dry toluene and the reading obtained deducted from that of the sample.

Calculation

$$\% \text{ water} = \frac{H \times F \times 100}{W}$$

where H = height of water column, in mm

F = capillary tube factor in g water/mm (0·00314 for a 2 mm capillary)

W = weight of sample in g

EMULSIFYING AGENTS IN POLVINYL CHLORIDE COMPOSITIONS

The following general notes on the extraction of emulsifying agents may be useful.

Polyvinyl chloride emulsion polymers may contain, in our experience, such materials as sulphonated fatty esters, alkyl sulphonate salts, octa-decanol and ethyl palmitate, and in order to obtain the emulsifying agent it is desirable to extract in the first place with methanol which extracts many emulsifying agents quite readily.

The methanol extraction does not, however, satisfactorily extract ethyl palmitate from a sample; if this substance is suspected, it is desirable to carry out an ether extraction after the methanol extraction.

In certain cases polyvinyl chloride emulsion polymers may be used in the subsequent preparation of polyvinyl chloride compositions; in the case of these latter compositions extraction with methanol may yield the emulsifying agent free from much contamination with plasticiser. In other cases it may be more desirable to extract the plasticiser first from the composition with ether, and then to dissolve the plasticiser free sample in peroxide free tetrahydrofuran and to precipitate the polymer with alcohol; this polymer is filtered off and the tetrahydrofuran solution evaporated carefully on a water bath. This procedure often yields the emulsifying or lubricating agent.

An example of the analysis of the emulsifying agent extracted from a polyvinyl chloride emulsion polymer is given on p. 134 and further examples are given below.

TESTS EMPLOYED TO ESTABLISH THAT AN EMULSION
POLYMER CONTAINED A SODIUM ALKYL SULPHONATE

Extraction of a sample of emulsion polymer, carried out in the course of general miscellaneous work, yielded $2·5\%$ of a soap-like solid.

Application of Wurzschmitt's classification procedure (p. 130) to the extract yielded the following evidence.

1. The material gave a positive test in the examination for anion active agents.
2. The material gave positive tests for the elements sulphur and sodium indicating that it belonged to Group 3 in Wurzschmitt's classification of anion active compounds.
3. In saponification tests no usage of alkali could be observed whether ethylene glycol monoethyl ether (b.p. $134°C$), ethylene glycol (b.p. $197°C$), diethylene glycol (b.p. $245°C$) or triethylene glycol (b.p. $276°C$) was used as solvent. This indicated that the compound was either the sodium salt of an alkyl sulphonic acid, or an alkyl aryl sulphonic acid, or a fatty sulphonic acid or the sodium salt of a sulphonated alkylene oxide addition product.
4. The material did not give a complex salt with phosphomolybdic acid and barium chloride, indicating the absence of alkylene oxide addition products.
5. The material did not reduce strongly alkaline potassium permanganate solution, indicating the absence of alkyl aryl sulphonates.
6. An aqueous solution of the material did not yield a precipitate when treated with Mohr's salt solution, sodium silicate solution, ammonium chloride solution or magnesium sulphate solution. This indicated that fatty sulphonic acids were absent and hence pointed to the probability that the material was the sodium salt of an alkyl sulphonic acid.

Accordingly, a fresh portion of the original extract was acidified with dilute sulphuric acid solution and then subjected to continuous extraction with ether.

On evaporation to dryness the ether extract yielded an oily liquid that darkened rapidly. The elementary composition of this acid corresponded

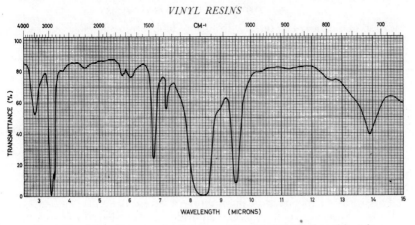

Fig. 3.11.   Infra-red spectrum of methanol extract from polyvinyl chloride emulsion polymer

with the empirical formula $C_{14}H_{29}SO_3H$ and infra-red examination of the original extract gave a spectrum closely resembling that of an alkyl sulphonate salt. This infra-red spectrum is shown in Figure 3.11.

TESTS REQUIRED TO ESTABLISH THAT A P.V.C. COMPOSITION CONTAINED STEARIC ACID AND A SULPHONATED FATTY ESTER

In the examination of polyvinyl compositions it may, on occasion, be desirable to proceed rather more directly to the identification of the emulsifying agent without applying the full Wurzschmitt classification procedure, e.g. a sample of rigid polyvinyl chloride foil, after preliminary removal of plasticiser by extraction with ether, was dissolved in peroxide free tetrahydrofuran and the polymer then precipitated by the

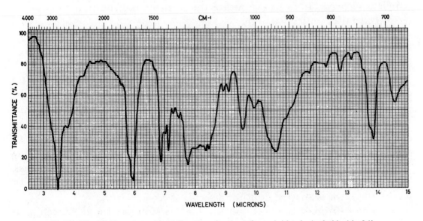

Fig. 3.12.   Infra-red spectrum of methanol extract from rigid polyvinyl chloride foil

175

addition of alcohol. The tetrahydrofuran-alcohol solution, on evaporation, then yielded a residue amounting to 2·3% of the deplasticised resin.

Chemically, it was shown by the tests given above that the material:

1. Contained sulphur.
2. Was anion active.
3. Was quite acid, having an acid value of 116 mg KOH/g.

Infra-red evidence indicated that the spectrum was quite consistent with the emulsifying agent being a mixture of stearic acid and a sulphonated fatty ester.

The chemical examination indicated that the amount of stearic acid corresponded with a figure of 59% of the original emulsifying agent.

The infra-red spectrum of the mixture is given in Figure 3.12.

Other samples of rigid polyvinyl chloride have yielded to this treatment and amongst them have been samples that contained alkyl sulphonate salts.

DETERMINATION OF A KNOWN SURFACE ACTIVE AGENT, I.E. CATANAC S.N., IN A VINYL CHLORIDE/VINYL ACETATE COPOLYMER

Catanac S.N. is stearamide-propyl dimethyl-hydroxyethyl ammonium nitrate, i.e.

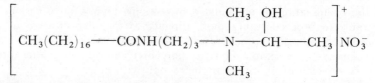

It is usually obtained in isopropanol solution and the first operation which is carried out is to evaporate the isopropanol solution to constant weight at 60°C under vacuum.

This yields dry Catanac S.N.

A solution of the titrating agent, dioctyl sulphosuccinate e.g. Manoxol O.T. is prepared as follows.

2·5 g of dioctyl sulphosuccinate are dissolved in distilled water and the solution diluted to 1 l.

In the test a known weight of the vinyl chloride/vinyl acetate copolymer (containing 0·025 to 0·04 g Catanac S.N.) is placed in a 250 ml conical flask and mixed with 10 ml of distilled water and 25 ml chloroform. 1 ml of 50% v/v sulphuric acid solution is added, then 1 ml of dimethyl yellow (0·01%) solution is added dropwise with continuous swirling of the flask. Standard dioctyl sulphosuccinate solution is run in slowly from a burette whilst the flask is swirled continuously.

At stages through the titration the flask is stoppered and shaken, care being taken not to shake the flask so vigorously that the contents form an emulsion. An emulsion can give rise to a false end-point by giving a red colour before the true end-point is reached. As the end-point is approached the contents of the flask appear orange or red during the shaking. From this point the dioctyl sulphosuccinate solution is added 0·1 ml at a time, stoppering and shaking the flask between additions. The end-point is indicated by the first appearance of a definite pink colour in the chloroform layer, 0·1 ml of the standard dioctyl sulphosuccinate solution is allowed for the development of the pink colour.

The dioctyl sulphosuccinate solution is standardised against the dry Catanac S.N. by the same procedure and, from the results of the two tests, the Catanac S.N. content of the copolymer is readily calculated.

INFRA-RED SPECTRA OF VINYL POLYMERS

SAMPLING

Many vinyl compositions, and particularly those based on polyvinyl chloride, contain plasticisers, fillers, lubricants or surface active agents. These additives have, on the whole, relatively strong infra-red absorption; in order to perform a useful infra-red examination of the resin these must be removed. As has been indicated, it is desirable first to extract with ether, followed (if polymeric plasticisers are expected) by extraction with hot methanol, or methanol-carbon tetrachloride mixtures. Ethylene dichloride is often used as a solvent for the preparation of thin films of vinyl resins for infra-red examination. If this solvent is not completely removed from the film, absorption bands occur at 11·3 μm and 14·1 μm. Similarly, samples prepared from tetrahydrofuran solution often show a spurious band at about 9·2 μm. This is frequently seen in published spectra of polyvinyl chloride.

POLYVINYL CHLORIDE

The spectrum of polyvinyl chloride (Spectrum 3.1) as encountered commercially has major absorption bands as listed in Table 3.3.

Table 3.3. Absorption bands in commercial polyvinyl chloride

Position	Group
3·4 μm	—CH$_2$— and —CH—
7·0 μm	—CH$_2$—
7·5 μm, 8·0 μm	—CH in —CHCl—
9·1 μm	—C—C—
10·4 μm	—CH$_2$—
14·5 μm	C—Cl

Apart from the rather broad absorption band at 14·5 μm attributed to C—Cl groups, no spectral features arise primarily from the presence of

177

chlorine in the molecule. This emphasises the necessity for conducting an elements test and in certain cases quantitative determination of elements, before considering the spectrum. Nevertheless the spectrum of polyvinyl chloride is quite characteristic, and there is no difficulty in distinguishing it from resins such as rubber hydrochloride and chlorinated rubber.

The spectrum of the syndiotactic species of this polymer (see discussion on stereoisomeric forms of polymers on p. 270) is reported by Hummel[22] No. 560; it is surprisingly different from that of the commercial (atactic) species.

POLYVINYL FLUORIDE

This relatively insoluble resin has, like most fluorinated materials, an intense infra-red spectrum (Spectra 7.4). This material is discussed in Chapter 7, together with other fluorinated hydrocarbon resins. Because prominent structure occurs in the 9 μm to 10 μm region, observation of the spectrum without a knowledge of the elements would suggest an aliphatic ether; this again shows the necessity for the determination of elements before considering the spectrum.

POLYVINYL ACETATE

This resin has a relatively simple spectrum (Spectrum 3.2) in which the majority of the strong bands are due to the acetate group.

A band near 5·75 μm is present in all esters, but a single C—O band occurs at 8·1 μm only in acetates and carbonates. The latter however do not exhibit the strong 9·8 μm band, and this serves to distinguish these two kinds of ester. Polyvinyl acetate has a spectrum rather similar to that of cellulose acetate (Spectrum 10.16), but these two materials can be distinguished by examination of the spectrum beyond 10 μm. This shows the importance of examining these resins in as clean a condition as possible, since the significant differences are largely obscured if additives such as plasticisers are present.

POLYVINYL ALCOHOL

Polyvinyl alcohol is occasionally encountered; the spectrum (Spectrum 3.3) shows clear evidence of the presence of hydroxyl groups from the relatively intense band near 3·0 μm (O—H stretching frequency). As polyvinyl alcohol is a secondary alcohol, the C—O frequency of the C—OH group occurs at 9·1 μm; this serves to distinguish the resin from cellulosic material and starch, the latter having a more complex C—O structure due to the presence of primary as well as secondary alcohol groups.

HIGHER ESTERS OF POLYVINYL ALCOHOL

Higher esters of polyvinyl alcohol are comparatively rare; the propionate, laurate, stearate and benzoate are sometimes encountered (Spectra 3.4, 3.5, 3.6 and 3.7). The complete spectrum of polyvinyl propionate in a pure state is quite characteristic but there is some difficulty in deducing the chemical nature of the material by examining the spectrum. This is clearly that of an ester, and the C—O structure includes an absorption band at 8·4 μm which is strong and distinct, and indicative of the propionate radical. The compound may be confused with an acrylate since these also show simple and strong C—O structure, although in acrylates the strongest C—O band lies between 8·5 μm and 8·6 μm rather than at 8·4 μm. To distinguish these compounds it is important to examine them in a sufficiently thin layer (0·0003 to 0·0004 in; 0·008 to 0·01 mm) to ensure that the wavelengths of the absorption maxima can be determined accurately.

In higher alkyl esters of polyvinyl alcohol the C—O structure remains substantially constant at about 8·6 μm, and the spectrum becomes less distinctive. It may well prove difficult to distinguish these higher vinyl ester polymers from polyacrylates and polyesters by their infra-red spectra, although the observations in Table 4.4 will be useful in some cases. The n.m.r. spectra are often more rewarding (see, for example, p. 204) and hydrolysis of the resin is a useful procedure to adopt in order to isolate an acid, alcohol or both. The monomeric products recovered may be:

1. A dibasic acid and a glycol. This will indicate that the unknown is a polyester.
2. An alcohol. This will suggest that the original is an acrylate, maleate or fumarate resin.
3. A monobasic acid. The original is then likely to be an ester of polyvinyl alcohol.

    It may be possible to identify the acid from its infra-red or n.m.r. spectrum. If this examination shows the unknown to be a linear carboxylic acid, it will be useful to convert the acid to its barium salt and examine the infra-red spectrum of this product. The barium salts of simple linear monobasic acids have a highly characteristic series of rather weak absorption bands between 8 μm and 9 μm[23] and identification will present no difficulty so long as only one acid is involved. If the infra-red or n.m.r. spectrum of the acid suggests that it is a mixture it will probably be better in this event to prepare the methyl esters of these mixed acids and to subject the mixture to gas liquid chromatographic separation and then attempt

179

the identification of the separated esters or the corresponding acids. It is possible that polyvinyl alcohol may be recovered in a sufficiently pure state for identification (Spectrum 3.3), but this is not always the case. The final identification will then depend upon a careful comparison of the spectrum of the unknown with that of the polyvinyl ester corresponding to the acid identified.

In work of this nature, the necessity for obtaining a pure specimen of the resin cannot be over-emphasised, since the isolation of hydrolysis products corresponding to impurities present in the resin will give false clues which are very troublesome and time consuming.

The only common polyvinyl aryl ester is polyvinyl benzoate (Spectrum 3.7). Recognition of this resin is comparatively simple, as its aromatic structure is evident at 14·1 μm.

VINYL ETHER POLYMERS

There are two distinct classes of vinyl ether polymers. One type has a side chain ether group

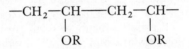

$$-CH_2-CH-\!\!\!-\!\!\!-CH_2-CH-$$
$$\qquad\quad |\qquad\qquad\quad\; |$$
$$\qquad\quad OR\qquad\qquad\;\; OR$$

while the other type is a cyclic ether

$$-CH_2-CH-CH_2-CH-$$
$$\qquad\quad |\qquad\qquad\quad |$$
$$\qquad\quad O-\!\!\!-CH-\!\!\!-O$$
$$\qquad\qquad\qquad |$$
$$\qquad\qquad\qquad R$$

Vinyl side-chain ether polymers

These compounds have simple spectra, characterised by an intense absorption band close to 9·0 μm wavelength, the rest of the structure being rather weak. In this respect, the infra-red spectra of vinyl side-chain ether polymers are similar to those of linear ether polymers such as polyethylene oxide and polypropylene oxide (Spectra 10.1 and 10.2) and the cellulose ethers (Spectra 10.19 and 10.20). Differentiation will depend upon examination of the resin as a thin layer ($\sim$0·0002 in; $\sim$0·005 mm) to observe the contour of the 9 μm band, and as a thick layer ($\sim$0·001 in; $\sim$0·03 mm) to examine the weaker structure. Further, the infra-red spectra of the common polyvinyl ethers, methyl, ethyl, *n*-butyl and isobutyl

(Spectra 3.8, 3.9, 3.10 and 3.11), are extremely similar, and are distinguishable mainly by the relative intensity and differing shape of the —CH$_2$— and —CH$_3$ group absorption bands between 6·8 μm and 7·5 μm. Minor differences in the shape of the ether band near 9 μm will assist identification. The n.m.r. spectra of these polymers show more pronounced differences.

Vinyl cyclic ether polymers

Cyclic ether polymers normally show a more complex C—O structure than the linear ether polymers just described. Their spectra are characterised by two or more rather broad bands of high intensity in the 8 μm to 11 μm region; as stated in Chapter 2, in compounds not containing elements other than C, H and O, this can be taken as indicative of an ether structure.

The common vinyl cyclic ether polymers are polyvinyl formal, acetal and butyral (Spectra 3.12, 3.13 and 3.14). In each case a weak ester carbonyl band is evident; this, together with the band near 8·0 μm, indicates residual polyvinyl acetate. These products are made by the hydrolysis of polyvinyl acetate to polyvinyl alcohol, and subsequent reaction with the appropriate aldehyde; it is reasonable therefore to expect minor and variable amounts of acetate and alcohol in the product.

The spectra of these materials are different from those of most other C, H and O containing resins, although there is some risk of confusion with cellulose ethers and esters (see Chapter 10).

OTHER STRAIGHT VINYL RESINS

The spectrum of polyvinyl cyclohexane (Nyquist[24]; Figure 109) is extremely uninteresting; the only band of consequence is a weak doublet at about 11·3 μm. The spectrum of poly-n-vinyl carbazole on the other hand (Nyquist; Figure 108) shows a useful pair of strong bands at about 13·4 μm and 13·9 μm and other aromatic structure.

In the spectrum of polyacrylonitrile (Spectrum 3.16) the 4·4 μm nitrile band is a very prominent feature.

POLYVINYLIDENE CHLORIDE

As with other chlorinated hydrocarbons, the spectrum of this substance (Spectrum 3.15) is quite characteristic, but does not directly reveal the presence of the CCl$_2$ group. The prominent and distinguishing features are the sharp doublet at 9·35 μm and 9·55 μm and the —CH$_2$—

band which by the influence of the adjacent chlorine atoms occurs at 7·15 μm.

Straight polyvinylidene chloride resin is partially crystalline at room temperature, and the doubling of the 9·4 μm band and the sharp bands at 11·3 μm and 13·3 μm arise from the crystalline material present.

Some variation in the intensity of these last two bands may be anticipated owing to variations in the sample preparation method which lead to varying crystallinity of the product. This will be accompanied by a change in shape of the 9·4 μm doublet, which, with decreasing crystalline content, gradually changes into a single broad band.

COPOLYMERS OF VINYL CHLORIDE

Copolymers of vinyl chloride are comparatively common, but it must not be assumed that absorption bands other than those of polyvinyl chloride found in the spectrum of a resin recovered from a polyvinyl chloride composition are necessarily indicative of the presence of a vinyl chloride copolymer.

The spectra of the common copolymers of vinyl chloride are described later in this chapter. If the unknown does not appear to match one of these exactly with regard to the wavelength of the absorption maxima it is possible that the resin is a mixture containing polyvinyl chloride, and in our experience the following possibilities should be considered before it is concluded that the unknown is a novel copolymer.

1. Weak ester bands at 5·8 μm

These often arise from a small residue of monomeric plasticiser. When such bands are found it is wise to submit the sample to a further solution and precipitation procedure and examine the resin again.

2. Substantial amounts of ester present

This may well indicate residual polymeric plasticiser, and the hot methanol extraction method (p. 144) should be followed. Both the methanol extracted material and the extracted polymer may then be examined. If the ester present is a polymeric plasticiser such as polyethylene adipate or sebacate it should be possible to identify this from the methanol extract. In some cases hot methanol treatment completely removes such plasticisers, but this is not always so and, where ester is still present after this treatment, the spectrum of the resin should be carefully examined to ascertain whether the absorption bands present in addition to those of the polyvinyl chloride are due to residual polyester plasticiser. If they do not coincide it is possible that the resin is a copolymer containing polyester plasticiser as an impurity.

Substantial amounts of ester may also arise from true resin additives, particularly methacrylate resins. The common mixed resins containing polyvinyl chloride are considered later in this chapter.

3. Foreign absorption bands other than those due to ester

A number of common fillers, such as carbonates and sulphates, have absorption spectra which appear remarkably like those of organic constituents. These fillers may be present in the composition in such a finely divided state that small amounts are contained in the resin after the precipitation procedure. If there is difficulty in identifying constituents in a resin, it is worth while obtaining the infra-red spectrum of any filler which has been removed during the precipitation process. Comparison of this spectrum with that of the resin may account for the additional absorption bands. The fillers which we have most frequently encountered in precipitated resins are calcium carbonate and china clay. (Spectra of common fillers are given in Reference 25.)

Additional absorption bands may also arise from residual solvents or decomposition products of solvents. In the latter case we have found tetrahydrofuran particularly troublesome, giving rise to a spurious band near 9·2 μm.

Polyvinyl chloride is occasionally blended with other resins, particularly rubbers (butadiene/acrylonitrile or ABS) and epoxy resins. These blends are considered later in this chapter.

VINYL CHLORIDE/VINYL ACETATE COPOLYMERS

The spectra of these copolymers are straightforward, and the major bands of both polyvinyl chloride and polyvinyl acetate are evident in the copolymer spectrum. The majority of commercial copolymers fall in the range up to 20% combined vinyl acetate; a selection of such copolymers is given in Spectra 3.17 to 3.20. Blends of polyvinyl chloride and polyvinyl acetate may be distinguished from copolymers by observing the structure near 10·5 μm (Spectrum 3.21).

At low acetate contents, in addition to the carbonyl absorption band at 5·8 μm, additional absorption is apparent at about 8·25 μm and 9·75 μm. These three bands, appearing simultaneously, are a very reliable indication of acetate when approximately 1% is present. Some other copolymerised constituents may give rise to structure resembling some, but not all of, these features. There is the possibility of confusion with a vinyl chloride/vinylidene chloride copolymer (producing additional absorption at 8·25 μm and 9·75 μm) which also contains a small amount of ester (producing absorption at 5·8 μm), but in such a case the ester absorption intensity will usually be out of proportion to the other two bands. If this is suspected, it will be useful to apply the morpholine test for the presence of polymerised vinylidene chloride (see p. 100), and, if this is positive, attempt to remove the ester, either by reprecipitation or solvent extraction of the resin, and determine the spectrum again on the purified sample.

With higher acetate contents, the ester carbonyl intensity increases, while the 8 μm band broadens and centres on 8·1 μm. The 9·75 μm band becomes prominent, while the methyl group band is seen at 7·3 μm.

The absorption bands of the acetate group are intrinsically much more intense than those of polyvinyl chloride. Hence at acetate concentrations above 50% there is difficulty in establishing the presence of copolymerised vinyl chloride, since it appears qualitatively in the spectrum only as the rather broad and weak C—Cl band at about 14·5 μm. It is therefore important that a test for chlorine be performed if there is reason to suspect that vinyl chloride may be present. However copolymers of this composition are rarely encountered as commercial samples.

Quantitative determination of free and combined vinyl acetate in vinyl chloride/vinyl acetate copolymer

Using solvent cast films of approx. 0·001 in (0·03 mm) thickness, the relative intensities of the three bands at 7·0 μm, 7·3 μm and 7·5 μm form a useful semi-quantitative test for acetate content, since the band intensity at 7·3 μm can be seen to increase relative to the intensities at 7·0 μm and 7·5 μm with increasing acetate content. The unknown sample can be matched against Spectra 3.17 to 3.20 and the acetate content judged to about $\pm 2\%$ on the composition.

A quantitative method for acetate content based on the measurement of the absorbance ratio 5·8 μm/7·0 μm (carbonyl to —CH$_2$— ratio) has been proposed by Weinberger and Kagarise[26] but we have no practical experience of this method.

The solution method described on p. 197 for quantitative determination of combined methyl acrylate in vinylidene chloride/methyl acrylate copolymer may also be applied to vinyl chloride/vinyl acetate copolymer, but a more convenient method is based upon the measurement of an absorption band due to the ester carbonyl group at about 2·15 μm.

This is a general method which may be applied, with appropriate calibration, to other ester copolymers of vinyl chloride and is furthermore a most useful method for the analysis of blends of polyvinyl chloride with ester polymers, such as polymethyl methacrylate.

For copolymers containing up to 20% acetate, a sample of about 0·4 in (10 mm) thickness gives a band of suitable intensity for measurement (Figure 3.13).

The resin is prepared by hot pressing to produce a sheet of $\sim\frac{1}{8}$ in (3 mm) thickness. Three pieces cut from this sheet are clamped together to form the test specimen.

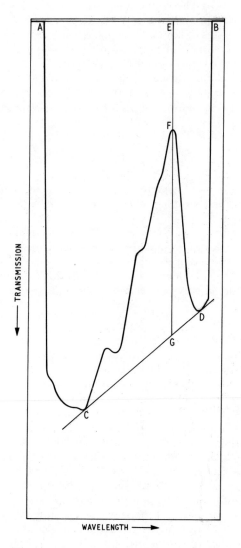

Fig. 3.13. Absorption band of acetate group at 2·15 μm

Calibration depends upon the assumption that the copolymer may be considered as a laminate of sheets of polyvinyl chloride and polyvinyl acetate. Thus a sheet of straight (ester free) polyvinyl chloride of about $\frac{1}{8}$ in (3 mm) thickness is prepared, and a number of sheets of straight polyvinyl acetate each of a different known thickness between 0·01 and 0·08 in (0·25 and 2 mm). A series of laminates is made, each with three layers of polyvinyl chloride but with a different thickness of polyvinyl acetate, and in this way a set of calibration samples is produced.

The spectra of these calibration samples are obtained in the vicinity of the 2·15 μm band, to give spectra similar to Figure 3.13.

Measurement of the absorbance of each calibration sample at 2·15 μm is made by drawing the straight line background CD under the band and calculating

$$\log_{10}\frac{EG}{EF}$$

The absorbance so measured is plotted against the thickness of polyvinyl acetate in the calibration samples. The graph should be linear and pass through the zero point.

This calibration curve may then be used to estimate the equivalent thickness of polyvinyl acetate in unknown test specimens.

If the thickness of the test specimen be T,

and the determined equivalent thickness of polyvinyl acetate is t,

then the w/w% combined vinyl acetate in the unknown is

$$\frac{1·20t \times 100}{1·40(T-t)+1·20t}\%$$

This assumes the densities of combined vinyl acetate and combined vinyl chloride are 1·20 and 1·40 g/cm^3 respectively.

It is also assumed that all the vinyl acetate in the specimen is combined and in many cases some free vinyl acetate (residual monomer) is present. Since the 2·15 μm band is due to the acetate group, both free and combined acetate absorb at this wavelength. However, vinyl acetate monomer may be conveniently observed by measuring the spectrum of a six-layer test specimen ($\sim$0·8 in or 20 mm thickness) at 1·63 μm (Figure 3.14). The band at 1·63 μm is due to the vinyl double bond ($-CH=CH_2$) and measurement of the absorbance at this wavelength will enable the monomer content of the test specimen to be determined.

For calibration a sample of copolymer containing no monomer is required. This may be prepared by the usual solution and precipitation method (see p. 147) after which the precipitated resin is dried in vacuo to constant weight at 100°C. The resulting resin is pressed into a $\frac{1}{8}$ in (3 mm) sheet, and six layers of this sheet are clamped together to form a calibration sample.

The calibration sample is placed in the spectrometer beam together with a glass absorption cell of known thickness (about 0·5 in or 10 mm path length) filled with carbon tetrachloride. The spectrum of this combination between 1·6 μm and 1·7 μm should reveal no absorption maximum at 1·63 μm if the calibration sample is indeed free of monomer, although a very weak monomer band (∼0·005 absorbance) may be tolerated. A series of solutions of vinyl acetate monomer in carbon tetrachloride is prepared,

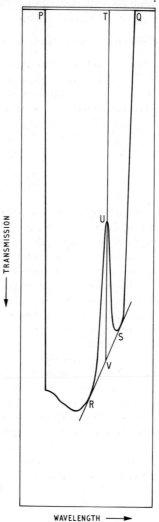

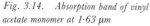

Fig. 3.14. *Absorption band of vinyl acetate monomer at 1·63 μm*

the concentrations being chosen to be equivalent to concentrations in the 0 to 4% range in the sheet.

The spectrum of the combination of the monomer-free copolymer specimen with each monomer solution in turn is recorded in the range 1·6 μm to 1·7 μm. The record should be similar to Figure 3.14. The

187

absorbance for each combination sample is measured from the record by drawing the straight line RS and calculating

$$\log_{10}\frac{TV}{TU}$$

This absorbance is plotted against the effective thickness of vinyl acetate monomer in the glass cell (If an $x\%$ v/v solution is placed in a cell X thick, the effective thickness of monomer is $\frac{x\ X}{100}$). The calibration graph should be linear, and pass through the zero point.

The effective thickness of vinyl acetate monomer in an unknown test specimen may then be determined by measuring its absorbance at $1\cdot63\ \mu m$ by the method described above and referring to the calibration graph.

If the thickness of the unknown specimen be M,

and the effective thickness of monomer be m,

then the w/w% monomer is given by

$$\frac{0\cdot94\ m \times 100}{1\cdot40\ M}\ \%$$

The densities of vinyl acetate monomer and the copolymer are assumed to be $0\cdot94$ and $1\cdot40$ g/cm^3 respectively.

It now remains to correct the absorbance measured at $2\cdot15\ \mu m$ for the monomer content of the test specimen. This may be done by making a three-layer laminate from the monomer-free copolymer used in the above monomer calibration, and placing this in the beam together with the glass absorption cell. The spectrum in the vicinity of $2\cdot15\ \mu m$ is first measured with the glass cell filled with carbon tetrachloride, and then with each of the monomer solutions in turn which were prepared for the monomer calibration.

By measuring the absorbance at $2\cdot15\ \mu m$ for each combination, the relationship between increase in absorbance at $2\cdot15\ \mu m$ and thickness of monomer may be determined. Since the equivalent thickness of monomer in an unknown test specimen may be determined by measurement at $1\cdot63\ \mu m$, the necessary correction to the $2\cdot15\ \mu m$ absorbance may be made before calculating the combined vinyl acetate content from this absorbance.

By this test, which in our experience is extremely good, the free and combined vinyl acetate content of a copolymer sample may be determined very rapidly.

TERPOLYMERS CONTAINING VINYL CHLORIDE AND VINYL ACETATE

Occasionally terpolymers of vinyl chloride and vinyl acetate with small amounts of carboxylic acid are encountered, for example Spectrum 3.22

shows a minor amount of copolymerised maleic acid. The acid carbonyl band appears as a very weak shoulder at 5·9 μm on the side of the strong ester band; rather better evidence of carboxylic acid is the appearance of a broad weak band underlying the 3·4 μm C—H band. This is due to the acid hydroxyl group. Such compositions often appear to contain some of the acid in the anhydride form, causing the shoulder near 5·6 μm. Although in these resins the spectrum gives evidence of the presence of carboxylic acid it is doubtful whether different copolymerised acids could be distinguished from the spectrum of the total resin.

The spectrum of a vinyl chloride/vinyl acetate/vinyl alcohol composition has been published[25]. The presence of secondary alcohol is evident from the O—H absorption band at 2·9 μm, and the enhanced intensity of the 9 μm band due to the presence of the C—O band of the secondary alcohol group.

OTHER ESTER COPOLYMERS OF VINYL CHLORIDE

Copolymers of vinyl chloride with acrylates may be encountered, the most common being those with methyl and ethyl acrylate (Spectra 3.23 and 3.25). Both these copolymers produce an ester carbonyl band at 5·8 μm, but the C—O structure near 8·5 μm is rather different in the two cases, and ethyl acrylate, even at low concentrations, shows a band at 9·75 μm not present in the methyl acrylate system.

Unfortunately the structure of the fumarate (maleate) copolymers is extremely similar to that of the acrylate copolymers and the infra-red spectrum does not enable the methyl acrylate copolymer to be distinguished from the dimethyl fumarate (maleate) copolymer (Spectrum 3.24) or correspondingly, the diethyl fumarate (maleate) copolymer (Spectrum 3.26) from that with ethyl acrylate. The apparent differences in the spectra reproduced here arise from the differing copolymer proportions, and examination of the spectra will show that the bands which appear in addition to those of polyvinyl chloride are substantially the same with either an acrylate or a fumarate (maleate). It has been reported[27] that pyrolysis/gas chromatography is a useful method of differentiating between those polymers containing acrylates and those containing fumarates (maleates). With copolymers of higher acrylates, such as 2-ethyl hexyl acrylate (Spectrum 3.27) evidence as to the nature of the acrylate is not easily obtained from the spectrum. In such cases it is better not to rely upon direct infra-red evidence; once the general nature of the resin has been ascertained by examining the infra-red spectrum, then it will be more useful to proceed to examination by n.m.r. or pyrolysis/gas chromatography.

Blends of polyvinyl chloride with polymethyl methacrylate (p. 202) have spectra which may be confused with those of vinyl chloride/acrylate or vinyl chloride/fumarate (maleate) copolymers. Fortunately the C—O

189

structure in the 8 μm to 9 μm region is somewhat different and serves as a distinguishing feature.

Quantitative analysis by infra-red spectroscopy of vinyl chloride/ester copolymers

The thick film (sheet) method for the analysis of vinyl chloride/vinyl acetate copolymers (p. 184) may be applied to the analysis of other ester copolymers of vinyl chloride, but in general it is desirable to calibrate with known copolymers rather than laminates of the appropriate homopolymers. Unfortunately the 2·15 μm ester band is broader and less well defined in ester copolymers other than acetates, hence the test is less satisfactory. Generally we prefer to apply the solution test described for ester copolymers of vinylidene chloride (see p. 197) which is very satisfactory provided the copolymer is soluble in tetrahydrofuran. Free ester monomer in the copolymer will interfere, and must first be removed by solvent extraction.

VINYL CHLORIDE/VINYL ETHER COPOLYMERS

The presence of aliphatic ether in vinyl chloride resins is evident from the presence of the relatively strong ether band near 9·0 μm. This is rather similar to a secondary alcohol absorption band, but the latter would be accompanied by a moderately intense O—H absorption band near 3 μm. In a vinyl chloride/vinyl isobutyl ether copolymer containing $\sim 5\%$ of the ether constituent (Spectrum 3.28) there is little indication of the minor component other than an intensification of the polyvinyl chloride band near 9·0 μm. At the 13% level of vinyl isobutyl ether (Spectrum 3.29) the rather broad ether band near 9·0 μm is now very prominent, and there is a doublet band due to —CH(CH$_3$)$_2$ at 7·3 μm. Even this however is not conclusive proof of the presence of the isobutyl group, and in general it is extremely difficult to ascertain the nature of ether constituents copolymerised with vinyl chloride from the infra-red spectrum. The n.m.r. spectrum is usually more useful (see p. 209).

Quantitative determination of combined vinyl isobutyl ether in vinyl chloride/ vinyl isobutyl ether copolymers

This is a general method which may be applied, with appropriate calibration, to the determination of any vinyl ether copolymerised with vinyl chloride, in amounts up to about 20%.

It relies upon the measurement of the absorption band at 11·9 μm for vinyl chloride and that at 9·1 μm for ether. The ether absorption has to be corrected for the overlapping vinyl chloride absorption.

A series of three or four standard copolymers are required for calibration. These may be small scale preparations with polymerisation taken
to completion, or samples calibrated from their n.m.r. spectra. For
approximate results, it may be adequate to use Spectra 3·28 and 3·29
together with that of straight polyvinyl chloride (Spectrum 3.1) as the
spectra of known compounds.

All standards, including a sample of pure polyvinyl chloride, and all
unknown samples are prepared as films approximately 0·002 in (0·05 mm)
thick. The exact thickness is not measured and is not critical, as this is an

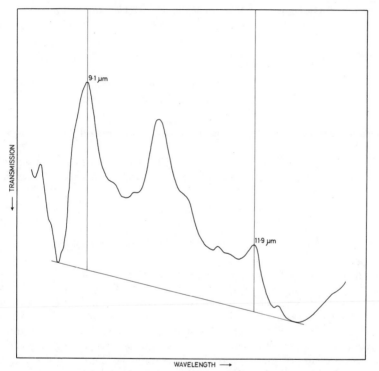

*Fig. 3.15. Determination of vinyl isobutyl ether in copolymer with vinyl chloride. Sample
contains 10% vinyl isobutyl ether*

absorbance ratio method. The films are pressed between polytetrafluoroethylene-coated plates (see p. 24).

The spectra of all samples are recorded over the range 7·5 μm to 13·0 μm
on a double beam spectrometer. The spectra of the samples will be similar
to Figure 3.15.

Straight line backgrounds are drawn as shown in the figure. The
absorbances are measured at the maxima at approximately 9·1 μm and
11·9 μm in the copolymer spectra. All absorbances, including those of the

straight polyvinyl chloride spectrum, must be measured at exactly the same wavelengths.

The absorbance ratio

$$\frac{\text{Absorbance at } 9 \cdot 1 \ \mu m}{\text{Absorbance at } 11 \cdot 9 \ \mu m} = R$$

for the polyvinyl chloride sample is first calculated. Then, for all other samples, the 'corrected' absorbance ratio

$$\frac{\text{Absorbance at } 9 \cdot 1 \ \mu m - R \times \text{Absorbance at } 11 \cdot 9 \ \mu m}{\text{Absorbance at } 11 \cdot 9 \ \mu m}$$

is calculated, and plotted against

$$\frac{\text{Weight ether}}{\text{Weight vinyl chloride}}$$

for each standard sample. The plot should be linear, and pass through the origin.

The composition of the unknown samples may now be determined from this calibration graph. The films from the standard samples are retained for future analyses.

VINYL CHLORIDE/VINYLIDENE CHLORIDE COPOLYMERS

The spectra of these copolymers (Spectra 3.30 to 3.34) are of some interest, because the copolymer spectrum is significantly different from that of the corresponding polymer mixture. In the case of copolymers containing vinylidene chloride as a minor constituent (Spectra 3.30 to 3.32) the copolymerised vinylidene chloride is non-crystalline, and the prominent polyvinylidene chloride band near $9 \cdot 4 \ \mu m$ is consequently broad. A change in shape of the $8 \cdot 1 \ \mu m$ polyvinyl chloride band is also seen; this may result from the changing environment of the —CHCl— groups due to adjacent CCl_2 units.

Resins containing 20 to 40% combined vinylidene chloride have spectra with rather broad bands, nevertheless these are quite satisfactory for qualitative purposes. With an increase in the combined vinylidene chloride content, the bands of crystalline vinylidene chloride, particularly the doublet at $9 \cdot 4 \ \mu m$, become prominent, and indeed, with more than 60% vinylidene chloride, vinyl chloride is difficult to detect. When the spectrum is similar to that of polyvinylidene chloride, and shows no sign of the presence of acrylonitrile or ester constituents, the spectral region $7 \ \mu m$ to $8 \cdot 5 \ \mu m$ must be examined. The bands near $7 \cdot 6 \ \mu m$ and $8 \cdot 3 \ \mu m$

arise from copolymerised vinyl chloride, although polyvinylidene chloride itself (Spectrum 3.15) appears to have weak bands in these positions.

VINYL CHLORIDE/ETHYLENE COPOLYMERS

Copolymers of vinyl chloride with ethylene are of some interest. At low ethylene levels (Spectrum 3.35) in addition to an extra —CH$_2$— band near 6·85 μm (on the edge of the polyvinyl chloride —CH$_2$— band) a new absorption band appears at 13·4 μm. This corresponds to the presence of the group —CH$_2$—CH$_2$—CH$_2$—, indicating that ethylene occurs only as isolated units. At a higher ethylene content (Spectrum 3.36) the 6·85 μm band is more pronounced, and the 13·4 μm band is accompanied by further absorption at 13·8 μm, showing the presence of —(CH$_2$)$_5$—, —(CH$_2$)$_7$— and so on, due to pairs and longer runs of ethylene units in the copolymer.

The spectra of these copolymers are similar to those of blends of polyvinyl chloride with chlorinated polythene (Spectrum 3.52) and in case of doubt, application of the solvent separation method suggested on p. 156 may be helpful.

Quantitative determination of ethylene content of vinyl chloride/ethylene copolymers

This method is applicable only to copolymers in which ethylene is present as isolated monomer units. This in our experience is the case with random copolymers containing up to 10% ethylene. The presence of adjacent ethylene units is revealed by the appearance of an additional band at 13·8 μm, and when this structure is present in the spectrum the method is not satisfactory. (Spectra 3.35 and 3.36.)

In this method the sample is used in the form of a thin film, hence it can be applied whether or not the polymer is soluble. Further, the method is not influenced by the presence of residues from the process provided these do not absorb at the wavelengths used in the analysis.

The composition of a number of samples is determined by n.m.r. measurements; these are used as calibration standards.

The standard samples, and the unknowns, are pressed to approx. 0·002 in (0·05 mm) thickness at 140°C between polytetrafluoroethylene-coated stainless steel plates (see p. 24). The spectra of all samples are measured on a double beam spectrometer over the range 8 μm to 14 μm. The reference beam attenuation is adjusted so that the maximum transmission point on the run is about 95%. The run will be similar to Figure 3.16, which is the spectrum recorded on a sample containing 3·5% w/w copolymerised ethylene.

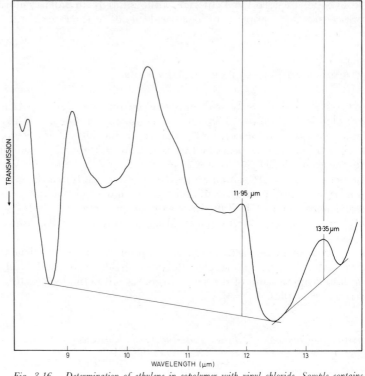

Fig. 3.16.   *Determination of ethylene in copolymer with vinyl chloride. Sample contains 3·5% ethylene*

Backgrounds are drawn as shown in the figure, and the absorbance measured at the maxima at approx. 12·0 μm (vinyl chloride) and 13·4 μm (ethylene).

From the series of standard samples

$$\frac{\text{Absorbance at 13·4 } \mu\text{m}}{\text{Absorbance at 12·0 } \mu\text{m}}$$

is plotted versus the concentration ratio

$$\frac{\text{Ethylene}}{\text{Vinyl chloride}}$$

The plot should be linear and pass through the origin. This calibration graph is used to determine the composition of the unknown samples.

Provided that the calibration curve is linear with no zero intercept, it is more convenient and more precise when the analysis is repeated to run one known standard sample on each occasion that a series of unknowns

194

is examined. The sample is retained in the form of a film, mounted ready to insert in the spectrometer.

Let the absorbance ratio of the standard be A_s,
and the concentration ratio of the standard be C_s.
If the absorbance ratio measured on an unknown is A_u
Then the concentration ratio of the unknown, C_u, is

$$\frac{A_u}{A_s} \times C_s$$

The w/w % combined ethylene in the unknown is

$$\frac{C_u}{1 + C_u} 100\%$$

OTHER VINYL CHLORIDE/OLEFINE COPOLYMERS

Vinyl chloride/propylene copolymer (Spectrum 3.37) is rather un-interesting because at low propylene levels the 6·85 μm band and the methyl band at 7·25 μm are the only indications of the presence of propylene. These features are seen also in the copolymers of vinyl chloride with higher α-olefins, but in addition there will usually be some extra structure in the spectrum between 12 μm and 14 μm indicative of runs of $-CH_2-$ groups.

COPOLYMERS OF VINYL ACETATE

Partially hydrolysed polyvinyl acetate (so called acetate/alcohol copoly-mer) has a spectrum containing the absorption bands of both these resins (Spectrum 3.38). There is some difficulty in distinguishing the spectrum of this resin from that of partially acetylated cellulose (Spectrum 10.15) but the alcoholic C—O structure (9 μm to 10 μm) in particular is different since vinyl alcohol is a secondary alcohol, and some residual primary and secondary alcohol groups are present in cellulose acetate.

Spectrum 3.39 is that of vinyl acetate/2-ethyl hexyl acrylate copolymer and the spectra of a number of copolymers of vinyl acetate with aliphatic esters appear in Reference 25. In each case a weak additional $-CH_2-$ band is evident at 6·8 μm and a C—O band near 8·5 μm. While this additional structure gives evidence of a second ester constituent, it would be insufficient, of itself, to offer a means of identifying the copolymerised ester.

The spectrum of vinyl acetate/vinyl stearate copolymer appears in the *Sadtler Commercial Spectra Series* D^{28} (No. D4158) but this spectrum shows remarkably little evidence of the presence of stearate.

195

Vinyl acetate/ethylene copolymers are of considerable interest. Copolymers with a major proportion of ethylene are considered in Chapter 6. Spectrum 3.40 is that of a copolymer containing 15% ethylene. Comparison with the spectrum of polyvinyl acetate shows enhanced absorption at 3·4 μm and a band at 6·8 μm. These bands originate from —CH$_2$— groups in alkyl chains, and hence these features would be expected in the infra-red spectra of other olefine copolymers of vinyl acetate. Quantitative analysis of these copolymers (of low ethylene content) is straightforward, based on the bands at 3·4 μm and 5·8 μm. Absorption of vinyl acetate at 3·4 μm is allowed for by correcting the absorbance as in the vinyl ether/vinyl chloride copolymer quantitative method (see p. 190).

COPOLYMERS OF VINYLIDENE CHLORIDE

COPOLYMERS WITH ACRYLONITRILE

Vinylidene chloride/acrylonitrile copolymer is easy to characterise from the infra-red spectrum: the absorption band due to copolymerised acrylonitrile at 4·4 μm is clearly visible even at low concentrations of acrylonitrile (Spectra 3.41 (a) and 3.41 (b)) while the remainder of the spectrum is essentially that of polyvinylidene chloride.

As with straight polyvinylidene chloride, the copolymers of this resin containing only minor amounts of a second component can vary considerably in crystallinity, and this may tend to cause confusion in identification. This effect 's demonstrated in Spectra 3.41(a) and 3.41(b), which are of the same copolymer, but in 3.41(a) the combined vinylidene chloride is essentially amorphous, while in 3.41(b) it is as crystalline as can be achieved.

Generally, vinylidene chloride copolymers containing less than about 10% of a comonomer can be either crystalline or amorphous. Films prepared by solvent casting tend to be crystalline, while those prepared by drying an aqueous latex tend to be amorphous, but the latter are readily crystallised by heat treatment, or even by standing at room temperature, when crystallinity develops after a few days.

The process of crystallisation can be followed by measuring the absorbance of the 16·7 μm band of crystalline vinylidene chloride, on films of known thickness. A similar test may be conducted on vinylidene chloride copolymer coatings on packaging films, but in this case the measurement is made with the KRS-5 prism or germanium multiple internal reflection element (see p. 29) and the contour of the 9·5 μm band is observed. This changes with increasing crystallinity from a broad single band in the amorphous resin to a well defined doublet, as seen in Spectra 3.41(a) and 3.41(b) respectively, and the same effect may be seen in the A.T.R. or multiple internal reflection measurement.

196

At higher combined acrylonitrile contents, the copolymer is likely to be largely amorphous. Thus the spectrum of a 20% acrylonitrile copolymer (Reference 25, Figure 294) shows that the sample is of low crystallinity. The degree of crystallinity which can be developed at a particular level of combined acrylonitrile is likely to depend, however, upon the amount of 'blocking' in the copolymer, more random copolymers having relatively lower crystallinity.

These copolymers are sometimes 'carboxylated' by the introduction of an acid monomer such as maleic, acrylic or methacrylic acid. Evidence for carboxyl groups is readily obtained from the presence of a carbonyl band at 5·9 μm, but we have no experience of differentiating different acidic components from the infra-red spectrum.

ESTER COPOLYMERS OF VINYLIDENE CHLORIDE

An important group of copolymers, used particularly as coating media, are the ester copolymers of vinylidene chloride. Here again care is required as variation in crystallinity of the combined vinylidene chloride produces changes in the spectrum but, since these are of the same nature as those observed in the vinylidene chloride/acrylonitrile copolymer, we have not included spectra of both crystalline and amorphous materials.

The ester copolymers commonly encountered are those with methyl, ethyl, butyl, and 2-ethyl hexyl acrylate (Spectra 3.42 to 3.45).

It will be observed that the C—O structure in the 8 μm to 9 μm region is considerably modified in these copolymers from its appearance in the straight acrylate resins (Spectra 4.1 to 4.3). If a resin is encountered which appears to be a vinylidene chloride/ester copolymer, it will not usually be possible to ascertain from the spectrum the nature of the ester constituent unless a known copolymer spectrum is available for comparison, and even then care is necessary, as the differences between the infra-red spectra of the higher acrylate copolymers are small and may be confused with effects due to change of crystallinity or degree of 'blocking' of the copolymer. Evidence from the n.m.r. spectrum, or pyrolysis/gas chromatography, will be particularly valuable.

Copolymers of vinylidene chloride with acrylonitrile and an ester constituent are also encountered e.g. the terpolymer with methyl acrylate. Caution is necessary in interpreting their spectra, especially as the materials may also be carboxylated (Spectrum 3.46).

Determination by infra-red spectroscopy of the combined ester content of vinylidene chloride/ester copolymers

This method is written for methyl acrylate/vinylidene chloride copolymer, but is applicable to other ester copolymers of vinylidene chloride. It is

197

particularly appropriate to these systems because it is conducted in solution, and therefore difficulties associated with crystallinity of the resin are overcome. However, the method may also be found convenient in other cases, particularly vinyl chloride/ester copolymers. Because esters in the monomeric state absorb at the wavelength used for the copolymerised ester, the sample must be free of ester monomer.

The band at approximately 5·8 μm is measured on a dilute solution of the polymer in tetrahydrofuran, in a cell of about 0·2 mm thickness. The cell thickness need not be precisely known, because a solvent band (5·092 μm) is measured for comparison. This also enables the absorption cell to be replaced with another of the same nominal thickness without recalibration.

One or more standard copolymers are required for calibration e.g. specially prepared copolymers or samples standardised by chlorine content or n.m.r. measurements.

Apparatus

Spectrometer. A double beam spectrometer is required having good resolution in the 5 μm to 6 μm region.
Absorption cell. Fixed path length in sodium chloride or potassium bromide, about 0·2 mm thick.
Mechanical shaker.

Solvent

Tetrahydrofuran.
Reflux 3 l of tetrahydrofuran over 100 g of caustic soda for 1 h. Distil, and reject the first 100 ml of distillate.

Procedure

Weigh out 0·5 g of the purified and dried copolymer into a 25 ml volumetric flask. The weight need only approximate to this figure, but must be known accurately. Add about 20 ml tetrahydrofuran, then place the flask on the mechanical shaker until dissolved (approx. 2 h). Fill up to the mark with solvent.

Fill the sample cell with solution and place in the sample beam of the spectrometer. Set the instrument to 4·9 μm and adjust the trimmer comb to bring the pen to 85% transmission. Record the spectrum between 4·9 μm and 6 μm at a speed of approx. 1 μm in 4 min. The recorder trace will be similar to Figure 3.17.

Draw the straight line background **XX'** to meet the curve tangentially at the absorption minima at about 4·9 μm and 5·35 μm and similarly **YY'** between about 5·6 μm and 5·9 μm. Draw lines through the maxima

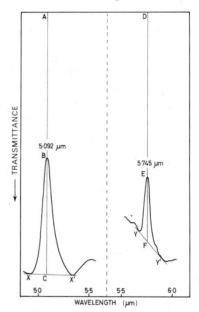

Fig. 3.17. Portions of infra-red spectrum of a vinylidene chloride/methyl acrylate copolymer solution in tetrahydrofuran, used for determining combined methyl acrylate content

B and E parallel to the absorption axis of the recorder paper. Following the lettering of Figure 3.17.

The tetrahydrofuran absorbance $= \log_{10}\dfrac{AC}{AB} = A_{THF}$

The methyl acrylate absorbance $= \log_{10}\dfrac{DF}{DE} = A_{MA}$

Calibration

Prepare solutions of one or more known copolymers of vinylidene chloride and methyl acrylate, to contain up to 5% w/v of methyl acrylate, related to the w/w % methyl acrylate in the copolymer as follows:

$$\% \text{ w/v methyl acrylate in solution} = \frac{Y \times Z}{25}$$

where Y = weight of copolymer in g
$\quad\quad Z$ = % w/w methyl acrylate in the copolymer
made up in 25 ml of tetrahydrofuran.

199

Examine these solutions by the procedure given above, then plot A_{MA}/A_{THF} against % w/v methyl acrylate.

Calculation

Calculate on all samples the absorbance ratio A_{MA}/A_{THF}, then read off the corresponding % w/v methyl acrylate from the calibration graph. From the % w/v methyl acrylate in the solution and the weight of copolymer taken,

$$\% \text{ w/w methyl acrylate} = \frac{25X}{Y}$$

where X = % w/v methyl acrylate in solution
Y = weight of copolymer taken in g
25 is the volume of the solution in ml

CHLORINATED POLYVINYL CHLORIDE

Chlorinated polyvinyl chloride is, structurally, not a well-defined product, and differences in the infra-red spectrum may be expected to occur owing either to differences in the chlorine content of the product, or to structural changes due to different manufacturing processes. Nevertheless, most commercial chlorinated polyvinyl chloride has an infra-red spectrum similar to Spectrum 3.47. Hummel[22] gives the infra-red spectra of a number of commercial samples, but many of the weak features in these spectra would seem to be due to impurities.

The major features of the spectrum are similar to those of polyvinyl chloride, but a pronounced shoulder is seen at about 7·8 μm, and the 8·3 μm band is intensified. These features are believed to arise from the changing environment of the —CHCl— group which is responsible for the 8·1 μm band of polyvinyl chloride. The increased absorption at 9·7 μm in the chlorinated polyvinyl chloride spectrum is probably analogous to the band at that wavelength in polyvinylidene chloride, and may arise from the group —CCl$_2$—. Additional bands in the 12 μm to 14 μm region are presumably C—Cl stretching bands.

BLENDS, BLOCK AND GRAFT COPOLYMERS OF POLYVINYL CHLORIDE WITH OTHER RESINS

It is now common practise to blend other resins with polyvinyl chloride to modify its physical properties. These blends, although physical mixtures, are often difficult to separate. Somewhat similar compositions (analytic-ally) are made by polymerising a second monomer in the presence of a polymer; this produces block or graft copolymers.

In the case of these mixtures or copolymers the spectrum is essentially that of a straight mixture of the two resins. Generally if the spectrum of the base resin can be recognised, consideration of the bands not thus accounted for will enable this second component to be identified, although minor constituents may well be missed in this simple examination.

Hence, if infra-red examination suggests that a material is a blend, or graft copolymer, further tests are desirable, including pyrolysis/gas chromatography (p. 154) and solvent separation (p. 156). The latter is usually followed by infra-red examination of the separated substances.

Blends of polyvinyl chloride with copolymers of butadiene, styrene and acrylonitrile

In the infra-red spectrum of a blend of polyvinyl chloride with a buta-diene/styrene/acrylonitrile copolymer (Spectrum 3.48), in addition to the bands due to polyvinyl chloride, those of copolymerised styrene are clearly seen at 13·2 μm and 14·3 μm. The nitrile band at 4·4 μm is sharp, but weak, and as the amount of acrylonitrile combined in these blends is often small, it may be necessary to examine the sample at increased thickness ($\sim$0·01 in or 0·3 mm pressing) to be certain that nitrile is present. If a polyvinyl chloride sample is found to contain styrene, this test should be applied if nitrile is not evident when the sample is run at the normal thickness of 0·001 to 0·002 in (0·025 to 0·050 mm), and a chemical test for nitrogen, by the oxygen flask method, should also be applied. The band of copolymerised butadiene, although prominent, is difficult to detect easily because it coincides in position with the strong polyvinyl chloride band at 10·3 μm. However a visual comparison of the intensity of the 10·3 μm band with the 7·0 μm polyvinyl chloride band will suffice in most cases to demonstrate the presence of butadiene.

Blends with butadiene/acrylonitrile copolymer are similar, but show no evidence of copolymerised styrene at 13·2 μm and 14·3 μm, while the 4·4 μm band is absent in blends of polyvinyl chloride with butadiene/styrene copolymer.

Blends of polyvinyl chloride with epoxy resin

The presence of epoxy resin in blends with polyvinyl chloride is some-times apparent from an examination of the ether extract, but in some cases the epoxy resin is so well cross-linked that little if any is removed by ether.

Even small amounts of epoxy resin in polyvinyl chloride are revealed by the presence of a sharp band at 6·65 μm (Spectrum 3.49). This appears to be an exceedingly sensitive test for epoxy resin, and if the band is found, a chemical test for epoxy resin should be performed. With higher concentrations of epoxy resin, the prominent bands at 9·7 μm and 12 μm may also be visible (Spectrum 10.9).

Blends of polyvinyl chloride with polymethyl methacrylate

Blends of polyvinyl chloride with polymethyl methacrylate (Spectra 3.50 and 3.51) (occasionally methyl methacrylate/acrylate copolymer) are likely to be confused with methyl acrylate and dimethyl fumarate copolymers of vinyl chloride (Spectra 3.23 and 3.24). However, the sharp C—O bands of polymethyl methacrylate at 8·5 μm and 8·7 μm (Spectrum 4.6(a)) lie at greater wavelengths than in the copolymers mentioned and the bands of polymethyl methacrylate at 11·9 μm and 13·3 μm are often visible.

There are so many possible combinations of polyvinyl chloride with these so-called 'Acrylic Modifiers' that the infra-red spectrum of the total composition is rarely conclusive, and complete identification will require solvent separation.

Blends of polyvinyl chloride with chlorinated polythene

The presence of chlorinated polythene in these blends is revealed by the appearance of a normal hydrocarbon type —CH$_2$— deformation band at 6·85 μm (Spectrum 3.52) often occurring as a shoulder on the shorter wavelength side of the strong 7 μm band of polyvinyl chloride. Unfortunately this feature is not conclusive evidence of chlorinated polythene, since it is present, for example, in vinyl chloride/olefine copolymers (see p. 195) and many possible additives contain normal hydrocarbon type —CH$_2$— groups. However if this additional feature is seen, without further evidence from the spectrum of other functional groups, the presence of chlorinated polythene may be suspected, and the solvent separation tests suggested on p. 156 should be carried out.

QUANTITATIVE ANALYSIS OF BLENDS OF POLYVINYL
CHLORIDE WITH OTHER POLYMERS BY INFRA-RED
SPECTROSCOPY

In principle, the determination of the proportions of a blend of two or more known resins is a simple infra-red analysis of a two or more component mixture, and depends for its success only upon the correct choice of suitable absorption bands for each constituent. Calibration samples are prepared by mill blending the known constituents together.

The major problem is inhomogeneity of the samples or phase separation of the constituents of the standards or samples, and for this reason it is preferable to work with thick samples whenever possible. Thus, for example, in the analysis of blends with acrylic modifiers, the thick film (sheet) method for vinyl chloride/vinyl acetate copolymers may be followed, but it is essential to calibrate with milled blends of the constituents; laminates of the constituent polymers are not satisfactory for calibration.

We have analysed blends of polyvinyl chloride with modifiers containing styrene copolymers by means of the absorption bands at 11·9 μm and 13·19 μm for polyvinyl chloride and polystyrene respectively, with samples prepared as hot pressed films of 0·0025 in (0·06 mm) thickness.

Similarly, a quantitative analysis of blends of polyvinyl chloride and chlorinated polythene may be based on measuring the absorbance of film samples at 6·85 μm and 7·0 μm, although films of only $\sim$0·001 in (0·02 mm) are required, and both samples and standards must be thoroughly homogenised for this test.

N.M.R. SPECTRA OF VINYL RESINS

IDENTIFICATION OF VINYL HOMOPOLYMERS BY N.M.R.

Identification of polymers by infra-red spectroscopy or pyrolysis/G.L.C. is, in many cases, such a simple procedure that one would not generally endeavour to identify, for example, polyvinyl chloride or polyvinyl acetate from their n.m.r. spectra. However, these identification methods all depend finally upon a direct comparison between the unknown and the behaviour of standard reference samples. In the case of more unusual resins, where reference samples or data are not available, n.m.r. examination is extremely valuable, since it is possible, in principle, to deduce the nature of an unknown substance from the n.m.r. spectrum, without reference to the spectrum of a pure sample of the material. In practice, it is preferred to examine the n.m.r. spectrum alongside other data, particularly the elementary analysis and the infra-red spectrum. Comparison with the n.m.r. spectrum of a known sample of the suspected material is clearly a valuable confirmatory test, although here some caution is required, particularly with vinyl resins, when the n.m.r. spectrum may well be sensitive to the stereochemical arrangement of the monomer units in the polymer chain.

To achieve a complete structural analysis from the n.m.r. spectrum, all the resonances must be accounted for and assigned, both in chemical shift value and area. Thus it is very important in identification work that the n.m.r. spectrum be recorded on the purest sample available.

The identification of unknown vinyl and acrylic resins by n.m.r.

This example shows how a vinyl resin and an acrylic resin of the same molecular formula were identified from their n.m.r. spectra. The Oxygen Flask Combustion test showed that resins A and B contained no 'additional elements'. Quantitative analysis showed that both had the approximate composition

C: 60%, H: 8%, O (difference): 32%, giving an empirical formula
$$C_5H_8O_2$$

Infra-red spectra showed that both resins were esters; no other functional groups such as ether or hydroxyl appeared to be present, and both materials were aliphatic in nature. On this account, it appeared that n.m.r. examination in o-dichlorobenzene solution would be preferred, as it would not be necessary to observe resonances of aryl protons, and the relatively high boiling point of this solvent ($\sim 180°C$) would enable the spectra to be recorded at high temperature, thus giving clearer and sharper resonances than would be expected at room temperature.

Figures 3.18 and 3.19 are the proton resonance spectra of resins A and B, recorded in o-dichlorobenzene solution at 140°C on the A 60D spectrometer. The solvent resonance was used as locking signal, and hexamethyl disiloxane as reference (9·94 τ). The resonances of the two polymers, and the corresponding relative areas measured from the integral curves in the figures, were as follows:

Polymer A	8·89 τ		8·11 τ	7·75 τ		4·95 τ
Areas	3	:	(4 total)		:	1
Polymer B	8·79 τ		8·22 τ	7·52 τ		5·88 τ
Areas	3	:	(3 total)		:	2

The resonances at 8·11 τ and 7·75 τ in the spectrum of polymer A overlapped as did those at 8·22 τ and 7·52 τ in the spectrum of polymer B, hence the total area of both resonances was measured in each case. The relative areas given above are the smallest values which give no fractional areas. The highest field resonance in each case is a triplet, and reference to the chemical shift tables[29, 30] shows that it may be assigned to a methyl group near (but not adjacent to) oxygen or carbonyl (since no other elements are present). Since the methyl resonance is a triplet in each case, it is adjacent to —CH$_2$—. The spectrum of resin A shows a quartet with

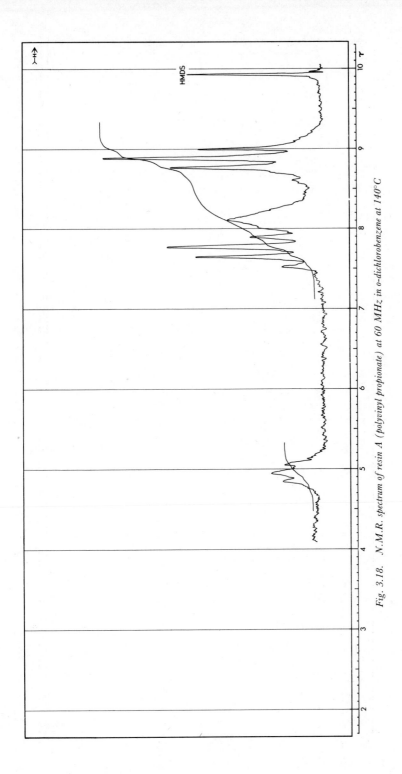

Fig. 3.18. N.M.R. spectrum of resin A (polyvinyl propionate) at 60 MHz in o-dichlorobenzene at 140°C

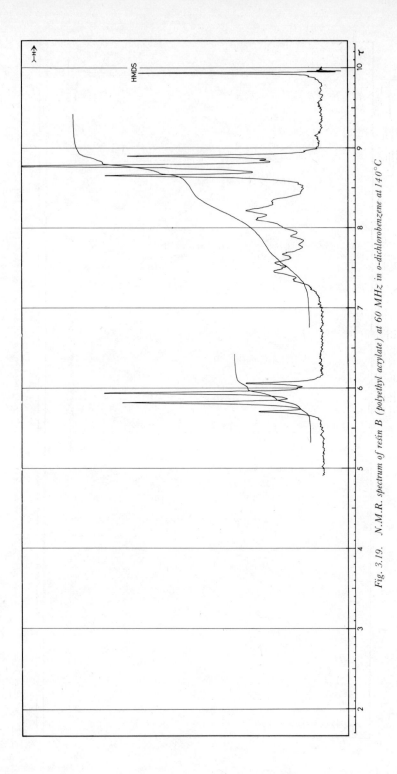

Fig. 3.19. N.M.R. spectrum of resin B (polyethyl acrylate) at 60 MHz in o-dichlorobenzene at 140°C

the same spin-spin splitting at 7·75 τ. According to the tables, this is reasonably assigned to

$$-\overset{\displaystyle ||}{\underset{\displaystyle O}{C}}-CH_2-$$

Hence resin A must contain the group

$$-\overset{\displaystyle ||}{\underset{\displaystyle O}{C}}-CH_2-CH_3$$

Since infra-red examination shows the resin to be an ester, the above group must be

$$-O-\overset{\displaystyle ||}{\underset{\displaystyle O}{C}}-CH_2-CH_3$$

In resin B, the corresponding resonance must be that at 5·95 τ since this is a quartet of the correct splitting. The chemical shift tables[29, 30] suggest

$$-\overset{\displaystyle ||}{\underset{\displaystyle O}{C}}-O-CH_2-$$

Thus the group

$$-\overset{\displaystyle ||}{\underset{\displaystyle O}{C}}-O-CH_2-CH_3$$

is present in resin B. This is consistent with the areas of the resonances.

If we assume that the repeat unit of each resin contains only one ester group, the area measurements above must represent correctly the total number of protons in each repeat unit of the polymer. Furthermore the molecular formula of the repeat unit must be $C_5H_8O_2$ in each case. We have accounted for $C_3H_5O_2$ in each resin, leaving C_2H_3, which must clearly be

$$-\overset{\displaystyle |}{CH}-CH_2-.$$

and the groups identified above must be the pendant groups along the vinyl chain in each case.

Hence the structures of the two resins must be
Resin A

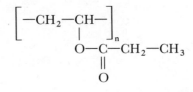

i.e. polyvinyl propionate.
Resin B

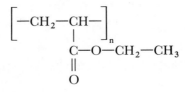

i.e. polyethyl acrylate.

Consideration of the positions of the remaining resonances shows that they are consistent with these structures.

QUALITATIVE IDENTIFICATION OF VINYL ETHER POLYMERS

Principles similar to the above may be employed to deduce the structure of vinyl ether polymers, the infra-red spectra of which are very similar to one another. The vinyl ethers are frequently encountered as copolymers, particularly with vinyl chloride, hence an example of the identification of a vinyl ether is considered in the discussion on identification of vinyl copolymers by n.m.r. on p. 209.

STEREO-ISOMERIC FORMS OF VINYL HOMOPOLYMERS

The n.m.r. spectrum has been extensively applied to the observation of the stereochemical arrangement of successive monomer units in the polymer chain, which are, in the simplest cases, the pure isotactic and pure syndiotactic forms. The general principles of this form of analysis, and the nomenclature, are described in Chapter 4 in the discussion of polymethyl methacrylate, as this is the best known example of this form of analysis.

The information available from the n.m.r. spectrum on stereochemical arrangement in vinyl homopolymers is included in Reference 31 where the τ values of the principle resonance lines and their sensitivity to stereochemical arrangement are noted. In most cases the lines are multiple because of spin-spin splitting, in addition to any complexity due to stereochemical arrangement, hence in general the measurement of stereochemical isomer content of vinyl polymers is a difficult problem.

QUALITATIVE ANALYSIS OF COPOLYMERS OF VINYL OR VINYLIDENE CHLORIDE BY N.M.R.

In a number of copolymers in which vinyl chloride or vinylidene chloride is the major constituent, while it may be possible to ascertain the general nature of the copolymerised constituent from the infra-red spectrum (e.g. ester, ether or olefin) it may be difficult to complete the identification. This is particularly so when the constituent concerned is the minor component of a random copolymer, since in such a case it is unlikely that the infra-red absorption pattern of the component will be identical with that of the corresponding homopolymer. Nevertheless, before proceeding to the identification of the minor constituent from the n.m.r. spectrum, it is advisable to test that the total material is a copolymer, as discussed earlier on p. 182. The following example will illustrate the general procedure when a copolymer is suspected.

Identification of an ether constituent in a vinyl chloride copolymer

In a particular case, a purified sample of a resin was found to contain chlorine as the only 'additional element'. The infra-red spectrum was generally very similar to that of polyvinyl chloride but the absorption band at 9 μm was enhanced in intensity. This suggested that the unknown was a vinyl chloride/vinyl ether copolymer, but there were no other features in the infra-red spectrum which would further identify this ether constituent.

The n.m.r. spectrum of the resin was recorded in p-dichlorobenzene solution at 140°C (Figure 3.20). The strongest resonances at about 5·5 τ and 8·0 τ arise principally from the protons in the vinyl chloride units —CH$_2$—$\underline{\text{CH}}$Cl and —$\underline{\text{CH}}_2$—CHCl respectively. The prominent doublet at 9·1 τ suggests a methyl group resonance split by a single adjacent proton, i.e.

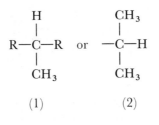

$$\text{(1)} \qquad\qquad \text{(2)}$$

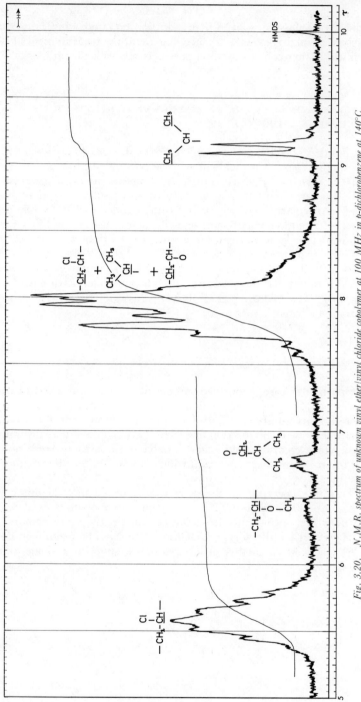

Fig. 3.20. N.M.R. spectrum of unknown vinyl ether/vinyl chloride copolymer at 100 MHz in p-dichlorobenzene at 140°C

The absence of other resonances in this region of the spectrum, such as might arise from normal alkyl $-CH_2-$ groups, strongly suggests structure (2) rather than structure (1). The resonance at $6\cdot7\ \tau$ must arise from the structure $-CH_2-O-$. This agrees with the infra-red observation that alkyl ether is present. Because this resonance is split into a doublet, the CH_2 group must be adjacent to a single proton, i.e.

$$\begin{array}{c} R \\ | \\ -O-CH_2-CH \\ | \\ R \end{array}$$

This is also in accordance with the relative areas of the resonances at $6\cdot7\ \tau$ and $9\cdot1\ \tau$ which are in the ratio $2:6$.

It follows that the unknown is most likely to be a vinyl chloride/vinyl isobutyl ether copolymer. The result was confirmed by direct comparison of the infra-red and n.m.r. spectra with those of a known sample.

QUANTITATIVE COMPOSITIONAL ANALYSIS OF VINYL COPOLYMERS BY N.M.R. SPECTROSCOPY

The n.m.r. spectrum is of particular value in compositional analysis of copolymers, because no reference is required to samples of known composition for calibration. Further, once an absolute calibration has been established, it will be possible to calibrate non-absolute methods, such as infra-red spectroscopy, which may be more appropriate in particular cases as general analytical methods. It is important, in these analyses, to ensure that all resonances are correctly identified, otherwise the result will be substantially in error. Similarly, all constituents in the resin must be known, and resonances of the solvent, or impurities in the solvent or resin, must be excluded from the area measurements. Thus both resin and solvent must be as pure as possible. The general procedure is illustrated by the following example.

Determination of the composition of a vinylidene chloride/ethyl acrylate copolymer from the n.m.r. spectrum

The copolymer was examined as a saturated solution in *o*-dichlorobenzene at 140°C. The copolymer contained the units

$$-CH_2-CCl_2- \text{ and } \begin{array}{c} -CH_2-CH- \\ | \\ C=O \\ | \\ O-CH_2-CH_3 \end{array}$$

211

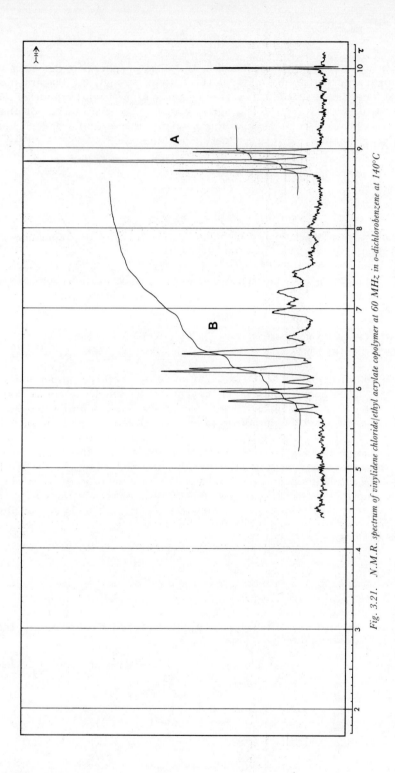

Fig. 3.21. N.M.R. spectrum of vinylidene chloride/ethyl acrylate copolymer at 60 MHz in o-dichlorobenzene at 140°C

The resonance of the terminal methyl group at $8 \cdot 8 \ \tau$ is well separated from the other resonances in the spectrum (Figure 3.21). Thus the area at $8 \cdot 8 \ \tau$, A, due to the 3 protons of the methyl group is compared with the area of all the other resonances, B, which are due to all the remaining protons in the ethyl acrylate unit and those in the vinylidene chloride unit.

Let N_a = No. of moles of ethyl acrylate in copolymer

N_v = No. of moles of vinylidene chloride in copolymer

Then

$$3N_a = KA \tag{1}$$

$$5N_a + 2N_v = KB \tag{2}$$

because ethyl acrylate contains 3 methyl protons, contributing to the $8 \cdot 8 \ \tau$ area, and 5 others contributing to the area at lower field, while vinylidene chloride contains 2 protons, both of which contribute to the low-field resonances. K is the proportionality constant. From (1) and (2), we require the molar ratio, eliminating K.

$$15N_a = 5KA$$

$$15N_a + 6N_v = 3KB$$

$$\therefore 6N_v = 3KB - 5KA$$

$$\therefore \frac{15N_a}{6N_v} = \frac{5KA}{3KB - 5KA}$$

$$\therefore \frac{N_a}{N_v} = \frac{2A}{3B - 5A}$$

Assuming the molecular weight of ethyl acrylate to be 100, and that of vinylidene chloride to be 97, we have

$$\frac{\text{Weight ethyl acrylate}}{\text{Weight vinylidene chloride}} = \frac{100 \ N_a}{97 \ N_v}$$

$$= \frac{100}{97} \left(\frac{2A}{3B - 5A} \right) = R$$

The weight percentage of ethyl acrylate is given by $100R/(1+R)\%$. In Figure 3.21 the relative areas from the integral are: $A = 4 \cdot 75$ and $B = 14 \cdot 65$.

$$\therefore \text{Ethyl acrylate} = 32 \cdot 7\% \text{ by weight.}$$

MONOMER SEQUENCE LENGTHS IN VINYL COPOLYMERS

In a copolymer of the vinyl monomers A and B it may be possible to identify resonances due to the interchange unit —AB— and distinguish these from —AA— and —BB— sequences. In favourable cases runs of four or even six monomer units can be distinguished, thus giving evidence on the distribution of monomer units within the polymer chains. The work on this form of analysis has been reviewed by Willis and Cudby[32].

REFERENCES

1. HASLAM, J., and SOPPET, W. W., *J. Soc. chem. Ind., Lond.*, **67,** 33 (1948)
2. HASLAM, J., and SQUIRRELL, D. C. M., *Analyst, Lond.*, **80,** 871 (1955)
3. ROBERTSON, M. W., and ROWLEY, R. M., *Br. Plast.*, **33,** 26 (1960)
4. HASLAM, J., and NEWLANDS, G., *J. Soc. chem. Ind., Lond.*, **69,** 103 (1950)
5. HASLAM, J., and SOPPET, W. W., *J. Soc. chem. Ind., Lond.*, **67,** 33 (1948)
6. HASLAM, J., and HALL, J. I., *Analyst, Lond.*, **83,** 196 (1958)
7. HASLAM, J., HAMILTON, J. B., and SQUIRRELL, D. C. M., *Analyst, Lond.*, **85,** 556 (1960)
8. HASLAM, J., HAMILTON, J. B., and SQUIRRELL, D. C. M., *J. appl. Chem., Lond.*, **10,** 97 (1960)
9. HASLAM, J., and SQUIRRELL, D. C. M., *J. appl. Chem., Lond.*, **9,** 65 (1959)
10. HASLAM, J., SQUIRRELL, D. C. M., and CLARKE, K. R., *J. appl. Chem., Lond.*, **10,** 93 (1960)
11. RUDDLE, L. H., SWIFT, S. D., UDRIS, J., and ARNOLD, P. E., (Ed. SHALLIS, P. W.), *Proceedings of the S.A.C. Conference, Nottingham 1965*, Heffer, Cambridge, 224 (1965)
12. GROSSMAN, S., and HASLAM, J., *J. appl. Chem., Lond.*, **7,** 639 (1957)
13. KARSTEN, P., KIES, H. L., and WALRAVEN, J. J., *Analytica. chim. Acta*, **7,** 355 (1952)
14. NIOLA, D., and DE SENA, C., *Rc. 1st sup. Sanità*, **27,** 168 (1964)
15. SQUIRRELL, D. C. M., (Ed. SHALLIS, P. W.), *Proceedings of the S.A.C. Conference, Nottingham 1965*, Heffer, Cambridge, 102 (1965)
16. CHAPMAN, A. H., DUCKWORTH, M. W., and PRICE, J. W., *Br. Plast.*, **32,** 78, Feb. (1959)
17. UDRIS, J., *Analyst, Lond.*, **96,** 130 (1971)
18. BELPAIRE, F., *Revue belge des Mater. Plast.*, **3,** 201 (1965)
19. TAUBINGER, R. P., and WILSON, J. R., *Analyt, Lond.*, **90,** 429 (1965)
20. STAPFER, C. H., and DWORKIN, R. D., *Analyt. Chem.*, **40,** 1891 (1968)
21. GROAGOVA, A., and PRIBYL, M., *Z. analyst. Chem.*, **234,** 423 (1968)
22. HUMMEL, D. O., *Infra-Red Analysis of Polymers, Resins and Additives, An Atlas*, **1,** Pt II, Wiley Interscience (1969)
23. MEIKLEJOHN, R. A., MEYER, R. J., ARONOVIC, S. M., SCHUETTE, H. A., and MELOCK, V. W., *Analyt. Chem.*, **29,** 329 (1957)
24. NYQUIST, R., *Infra-Red Spectra of Polymers and Resins*, The Dow Chemical Co., Midland, Mich., 2nd edn (1961)
25. *Infra-Red Spectroscopy. Its Use in the Coatings Industry*, Federation of Societies for Paint Technology, Philadelphia (1969)
26. WEINBERGER, L. A., and KAGARISE, R. E., *OTS Bull* No. PB 111438 (1954)
27. NOFFZ, D., and PFAB, W., *Z. analyt. Chem.*, **228,** 188 (1967)
28. *Sadtler Commercial Spectra Series D*, The Sadtler Research Laboratories, Philadelphia 2
29. JACKMAN, L. M., *Applications of Nuclear Magnetic Resonance Spectroscopy in Organic Chemistry*, Pergamon, Oxford (1959)
30. BIBLE, R. H., *Interpretation of n.m.r. Spectra. An Empirical Approach*, Plenum Press, New York (1965)
31. RITCHEY, W. M., and KNOLL, F. J., *Jnl Polym. Sci., Pt. B*, **4,** 853 (1966)
32. WILLIS, H. A., and CUDBY, M. E. A., *Appl. Spectrosc. Rev.*, **1,** 237 (1968)

4

POLYESTERS AND ESTER POLYMERS

Most of the problems connected with the examination of ester polymers and polyesters may be solved by taking account of the methods that have been worked out for the analytical examination of four rather different types of polymer, viz.

1. Polymethyl methacrylate and related polymers and copolymers— Acrylics
2. Polyethylene terephthalate and related polymers and copolymers
3. Polycarbonates
4. Polyesters of acids such as phthalic, fumaric and maleic acids.

The allocation of a sample under test to one or other of the above four classes may often be accomplished after carrying out some of the simple preliminary qualitative tests described in Chapter 2 and after determination of the infra-red spectrum of the sample.

A considerable part of this chapter is concerned with item 1 in the above list e.g. with polymethyl methacrylate and related polymers, and observations are made on such points as the removal and determination of plasticiser, and the use of pyrolysis/gas chromatography, gas liquid chromatography and vacuum depolymerisation in the identification of this type of polymer. Determinations of many of the additives which may be present in polymer preparations are discussed in this section.

The hydrolysis of polyethylene terephthalate and related polyesters is discussed and the chemical identification of a complex polycarbonate is dealt with at length because the principle of the procedure may be readily adapted to other identifications.

In discussing the analytical examination of polyesters of such acids as phthalic, fumaric and maleic acid, the opportunity has been taken to make some general observations on the identification of esters derived from various acids and polyols, and these remarks may have some significance in other fields of polymer identification.

Throughout the chapter the application of infra-red methods to the qualitative identification of the principal ester and polyester resins is described. In many ways, emphasis is laid on the limitations of the infra-red method in the polyester field, since the spectra of a number of these resins are very similar. Subtle properties which may be imparted to a resin, e.g. by the copolymerisation of minor constituents, in many cases will be missed by a simple examination of the infra-red spectrum of a resin.

Nevertheless it is made quite clear that valuable information may be obtained by infra-red examination of products obtained firstly by chemical attack on polymers and particularly by hydrolysis followed by derivative preparation, and secondly by pyrolysis and vacuum depolymerisation. Both these procedures are shown to be even more useful when combined with efficient separation procedures.

The increasing importance of n.m.r. methods in this field is seen in the descriptions of

1. Quantitative infra-red and n.m.r. methods for the examination of copolymers of methyl methacrylate.
2. Qualitative observations made on the n.m.r. spectra of isotactic and syndiotactic forms of polymethyl methacrylate.
3. N.M.R. observations made on glycol-phthalate resins and particularly on
   (a) Resins based on mixed glycols
   (b) Resins containing mixed acids
   (c) Resins containing substantially *o*-phthalic acid in which the differentiation of maleic and fumaric acid units is required, on both qualitative and quantitative bases.

POLYMETHYL METHACRYLATE AND RELATED POLYMERS AND COPOLYMERS

The polymethyl methacrylate of commerce is produced by the polymerisation of methyl methacrylate, usually by one or other of two processes, viz. (1) a procedure in which the polymerisation is carried out in bulk, in which case the final polymerisation is effected in cells formed from glass plates and the polymer is obtained in the form of sheets, and (2) a process in which the polymerisation is carried out in the presence of colloidal material, the product being obtained in the form of granules.

The polymer obtained in (2) is usually of a much lower molecular weight than that obtained in (1).

Polymethyl methacrylate polymers of this type are usually considered to have the following formulae:

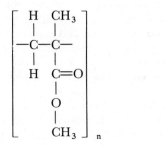

In practice, copolymers of methyl methacrylate and many other monomers are encountered, among the principal ones being copolymers of methyl methacrylate with such substances as ethyl acrylate, methyl acrylate, styrene, allyl methacrylate, cyclohexyl methacrylate, acrylonitrile and methacrylic anhydride.

Experience has indicated that samples of polymethyl methacrylate and related polymers and copolymers are submitted for examination in many different forms. Clear sheet may be received with and without added plasticiser (which is often dibutyl phthalate) or colouring agent. Further, opal material may be submitted containing opacifier and pigment. Occasionally, so-called pearlescent material, containing such substances as lead pyrophosphate and guanin, may have to be examined.

The most important question that has to be answered about an acrylic polymer concerns the nature of the polymer or copolymer. It is necessary to ensure, in the first place, that the polymer is ready for examination. In many cases this involves the isolation of the plasticiser-free polymer.

This isolation forms part of the ordinary method of determination of plasticiser and lubricant in a sample and this is described below.

THE DETERMINATION OF TOTAL PLASTICISER/LUBRICANT AND ISOLATION OF THE POLYMER

Solvents

Acetone.
Petroleum ether (b.p. 40–60°C).

Procedure

1 to 2 g of the sample, in the form of finely divided drillings, is weighed into a beaker (250 ml), 40 ml of acetone added, and the mixture stirred,

with occasional warming on a water bath. Alternatively, the beaker is covered and the mixture allowed to stand overnight.

If a clear solution is obtained and no insoluble gels are present, then the polymer is not cross-linked and it is unlikely to contain modifiers such as butadiene/styrene copolymer or acrylonitrile/butadiene/styrene terpolymer. If the sample simply swells then cross-linking may be suspected. If the solution is milky, or insoluble matter is present, then the presence of a modifier, additive or emulsifier may be suspected.

1. If the sample dissolves completely in the acetone, the volume of the solution in the beaker is reduced to 20 ml by evaporation on the water bath under a jet of air. The cooled solution is stirred vigorously and 100 ml of petroleum ether (b.p. 40–60°C) are added slowly. The beaker is then covered and the precipitated polymer is allowed to settle. After standing for 1 h the precipitate is filtered off on a No. G3 sintered glass crucible and the filtrate is collected in a beaker (250 ml). The polymer is washed on the filter with three separate 20 ml portions of petroleum ether, on each occasion squeezing the polymer with a glass rod to ensure penetration by the solvent. The washings and main filtrate are collected and retained. The polymer is dried at 60°C under vacuum for at least 24 h.

2. If the acetone solution is turbid, indicating the presence of insoluble material, the suspension is transferred, quantitatively, to a tared 100 ml centrifuge tube and centrifuged at 3000 rev/min for 45 min. If separation of the insoluble matter is poor the centrifuging is repeated at 17000 rev/min. The clear supernatant solution is decanted into a 250 ml beaker. The residue in the tube is washed with 20 ml warm acetone, the mixture re-centrifuged, and the acetone washings are added to the main acetone solution. The residue in the tube is dried and weighed. This residue usually consists of inorganic filler and may be examined as described in Chapter 11.

   The combined acetone solution and washing are evaporated to about 20 ml after which the precipitation and separation of the polymer are carried out exactly as described in Procedure 1 above.

3. If the polymer simply swells in the acetone indicating that the polymer is cross-linked, the volume of acetone is increased to 80 ml and warming and stirring carried out so as to induce as much swelling as possible. An excess of petroleum ether is added to precipitate any polymer that has dissolved and the procedure for isolation of the polymer then carried out as in Procedure 1, and the polymer dried as described. The extracts and washings are combined and reserved.

Sometimes the polymer disperses in the acetone and forms a large gel-like globule on the addition of the petroleum ether. These polymer aggregates should then be cut into small pieces during the washing with petroleum ether. It is advisable to repeat this acetone-petroleum ether extraction process to ensure complete extraction of the additives.

RECOVERY OF THE PLASTICISER/LUBRICANT

In all cases i.e. Procedures 1, 2 and 3, the acetone-petroleum ether filtrate is evaporated and the residue is extracted with three separate 10 ml portions of petroleum ether. Each extract is filtered through a filter into a 50 ml extraction flask and the petroleum ether is evaporated off from the filtrate on a water bath. The residual plasticiser/lubricant is dried to constant weight at 105°C. From the weight of residue the % plasticiser/lubricant is calculated. This plasticiser/lubricant is reserved for examination as described in Chapter 11. The material insoluble in the petroleum ether is retained and its infra-red spectrum is examined. This insoluble matter is normally low molecular weight polymer but occasionally it contains other substances such as emulsifying agents.

RECOVERY OF VOLATILE CONSTITUENTS FROM ACRYLIC COMPOSITIONS

On occasion an acrylic composition is encountered which contains a constituent which may be lost by volatilisation or decomposition when attempts are made to extract it from the polymer either by the method just described or by ether extraction. In such cases the principle of the procedure described below, which we have used successfully for the recovery of triethyl phosphate from a methacrylate type polymer containing 3% phosphorus, may be applied with advantage.

The apparatus used consists of a boiling tube (25×200 mm) provided with two side arms, Figure 4.1(a).

Tube A is filled with finely divided sample to a height of about 20 mm from the side arm and a plug of glass wool is then placed on top of the sample. Tube A is then sealed off at the top, after which the apparatus is evacuated prior to sealing at C (see Figure 4.1(b)). The sample is now heated at between 100 and 120°C for 30 min by immersion in a heated metal bath, the side tube B being immersed in a solid carbon dioxide-methanol bath at about -80°C. The freezing mixture is then removed and the water that has distilled is allowed to melt and run to the bottom of B. When all the water has been transferred the freezing bath is replaced and the temperature of the metal bath brought to between 230 and 250°C at which temperature it is maintained for about 30 min.

The heating is then stopped and the vertical portion of B is cut off. The liquid on top of the ice is drawn off by means of a capillary pipette and dried over anhydrous sodium sulphate prior to infra-red test.

219

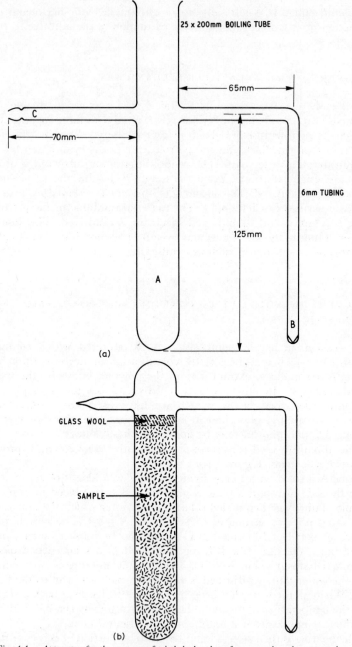

25 x 200mm BOILING TUBE

65mm

C

70mm

6mm TUBING

125mm

A

B

(a)

GLASS WOOL

SAMPLE

(b)

Fig. 4.1.   Apparatus for the recovery of triethyl phosphate from a methacrylate type polymer
(a) before loading with sample and (b) after loading with sample

If any doubt remains the liquid may be given a final purification by gas liquid chromatographic test.

The following conditions will be found suitable:

Column length	6 ft (1·83 m)
Composition of column	30% w/w silicone grease on Celite
Temperature of column	130°C
Flow rate	2 l/h of nitrogen.

THE EXAMINATION OF PLASTICISER-FREE POLYMERS

Various methods of test are available for the examination of the plasticiser-free polymer, the more important being the following:

1. Pyrolysis/G.L.C.
2. Chemical attack on the polymer followed by G.L.C.
3. Vacuum depolymerisation.
4. Infra-red spectroscopy.
5. Nuclear magnetic resonance spectroscopy.

Observations on the first three methods are given below but the sections on infra-red and n.m.r. spectroscopy are given later in this chapter. The reason for this is that it is more convenient to precede the remarks on individual types of ester polymer by some general observations on infra-red and n.m.r. spectra which are applicable to all types of ester polymer.

PYROLYSIS/GAS CHROMATOGRAPHY OF ACRYLIC RESINS

In particular cases, the identification of acrylic polymers or copolymers by infra-red or n.m.r. spectroscopy may be inconclusive, and it will be valuable to have a completely different form of test procedure available. Pyrolysis/gas chromatography may well provide the necessary information. The general qualitative procedure used has been detailed in Chapter 1, and typical pyrograms are given in Figure 4.2. Figure 4.3 shows the quantitative character of pyrograms of methyl methacrylate/ethyl acrylate copolymers and a further quantitative application is given below.

The determination of 2-ethyl hexyl acrylate in a copolymer of methyl methacrylate and 2-ethyl hexyl acrylate

No satisfactory chemical or infra-red method has been found for the analysis of copolymers of methyl methacrylate and 2-ethyl hexyl acrylate. Pyrolysis/gas chromatography has however been used with success, and

221

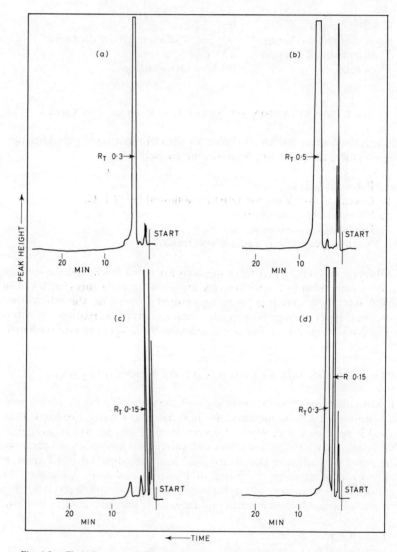

Fig. 4.2. Typical pyrograms of acrylic polymers: (a) polymethyl methacrylate, (b) polyethyl methacrylate. (c) polymethyl acrylate, (d) methyl methacrylate/methyl acrylate copolymer

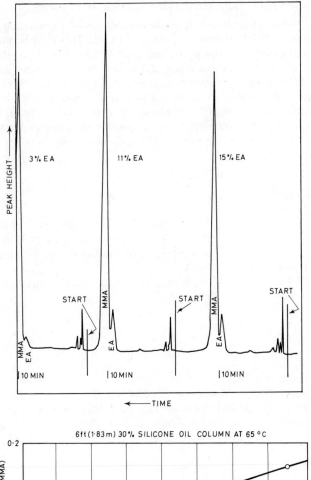

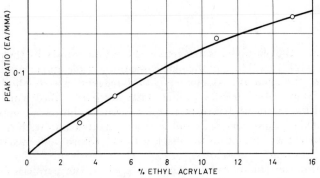

Fig. 4.3. *Pyrolysis/G.L.C. analysis of methyl methacrylate/ethyl acrylate copolymers of various compositions: pyrograms (above) and EA to MMA peak ratios (below)*

since the procedure demonstrates a novel method of internal standardisation using added polystyrene it is detailed in full. It should be noted that a simple comparison of the pyrogram peak heights due to 2-ethyl hexyl acrylate and methyl methacrylate monomers is in this case impossible since at the sensitivity required to produce a measurable 2-ethyl hexyl acrylate peak the methyl methacrylate peak is off scale. Another important point is that copolymers and not blends must be used as standards since 2-ethyl hexyl acrylate monomer is only produced in the pyrolysis of the copolymer and *not* in the pyrolysis of poly-2-ethyl hexyl acrylate.

Gas chromatographic conditions

Column	2 ft (0·6 m) 30% w/w silicone elastomer on Celite (30–80 mesh) and 6 in (150 mm) 25% w/w Apiezon 'L' on Celite (30–80 mesh)
Temperature of column	150°C
Carrier gas rate	50 ml/min hydrogen/nitrogen mixture, 1 : 1
Air	800 ml/min

Reagents

Chloroform. Analar.
Polystyrene, 0·2% w/v solution in chloroform.
Copolymers of methyl methacrylate and 2-ethyl hexyl acrylate of known 2-ethyl hexyl acrylate contents, 6 to 10%.

Procedure

If necessary, the plasticiser is first removed from the sample by the method given on p. 217. 1±0·01 g of the plasticiser-free sample and corresponding weights of the standard copolymers (third reagent) are weighed into a series of 50 ml volumetric flasks. The flasks are partly filled with chloroform and the mixtures shaken mechanically until solution is complete.

5 ml of the polystyrene solution are then pipetted into each flask and the solutions then diluted to the mark with chloroform and mixed. Each solution is examined as follows:

0·02 ml of the chloroform solution is placed on the pyrolysis coil and the solvent evaporated off by heating the coil in air at 150°C for 1 min. The pyrolysis head is fitted into the gas chromatographic column and the stream of carrier gas is turned on. The detector flame is lit, the flame cap is placed in position and the ionisation potential switched on. The sensitivity is adjusted to the correct value and the carrier gas flow is regulated to 50 ml/min. The column is then purged for 1 min whilst the recorder base line is adjusted.

The sample is pyrolysed at 550°C for 15 s and the trace given by the

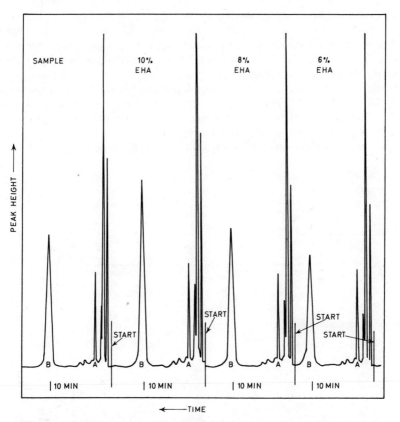

Fig. 4.4. Pyrograms of methyl methacrylate/ethyl hexyl acrylate copolymers; A are peaks due to styrene (internal standard) and B peaks due to ethyl hexyl acrylate; the sample contains 7% ethyl hexyl acrylate

pyrolysis products is recorded. The procedure is repeated with the standard polymers.

Calculation

The heights of the 2-ethyl hexyl acrylate and styrene peaks are measured and the 2-ethyl hexyl acrylate/styrene peak height ratios are calculated.

The ratio obtained from the sample is compared with those ratios

225

obtained from the standard copolymers. The percentage of 2-ethyl hexyl acrylate in the sample is deduced from this comparison.

Some characteristic pyrolysis chromatograms are given in Figure 4.4.

VACUUM DEPOLYMERISATION

The method of pyrolysis/gas chromatography is now used extensively for the examination of acrylic polymers. There are occasions, however, when the older method of vacuum depolymerisation, which was developed by Haslam and Soppet[1] may be used with advantage. In their original work the above authors had to rely on a critical examination of the yield and the physical and chemical properties of the depolymerisation products.

Nowadays, however, this older method may be used in conjunction with gas chromatography and infra-red spectrophotometry: this is

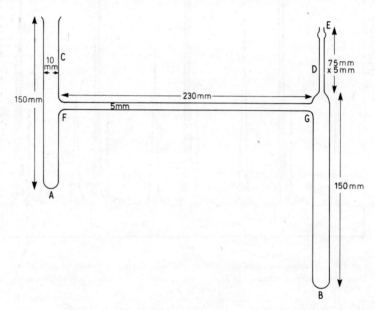

Fig. 4.5. Vacuum depolymerisation apparatus in Pyrex glass

particularly useful in the examination of unfamiliar polymers. The procedure, *which must be carried out behind a safety screen,* is as follows: 1 g of the plasticiser-free sample is weighed into the tube A of the Pyrex glass vacuum distillation apparatus shown in Figure 4.5. The upper portion of tube A is then sealed off at C in the blowpipe flame leaving a capillary tip. Tube B is now connected at E to a high vacuum pump, and when the pressure registered on the manometer connected to the pump has been reduced to 1 torr (133 N/m^2), the apparatus, with the pump still connected, is sealed off in the blowpipe flame at D.

The apparatus is now assembled with tube A immersed in a bath of Wood's metal (an alloy of 4 parts of bismuth, 2 parts of lead, 1 part of tin and 1 part of cadmium, which melts at 60·5°C) at 100 to 120°C. The Wood's metal is contained in a small metal bath, 1 in (25 mm) internal diameter and $3\frac{1}{4}$ in (82·5 mm) long.

Tube B is immersed in a solid carbon dioxide-methanol mixture at -70 to $-80°C$ contained in a Dewar flask (capacity 1 l) and the levels of A and B are so arranged that the point G is at a higher level than point F. The temperature of the Wood's metal bath is now raised to 340–350°C over a period of 15 min and is maintained at this temperature for 1 h. It may be noted that the depolymerisation only takes place to the extent of about 10% at 250°C.

The apparatus is then withdrawn from the hot Wood's metal bath and cold solid carbon dioxide-methanol mixture, the capillary tip at C is broken and the tube B is cut off near the top, i.e. at a point G.

The analytical examination of the monomer in tube B is carried out without delay, in order to avoid polymerisation.

In their original work, Haslam and Soppet showed that when the depolymerisation is carried out on plasticiser-free polymer in vacuo the product obtained is of very high quality as the following figures show.

	Monomer obtained by depolymerisation of plasticiser-free polymer in vacuo	Pure Monomer
Recovery	90 to 97%	
Ester value mg KOH/g	556 to 560	560
Refr. index 20°C	1·4143 to 1·4144	1·4144
Aldehydes	nil	nil
Colour	water-white	water-white

If the polymer itself is depolymerised in air instead of in vacuo, the yield of monomer is less and the product is poorer in quality. If the plasticised polymer is depolymerised in air then the product obtained is poorer still and contains aldehydes as the following figures show.

	Monomer obtained by depolymerisation of plasticiser-free polymer in air	Monomer obtained by depolymerisation of plasticised polymer in air
Ester value mg KOH/g	540	527
Refr. index 20°C	1·416	1·420
Aldehydes as H·CHO	nil	0·11%
Colour	yellow	yellow

227

Over the years this test has proved to be of great value and particularly so when comparing one polymer with another, when the differences in composition may be small, because many delicate tests may be carried out on the depolymerisation products.

If the analytical chemist is prepared to provide himself with pure polymers and known copolymers and establish for himself data on the quality of the depolymerisation products under precise conditions of depolymerisation, he will find this library of knowledge exceedingly valuable.

Some idea of the differences that may readily be brought out may be seen from the following table which gives information about the quality of the depolymerisation products obtained on application of the depolymerisation test to known copolymers of methyl methacrylate with other monomers.

	Copolymer from methyl methacrylate with 10% styrene	Copolymer from methyl methacrylate with 10% ethyl acrylate	Copolymer from methyl methacrylate with 46% cyclohexyl methacrylate
Recovery %	70 to 75	50 to 60	40
Ester value mg KOH/g	510	548	120 to 200
Refr. index 20°C	1·4242	1·4141	1·42 to 1·43
Colour	water-white	water-white	
Acid value mg KOH/g			3
Water (%) by Fischer test			7 to 10

Haslam and Soppet[1] have provided some useful chemical tests that may with advantage be applied to the depolymerisation products of the three copolymers mentioned above, and in favourable cases, to other copolymers.

The above figures are quoted in order to illustrate the behaviour of certain polymers and copolymers in a vacuum depolymerisation, when only conventional methods of examination of the products are available.

It should, of course, be realised that nowadays one of the most fruitful lines of attack on the depolymerisation products of polymers and copolymers is by means of gas chromatographic separation, with infra-red examination of the separated components. Figure 4.6 for example illustrates the behaviour on gas chromatographic test of the depolymerisation products of three copolymers, viz.

1. A copolymer of methyl methacrylate and ethyl methacrylate
2. A copolymer of methyl methacrylate and methyl acrylate
3. A copolymer of methyl methacrylate and styrene.

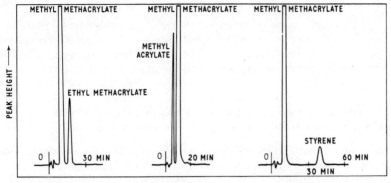

Fig. 4.6. *Chromatograms of vacuum depolymerisation products of copolymers, all using 6 ft (1·83 m) of 30% w/w dinonyl phthalate on celite with:*

Column temperature	*105°C*
Nitrogen flow rate	*1·5 l/h*
Bridge current	*140 mA*

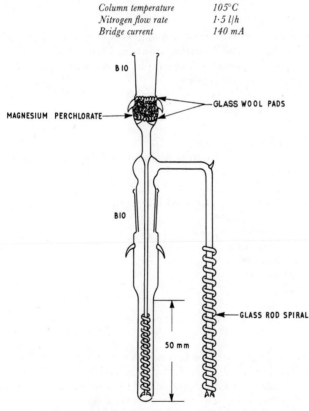

n−HEPTANE TRAP WITH EXTENSION FOR A BUFFERED BROMINE ABSORBER

Fig. 4.7. *Absorber for the determination of combined ethyl esters in methyl methacrylate copolymers*

The fractions may be collected in cold traps and subsequently examined as liquids or gases by infra-red spectroscopy[2]. It is more convenient, however, and more efficient, to collect the fractions directly in a heated gas cell for infra-red examination, with the apparatus described on p. 68.

GAS LIQUID CHROMATOGRAPHIC EXAMINATION OF THE PRODUCTS OF CHEMICAL ATTACK ON ACRYLATES

This is well illustrated by the method developed here for the determination of combined ethyl esters in methyl methacrylate copolymers[3].

The alkoxyl groups in the sample are converted to the corresponding iodides which are then determined by gas liquid chromatographic test.

Apparatus

The reaction flask, spiral scrubber and glass spoon as used by Easterbrook and Hamilton[4], and the absorber shown in Figure 4.7.
Conventional gas liquid chromatographic apparatus.
Agla micrometer syringe pipette.

Reagents

Hydriodic acid, sp. gr. 1·17.
The M.A.R. reagent, which is supplied in 6 ml ampoules.
Phenol. Analar.
n-Heptane.
The pure synthetically prepared material.
Sodium acetate solution.
A 25% w/v solution in water.
Methylene dichloride.
Ethylidene dichloride.
Methyl iodide.
Ethyl iodide.

Calibration

The column used consisted of a U tube having a total length of 6 ft (1·83 m) and a nominal bore of 0·25 in (6 mm), packed with a 30% w/w mixture of dinonyl sebacate on Celite 545 that had been graded as described by Martin and James[5]. The conditions were as follows:

Column temperature	75°C
Nitrogen flow-rate	3·0 l/h
Bridge current	140 mA
Katharometer temperature	Room temperature (22°C)

Add 0·500 ml of methylene dichloride and 0·250 ml of ethylidene dichloride accurately to *n*-heptane contained in a 100 ml calibrated flask by means of an Agla micrometer syringe pipette. Dilute the contents of the flask to the mark with *n*-heptane and shake thoroughly.

1 ml of this *n*-heptane solution contains 0·005 ml of methylene dichloride and 0·0025 ml of ethylidene dichloride.

To 10·0 ml aliquots of the *n*-heptane solution add various known volumes of methyl and ethyl iodides accurately by micrometer syringe pipette (see Table 4.1). Subject 2 to 5 drops of each of these solutions to gas liquid chromatographic test under the above-mentioned conditions (the actual amount depending on the iodide content). From the chromatograms determine the ratios (a) peak height of methyl iodide to the peak height of 0·005 ml of methylene dichloride, and (b) peak height of

Table 4.1. Results obtained in the construction of the calibration curves

Standard No.	Volume of methyl iodide added, ml	Weight of methyl iodide mg/ml of solution	Volume of ethyl iodide added, ml	Weight of ethyl iodide mg/ml of solution	Ratio (a)	Ratio (b)
1	0·170	38·63	—	—	5·79	—
2	—	—	0·170	32·86	—	7·95
3	0·100	22·79	0·010	1·93	3·33	0·42
4	0·080	18·23	0·030	5·79	2·62	1·40
5	0·050	11·40	0·060	11·58	1·66	2·87
6	0·020	4·56	0·100	19·33	0·67	4·68

ethyl iodide to peak height of 0·0025 ml of ethylidene dichloride, and construct curves, with the weights of iodide as ordinates and these ratios as abscissae. Our results are summarised in Table 4.1 and the calibration curves are shown in Figure 4.8.

Procedure

Clean the reaction flask, spiral scrubber and the *n*-heptane absorber assembly with chromic-sulphuric acid mixture and wash them thoroughly with distilled water. Dry the apparatus by washing it with acetone and allowing residual acetone to evaporate. Place 2·5 g of phenol and the contents of a 6 ml ampoule of hydriodic acid in the reaction flask. Put 4 ml of 25% w/v sodium acetate solution in the spiral scrubber and place the stopper in the open end. Connect the spiral scrubber to the reaction flask with suitable tension springs. With water flowing through the condenser and a nitrogen flow of 6 ml/min passing through the apparatus, heat the contents of the reaction flask to boiling and maintain gentle boiling for 30 min. This serves to condition the apparatus for the test. Withdraw the burner and allow the reaction flask to cool. Measure

231

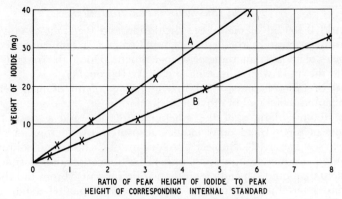

Fig. 4.8. *Calibration curves for the determination of combined ethyl esters in methyl methacrylate copolymers: A is for methyl iodide and B for ethyl iodide*

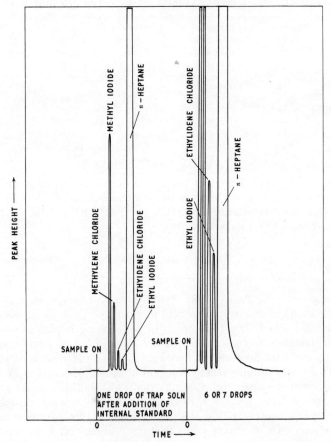

Fig. 4.9. *Chromatograms of the iodides evolved from a methyl methacrylate/ethyl acrylate (91:9) copolymer*

accurately 0·500 ml of *n*-heptane into the absorber by micrometer syringe pipette and fit the absorber assembly to the spiral scrubber by means of tension springs. Weigh out accurately about 20 mg of copolymer into the glass spoon and lower the spoon plus sample carefully down through the condenser into the reaction flask. Reconnect the spiral scrubber and when assembled submerge the *n*-heptane absorber in solid carbon dioxide-methanol mixture. Heat the contents of the reaction flask to boiling and maintain boiling for 1 h. Disconnect the absorber assembly. Add 0·500 ml of a *n*-heptane solution containing 1% v/v of methylene dichloride and 0·5% v/v of ethylidene dichloride (made up accurately by micrometer syringe pipette) to the contents of the absorber by micrometer syringe pipette and mix the solutions. Hence the absorber now contains 0·005 ml of methylene dichloride, 0·0025 ml of ethylidene dichloride, *n*-heptane up to 1 ml and the evolved iodides. Subject 1 to 6 drops of this heptane solution to chromatographic test under the conditions of calibration.

With copolymers having a low combined ethyl acrylate content, it is necessary to run 2 chromatograms to find the ratios (a) and (b), i.e. 1 drop of solution to obtain ratio (a) and 5 or 6 drops of solution to find ratio (b) (Figure 4.9). This would be unnecessary with a gas liquid chromatograph fitted with an attenuator. Calculate the ratios (a) and (b) and from the calibration curves find the weights of each iodide evolved from the polymer. Calculate the percentage of combined ethyl acrylate in the copolymer from the yield of the two iodides.

Results

A 19·9 mg sample of a copolymer known to consist of 91% of combined methyl methacrylate and 9% of combined ethyl acrylate was subjected to the procedure described above. The chromatograms obtained are shown in Figure 4.9. The peak heights of methyl iodide, methylene dichloride, ethylidene dichloride and ethyl iodide were measured and the ratios (a) and (b) were calculated. These ratios were 3·75 and 0·66, respectively. By means of the calibration curve it was found that the trap therefore contained:

25·6 mg of methyl iodide = 18·05 mg of combined methyl methacrylate
and
2·8 mg of ethyl iodide = 1·80 mg of combined ethyl acrylate.

(*Note:* Methyl methacrylate and ethyl acrylate are isomeric and have a mol. wt. of 100·11.)

∴ Recovery (in terms of original polymer) = 19·85 mg

∴ Combined ethyl acrylate in copolymer = $\dfrac{1·80}{19·85} \times 100 = 9·07\%$

Other results found for this copolymer were 9·04 and 9·09%.

Similarly, for a copolymer known to consist of 97% of combined methyl methacrylate and 3% of combined ethyl acrylate, the results were 2·96, 3·15 and 3·09% ethyl acrylate.

Incidentally, work on higher iodides by the above procedure has led to some doubt as to the stability of secondary butyl iodide in the presence of boiling phenol and hydriodic acid mixture.

THE DETECTION AND DETERMINATION OF ADDITIVES IN ACRYLIC POLYMERS

The determination of the total petroleum ether soluble matter has been dealt with on p. 217. The qualitative analysis of this extract may be carried out by the methods given in Chapter 11, which deals with plasticisers since, in many cases, the extract consists of plasticiser.

Certain tests for additives, however, not dealt with in the plasticiser chapter, are carried out either directly on the polymer composition or directly on the petroleum ether extract and these tests are more conveniently discussed here as follows:

DETERMINATION OF PHTHALATE PLASTICISER

If the nature of the phthalate plasticiser is known it may be determined directly by measurement of the U.V. absorption, at 276 nm, of a 0·5% w/v solution of the sample in chloroform.

When the sample is cross-linked and only swells in the solvent, the suspension is stirred frequently and allowed to stand overnight to equilibrate with respect to the phthalate distribution. The solution is then filtered for U.V. measurement. The $E_{1\,\text{cm}}^{1\%}$ values for two common phthalate plasticisers are: dibutyl phthalate 46·0, dioctyl phthalate 31·8.

DETERMINATION OF LONG-CHAIN ALCOHOL LUBRICANTS E.G. STEARYL ALCOHOL

If a long-chain alcohol such as stearyl alcohol has been detected in the petroleum ether extract then this substance may be determined as follows: A duplicate petroleum ether extract is prepared and this duplicate is dried for only 10 to 15 min. It is then dissolved in a known volume of carbon disulphide and the solution is examined over the spectral range 2·6 μm to 2·9 μm.

The concentration of the long-chain alcohol is determined from the height of the absorption band at 2·76 μm, taken in conjunction with calibration data obtained from standard solutions of the alcohol. A concentration range of 0·005 to 0·01% w/v stearyl alcohol in carbon disulphide is suitable for measurement.

DETERMINATION OF LONG-CHAIN ACID LUBRICANTS E.G. STEARIC ACID

If a long-chain acid such as stearic acid has been detected in the petroleum ether extract then this acid may be determined quantitatively by titration of an ethanol solution of the petroleum ether extract with standard alkali using phenolphthalein as end-point indicator.

Alternatively, a solution of the original sample in acetone may be titrated with 0·02 N tetrabutyl ammonium hydroxide solution. This method gives a measure of the total acidity of the sample and it may also be used to determine copolymerised methacrylic acid.

DETERMINATION OF LUBRICANTS OTHER THAN STEARYL ALCOHOL AND STEARIC ACID

Long-chain ester lubricants, which cannot readily be determined in the presence of a phthalate plasticiser by means of a reaction or absorption specific to the lubricant, may often be determined by a difference method. The phthalate is determined quantitatively from the U.V. absorption of the petroleum ether extract in alcohol. The long-chain ester is estimated by weight difference from the total weight of the petroleum ether extract. Alternatively, the thin-layer separation described in Chapter 11 may be applied quantitatively and the constituents weighed after separation and drying. A semi-quantitative analysis of the mixture may be obtained in this way.

DETERMINATION OF FLAME RETARDANTS

The presence of a flame-retardant additive is usually indicated by the presence, in the original compound, of a relatively large proportion of phosphorus together with chlorine or bromine. Quantitative determinations of these elements are therefore required. Phosphorus is determined by wet digestion followed by precipitation of the phosphate as ammonium phosphomolybdate. The precipitate, after washing, is dissolved in a known volume of standard alkali and the excess alkali is back titrated with standard acid. Chlorine and bromine are determined by Oxygen Flask Combustion methods or after sodium peroxide fusion.

235

However, the three elements chlorine, bromine and phosphorus are much more conveniently determined by X-ray Fluorescence on disks cut or moulded from the original sample. Intensity measurements are made at the Kα lines in each case and calibration effected with standard disks containing known amounts of flame-retardant additives. The calibration for bromine curves seriously above 4% bromine due to self-absorption of the fluorescent radiation, and for samples containing higher concentrations a dilute solution in acetone is examined.

DETERMINATION OF ULTRA-VIOLET ABSORBERS

U.V. absorbers are often encountered in polymethylmethacrylate and related polymers, and in clear sheets their presence can often be detected by direct measurement of the absorption spectrum of the sample itself. Many U.V. absorbers have been encountered including phenyl and methyl salicylates, 2:4-dihydroxybenzophenone and other substituted benzophenones, resorcinol and derivatives of resorcinol, stilbene benzoin and triazole derivatives.

In 1953, Haslam, Squirrell, Grossman and Loveday[6] described in some detail chemical and U.V. spectrophotometric methods for the identification and determination of some of these compounds and for some years it was our practice to use both methods of analysis concurrently for confirmation purposes. Improvements in instrumentation and development of more specific U.V. absorption methods (see p. 335) have meant that the chemical tests are seldom used nowadays.

For most purposes it is sufficient to measure the U.V. absorption spectrum over the spectral range 250 nm to 400 nm of a 0·5% w/v solution of

Table 4.2. Spectral characteristics of some common U.V. absorbers in chloroform solution

Substance	λ max. nm	λ min. nm	$E^{1\%}_{10\,mm}$ at highest λ max. quoted
Methyl salicylate	308	265	285
Phenyl salicylate	312	275	260
2-Hydroxy-4-methoxy benzophenone	328, 288	308, 264	420
2 : 4-Dihydroxy benzophenone	325, 288	310, 264	465
2 : 2(2'Hydroxy-5'-methyl phenyl benzotriazole)	341, 299	315, 261	770
2 : 2'-Dihydroxy-4-n-octoxy benzophenone	352, 299	314, 278	430

the sample prepared as described for the determination of phthalate. The nature of the constituent is deduced by reference to spectra of known U.V. absorbers and the concentration calculated by reference to a

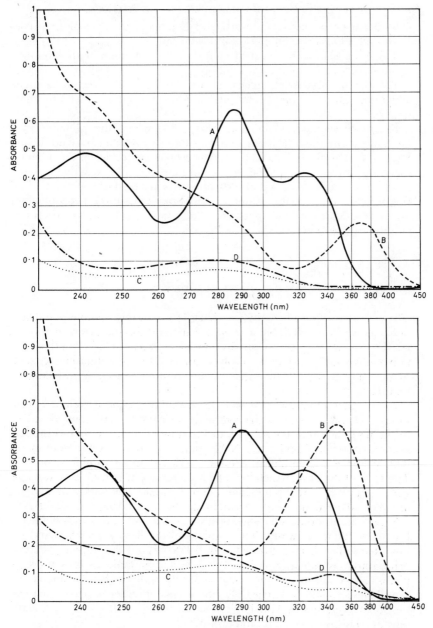

Fig. 4.10. Spectra of 2-hydroxy-4-methoxy benzophenone (upper) and 2:4-dihydroxy benzophenone (lower). Curve A, ethanolic solution; curve B, alkaline ethanolic solution (10 ml of ethanolic solution plus 1 ml of water plus 2 ml of ethanolic potassium hydroxide solution); curve C, ethanolic solution after reaction with nickel peroxide; curve D, nickel peroxide reaction product made alkaline with 2 drops of ethanolic potassium hydroxide solution

calibration graph relating absorbance at a specific wavelength to concentration of absorber in chloroform solution.

Table 4.2 shows the spectral characteristics of some of the common U.V. absorbers in chloroform solution.

Polystyrene opacifying agent absorbs in the U.V. region (λ max. 269·5 nm) and may be determined in the absence of other absorbing materials.

Reference should also be made to p. 335 where the principle of a method for the identifications of antioxidants developed by Ruddle and Wilson[7] is described. This method, which takes advantage of the different spectra given by a compound in each of four solvent or reaction systems, is also applicable to U.V. absorbers and is particularly useful for distinguishing between, for example, 2-hydroxy-4-methoxy benzophenone and 2 : 4-dihydroxy benzophenone which have very similar spectra in chloroform solution (see Figure 4.10).

DETERMINATION OF RESIDUAL MONOMER

The polymerisation of methyl methacrylate is rarely complete and it is desirable to have available a chemical method for the determination of residual monomer in polymethyl methacrylate. In this method the polymethyl methacrylate is dissolved directly in the iodinating reagents so that any monomer liberated in the solution process is immediately iodinated.

In this determination[8] 1·0 g of the granular sample is weighed accurately and transferred to a dry iodine flask (500 ml). Care must be taken, in the preparation of the sample, to avoid depolymerisation and production of monomer. 50 ml of 0·2 N Marshall's solution (a solution of iodine monochloride in chloroform) are then added from an automatic burette. The neck of the flask, which must be a very good fit indeed, is now sealed with 1 ml of 20% w/v potassium iodide solution. The flask and contents are then allowed to stand overnight in the dark. About 25 ml of 20% w/v potassium iodide solution are now added carefully to the flask followed by 100 ml of water. The contents of the flask are shaken and the liberated iodine is then titrated with 0·1 N sodium thiosulphate solution using starch as indicator. A blank determination, omitting only the sample, is carried out at the same time under the same conditions. Blank and sample tests should be carried out side by side in order to avoid inequalities in the volumes of Marshall's solution delivered from the automatic burette. This is exceedingly important.

The iodine which is used up is calculated in terms of methyl methacrylate monomer.

The above method is one based on the determination of total unsaturation which is then calculated in terms of monomer. Infra-red methods are much faster and more convenient for the determination of free monomer in acrylic sheet (see p. 257).

DETERMINATION OF WATER

Two methods of determining water are in common use. In the case of powder samples where the particle diameter is less than 150 μm the Karl Fischer method[9] may be used. In this instance the surface moisture approximates closely to the water content of the sample. A direct infra-red absorption method may be applied to acrylic sheet, and an infra-red method may also be used in the case of samples in granule or powder form (p. 257). The gas chromatographic method (p. 51) has also been used successfully with powder samples.

DETERMINATION OF FREE MERCAPTAN

The analyst concerned with polymethyl methacrylate and related polymers may encounter problems connected with mercaptans, e.g. lauryl mercaptan, occasionally used as a chain transfer agent in polymerisation processes.

The test for mercaptan is made with an apparatus similar to that devised by Kolthoff and Harris[10], namely containing a rotating platinum wire indicator electrode and a reference half cell containing mercury in contact with a potassium iodide-mercuric iodide-potassium chloride solution connected to the test solution by means of a salt bridge and short circuited through a micro-ammeter.

2 g of the sample of polymethyl methacrylate is dissolved in 100 ml of acetone, with magnetic stirring. 5 ml of 5 N ammonia solution and 5 ml of 1 M ammonium nitrate solution are added, stirring for a further 20 min, when the polymer redissolves.

The amperometric titration of the mercaptan is now carried out with 0·005 N silver nitrate solution. The end-point is quite sharp.

DETERMINATION OF SMALL AMOUNTS OF SULPHUR, NITROGEN AND PHOSPHORUS

Small amounts of these elements arise from various sources in acrylic polymers and it is useful to have methods available for their accurate determination.

SULPHUR

Haslam and Squirrell[11] have developed a simplified and speedy method for this determination in which the sample is combusted at atmospheric pressure in oxygen in a 2 l Pyrex oxygen flask, with electrical ignition. In this procedure, the combustion products are absorbed in alkaline hydrogen peroxide solution in the flask, and the sulphate is finally determined turbidimetrically as barium sulphate.

NITROGEN

Small amounts of nitrogen are best determined in polymethyl methacrylate and related polymers by wet digestion with sulphuric acid prior to direct application, without distillation, of the colorimetric phenol hypochlorite reaction of Crowther and Large[12] to the reaction products.

PHOSPHORUS

Small amounts of phosphate in polymethyl methacrylate and related polymers, possibly due to the presence of residual phosphate from granulating agents, may be determined by preliminary wet digestion with sulphuric acid, followed by formation of phosphomolybdic acid and reduction of this with ascorbic acid to a molybdenum blue complex; the blue colour obtained is suitable for colorimetric measurement.

X-RAY FLUORESCENCE DETERMINATION OF SULPHUR AND PHOSPHORUS

Both sulphur and phosphorus can also be determined with adequate sensitivity by X-ray Fluorescence, measurements being made at the $K\alpha$ lines with maximum sensitivity conditions (chromium X-ray tube, pentaerythritol crystal) and calibration against acrylic disks polymerised with known additions of phosphorus- and sulphur-containing additives.

POLYETHYLENE TEREPHTHALATE

Polyethylene terephthalate is usually prepared from dimethyl terephthalate by reaction with ethylene glycol. A process of ester interchange takes place with the production of a polymer of general formula

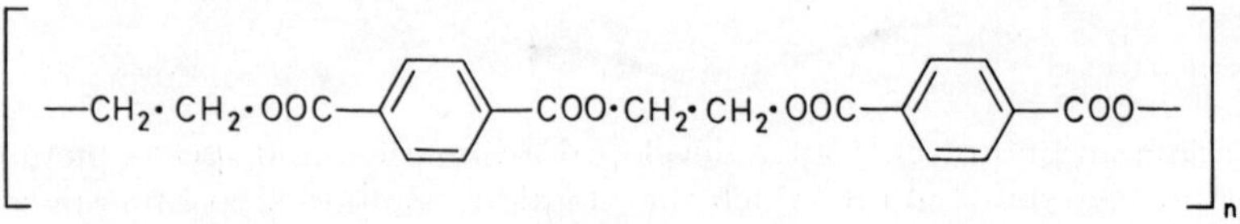

$$\left[-CH_2\cdot CH_2\cdot OOC - \left\langle\bigcirc\right\rangle - COO\cdot CH_2\cdot CH_2\cdot OOC - \left\langle\bigcirc\right\rangle - COO - \right]_n$$

The ivory coloured polymer, of high molecular weight, has a melting point range within a few degrees of 265°C.

The polymer is usually met with as such or as film or filament. The filament is usually produced by melt spinning, whilst the film is produced by melt casting, i.e. by forcing the filtered molten polymer through a long narrow slit and quenching of the curtain of melt so formed in water.

The carbon, hydrogen and oxygen contents of pure polyethylene terephthalate are as follows:

Carbon	62·50%
Hydrogen	4·20%
Oxygen	33·30%

In practice, it is found that in most laboratories the tests required on polyethylene terephthalate involve such determinations as that of intrinsic viscosity in *o*-chlorophenol solvent, of melt viscosity, of colour, of softening point, of residual catalyst and of carboxyl end-groups. Nevertheless, the occasion will arise in most laboratories when sound chemical tests for polyethylene terephthalate may be required.

Polyethylene terephthalate samples are often received in a form in which identification work can be carried out directly on the samples as received, that is the amount of contamination by additives is insufficient to interfere in any tests on the polymers. In some cases however fibre samples may be coated with a spin finish and removal of this by petroleum ether or benzene-methanol extraction may be necessary before analysis. The sample may contain delustrants or inorganic filler but in small amounts these do not usually complicate the analysis. In the case of laminated films separation of the laminates is required before identification is sought, as in Chapter 2.

A very useful test for the detection of polyethylene terephthalate is that involving aminolysis[13]; this procedure hydrolyses the ester and converts the liberated terephthalate to a crystalline derivative at the same time.

Approximately 1 g of the finely ground sample is accurately weighed into a 50 ml round bottomed flask and approximately 3 g of redistilled ethanolamine, also accurately weighed, are added. The flask is connected to a reflux condenser and the contents of the flask are boiled for 2 h and then allowed to cool. A blank test is carried out on a known weight of the ethanolamine used. When cool, the contents of the flask solidify if poly-ethylene terephthalate is being tested. This residue is warmed with a small amount of water and ground finely with a glass rod before being transferred quantitatively to a 100 ml beaker, a total of 20 ml of distilled water being used for washing the flask and condenser. The slurry is now titrated, preferably automatically, to pH 6·0; it is usual to employ an automatic titrimeter and to slow down the titration at an appropriate distance from the end-point.

The blank is titrated similarly, and the ester value of the sample is calculated in appropriate terms. The equation of the reaction is

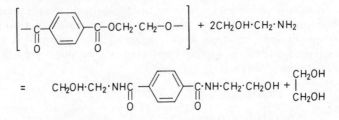

After neutralisation, the precipitated derivative i.e. terephthalic acid-bis-2-hydroxy-ethylamide from polyethylene terephthalate is dried at 100°C and its m.p. is determined. With polyethylene terephthalate the ester value corresponds to about 99·0 to 99·5% hydrolysis. The m.p. of the derivative is 234·5°C.

POLYESTERS BASED ON TEREPHTHALIC AND
OTHER ACIDS

There are cases where polyesters of acids other than terephthalic acid may be present, and the following method has been applied with success to glycol terephthalate/glycol sebacate copolymer. It illustrates the method of isolation of the total acids and the separation of terephthalic acid and sebacic acid.

0·2 g of the sample are weighed into a 250 ml saponification flask and refluxed with 25 ml of N alcoholic KOH solution and 25 ml of water until solution is complete. This potash hydrolysis is difficult; very finely ground material and film usually take about 3 h and chips may require as long as 8 h.

The solution is cooled, the condenser washed down with water and the whole transferred to a 500 ml beaker. Neutralisation with N hydrochloric acid solution is now carried out using methyl orange as indicator, after which 5 ml of N sodium hydroxide solution is added and the alcohol is boiled off. The solution is again cooled, then transferred to a separating funnel and extracted with three lots of 20 ml, 10 ml and 5 ml of ether respectively. The combined ether extracts are extracted with two lots each of 10 ml of water, and these extracts are added to the main aqueous solution. The ether remaining in this solution is boiled off on the water bath, then 20 ml of 25% v/v hydrochloric acid solution (1 volume of conc. hydrochloric acid solution sp. gr. 1·18 to 3 volumes water) are added with stirring. The solution is then cooled and the precipitated acids are filtered off on a sintered glass crucible. The precipitate is carefully washed with 50 ml of water and the crucible and contents then dried at 100°C.

The combined filtrate and washings are neutralised, then made alkaline with 5 ml N sodium hydroxide solution, and boiled down to 50 ml, then cooled and acidified with 25% v/v hydrochloric acid solution, prior to transfer to a liquid extractor and extraction with ether for 5 h. The ether extract is evaporated to dryness in a tared extraction flask and the flask and contents finally dried to constant weight in an oven at 100°C.

The total weight of terephthalic and sebacic acids is now calculated.

The sintered glass crucible and contents are placed in a lipless beaker (100 ml) after which 25 ml of ether are placed in the crucible and, after breaking up the precipitate with a small glass rod, a stopper is inserted firmly in the mouth of the beaker and the whole allowed to stand overnight. 10 ml of ether are placed in the tared extraction flask, which is stoppered, and allowed to stand overnight.

With careful washing down, the various ether extracts are filtered through the sintered glass crucible, collected, then evaporated to dryness in a tared beaker (100 ml) and dried to constant weight at 100°C.

In this way the weight of ether soluble sebacic acid is obtained. Subtraction of this weight from the weight of total acids yields the terephthalic acid.

This method has been shown to give excellent results in the examination of known glycol terephthalate/glycol sebacate copolymers. The principle of this test may be extended to the examination of other polyethylene terephthalate-like polymers, although special precautions will have to be taken if esters of volatile acids are involved.

PYROLYSIS/GAS CHROMATOGRAPHY OF POLYETHYLENE TEREPHTHALATE

Polyethylene terephthalate gives a characteristic pyrogram which readily distinguishes it from other polyesters. This method is particularly valuable when the amount of sample is limited as for example in the identification of a single fibre isolated from a mixed fibre blend. The presence of inorganic matter does not interfere with the identification.

THE DETECTION AND DETERMINATION OF CATALYST RESIDUES AND INORGANIC ADDITIVES

The inorganic elements present in polyethylene terephthalate samples are detected by spectrographic examination of the ash. Quantitative X-ray Fluorescence determinations can be made directly on $\frac{1}{8}$ in (3 mm) thick moulded disks, for the elements listed in Table 4.3, which also gives typical calibration ranges and appropriate measurement conditions. Alternatively, atomic absorption measurements made on the solution obtained after wet oxidation of the sample (p. 5) may be used for those elements

Table 4.3. X-ray fluorescence analysis of polyethylene terephthalate samples

Element	Range	X-ray Tube. Anode, kV/mA	Crystal and Order	Lines used	Background ± °2θ	Automatic Pulse Height Analyser	Instrument Standard (IS)	Counter	Calibration Graph
Antimony	20–500 p.p.m.	W 40/32 Cr 60/32	LiF, 1st LiF, 1st	$Sb_{L\alpha}$ $Sb_{L\alpha}$	+2 +2	in in		F.C. F.C.	% Sb v. ratio to I.S. % Sb v. ratio to I.S.
Zinc	10–250 p.p.m.	W 40/32 Cr 60/32	LiF, 1st	$Zn_{K\alpha}$	(−0.7)	in	Borax Glass Disk containing 0.025% Sb 0.015% Zn 0.020% Mn 0.10% Ti 0.02% Ca 0.20% Si	F+S	% Zn v. ratio to I.S.
Manganese	10–300 p.p.m.	W 40/32	LiF, 1st	$Mn_{K\alpha}$	—	in		F+S	% Mn v. ratio to I.S.
Titanium	0.01–1.0%	Cr 40/8 W 40/20	LiF, 1st	$Ti_{K\alpha}$	—	in		F.C.	% Ti v. ratio to I.S.
Calcium	10–250 p.p.m.	W 60/32 Cr 60/32	PE, 1st	$Ca_{K\alpha}$	+2	in		F.C.	% Ca v. ratio to I.S.
Silicon	0.02–0.5% SiO$_2$	Cr 60/32	PE, 1st	$Si_{K\alpha}$	+2	in		F.C.	% SiO$_2$ v. ratio to I.S.
Phosphorus	10–200 p.p.m.	Cr 60/32	PE, 1st	$P_{K\alpha}$	+1	in	Borax Glass Disk containing 2·2% P	F.C.	% P v. ratio to I.S.

for which hollow-cathode lamps are available e.g. zinc, manganese and calcium, or conventional chemical methods similar to those described on p. 158 may be applied.

DETERMINATION OF CARBOXYL END-GROUPS

The determination of the carboxyl end-groups gives a measure of the chain degradation of a polyethylene terephthalate sample, and two methods have found favour. For accurate work the sample is dissolved by stirring at 90°C with a solvent mixture of 70% v/v *o*-cresol in chloroform. A known volume of standard alkali is added slowly so as to avoid precipitation of the polymer. The solution is now titrated potentiometrically with standard hydrochloric acid, standard pH electrodes being used. Two end-points are observed, the first due to neutralisation of the excess alkali and the second due to the titration of the sodium carboxylate; the difference between the two end-points corresponds to the carboxyl groups present. After appropriate corrections for any blank on the reagents and standardisation of the titrant with benzoic acid put through the whole procedure, the carboxyl value can be calculated. Full-scale recording titration is recommended.

The second, more direct method is simpler and although not so accurate as the first method has been found useful for routine and comparative tests. 2 g sample, dissolved in 70 ml mixed solvent (1 kg phenol, 1 l *m*-cresol, 600 ml chloroform and 11·0 ml water) is titrated with dilute standard alkali to a bromophenol blue end-point. A comparison end-point is prepared by placing the indicator in a buffer solution at pH 3·35 in a separate beaker. Appropriate blank and standardisation titrations are performed. A direct conductometric titration with standard alkali has also been used.

DETERMINATION OF HYDROXYL END-GROUPS

An indirect method based on the work of Conix[14] is used for the determination of hydroxyl groups in polyethylene terephthalate. The sample is dissolved in nitrobenzene at 180°C and treated with succinic anhydride for 3 h to produce an adduct which is precipitated by pouring the solution into a large volume of acetone. The precipitate is filtered and reprecipitated again with acetone from nitrobenzene solution. The filtered and dried product is washed free of residual nitrobenzene with warm ether and finally re-dried.

The total carboxyl groups in the adduct are now determined by the first of the methods previously described and the difference between the two values obtained before and after treatment with succinic anhydride calculated as gramme equivalents hydroxyl groups/10^6 g polymer.

POLYETHYLENE TEREPHTHALATE YARN

DETERMINATION OF KNOWN SPIN FINISHES

These spin finish preparations are often proprietary compositions containing hydrocarbon oils, fatty acids, esters of fatty acids, surface active agents, sulphonated oils, basic constituents and water.

Usually the proportion of spin finish may be determined by direct ether extraction. The method gives a measure of the total solids, i.e. constituents of the spin finish other than water remaining on the yarn and is useful as a control test.

Procedure

A plug of loose cotton wool is placed in a Soxhlet extraction apparatus, i.e. at the place normally occupied by the base of the thimble. This cotton wool is covered with 10 g of the yarn.

The yarn is now extracted for 3 h on the boiling water bath with 150 ml of ether, the extract being received in a 250 ml flask, previously dried at 100°C and weighed.

The ether is then allowed to drain into the extraction flask, after which the flask is removed and the ether is removed by evaporation on the water bath. The flask and contents are dried in the oven at 100°C for 1 h, then cooled in a desiccator for at least 30 min, and re-weighed.

The weight of ether extract is then calculated as a percentage of the yarn taken.

In cases, however, where the known spin finish contains a surface active agent of the polyethylene oxide type i.e. that gives a weighable complex with phosphomolybdic acid and barium chloride, it may be better to proceed as follows. 5 g of the yarn are first extracted with ether as in the determination of total solids given above.

50 ml of water are added to the ether extract in the extraction flask and the solution is brought to the boil on an electric hot plate and swirled at the boil for 5 min. Any residue adhering to the sides of the flask is removed with a rubber 'policeman'.

Using suction, the solution is filtered through a No. 1 grade 3 sintered glass crucible, and the insoluble matter is washed with hot water until about 120 ml of filtrate have been collected.

5 ml of 25% v/v hydrochloric acid solution are added, followed by 5 ml of 10% w/v barium chloride ($BaCl_2 \cdot 2H_2O$) solution and 2 ml of phosphomolybdic acid reagent (10 g Analar phosphomolybdic acid dissolved in 100 ml water) in this order.

The mixture is diluted to 150 ml with water, then raised to the boil, after which it is cooled, and allowed to stand overnight.

The precipitate is filtered off on a No. 1 grade 4 sintered glass crucible which has been dried and weighed along with a small glass rod about 3 in (80 mm) long.

The precipitate in the crucible is washed with a minimum of 100 ml of water, followed by 50 ml of alcohol and then by 100 ml of ether. Additions of alcohol and ether to the crucible are made 10 to 20 ml at a time and the filtration must be dropwise. The precipitate is stirred frequently by means of the glass rod throughout the washing process. On completion of the filtration and removal of the ether by suction the crucible, precipitate and glass rod are dried to constant weight at 100°C.

A calibration curve is prepared by using known amounts of the particular spin finish or active principle of the spin finish in water, and determining the corresponding amounts of phosphomolybdic acid complex; the proportion of spin finish on the yarn is then readily calculated.

POLYCARBONATES

These are exceedingly interesting substances to the analytical chemist. It will probably be helpful, in the first place, to realise how the substances are prepared.

It is usual to prepare a 4 : 4'-dihydroxy-diphenyl-alkane initially by a procedure such as the following:

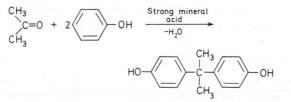

This 4 : 4'-dihydroxy-diphenyl-alkane is then allowed to react with phosgene in the presence of alkali to yield a polycarbonate, thus:

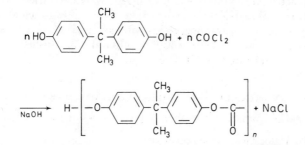

Various 4 : 4'-dihydroxy-diphenyl-alkanes may be used, but it may be useful to the analytical chemist to give the steps that led to the positive

247

identification of an unknown substance as the polycarbonate obtained from Bisphenol A i.e. 4 : 4′-dihydroxy-diphenyl 2 : 2′-propane and phosgene.

All the evidence suggests that the information is likely to be useful in other cases.

With completely unknown substances there is still a lot of analytical virtue in carrying out elementary determinations of carbon, hydrogen and oxygen. These were carried out on the original polymer with the following results. Comparative results are given for the pure polycarbonate derived from Bisphenol A.

	Original polymer %	Theoretical for pure polycarbonate derived from Bisphenol A %
Carbon	75·2	75·57
Hydrogen	5·7	5·55
Oxygen	18·6	18·88

Preliminary infra-red tests indicated that the unknown substance was an aromatic ester of some kind.

Although potash hydrolysis of the ester was reasonably satisfactory and gave evidence of the potassium derivative of a phenol in the hydrolysis product, this hydrolysis gave no evidence of the presence of the potassium salt of an organic acid in the hydrolysis product. Eventually it was agreed that preliminary ethanolamine hydrolysis of the original substance, carried out as follows, was an extremely efficient process.

ETHANOLAMINE HYDROLYSIS

1 g of the unknown substance was boiled under reflux for 30 min with 3 ml redistilled ethanolamine in a 50 ml round bottomed flask. The solution was cooled, acidified with 2 N hydrochloric acid solution and the acid solution extracted three times with 15 ml ether in a separating funnel. The combined ether extracts were washed with water, dried over anhydrous sodium sulphate, then evaporated to dryness to yield a white crystalline phenolic substance. This phenol was recrystallised from boiling 40% v/v acetic acid solution, filtered off, washed with cold distilled water and dried at 100°C for 2 h.

CHEMICAL EXAMINATION OF PHENOLIC COMPONENT

On chemical examination the phenol gave the following results:

1. Gibbs Indophenol Test: positive.
2. Coupling with diazotised sulphanilic acid in alkaline media gave a deep orange dyestuff.

3. Reaction with phosphomolybdic acid and ammonia: pale blue solution characteristic of a phenol.
4. Reaction with neutral ferric chloride in neutral acqueous solution: very pale violet colour.

The phenol was then submitted to elementary analysis with the following results. The figures for pure Bisphenol A are given for comparison purposes.

	Unknown Phenol %	Pure Bisphenol A %
Carbon	79·6	78·92
Hydrogen	7·0	7·06
Oxygen	13·7	14·02
Mol. wt	approx. 216	228
M.p.	154 to 159°C	153 to 155°C

These figures seemed to indicate that the phenolic component had a probable molecular formula of $C_{15}H_{16}O_2$ and it looked strongly probable that we were dealing either with a compound containing two phenolic residues connected directly, or connected through an intermediate or substituted —CH_2— linkage.

Spectrophotometric examination in the U.V. region of the phenol we had isolated, together with that of known substances viz. 4 : 4′-dioxy-3 : 3′-dimethyl-β : β'-diphenyl propane, 4-hydroxydiphenyl and 4 : 4′-dihydroxydiphenyl left no doubt that we were *not* dealing with a substance containing two conjugated phenolic residues directly connected and therefore, as it were, reinforcing each other in their conjugation. On the contrary, it seemed reasonable to believe we were dealing with a dihydric phenol, the phenolic residues of which were connected through a —CH_2— or substituted —CH_2— grouping.

Search of the literature revealed two possible substances, both of which were prepared, then converted to their tetrabromo derivatives.

These were:

1. A phenolic substance prepared as follows: 2 : 2′-dimethyl-5 : 5′-dioxy diphenyl methane was synthesised by reaction of *p*-nitrotoluene with formaldehyde to yield 2 : 2′-dimethyl-5 : 5′-dinitrodiphenyl methane. This was reduced with tin and HCl to yield 2:2′-dimethyl-5:5′-diamino diphenyl methane. This amine was diazotised and converted to the corresponding phenol.
2. 4:4′-dioxyl-β:β'-diphenyl propane. This was prepared by the reaction of phenol with acetone in the presence of hydrochloric acid.

Examination of the phenol isolated from the unknown substance and the two phenols that had been synthesised, i.e. carrying out m.p. tests and mixed m.p. tests on the parent phenols and also on their tetrabromo derivatives, left no doubt that we were dealing with 4:4'-dioxy-β:β'-diphenyl propane (Bisphenol A). All this was confirmed by infra-red and U.V. tests on both parent phenols and the corresponding tetrabromo derivatives.

Moreover, with this evidence, it seemed likely that, in the original substance, two molecules of Bisphenol A were linked together by carbonate linkages and it seemed desirable to prove this by a carbonate determination.

CARBONATE DETERMINATION

Reagents

Alcoholic potassium hydroxide solution N.
Prepared 16 h before use, stored in a refrigerator and filtered immediately prior to use.
Baryta solution 0·5 N.
Hydrochloric acid solution 50% v/v.
Hydrochloric acid solution N.
Hydrochloric acid solution 0·1 N.
Sodium hydroxide solution 0·1 N.

Procedure

0·8 g of the original substance under test, accurately weighed, were now boiled under reflux with N alcoholic potash for 2 h, the solution being protected from atmospheric CO_2 by a 'Carbosorb' tube placed on the top of the condenser. The solution was cooled and washed into a 50 ml standard flask, then diluted to the mark with cold water, previously boiled free of CO_2.

The carbonate was then determined in a 20 ml aliquot of this solution by the vacuum evolution method described in detail in BS 2455, using 0·1 N hydrochloric acid to dissolve the precipitated barium carbonate followed by back titration of the excess hydrochloric acid with 0·1 N sodium hydroxide solution.

By this procedure the carbonate content, as CO_3^{2-}, of the unknown substance proved to be 24·5%, in good agreement with the theoretical carbonate content of 23·7% of the polycarbonate produced by reaction of Bisphenol A with phosgene.

All this evidence, presented in relation to a particular case, should, one feels sure, be of general value in the examination of polycarbonates and related substances of many types.

THE EXAMINATION OF POLYESTER RESINS OF ACIDS SUCH AS PHTHALIC, MALEIC AND FUMARIC ACIDS

The analyst in the plastics industry is often called upon to examine unknown polyester resins. These resin preparations may contain free monomers, and the esters may have been derived from various glycols and acids. The analyst is therefore in need of an all-round method, capable of dealing with all kinds of polyester identification. One of us (J.H.) is extremely indebted to Messrs Partridge, Traxton, and Karley of Scott Bader and Co. Ltd for most of what follows and for many profitable discussions on this and allied topics.

In the identification of unknown polyesters, the above authors use the principle of Swann's method[15, 16] in making the acetates of the polyols, but substitute ethanolamine for Swann's benzylamine and butylamine. They carry out G.L.C. examination of the polyol acetates using two columns:

1. Carbowax 20 M (20 000 mol. wt)
2. Methyl silicone gum E30

All elution times are determined against knowns and the authors, being great believers in trapping fractions as well, also obtain the infrared spectra of all the trapped acetates from the G.L.C. columns.

They find that the acetates of ethylene and propylene glycols are not well separated on Carbowax and that a much better separation is achieved on the silicone gum rubber.

In searching for acids such as fumaric, phthalic and maleic acids the above authors have had only half-hearted success with the lithium methoxide method of preparation of the methyl esters. They prefer the method of esterification using boron trifluoride, methanol and sulphuric acid. Incidentally, if the preparation contains monomers such as styrene, then it is treated with a small amount of hydroquinone prior to removal of the monomer in a vacuum oven as a preliminary to the esterification procedure.

The esters prepared as above are separated by G.L.C. and the elution times and the infra-red spectra of all trapped fractions are obtained.

The above authors have examined a very large number of polyester resins, made to their own specifications, whether carbon dioxide is lost in the preparation or not, and have thus collected a library of data on known resins, produced under known conditions giving e.g.

1. The infra-red spectrum of the resin.
2. The refractive index of the preparation.
3. The ester value of the preparation.
4. The results of the application of the modified Swann method, with both the elution times and the infra-red spectra of the corresponding acetates.
5. The results of the application of the methanolysis method, with both the elution times and the infra-red spectra of the methyl esters of the acids.

Corresponding information is obtained about any unknown polyester submitted for test, and there must be a perfect match at all points with a known polyester before it is concluded that a satisfactory identification has been achieved.

In dealing with the polyester resins of such acids as phthalic, maleic and fumaric acids, Traxton's method[17] is of particular value. In this method, the polyester is hydrolysed with potassium hydroxide solution in the presence of acetone, when neglible isomerisation of the unsaturated acids takes place. Acetone and any monomeric materials are evaporated off and the hydrolysis product is neutralised with hydrochloric acid prior to dilution to volume.

An aliquot of this solution, in the presence of ammonium chloride and hydrochloric acid and after deoxygenation with nitrogen, is polarographed to yield

1. A wave due to maleic + fumaric acids at -0.8 V v. the S.C.E.
2. A wave due to phthalic acid at -1.3 V v. the S.C.E.

Isophthalic, succinic, glutaric, sebacic, crotonic and itaconic acids do not interfere. Mesaconic and citraconic acids *do* interfere with the maleic-fumaric acid determination, but not with the phthalic acid determination. (In the absence of maleic and fumaric acids mesaconic and citraconic acids may be determined simultaneously with phthalic acid, according to Novak[18].) Ethylene glycol and propylene glycol do not interfere but low molecular weight polyethylene glycol does interfere with the maleic-fumaric acid determination (see below).

The maleic and fumaric acids may, however, be determined separately in the original hydrolysis product as follows.

An aliquot of the solution, in the presence of ammonium chloride, and with the pH adjusted to 8·2, is deoxygenated with sodium sulphite and polarographed. The wave due to maleic acid is at about -1.4 V v. S.C.E.

The determination of fumaric acid is carried out similarly, but the pH is adjusted to 9·7. The wave due to fumaric acid occurs at about -1.6 V v. S.C.E.

Low molecular weight polyethylene glycol does not interfere with the above two tests, as these acids are reduced at more negative potentials than in the ammonium chloride-hydrochloric acid supporting electrolyte.

Incidentally, in recovering monomers and solvents from polyester preparations, the above authors have found the H-tube method developed by Haslam and Jeffs[19] to be of particular value.

Occasionally samples of polyesters may be received for test which contain not only polyesters derived from such acids as phthalic, fumaric and maleic acids, but also constituents derived from linseed oil. It is likely that the methods outlined above will prove useful in the analytical examination of such samples.

INFRA-RED SPECTRA OF ESTER RESINS

The prominent absorption bands of all ester resins occur near 5·8 μm (C=O stretching band) and between 7·5 μm and 9·5 μm (C—O stretching bands). In using the infra-red spectrum for qualitative analysis, it is important to remember that other carbonyl containing compounds, particularly acids, ketones, aldehydes and imides, show rather similar characteristics. However, acids have additional structure due to the hydroxyl group which appears as a broad band underlying the 3·4 μm C—H structure, and aldehydes have —CH bands near 3·7 μm which, although weak, are easily seen because this spectral region is usually clear of absorption. Thus both acids and aldehydes may be distinguished from esters. Although esters contain C—O groups which are not present in ketones, this unfortunately does not enable them to be distinguished from the spectrum, because in ketones the carbon-carbon linkages adjacent to the carbonyl group produce bands of considerable intensity in the 8 μm to 9 μm region which, in both shape and intensity, can be mistaken for C—O structure. Thus the infra-red spectrum does not provide a reliable method of distinguishing esters and ketones; fortunately ketone polymers are encountered very rarely, and it will be reasonable to assume initially that a resin is an ester rather than a ketone.

The diaryl ether group shows a strong band at about 8·0 μm which is very similar in appearance to the C—O band of an aryl ester at this wavelength, and aryl ether polymers which contain carbonyl groups are easily mistaken for aryl polyesters. Polyimides based on *pp'*-diaminodiphenyl ether are of this nature, but may be distinguished by the presence in their spectra of a doublet C=O band at 5·6 μm and 5·8 μm. If this is suspected it will be important to perform a qualitative test for nitrogen.

CLASSIFICATION OF ESTER RESINS ACCORDING TO HETERO-ELEMENTS PRESENT

NO ELEMENTS DETECTED OTHER THAN CARBON AND HYDROGEN

These materials are likely to be homopolymers or copolymers of vinyl or acrylic esters, esters of cellulose, or polyesters. The vinyl esters are described in Chapter 3, and the cellulose esters in Chapter 10. The remainder are described in this chapter, except for copolymers of esters with hydrocarbon resins containing only a minor proportion of ester. The latter are described in Chapter 6.

ESTER RESINS CONTAINING CHLORINE

By far the most common chlorine-containing resins which have spectra with the characteristics of esters are copolymers of chlorine-containing monomers namely vinyl chloride or vinylidene chloride with vinyl or acrylic esters, and blends of acrylic resins with polyvinyl chloride (or vinyl chloride copolymers). There is also the possibility that the material is ester-plasticised polyvinyl chloride. These relatively common resins have been described in Chapter 3. If such resins seem unlikely, the material may be an ester based on a chlorinated acid, of which the most common are those based on HET acid (p. 269), and those derived from a chlorinated acrylic acid e.g. polymethyl α-chloroacrylate (p. 269).

ESTER RESINS CONTAINING NITROGEN

The key to the identity of these resins is to establish how the nitrogen is combined; in many cases this should be possible using Table 2.8 (p. 120).

Resins containing nitrile groups are likely to be copolymers of acrylo-nitrile or methacrylonitrile with esters; in these compositions the spectrum of the ester is modified only to a minor extent by the presence of the nitrile-containing monomer, and it should be possible to proceed with the identification of the ester constituent as if it were a straight polymer. The same observations apply to resins found to contain primary amide groups; these may well arise from the copolymerisation of acryla-mide or methacrylamide.

The other important class of ester resins containing nitrogen are the 'ester urethane' rubbers. These are products in which the terminal hydroxyl groups of a relatively low molecular weight polyester are reacted with an aryl di- (or poly)-isocyanate to produce chain length-ening and cross-linking. This reaction with an isocyanate introduces the urethane or N-substituted carbamate group

$$-NH-\underset{\underset{O}{\|}}{C}-O-$$

This group shows strong absorption due to the N—H group at about 6·5 μm and a carbonyl band at 5·9 μm, often to be seen as a shoulder on the long-wave side of the main carbonyl absorption band. Additional absorption due to N—H groups also appears at 3·0 μm, but this may be difficult to distinguish from hydroxyl absorption which often occurs due to the terminal groups in a polyester. The chemistry of this process is given more fully on p. 452, in the description of 'polyether urethanes'.

The polyester urethanes have spectra somewhat similar to the true polyurethane resins, made by the condensation of alkyl or aryl di-isocyanates with glycols. The latter show bands at 6·5 μm and 3·0 μm which are relatively stronger, compared with the carbonyl intensity, than in polyester urethanes, since they contain no true ester groups. Furthermore the carbonyl band is now at 5·9 μm rather than 5·8 μm.

The polyimide resins must also be considered, since they may be confused with ester polymers. These resins show a double carbonyl band, one component being in the normal ester position of 5·8 μm and a weaker maximum at 5·6 μm.

All these nitrogen-containing resins are considered in this chapter, except for polyimides and the true polyurethanes, which are described in Chapter 5.

CHARACTERISTIC ABSORPTION BANDS FOR ESTERS

The number of ester resins in existence is very large, and before resorting to the comparison of an unknown spectrum with a wide range of standard

Table 4.4. Characteristic absorption patterns of esters

Ester	C=O band (μm)	C—O bands (μm)	Other spectral features (μm)
o-Phthalates	5·80	7·8, 8·9	6·23, 6·35 (doublet) 9·35, 13·45, 14·25
iso-Phthalates	5·85	7·7, 8·1	6·23, 6·32 (doublet) 8·6, 8·85, 9·15, 9·3, 13·7
Terephthalates	5·82	7·9, 9·0	9·8, 11·5, 13·7
Benzoates	5·85	7·9, 9·05	9·4, 9·8, 14·1
Methacrylates	5·78	7·9, 8·1 8·4, 8·7	13·4
Acrylates	5·78	7·9, 8·55	—
Alkyl carbonates	5·73	8·0	—
Aryl carbonates	5·65	8·1, 8·4, 8·6	Usually a band near 12 due to use of p-substituted phenols in the condensation
Acetates	5·78	8·0, 9·8	—
Propionates, butyrates	5·80	8·4	9·25
Esters of long chain aliphatic acids or alcohols	5·75	8·6	{ 13·9 or, if crystalline, doublet at 13·7 and 13·9 in many cases
Adipates, azeleates, sebacates	5·80	8·05, 8·5	—
*HET acid esters	5·70	8·35	6·05 (weak), 6·22 (sharp, medium), 10·9
Abietic acid esters	5·80	8·15, 8·55	Many weak bands

*1 : 4 Endodichloromethylene tetrachlorophthalic acid

spectra of ester resins, it is better to attempt a preliminary classification of the type of ester by considering both the elements present and the features of the spectrum. Fortunately many esters have band patterns which are extremely useful and characteristic.

These patterns are summarised in Table 4.4. The data refer equally to monomeric and polymeric systems, except in the case of acrylates and methacrylates, where the figures given apply to the polymer only (in which the double bond of the acid is removed).

The bands noted in Table 4.4 appear in straight resins, block copolymers and mixtures. In random copolymers some modification may be expected, but this is not important except in acrylates and methacrylates, where a significant modification of the C—O structure occurs; the relatively sharp C—O pattern found in the straight polymer becomes increasingly broad and ill defined with increasing amount of the copolymerised constituent. The exception to this is in copolymers of methacrylates with methacrylates where the characteristic C—O pattern of the homopolymer is retained.

SIMPLE ACRYLATE AND METHACRYLATE RESINS

Most acrylate resins have a simple and characteristic C—O structure consisting of a pair of bands at 7·9 μm and 8·55 μm, the latter being considerably the stronger. Polymethyl acrylate (Spectrum 4.1) is exceptional in having an additional band at 8·35 μm. This, together with the 12·0 μm band, makes this resin easily recognisable from its spectrum.

Other common acrylate resins are the ethyl, n-butyl, isobutyl and 2-ethyl hexyl esters (Spectra 4.2 to 4.5). The ethyl compound is recognised from the relatively strong bands at 9·75 μm and 11·75 μm; the spectrum of poly-n-butyl acrylate, on the other hand, is rather complex, and the only band of interest, other than C=O and C—O structure, is the band at 13·6 μm due to the n-butyl group. The isobutyl ester shows a useful band pair at 10·1 μm and 10·55 μm. Spectra of the other isomeric poly-butyl acrylates have been published[20]. The bands at 13·0 μm and 13·7 μm in the spectrum of poly-2-ethyl hexyl acrylate (Spectrum 4.5) arise from the ethyl and butyl groups respectively, making this resin comparatively easy to recognise from the infra-red spectrum.

A large number of methacrylate resins exist, and the spectra of those commonly encountered are given (Spectra 4.6(a) to 4.10). All the straight resins show the series of four C—O bands (7·8 μm, 8·55 μm, 8·7 μm) which seems to be a characteristic of the methacrylate structure, together with the band at 13·35 μm. To observe this structure, a thin film (0·0005 in or 0·01 mm thick) is required. Having established that the unknown is a methacrylate, it is necessary to identify the individual esters by matching their spectra since, although the remaining structure in the spectra is distinctive, it is not in most cases characteristic for particular groups. An exception is polycyclohexyl methacrylate (Spectrum·4.10) which shows in its spectrum a relatively large number of sharp bands, a feature of the spectra of most cyclohexyl compounds. In the case of polymethyl methacrylate the substantially isotactic species has been prepared and its spectrum (Spectrum 4.6(b)) is markedly different from that of the commercial resin. There are however no features by which this spectrum could be recognised as that of an isotactic polymer; in this case the isotactic and syndiotactic species may be characterised from their n.m.r. spectra (p. 270).

POLYHYDROXY METHACRYLATES

These materials are hydroxy esters; thus, in addition to the usual absorption bands associated with the acrylate or methacrylate, a prominent hydroxyl band is seen at 2·9 μm (Spectra 4.11 and 4.12). The spectra are therefore quite satisfactory for qualitative identification of these substances.

INFRA-RED DETERMINATION OF MONOMER AND WATER CONTENT OF METHACRYLATE RESINS

An infra-red method for the determination of the monomer content of methacrylate sheet is given by Miller and Willis[21]. The absorption band due to the unsaturated group $C{=}CH_2$ at 1·63 μm is used to estimate monomer contents above 0·1%. The test is not suitable for opaque sheet, as such specimens do not transmit sufficient radiation in the 1 μm to 2 μm wavelength range at the sample thickness of $\frac{1}{2}$ in (12 mm) which is required for the test.

An infra-red method may also be used to estimate the water content of polymethyl methacrylate sheet in thicknesses of $\frac{1}{16}$ in (2 mm) and above, using the absorption band due to water at 1·93 μm. A $\frac{1}{8}$ in (3 mm) sheet of polymethyl methacrylate is first dried for 4 weeks in a vacuum dessicator. The sheet is weighed and the infra-red spectrum between 1·8 μm and 2·1 μm recorded on this dried sample. The sheet is then allowed to stand in the atmosphere, and at suitable intervals of time it is weighed and the spectrum recorded over the same wavelength range.

The water content of the sheet is calculated from the increase in weight, and a calibration graph of water content versus absorbance at 1·93 μm is made. It is convenient to plot the water content of the sheet as mg/cm^2, since the resulting calibration curve may then be applied directly to sheets of other thicknesses up to $\frac{1}{2}$ in (12 mm).

The following infra-red absorption method may be applied to the determination of water in acrylic moulding powders which do not contain opaque fillers. The method covers the range 0·02 to 0·3% water, and is based on measurement of the absorption band of water at 1·90 μm and a band of the resin at 1·95 μm.

The sample is examined in solution in a glass cell 25 mm thick. To avoid the labour of cleaning glass cells and handling viscous solutions, we use flat-bottomed glass sample tubes 25 mm in diameter by 75 mm long (Catalogue No. TW–600, A. Gallenkamp and Co. Ltd, London). When the measurement has been made, the sample and tube are thrown away. A mounting block is prepared to hold the sample tube vertically as close as possible to the entrance slit of the spectrometer.

Preparation of solvent

Mix 500 ml each of chloroform and carbon tetrachloride (spectroscopic grade reagents) in a 2 l stoppered glass container. Add 200 g molecular sieve (Type 5A, $\frac{1}{8}$ in pellets, British Drug Houses Ltd, London) to dry the solvents. Shake well, and allow to stand for 2 d before use. This solvent mixture decomposes on prolonged standing, and is to be rejected after 2 weeks.

Preparation of solution

Weigh 9·0 g polymer into a sample tube, add 18 ml solvent and stopper with a polythene stopper as quickly as possible. Shake the tube (mechanically if possible) until solution is complete.

Procedure

Mount the sample tube in the spectrometer in the sample holder. Record the spectrum over the range 1·8 μm to 2·0 μm with a suitably slow traverse speed. The spectrum will be similar to Figure 4.11.

Draw linear backgrounds as in the figure, and measure the absorbances at the absorbance maxima 1·90 μm and 1·95 μm.

Calculate

$$\frac{\text{Absorbance of water}}{\text{Absorbance of polymer}} = \frac{\text{Absorbance at } 1\cdot90 \ \mu\text{m}}{\text{Absorbance at } 1\cdot95 \ \mu\text{m}}$$

This ratio is converted to % w/w water from a calibration graph prepared in the following way. Dry a sample of the moulding powder in a vacuum oven at 60°C until no water is present as determined by the absorbance at 1·90 μm. Then weigh out 9·0 g samples of the dried polymer into sample tubes. Add a known amount of water to each tube from a micro-pipette to give solutions containing 0·01% to 0·2% water, calculated on the 9 g polymer. This operation must be conducted rapidly, as dry polymer picks up water very rapidly from the air.

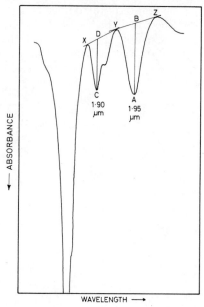

Fig. 4.11.   Infra-red spectrum of acrylic moulding powder (Diakon) containing water (0·10%)

After solution is complete, examine these standard samples in the same way as described above, calculate the absorbance ratio in the same way and plot this against the known water content of the individual tubes. This calibration graph is then used to determine the water content of the unknown samples.

COPOLYMERS OF METHYL METHACRYLATE WITH ACRYLATES

These copolymers are commonly encountered as moulding powders, and some care is required to differentiate between these and straight polymethyl methacrylate. The resin, preferably precipitated and thoroughly dried to remove residual solvents, is hot pressed to 0·002 to 0·003 in (0·05 to 0·08 mm) thickness. Solvent casting is unsatisfactory, as solvent is extremely difficult to remove from cast methacrylate films.

Copolymerised ethyl acrylate is apparent by the appearance of weak additional bands at 9·6 μm and 9·8 μm (Spectrum 4.13). The presence of copolymerised methyl acrylate is shown by an additional band at 9·6 μm and extra absorption at 12·0 μm (Spectrum 4.14). At low concentrations (below 5%) of methyl or ethyl acrylate, these differences are small, and with some other copolymerised esters the differences are even less readily seen.

259

Thus, when the spectrum suggests unmodified polymethyl methacrylate, and there is reason to suspect that the sample is a copolymer, a difference run between the unknown sample (pressed to about 0·003 in or 0·08 mm thickness) and a sample of straight polymethyl methacrylate in the comparison beam of the instrument is particularly useful for showing up the

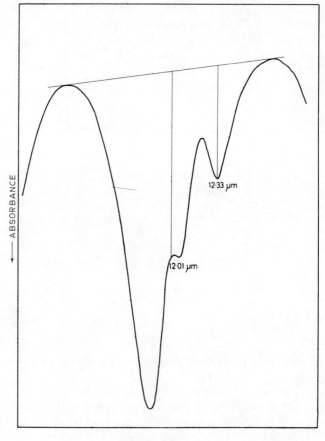

WAVELENGTH ⟶

Fig. 4.12. Infra-red spectrum of methyl methacrylate/methyl acrylate (90 : 10) copolymer

small differences in the spectra. If the comparison sample of polymethyl methacrylate is prepared as a wedge[22] tapering from 0·004 to 0·001 in (0·1 to 0·025 mm), a good balance for the polymethyl methacrylate absorption can be obtained by pushing the wedge across the comparison beam until the 13·3 μm band of polymethyl methacrylate is exactly balanced. It is advisable to prepare some specimen spectra by this method on known copolymers for comparison purposes, especially as the spectrum of isolated units of a second ester distributed in a methyl methacrylate chain is usually quite different from that of the homopolymer of the second ester.

It is important that the sample be free from plasticiser and from any solvents used in preparing the sample.

Although this is a rapid test, it is not necessarily conclusive, as it may not reveal the presence of small amounts of the higher acrylates, such as 2-ethyl hexyl acrylate. The pyrolysis/G.L.C. test can be used to great advantage with such copolymers (p. 221).

QUANTITATIVE DETERMINATION OF THE METHYL ACRYLATE
CONTENT OF METHYL ACRYLATE/METHYL METHACRYLATE
COPOLYMER BY INFRA-RED SPECTROSCOPY

This method is suitable for copolymers containing 2% to 20% methyl acrylate. Standard samples of known composition are required for calibration.

Prepare all standards and samples as films of 0·002 in (0·05 mm) thickness by pressing between polytetrafluoroethylene coated plates (p. 24). Record the spectra of all samples over the range 11 μm to 13 μm. Record also the spectrum of a similar sample of methyl methacrylate homopolymer. The spectrum of a copolymer will be similar to Figure 4.12.

Draw the straight line background as shown in the figure. Measure the absorbance at the methyl acrylate wavelength (12·01 μm) and at the methyl methacrylate wavelength (12·33 μm).

Plot the ratio

$$\frac{\text{Absorbance at } 12\cdot01 \ \mu\text{m}}{\text{Absorbance at } 12\cdot33 \ \mu\text{m}}$$

against the concentration ratio

$$\frac{\text{Methyl acrylate}}{\text{Methyl methacrylate}}$$

for all standards, including the methyl methacrylate homopolymer. The plot should be linear, but will intercept the absorbance axis at the value for the methyl methacrylate homopolymer. This graph is used to determine the composition of the unknown samples from the appropriate absorbance ratios.

261

For semi-quantitative work, absorbances measured from Figure 4.12 may be used, together with the corresponding data on methyl methacrylate homopolymer, to construct a calibration graph.

QUANTITATIVE DETERMINATION OF THE ETHYL ACRYLATE CONTENT OF ETHYL ACRYLATE/METHYL METHACRYLATE COPOLYMERS

The method above may be adapted for ethyl acrylate copolymers. Samples are prepared for examination in the same way. Appropriate reference standards are required.

Spectra of all samples are recorded over the range 8·5 μm to 11·5 μm. The band measured for methyl methacrylate is at 10·95 μm and ethyl

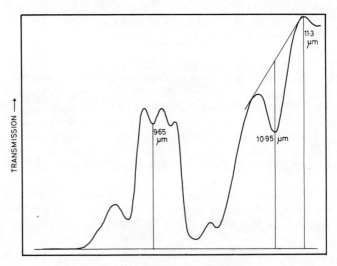

Fig. 4.13. Infra-red spectrum of methyl methacrylate/ethyl acrylate (90 : 10) copolymer

acrylate is measured at 9·65 μm. The base line is drawn as in Figure 4.13 for the 10·95 μm band, but as no reasonable background can be drawn for the 9·65 μm band, the procedure is as follows

Calculate

$$\log_{10}\frac{\text{Transmission at }11\cdot3\ \mu\text{m}}{\text{Transmission at }9\cdot65\ \mu\text{m}} = E$$

Then calculate the ratio

$$\frac{E}{\text{Absorbance at }10\cdot95\ \mu\text{m}}$$

262

Plot this ratio against concentration ratio for all standards, including methyl methacrylate homopolymer. As in the previous case, the graph should be linear with a zero intercept. The composition of all unknowns is determined from this calibration graph.

COPOLYMERS OF METHYL METHACRYLATE WITH OTHER METHACRYLATES

Methyl methacrylate/cyclohexyl methacrylate copolymer, occasionally encountered, is of some interest because, unlike most methacrylate copolymers, the spectrum of a 50/50 copolymer is almost identical with that of the corresponding polymer mixture (Spectrum 4.15). This is presumably because copolymers of methacrylates with methacrylates as prepared commercially consist substantially of syndiotactic monomer sequences so that the position and appearance of all the prominent bands is unaffected by copolymerisation.

Allyl methacrylate, glycol dimethacrylate and triethylene glycol dimethacrylate are used in small concentrations ($\sim 1\%$) as cross-linking agents for polymethyl methacrylate. These copolymers have spectra which are extremely similar to that of straight polymethyl methacrylate, but occasionally the presence of such a cross-linking agent may be revealed by a difference spectrum as described above, comparing the difference spectrum with those obtained under the same conditions from known copolymers. However, the infra-red method is of limited value in the analysis of these compositions and pyrolysis/G.L.C. is the preferred method of analysis.

COPOLYMERS OF METHYL METHACRYLATE WITH STYRENE AND α-METHYL STYRENE

The random copolymers of methyl methacrylate with styrene and α-methyl styrene have rather similar spectra (Spectra 4.16 and 4.17) as the spectra of polystyrene and poly-α-methyl styrene are themselves rather similar (Spectra 6.37 and 6.38). In each case the presence of the mono-substituted benzene ring is evident from the bands at $13 \cdot 3$ μm and $14 \cdot 3$ μm and a further weak but sharp aromatic band appears at $6 \cdot 25$ μm. The presence of ester is evident from the $5 \cdot 8$ μm band, and the C—O structure has the characteristic methacrylate pattern ($7 \cdot 9$ μm, $8 \cdot 1$ μm, $8 \cdot 4$ μm and $8 \cdot 7$ μm) although the C—O bands are broader and less well defined than in polymethyl methacrylate. The major differences in the spectra of these copolymers lie between 9 μm and 10 μm and beyond 12 μm.

Block copolymers of methyl methacrylate and styrene and mixtures of these two resins may be distinguished from the equivalent random copolymers. When the amount of polystyrene is in the range 0 to 20%, there is a significant difference in the position of the styrene absorption band near 14·3 μm. In block copolymers and blends this occurs at 14·33 μm, while in random copolymers it moves to 14·25 μm. This difference is not apparent at high styrene contents, but here the broadening of the C—O structure of the methyl methacrylate which is evident in random copolymers does not occur in blends or block copolymers.

METHYL METHACRYLATE/ACRYLONITRILE COPOLYMERS

Copolymers of methyl methacrylate containing relatively small amounts of combined acrylonitrile (up to 20%) show a comparatively unmodified pattern of polymethyl methacrylate, the only evidence of acrylonitrile being the appearance of the nitrile group absorption band at 4·4 μm (Spectrum 4.18). Increasing concentration of combined acrylonitrile has the somewhat surprising effect of causing some polymethyl methacrylate absorption bands to disappear and others to drift to shorter wavelength (Spectrum 4.19) although there is still no evidence, other than from the 4·4 μm band, of the presence of acrylonitrile.

METHYL METHACRYLATE/BUTADIENE COPOLYMERS

The effect of copolymerisation on the C—O structure of methyl methacrylate is well demonstrated in methyl methacrylate/butadiene copolymers. With about 40% butadiene (Spectrum 4.20) the structure between 8 μm and 9 μm is reduced to two broad bands at 8·4 μm and 8·8 μm, while at about 70% butadiene it appears as a single broad band (Spectrum 4.21). In all cases the butadiene absorption bands at 10·3 μm and 10·9 μm are evident, but the rest of the spectrum beyond 11·0 μm is rather ill defined.

BLENDS OF POLYMETHYL METHACRYLATE AND COPOLYMERS WITH OTHER RESINS

A number of complex polymer compositions are encountered in which the major constituent is polymethyl methacrylate. Polymethyl methacrylate may be blended with polyethyl acrylate (Spectrum 4.22). It is interesting to note that this spectrum corresponds with those of the homopolymers, and is considerably different from that of the copolymer (Spectrum 4.13). Other common blends include those with butadiene/styrene rubber, butadiene/ acrylonitrile rubber, or an ABS rubber (Spectra 4.23, 4.24 and 4.25). Similar blends may be made in which the base polymer is a methacrylate/styrene copolymer, rather than straight polymethyl methacrylate.

The presence of styrene is clear, even at low concentrations, from the 13·3 μm and 14·3 μm bands; acrylonitrile in such compositions is evident from the appearance of the nitrile band at 4·4 μm. Butadiene may be detected by means of the strong band at 10·3 μm, but this is sometimes difficult to observe as methyl methacrylate resin itself has a moderately intense band this wavelength. Additional information may be obtained from a difference spectrum in which a specimen of straight polymethyl methacrylate is placed in the comparison beam of the spectrometer as described on p. 260. In these compositions it is sometimes difficult to decide which components are copolymerised and which are mixed. The observations in the preceding sections (p. 263) may be helpful in deciding whether methyl methacrylate is copolymerised with styrene or acrylonitrile, and further evidence may be obtained in favourable cases by solvent separation of the resin.

o-PHTHALATE RESINS

Resins based on *o*-phthalic acid are very common, and are easily recognised by the presence in the spectrum of the series of bands noted in Table 4.4 (Spectrum 4.26). There is the possibility of confusion with compositions, particularly vinyl chloride resins, which are heavily plasticised with *o*-phthalate plasticisers. Advantage may be taken however of the solubility of *o*-phthalate plasticisers in solvents such as ether, and in case of doubt the examination of the spectrum of the resin after solvent extraction, and that of the extracted material, should decide between these possibilities.

These resins are often cross-linked and insoluble, and may be heavily filled. When doubt arises as to which bands in the spectrum are due to resin and which to filler, the sample may be subjected to vacuum distillation at the lowest possible temperature. The residual filler may then be examined (as a halide disk) and the true spectrum of the resin deduced.

In most cases, it will not be possible to decide from the spectrum of the resin the nature of the glycol (or polyhydric alcohol) on which the resin is based, as the characteristic bands of the alkyl groups are submerged in the relatively intense absorption of the phthalate nucleus. However, the n.m.r. spectrum will usually give valuable evidence on any soluble resin. Alternatively the resin may be hydrolysed, and the constituent glycol recovered for identification. The spectra of the glycols commonly used are presented in Spectra G.1 to G.11. Additional glycol spectra appear in the *Sadtler Standard Spectra*[23], and in Hummel[24] (Figures 1666 to 1683).

STYRENE MODIFIED *o*-PHTHALATE RESINS

O-phthalate resins are often modified with styrene, and a sample may contain styrene as monomer, polymer or both. If bands are found in the

265

spectrum at 14·3 μm and 13·3 μm (styrene polymer), or 14·4 μm and 13·9 μm (styrene monomer), the presence of styrene may be suspected. Styrene monomer may be removed by vacuum distillation, or, if the resin is soluble, by solution and precipitation of the resin. Polymerised styrene is much more difficult to remove, especially as it may be copolymerised (as a block copolymer) with the base resin. However, the presence of polystyrene will not prevent the identification of a material from its spectrum as an *o*-phthalate resin (Spectrum 4.27) and the hydrolysis procedure to recover the glycol for identification is possible in the presence of polystyrene.

'OIL-MODIFIED' *o*-PHTHALATE RESINS

The spectra of *o*-phthalate resins modified with aliphatic ester oils show additional absorption near 8·5 μm due to the aliphatic C—O group (Spectrum 4.28). This additional absorption is relatively weak, and furthermore the nature of the aliphatic ester present cannot usually be ascertained; if further information is required, it is necessary to resort to hydrolysis and the recovery and identification of acids, alcohols and glycols.

The spectra of a number of these 'oil-modified' resins of stated composition have been published[25].

COPOLYESTERS OF GLYCOLS, *o*-PHTHALIC ACID AND UNSATURATED DICARBOXYLIC ACIDS

Small amounts of unsaturated aliphatic acids, such as maleic, may be condensed with *o*-phthalic acid and glycols to form copolyesters. The infra-red spectrum gives no satisfactory evidence of the nature of these additives unless substantial amounts are present, and n.m.r. examination or hydrolysis is preferred for their examination.

POLYESTERS OF TEREPHTHALIC ACID

These polyesters are easily recognised from the C—O absorption bands at 8·0 μm and 9·0 μm, and the benzene ring substitution band which occurs at 13·8 μm. The last band differs markedly in position from the *p*-substitution band in aromatic hydrocarbons which is near 12 μm; this large shift may arise from extended conjugation between the C═O groups and the benzene nucleus since a similar shift to greater wavelength occurs with the benzoates.

Although in principle a large number of terephthalate resins can be made, in fact only two, polyethylene terephthalate, and polybis-1 : 4-hydroxymethyl cyclohexane terephthalate (Spectra 4.29(a), 4.29(b) and 4.30) are now commonly encountered.

Bis-1 : 4-hydroxy-methyl cyclohexane has *trans-* and *cis-*isomers. The spectra of the terephthalate polymers from each of the glycols in a pure state, in both crystalline and amorphous conditions, are given by Boye[26].

Comparison of the spectra shows that, as with the *o*-phthalate resins, the spectrum is dominated by the absorption of the phthalate group, and the modification of the spectrum with different glycols, although well marked, is largely confined to the medium-intensity bands in the 9 μm to 13 μm region.

An additional problem with terephthalate ester resins is that many of them can crystallise, and with some fabrication methods, particularly those employed in the manufacture of film and fibre, molecular orientation may be introduced. The spectra of the crystalline and amorphous resins differ, particularly in the 9 μm to 13 μm region, while the relative intensity of the absorption bands is influenced by orientation. Examples of these effects are given in Spectra 4.29(a) and 4.29(b), which are spectra of straight polyethylene terephthalate in different physical states. Daniells and Kitson[27] have studied the variation in the spectrum of polyethylene terephthalate due to crystallinity, while Krimm[28] shows the relation between band intensity and orientation.

A quantitative method for the measurement of the amorphous content of polyethylene terephthalate samples is given by Miller and Willis[29].

If it is desired to examine polyethylene terephthalate fibre in its original form the 'fibre grid' method described by Beaven *et al.* is useful[30].

COPOLYESTERS OF GLYCOLS WITH TEREPHTHALIC AND OTHER ACIDS

The spectrum of the condensate of terephthalic acid with mixed ethylene and diethylene glycols shows rather small differences from that of the simple ethylene glycol ester, although the resin is evidently amorphous. The additional band at 10·6 μm arises from the presence of the diethylene glycol (Spectrum 4.31).

Polyesters formed from a glycol with terephthalic acid and another dibasic acid are of some importance. The principal representatives here are poly(ethylene terephthalate/sebacate) (Spectrum 4.32) and poly-(ethylene terephthalate/isophthalate) (Spectrum 4·33). Aliphatic ester groups are evident in the spectrum of the first material at about 8·5 μm, while additional absorption bands at 6·2 μm and 10·7 μm are evident in the latter. Clearly the possibilities in this field are very large, and unless the spectrum corresponds exactly to that of a straight terephthalate ester, such as polyethylene terephthalate, the features of the spectrum must be carefully considered, since in many cases the spectrum will not give an unambiguous answer. For example, the additional absorption band near 8·5 μm in Spectrum 4.32 could be due to a dibasic aliphatic acid other than sebacic.

267

Interpretation of these spectra must therefore be done with particular care, and in any case of doubt the material should be hydrolysed and the acids and glycols recovered for identification. The spectra of the glycols commonly used are given in Spectra G.1 to G.11.

N.m.r. examination is very useful with soluble resins.

CARBONATE RESINS

Carbonates, whether aliphatic or aromatic, have distinctive and easily recognised spectra. The carbonyl band occurs at relatively short wavelength (near 5·7 μm) while the C—O absorption band is close to 8 μm.

The most common alkyl carbonate is poly-(diethylene glycol bis-allyl) carbonate (Spectrum 4.34). Other than the C=O and C—O bands mentioned above, the spectrum is rather weak and broad in character. This resin is completely insoluble, and the best spectrum is obtained by powdering the sample with a diamond faced spatula and using the alkali halide disk method.

A number of aromatic carbonate resins are available commercially. These products are all based on p-substituted biphenols, and their spectra are rather similar. The most common is the carbonate from so-called diphenylol propane.

Spectrum 4.35 is recognised without difficulty as that of an aromatic carbonate from the exceptionally short wavelength carbonyl band (5·65 μm) and the intense C—O band at 8·1 μm. The p-substituted benzene nucleus causes the appearance of the relatively strong band near 12 μm.

The spectra of polycarbonates from different phenols are chiefly distinguished in the 9 μm to 11 μm region[31].

Unless the spectrum of an aromatic polycarbonate can be matched directly against Spectrum 4·35 or one of the spectra reproduced in Ref. 31, identification is best performed by examination of the phenol recovered by hydrolysis (p. 248) by chemical, U.V. and infra-red tests.

ALIPHATIC POLYESTERS AND ESTER URETHANE RUBBERS

Polyesters from glycols, principally ethylene and propylene glycol with dibasic acids, are used as plasticisers, particularly in polyvinyl chloride compositions. They are discussed in Chapter 11. Another important use of these materials is in the formation of the polyester urethane rubbers.

In these materials, the terminal hydroxyl groups of a relatively low molecular weight polyester are reacted with an aromatic di- or poly-isocyanate, to produce a chain lengthened and cross-linked rubbery product. The chemistry of the cross-linking process is entirely analogous to that described for the polyether urethanes (p. 452), and in the spectrum of a polyester urethane therefore, one may expect the absorption bands of an ester, together with the bands of urethane and substituted urea groups.

The spectrum of polyethylene adipate cross-linked with naphthalene di-isocyanate is typical of the spectra of these products (Spectrum 4.36). The presence of aliphatic ester is evident from the absorption bands at 5·8 μm and 8·5 μm, while the carbamate structure gives rise to bands at 5·9 μm and 6·5 μm with a weak urea band at 6·1 μm. Occasionally, in an imperfectly cured resin, residual isocyanate absorption appears at 4·4 μm. This may cause confusion with the nitrile band in, for example, acrylo-nitrile copolymers, but the latter will usually not contain carbamate (urethane) groups.

The moderately intense band at 12·8 μm arises from the naphthalene nucleus, and gives a useful clue as to the cross-linking agent present. Other cross-linking agents also show useful structure in this region[32]. However, experience suggests that it is unwise to attempt to deduce the type of ester directly from the spectrum of the resin. Hydrolysis of the material, with the recovery for identification of the glycol and acid (or acids), is the only reliable method. In the hydrolysis, the diamine (or polyamine) corresponding to the isocyanate used may be recovered from the products for identification[32].

POLYESTERS OF CHLORINATED ACIDS

POLYESTERS OF HET ACID AND TETRACHLOROPHTHALIC ACID

Polyesters based on HET acid are comparatively easy to recognise from their infra-red spectra. A useful feature is the medium-strength sharp band at 6·25 μm, and there is a series of weak rather broad bands in the 10 μm to 15 μm region (see Hummel[24] Figures 819 and 820). Polyesters of tetrachlorophthalic acid have less satisfactory infra-red spectra for identification, but the rather sharp band of medium intensity at about 6·5 μm seems to be of some value in recognising these materials (Hummel Figure 818). It is very probable that these acids will be used together with o-phthalic and possibly other acids, and this can usually be confirmed by hydrolysis and the identification of hydrolysis products. These resins are sometimes encountered mixed with either monomeric or polymerised styrene, from which they may be separated, in suitable cases, by solvent extraction.

POLYMETHYL α–CHLOROACRYLATE

This resin is recognised fairly easily from its spectrum (Spectrum 4.37) as it has a double carbonyl band at about 5·8 μm and 5·85 μm. The C—O structure is also rather unusual in appearance. The split carbonyl band is thought to arise from the presence of a chlorine atom on the carbon α to the carbonyl group, and is presumably a feature of the spectra of other esters of chloroacrylic acid, although we have no experience of such compounds.

OTHER ESTER RESINS

Although the spectra of the most commonly encountered ester resins have been described in some detail above, there are so many possibilities in this field that inevitably others will be encountered. Hummel[24] gives the spectra of a number of ester resins, while further examples appear in Ref. 25 and in *Sadtler Commercial Spectra Series D*[33]. In using these compilations for identifying ester resins, it is particularly important to consider how far the spectrum is really characteristic of the substance sought, as opposed to the spectra of other similar substances. Generally speaking it is a more profitable procedure to identify an ester from hydrolysis products than to attempt direct identification based on weak features of the spectrum.

THE STUDY OF STEREOREGULARITY IN POLYMERS BY N.M.R. SPECTROSCOPY

Polymers of the general structure

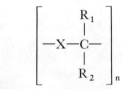

where X is any bifunctional atom or group, e.g. $-CH_2-$ or $-O-$ and R_1 and R_2 are different.

can in principle exist in different stereo-isomeric forms, which are distinguished by differences in the relative configurations of the successive repeat units along the polymer chains.

Examples of such polymers are:

X = CH$_2$	R$_1$ = CH$_3$	R$_2$ = H	Polypropylene
X = CH$_2$	R$_1$ = CH$_3$	R$_2$ = C—O—CH$_3$ (with C=O)	Polymethyl methacrylate
X = CH$_2$	R$_1$ = H	R$_2$ = Cl	Polyvinyl chloride
X = O	R$_1$ = H	R$_2$ = CH$_3$	Polyacetaldehyde

Stereoregular polymers are those in which repeating units follow one another in the polymer chains in a regular configurational sequence. The simplest stereoregular sequence is that in which the successive repeat units all have the same configuration. This is known as the isotactic species (see Figure 4.14).

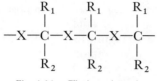

Fig. 4.14. *The isotactic species*

This graphical presentation follows the convention for describing simple stereoisomeric molecules. The backbone of the polymer is considered to be fully extended into a regular zig-zag and a projection is drawn into a plane at right angles to that containing the zig-zag.

The other common regular form is that in which successive repeat units alternate in opposite configurations. This is the syndiotactic species which, by the same convention, is as in Figure 4.15.

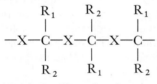

Fig. 4.15. *The syndiotactic species*

Examples of polymers which have been prepared as both the substantially isotactic and syndiotactic species include polymethyl methacrylate and polypropylene. The pure isotactic species has been prepared in the case of a number of the higher α-olefin polymers.

The backbone atoms of many substantially pure stereoregular polymers tends to assume a preferred conformation (i.e. the molecule has a preferred shape in space). For example, the isotactic species of polypropylene tends to assume the conformation in which the backbone carbon atoms lie on a regular helix, in which three monomer units form a complete turn. In these regular conformations the polymer chains are able to pack together in a three-dimensional crystalline arrangement, and in this state examination by X-ray diffraction enables both the conformation and the configuration of the polymer to be determined. This is generally the case with those α-olefin polymers which are encountered commercially as the substantially pure isotactic species.

The general definition given above of potentially stereoregular polymers includes all the well-known vinyl and acrylic polymers, but as commonly available commercially, they consist of mixed so-called atactic

271

forms, which are either completely amorphous, or show only very low crystallinity. In these cases X-ray diffraction examination gives no evidence on the overall stereochemical composition of the sample. However, the n.m.r. spectrum gives evidence of short-period regularity in the polymer chains, and the ability to investigate the nature of configurational sequences in polymers containing mixed sequences is of considerable interest to the polymer chemist. It is also important for the analyst who is using the n.m.r. spectrum for qualitative and quantitative analysis to appreciate the effects of configurational differences which may occur in the spectra of a number of common acrylic and vinyl polymers and copolymers.

THE N.M.R. SPECTRUM OF POLYMETHYL METHACRYLATE

Polymethyl methacrylate was the first polymer whose stereochemical structure was deduced from the n.m.r. spectrum. It is a particularly simple example of this form of analysis, because the α-carbon atom of the repeat unit is doubly substituted, hence there is no spin-spin coupling between the α and β protons, and the effects of stereochemical arrangements are seen clearly without decoupling the resonances. The original work on this polymer was done by Bovey and Tiers[34], and the following observations follow the general line of their argument.

The proton resonance spectra of three stereochemically distinguishable forms of polymethyl methacrylate are shown in Figures 4.16, 4.17 and 4.18. The spectra are recorded at 100 MHz in dichlorobenzene solution at 150°C.

A sharp resonance at 6·46 τ is assigned to the —O—CH$_3$ protons i.e. the resonance is shifted well downfield by the influence of the adjacent oxygen atom. This resonance is identical in all three spectra, and is not shown so that the significant bands may be presented on as large a scale as possible.

However there are substantial differences in the spectra in the region of 9 τ. The substantially isotactic sample (see Figure 4.16) shows a strong resonance at 8·68 τ with only weak features at 8·82 τ and 8·93 τ. The syndiotactic sample (see Figure 4.17) shows a strong resonance at 8·93 τ and weak resonances at 8·68 τ and 8·82 τ. The third sample, which is that of a commercial sample of the polymer, shows resonances in the same positions, but here 8·93 τ is strong, 8·82 τ is appreciable in intensity, while 8·68 τ is extremely weak (see Figure 4.18).

Resonances in this region of the spectrum may be assigned to the protons of the α-CH$_3$ group and the evidence is that this must exist in three different chemical environments, each of which corresponds to a particular chemical shift of the resonance of the α-CH$_3$ protons. If the possible environments of the α-CH$_3$ group are considered, it is found that, if account is taken of the configuration of the monomer unit on either side of a particular monomer unit in the structure, there are three (and only

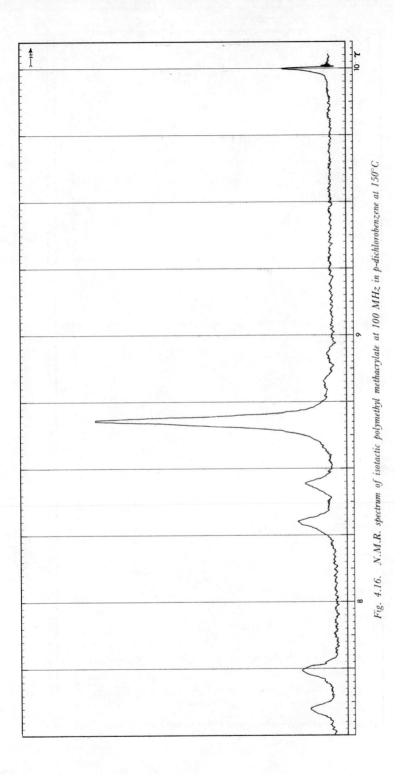

Fig. 4.16. N.M.R. spectrum of isotactic polymethyl methacrylate at 100 MHz in p-dichlorobenzene at 150°C

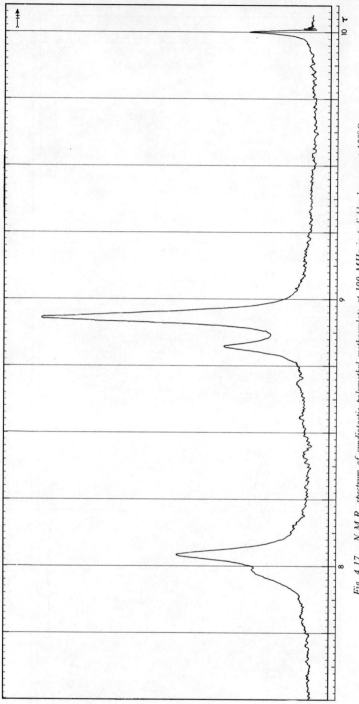

Fig. 4.17. N.M.R. spectrum of syndiotactic polymethyl methacrylate at 100 MHz in p-dichlorobenzene at 150°C

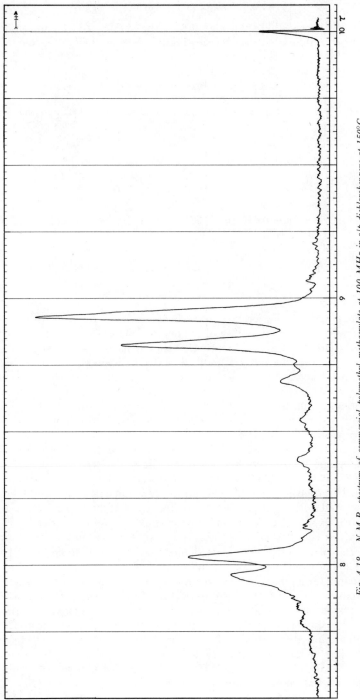

Fig. 4.18. N.M.R. spectrum of commercial polymethyl methacrylate at 100 MHz in o/p-dichlorobenzene at 150°C

three) distinguishable environments for the central α-CH$_3$ protons. This group is represented by R$_1$ in Figure 4.19 and R$_2$ represents the (—C=O)—O—CH$_3$ group.

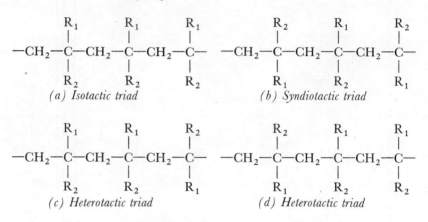

(a) Isotactic triad   (b) Syndiotactic triad

(c) Heterotactic triad   (d) Heterotactic triad

Fig. 4.19. The four triad structures

These are called triad structures because they consist of runs of three monomer units. Examination of the environment of the central R$_1$ group in the triad structures of Figure 4.19 shows that three different environments can exist (the forms (c) and (d) are not chemically distinguishable). It is therefore reasonable to suppose that the three resonances observed in the α-CH$_3$ region relate to the presence of these distinguishable triads of monomer units, and we can conclude that in principle it should be possible to observe effects due to these triad structures in any polymer of this general type.

Because Figures 4.16 and 4.17 are known to be the spectra of the isotactic and syndiotactic species, it may be concluded that the resonance at 8·68 τ is due to the isotactic triad, and that at 8·93 τ to the syndiotactic triad. It is then reasonable to assign the resonance at 8·82 τ to the heterotactic triad as this is prominent in the spectrum of the commercial sample, which is expected to have the most irregular structure. Indeed, in triad terms the commercial resin may be described as runs of syndiotactic material, with faults, which are demonstrated by the heterotactic triads, there being very little isotactic material.

Further differences are observed in the spectra near 8 τ, where the resonance of the —CH$_2$— group occurs. Figure 4.16 shows a quartet in this region while Figure 4.17 shows a singlet. Evidently the quartet arises by spin-spin coupling because the two protons of the —CH$_2$— group are in a different chemical environment, whereas in the form showing a singlet these two protons are equivalent. If, as in the treatment above, we take nearest neighbours into account, we can describe only two forms (see Figure 4.20(a) and (b)).

In Figure 4.20(a), one proton of the marked —CH_2— group lies between two R_1 groups while the other lies between two R_2 groups. Thus in principle the two protons of this CH_2 group are distinguishable. However, in Figure 4.20(b), each proton of the marked CH_2 group lies between R_1 and R_2 and evidently these two protons should be equivalent.

In accordance with the usual nomenclature for simple substances containing two asymetric centres, these forms are described as meso and racemic diads respectively. The isotactic polymer contains only the meso diad, while the syndiotactic polymer contains only the racemic diad. Thus

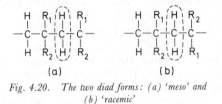

(a)   (b)

Fig. 4.20.   The two diad forms: (a) 'meso' and
(b) 'racemic'

the syndiotactic polymer should show a singlet resonance, and the isotactic polymer a quartet, in the 8 τ region, and this is found in practice. It has therefore been possible to differentiate and identify the stereochemically pure forms of polymethyl methacrylate purely from the features of the n.m.r. spectrum, and in principle similar observations can be made

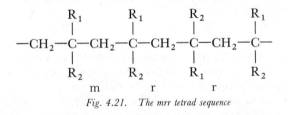

Fig. 4.21.   The mrr tetrad sequence

on other polymers. In practice most of the common polymers which could exist in pure isotactic and syndiotactic forms are not doubly substituted on the α-carbon atom, and the methylene proton pattern is very complex because of spin-spin splitting between the protons on the α- and β-carbon atoms.

Turning now to the spectrum of the commercial resin (Figure 4.18) the —CH_2— resonance shows differences from that in either of the other two spectra. The singlet present in the spectrum of the syndiotactic polymer has a second resonance appearing on the low-field side. Following the argument that the resonance of the —CH_3 group shifts with configurational changes in adjacent units in the chain, the —CH_2— might be expected to be similarly affected. Thus the singlet resonance of the racemic diad (r diad) should differ in the tetrad sequences rrr, mrr (or rrm, which is presumed equivalent) and mrm. An example of a tetrad sequence is shown in Figure 4.21.

277

The commercial polymer, being principally syndiotactic, would contain a comparatively large number of rrr sequences, thus accounting for the methylene singlet resonance coincident with that in the syndiotactic polymer. However, some mrr (or rrm) units would be expected, and this accounts for the additional singlet resonance at a lower field. The resonance of the mrm tetrad is presumably again at a lower field, but evidently there are not sufficient of these tetrads for the resonance to be observed. The situation with the m diad quartet will be more difficult to observe. Evidently each of the three distinguishable tetrads mmm, mmr (rmm) and rmr will produce a quartet, and in practice this will produce a large number of weak resonances which will be partially obscured by the stronger resonances of the r-centred tetrads. This accounts for the apparently randomly-spaced weak bands near 8 τ.

It is apparent that the same argument can be extended to the possibility of observing structure on the three α-CH$_3$ resonances due to pentad effects, and indeed there is evidence of this from 220 MHz spectra of polymethyl methacrylate. So far there has been no indication in the 8 τ region of structure due to hexad sequences; on the other hand sequence effects involving hexads have been observed in copolymers of vinylidene chloride and vinyl acetate[35].

QUALITATIVE IDENTIFICATION OF ACRYLIC RESINS FROM THE N.M.R. SPECTRUM

An example of the identification of an acrylic ester homopolymer is given on p. 204, where the spectrum is compared with that of a vinyl ester resin. The identification of copolymers of acrylates or methacrylates with minor amounts of aliphatic monomers is not very satisfactory. On the other hand the n.m.r. spectrum gives good evidence of the presence of unsaturated components because the resonances of protons on double bonds and aromatic rings fall well to low field of those of saturated esters. An example is methyl methacrylate/styrene copolymer, which is now considered as an example of quantitative analysis of an acrylic copolymer from the n.m.r. spectrum.

QUANTITATIVE ANALYSIS OF ACRYLIC COPOLYMERS FROM THE N.M.R. SPECTRUM

The n.m.r. spectrum of a methyl methacrylate/styrene copolymer is as shown in Figure 4.22. The resonances of the aryl protons are seen near 3 τ. The methoxyl resonance is of particular interest as it appears within the rather wide range of 6·5 τ to 7·6 τ depending upon the environment of the group. It is possible to calculate from the reactivity ratio of the two monomers the relative proportions of the triad sequences MMM, MMS

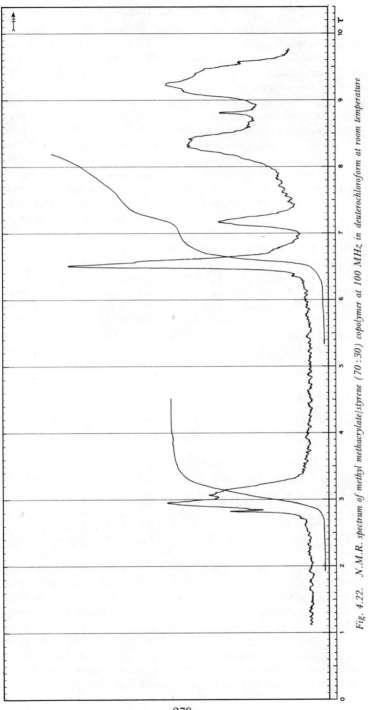

Fig. 4.22. N.M.R. spectrum of methyl methacrylate/styrene (70 : 30) copolymer at 100 MHz in deuterochloroform at room temperature

279

and S$\underline{M}$S in a copolymer and to relate these to the relative areas of resonances in this region of the spectrum. From this it may be deduced[36] that the resonance near 6·5 τ arises from the methoxyl protons of the central $\underline{M}$ unit in the sequence M$\underline{M}$M, that near 7·2 τ from M$\underline{M}$S and finally that a resonance near 7·6 τ (which is very weak in Figure 4.22) arises from the $\underline{M}$ unit in the triad S$\underline{M}$S. This assignment appears reasonable, as one would expect the resonance to move to higher field with decreasing polarity of the environment. Additional structure appears on each of these resonances because isotactic, syndiotactic and heterotactic arrangements are possible in each case and these have a small but observable effect on the positions of the resonances.

The quantitative analysis of such a copolymer is relatively straightforward. The low-field resonance near 3 τ is due to the 5 protons of the benzene ring of styrene. For these 5 protons there are 3 protons in the aliphatic portion of the styrene molecule.

If the area of the resonance at about 3 τ be A,

and the area of the remaining resonances due to the copolymer be B,

Then, of the latter area, $3A/5$ is due to styrene.

Hence $(5B-3A)/5$ is due to the resonance of the protons of methyl methacrylate. For each $A/5$ molecules of styrene combined in the copolymer there are $(5B-3A)/(5 \times 8)$ molecules of methyl methacrylate (because there are 8 aliphatic protons in the methyl methacrylate repeat unit in the copolymer).

Hence,

$$\frac{\text{Molar concentration of styrene}}{\text{Molar concentration of methyl methacrylate}} = \frac{A}{5} \times \frac{5 \times 8}{5B - 3A}$$

$$= \frac{8A}{5B - 3A}$$

from which the w/w ratio may be calculated.

QUALITATIVE AND QUANTITATIVE ANALYSIS OF COPOLYESTERS FROM THE N.M.R. SPECTRUM

Provided that a copolyester sample is soluble, a most satisfactory qualitative and quantitative analysis is possible from the n.m.r. spectrum. The spectrum of the polyester from ethylene glycol and terephthalic acid and the spectra of a number of copolyester samples derived from mixed aryl, alkyl and unsaturated acids and commonly encountered glycols are given in Figures 4.23 to 4.28, and Figure 4.29 gives a key to the origin of the resonances. Once the resonances have been identified, the quantitative analysis is straightforward from band area measurement, but the integral curves have been omitted from the figures for clarity.

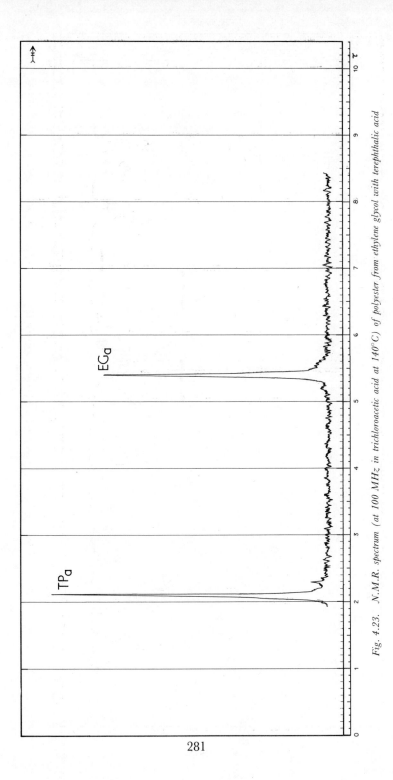

Fig. 4.23. N.M.R. spectrum (at 100 MHz in trichloroacetic acid at 140°C) of polyester from ethylene glycol with terephthalic acid

281

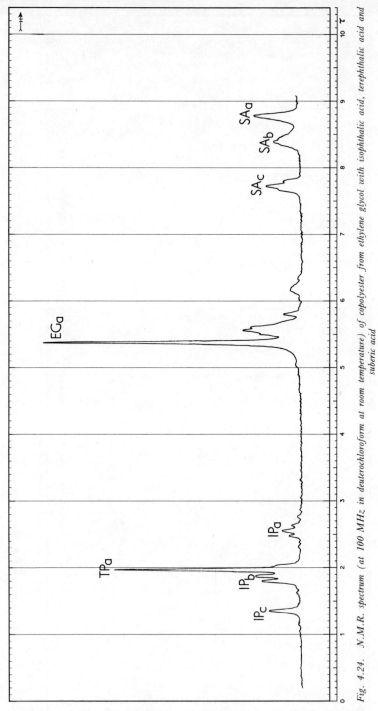

Fig. 4.24. N.M.R. spectrum (at 100 MHz in deuterochloroform at room temperature) of copolyester from ethylene glycol with isophthalic acid, terephthalic acid and suberic acid

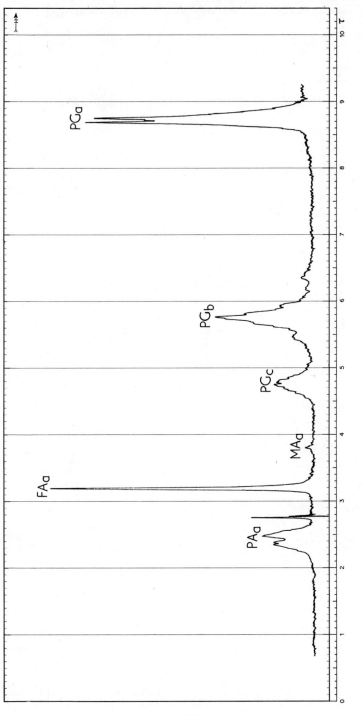

Fig. 4.25. N.M.R. spectrum (at 100 MHz in deuterochloroform at room temperature) of copolyester from propylene glycol with o-phthalic acid, fumaric acid and maleic acid

283

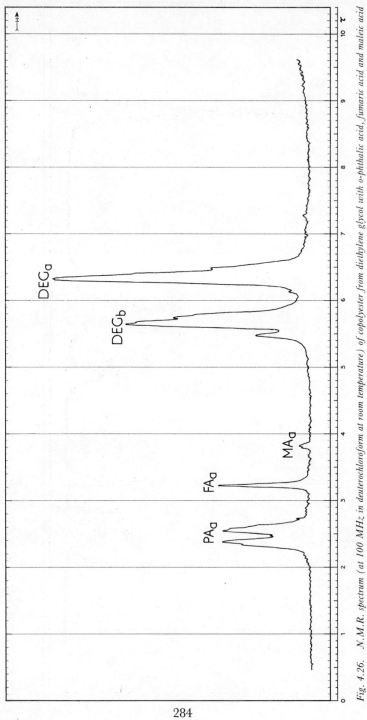

Fig. 4.26. N.M.R. spectrum (at 100 MHz in deuterochloroform at room temperature) of copolyester from diethylene glycol with o-phthalic acid, fumaric acid and maleic acid

284

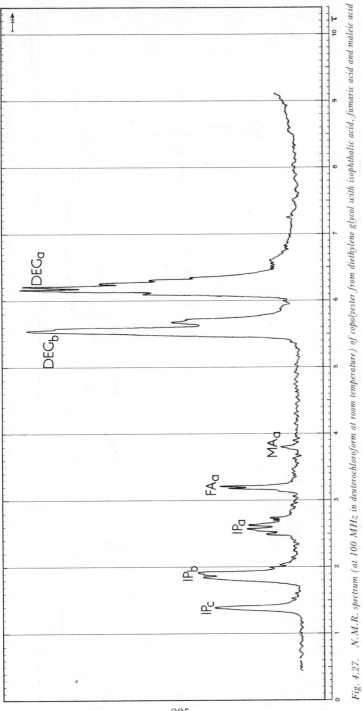

Fig. 4.27. N.M.R. spectrum (at 100 MHz in deuterochloroform at room temperature) of copolyester from diethylene glycol with isophthalic acid, fumaric acid and maleic acid

285

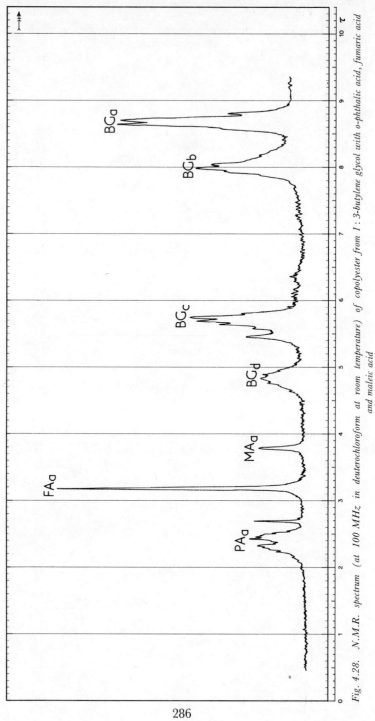

Fig. 4.28. N.M.R. spectrum (at 100 MHz in deuterochloroform at room temperature) of copolyester from 1 : 3-butylene glycol with o-phthalic acid, fumaric acid and maleic acid

286

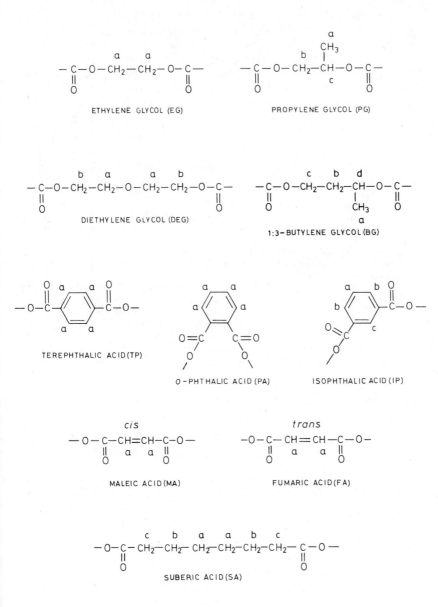

ETHYLENE GLYCOL (EG)

PROPYLENE GLYCOL (PG)

DIETHYLENE GLYCOL (DEG)

1:3-BUTYLENE GLYCOL (BG)

TEREPHTHALIC ACID (TP)

O-PHTHALIC ACID (PA)

ISOPHTHALIC ACID (IP)

MALEIC ACID (MA)

FUMARIC ACID (FA)

SUBERIC ACID (SA)

Fig. 4.29.   *Key to origin of the resonances in Figs. 4.23 to 4.28*

In particular the resonances in the 1τ to 3τ region arise from aryl protons and the patterns from the different phthalic acids are easily distinguished. The resonances of protons in the glycol residues all appear to high field, together with those of any alkyl acid constituents. The most striking feature however is the ease with which copolymerised maleic and fumaric acid residues may be observed in the 3τ to 4τ region, and this is a very useful test for the presence of these unsaturated constituents in copolyesters.

REFERENCES

1. HASLAM, J., and SOPPET, W. W., *Analyst, Lond.*, **75**, 63 (1950)
2. HASLAM, J., JEFFS, A. R., and WILLIS, H. A., *Analyst, Lond.*, **86**, 44 (1961)
3. HASLAM, J., HAMILTON, J. B., and JEFFS, A. R., *Analyst, Lond.*, **83**, 66 (1958)
4. EASTERBROOK, W. C., and HAMILTON, J. B., *Analyst, Lond.*, **78**, 551 (1953)
5. MARTIN, A. J. F., and JAMES, A. T., *Biochem. J.*, **50**, 679 (1952)
6. HASLAM, J., SQUIRRELL, D. C. M., GROSSMAN, S., and LOVEDAY, S. F., *Analyst, Lond.*, **78**, 92 (1953)
7. RUDDLE, L. H., and WILSON, J. R., *Analyst, Lond.*, **94**, 105 (1969)
8. HASLAM, J., (Ed. KAPPELMEIER, C. P. A.), *Chemical Analysis of Resin-Based Coating Materials*, Interscience, 354 (1959)
9. SQUIRRELL, D. C. M., *Automatic Methods in Volumetric Analysis*, Hilger & Watts, London, 27 (1964)
10. KOLTHOFF, I. M., and HARRIS, W. E., *Ind. Engng Chem. analyt. Edn*, **18**, 161 (1946)
11. HASLAM, J., and SQUIRRELL, D. C. M., *J. appl. Chem., Lond.*, **11**, 244 (1961)
12. CROWTHER, A. B., and LARGE, R. S., *Analyst, Lond.*, **81**, 64 (1956)
13. HASLAM, J., and SQUIRRELL, D. C. M., *Analyst, Lond.*, **82**, 511 (1957)
14. CONIX, A., *Makromolek. Chem.*, **26**, 226 (1958)
15. ESPOSITO, G. G., and SWANN, M. H., *Analyt. Chem.*, **33**, 1854 (1961)
16. ESPOSITO, G. G., and SWANN, M. H., *Analyt. Chem.*, **34**, 1173 (1962)
17. TRAXTON, M. E., *Chemy Ind.*, 1613 (1966)
18. NOVAK, V., *Plaste Kautsch.*, **12**, 202 (1965)
19. HASLAM, J., and JEFFS, A. R., *Analyst, Lond.*, **83**, 455 (1958)
20. KAWASAKI, A., FURAKAWA, J., TSURUTA, T., WASAI, G., and MAKIMOTO, T., *Makromolek. Chem.*, **49**, 76, 1961
21. MILLER, R. G. J., and WILLIS, H. A., *J. appl. Chem., Lond.*, **6**, 385 (1956)
22. HARVEY, M. C., and PETERS, L. L., *Analyt. Chem.*, **32**, 1725 (1960)
23. *Sadtler Standard Spectra*, The Sadtler Research Laboratories, Philadelphia 2
24. HUMMEL, D. O., *Infra-Red Analysis of Polymers, Resins and Additives, An Atlas*, **1**, Pt II, Wiley Interscience (1969)
25. *Infra-Red Spectroscopy. Its Use in the Coatings Industry*, Federation of Societies of Paint Technology, Philadelphia (1969)
26. BOYE, C. A., *J. Polym. Sci.*, **55**, 263 (1961)
27. DANIELLS, W. W., and KITSON, R. E., *J. Polym. Sci.*, **33**, 161 (1958)
28. KRIMM, S. *Adv. Polym. Sci.*, **2**, 51 (1960)
29. MILLER, R. G. J., and WILLIS, H. A., *J. Polym. Sci.*, **19**, 485 (1956)
30. BEAVEN, G. H., and JOHNSON, E. A.; WILLIS, H. A., and MILLER, R. G. J., *Molecular Spectroscopy*, Heywood, London (1961)
31. SCHNELL, H., *Trans. J. Plast. Inst.*, **28**, 143 (1960)
32. CORISH, P. J., *Analyt. Chem.*, **31**, 1298 (1959)
33. *Sadtler Commercial Spectra Series D*, The Sadtler Research Laboratories. Philadelphia 2
34. BOVEY, F. A., and TIERS, G. V. D., *J. Polym. Sci.*, **44**, 173 (1960)
35. JOHNSEN, V., and KOLBE, K., *Kolloidzeitschrift und Z. fur Polym.*, **220**, 145 (1967)
36. ITO, K., and YAMASHITA, Y., *Jnl. Polym. Sci., Pt. B.*, **3**, 625 (1965)

5

NYLON AND RELATED POLYMERS

In the first part of this chapter attention is drawn to the determination of plasticiser in nylon preparations and to the chemical hydrolysis of nylon and related polymers by means of 50% v/v hydrochloric acid solution.

The chemical examination of the hydrolysis products of polymers, copolymers and terpolymers is discussed and information is provided on the chromatographic examination of these products.

Other tests that are dealt with include the pyrolysis/gas chromatographic identification of nylons, and determinations of amine end-groups, carboxyl end-groups, lubricants, relative viscosity, spin finishes, water, delustrants, fillers and stabilising elements.

Some observations are made on the identification of polyimides.

In the second part of the chapter the infra-red spectra of many nylon polymers, and of some polyimides and primary and tertiary amide resins, are described and discussed. Attention is drawn to the distinguishing features in the spectra of simple nylon polymers, and also to the differences which occur in the spectra of some nylon resins due to variations in the crystalline form of the sample. Identification of nylon copolymers directly from their infra-red spectra is not recommended, and it is suggested that with these materials hydrolysis and identification of the hydrolysis products is a much more reliable procedure.

Information is provided on the n.m.r. spectra of some nylon polymers, and it is shown that the n.m.r. procedure is a useful method for the differentiation of e.g. 6 : 6 / 6 and 6 : 6 / 6 : 10 copolymers and for the determination of the proportions of the constituents in them.

NYLON

Nylon was discovered by W. H. Carothers in America in 1928.

The method of production was to prepare a combination of a diamine and a dibasic acid to give a salt, that, on heating, lost water to yield a linear, long-chain polymer. Since there are many diamines and many dibasic acids that lend themselves to this reaction, the theoretical number of possible nylon polymers is considerable, but, in fact, it is probable that only two types are in large-scale production by this method. One of these was the first nylon type to be developed commercially and was made from hexamethylene diamine $H_2N \cdot (CH_2)_6 \cdot NH_2$ and adipic acid $HOOC \cdot (CH_2)_4 \cdot COOH$. This was called type 6 : 6 (six : six) nylon from the number of carbon atoms in the parent diamine and dibasic acid respectively.

Subsequently type 6 : 10 (six : ten) nylon was developed from hexamethylene diamine and sebacic acid $HOOC \cdot (CH_2)_8 \cdot COOH$.

A different method of making nylon polymers by the condensation of certain amino acids was discovered later and is now in use in certain parts of the world. Types of nylon so made are described by a single number derived from the number of carbon atoms in the amino acid from which they are formed. Two of those in commercial manufacture are type 6 (six) nylon and type 11 (eleven) nylon made from 5-amino caproic acid and ω-amino undecanoic acid respectively.

In addition to straight nylon polymers it is possible to make copolymers by heating together different nylon salts or nylon salts and appropriate amino acids; many combinations of polymers and copolymers are prepared as blends, and polymers, copolymers and blends can all be plasticised to modify their properties.

The analytical chemist is called upon to deal with substances of this type of widely differing properties.

He quite readily may meet, in his general work, preparations derived from the following:

Nylon 6 : 6
Nylon 6 : 10
Nylon 6
Nylon 11
Methoxy-methyl Nylon 6 : 6
Igamid U. Not strictly a nylon, being derived from a di-isocyanate and a diol
Nylon 6 : 6 / 6 : 10
Nylon 6 : 6 / 6
Nylon 6 : 6 / 6 : 10 / 6
Nylon 6 : 6 / 6 / PACM : 6. Here a new amine is involved, so-called
    PACM, *pp'*-diamino-dicyclo-hexyl methane.

Methoxy-methyl nylon is 6 : 6 nylon polymer that has been modified by treatment with formaldehyde and methanol. This treatment introduces methoxy-methyl groups probably as follows:

$$-R-CO-N-R-$$
$$\underset{\displaystyle CH_2OCH_3}{\overset{\displaystyle |}{}}$$

and renders the nylon polymer soluble in alcohol.

There is a certain amount of controversy about the melting points of samples of nylons and the figures depend, to a certain extent, on the methods of test. In our experience, the most satisfactory laboratory method is a birefringence method due to the Koflers (A.S.T.M. D-2117-64) in which a Kofler hot-stage microscope is used. In this method, crystalline

Table 5.1. Melting points of nylons

Nylon	m.p.
Nylon 6	220–230°C
Nylon 6 : 6	258–268°C
Nylon 6 : 10	218–228°C
Nylon 11	189–191°C
Nylon 12	177–180°C

melting point is defined as the temperature at which all traces of double refraction vanish. Using this method the figures shown in Table 5.1 have been obtained on various nylons.

REMOVAL AND DETERMINATION OF PLASTICISER

When a nylon composition is received for examination it is first necessary to isolate the polymer, free from plasticiser, before seeking to identify it. A solution and precipitation method is employed and this is illustrated by the method used to determine N-ethyl *p*-toluene sulphonamide plasticiser in a 6 : 6 / 6 : 10 / 6 preparation.

Procedure

1 g of the sample is weighed into a beaker (100 ml) containing a glass rod. 5 ml of 90% w/w formic acid solution are added and the mixture is allowed to stand overnight in order to effect solution of the sample. If the mixture is stirred frequently, most samples dissolve in about 4 h. Nylon 6 : 10 preparations may require heating on the water bath in order to effect solution. 10 ml of methyl alcohol are now added to the viscous solution of the polymer sample and the mixture is stirred thoroughly.

Most of the nylon copolymers will remain in solution at this stage but some precipitation will take place with the straight polymers and particularly so with nylon 6 : 10.

The main purpose of the addition of methyl alcohol is to dilute the formic acid solution.

70 ml of ether are now added slowly with stirring; at this stage most polymers yield granular precipitates but copolymers generally pre-cipitate as viscous gels. In such cases the precipitate, as it forms, is spread

as thinly as possible around the inside of the beaker and care is taken to avoid the formation of a solid mass of precipitated nylon.

After standing for a few minutes the precipitated nylon becomes more brittle and is easily removed from the sides of the beaker by means of the glass rod. After standing for at least 30 min, the ethereal solution is filtered off through a No. 1 porosity grade 3 crucible using suction; the beaker is carefully washed out with ether, and the precipitated nylon then transferred to the sintered glass crucible. The crucible and contents are then washed with four successive quantities of ether each of 20 ml, stirring the nylon with a glass rod after each addition of ether. The filtrate and washings are collected in a 500 ml beaker and the solution evaporated to dryness on a water bath, i.e. until there is no longer any odour of formic acid.

The beaker is cooled, 20 ml of ether are added and the plasticiser residue is dissolved by swirling and stirring with a glass rod. The solution is filtered through a No. 1 porosity grade 3 sintered glass crucible and the beaker washed with three successive quantities each of 20 ml of ether. The filtrate and washings are collected in a tared extraction flask and the ethereal solution evaporated to dryness on the boiling-water bath. The flask and contents are dried by heating for 1 h at 100°C and then for further periods of 30 min at 100°C until constant weight is obtained. The weight of plasticiser thus obtained is calculated as a percentage of the original sample.

The method has been shown to give excellent results in the determination of N-ethyl *p*-toluene sulphonamide in nylon 6 : 6 / 6 : 10 / 6, (40 : 30 : 30) N-ethyl *p*-toluene sulphonamide preparations at 8 and 18% levels.

The plasticiser-free nylon left in the crucible is dried by air suction and is then available for further examination.

HYDROLYSIS OF NYLONS AND RELATED POLYMERS WITH 50% v/v HYDROCHLORIC ACID SOLUTION

This test, applied to nylon samples from which the plasticiser has been removed or directly to plasticiser-free samples, is exceedingly important.

In the case of e.g. 6 : 6 nylon, the purpose of the test is to hydrolyse a known weight of the sample with 50% v/v hydrochloric acid solution (100 ml of this solution contain 50 ml of conc. hydrochloric acid solution of sp. gr. 1·18) and to recover, quantitatively, the adipic acid and hexamethylene diamine dihydrochloride produced as a result of the hydrolysis process. The weights obtained are noted.

6 : 10 nylon yields hexamethylene diamine dihydrochloride and sebacic acid in known proportions and nylon 6 yields only 5-amino caproic acid hydrochloride.

Methoxy-methyl 6 : 6 nylon loses the elements of methyl alcohol and formaldehyde in the hydrolysis process and consequently yields lower proportions of hexamethylene diamine dihydrochloride and adipic acid than the parent 6 : 6 nylon.

Copolymers, according to their composition, yield mixtures of hydrochlorides and mixtures of acids. The hydrolysis procedure has been studied in considerable detail by Haslam and Swift[1].

They hydrolysed samples of nylon in sealed tubes with hydrochloric acid solution for definite periods of time. The hydrolysis products were titrated potentiometrically with standard alkali. The first end-point gave a measure of the free hydrochloric acid, whilst e.g. the adipic acid resulting from the hydrolysis of 6 : 6 nylon was titrated between the first and second end-points. This difference between the first and second end-points gave a measure of the extent of the hydrolysis.

The results indicated that nylon 6 is hydrolysed with great ease, that the hydrolysis of nylon 6 : 6 proceeds to completion with greater difficulty and that nylon 6 : 10 is difficult to hydrolyse completely.

It is for reasons such as these that the routine hydrolysis procedure for nylons involves a prolonged boiling with hydrochloric acid solution.

The procedure is carried out as follows[2, 3]:

1 g of the sample is hydrolysed by boiling under reflux for 40 h with 20 ml of 50% v/v hydrochloric acid solution.

The hydrolysis product is transferred to a continuous liquid-liquid extractor. In their original work Haslam and Clasper used a particular form of apparatus[2, 3], although other forms of liquid-liquid extractor, available commercially, may prove to be equally reliable e.g. the 100 ROLUS made by Quickfit and Quartz Ltd, Stone, Staffordshire, England.

Extraction of the hydrolysis product with ether is carried out for 6 h. The ether extract is then evaporated to dryness on a water bath and the residue is dried to constant weight at 100°C.

The proportion of acid or mixed acids thus recovered is calculated as a percentage of the original nylon.

The aqueous solution remaining in the extractor is transferred to a tared basin and evaporated to dryness on the water bath. A small amount of water is then added and the solution is again evaporated to dryness. This process is repeated in order to remove the last traces of hydrochloric acid. The residue obtained is dried to constant weight at 100°C.

The proportion of hydrochloride or mixed hydrochlorides thus obtained is calculated as a percentage of the original nylon.

The following figures, quoted directly from the paper by Clasper and Haslam[2], illustrate how the application of the above method readily distinguishes e.g. between nylon 6 : 6, nylon 6 : 10 and nylon 6.

Results obtained on hydrolysing nylon 6 : 6, nylon 6 : 10 and nylon 6 with 50% v/v hydrochloric acid solution.

The results are calculated as percentages of the original commercial nylons.

293

Sample	Unhydrolysed Material %	Acid %	Base Hydrochloride %
Nylon 6 : 6 (a)	Nil	62·1	81·3
(b)	Nil	61·8	82·3
Nylon 6 : 10 (a)	0·32	71·1	65·8
(b)	0·36	69·5	67·3
Nylon 6 (a)	0·43	2·2	139·6
(b)	0·88	2·9	137·8

TESTS ON THE HYDROLYSIS PRODUCTS

ACID

The acid isolated on hydrolysis of most commercial nylons is usually either adipic acid or sebacic acid and in certain cases, i.e. with copolymers, it consists of a mixture of the two acids.

Two useful figures to determine on the acid obtained on hydrolysis are:

1. The melting point in a capillary tube.
2. The equivalent weight of the acid by titration of the alcohol solution of the acid with standard alkali.

Typical figures obtained on the acids produced on hydrolysis of commercial nylons are given below, alongside the generally accepted figures for the pure acids.

Acid produced on hydrolysis of 6 : 6 nylon

m.p. of acid	m.p. of pure adipic acid	Equivalent weight of acid	Equivalent weight of pure acid
148 to 149·5°C	150°C	73·6 to 74·5	73·0

Acid produced on hydrolysis of 6 : 10 nylon

m.p. of acid	m.p. of pure sebacic acid	Equivalent weight of acid	Equivalent weight of pure acid
132°C	132°C	102·9 to 103·5	101·0

Nylon 6, of course, produces negligible acid in the hydrolysis test.

If, from the above preliminary tests on the acid it is suspected that a mixture of acids is present, then it is necessary to determine the proportions of the two acids in the mixture, and moreover, to isolate samples of the individual acids.

DETERMINATION OF THE PROPORTIONS OF ADIPIC AND SEBACIC ACIDS IN MIXTURES OF THE TWO ACIDS

0·500 g of the mixed acids is equilibrated in a glass-stoppered bottle at 20°C for 30 min with 50 ml of water and 35 ml of ether. 25 ml of the aqueous layer are withdrawn and titrated very accurately with N sodium hydroxide solution delivered from a microburette using 1 drop of phenolphthalein solution as indicator. The proportion of adipic acid (and hence of sebacic acid) in the original mixed acids is then deduced from the volume of N sodium hydroxide solution used up in the above test and a graph drawn from the figures given in Table 5.2, which were obtained by carrying out the above test on known mixtures of the two pure acids.

Table 5.2

Adipic acid added g	Sebacic acid added g	Adipic acid in prepared mixture %	Titration in ml N NaOH of 25 ml of aqueous layer in equilibration test
0·5001	Nil	100	2·41
0·4000	0·1001	79·99	1·94
0·3501	0·1499	70·02	1·70
0·3001	0·2000	60·01	1·45
0·2002	0·3001	40·02	0·98
0·1498	0·3503	29·96	0·76
0·0999	0·3998	19·99	0·52
Nil	0·5005	Nil	0·07

The equilibration test, of course, is based on the preference, in the test, of the sebacic acid for the ether phase and the adipic acid for the aqueous phase.

ISOLATION OF THE INDIVIDUAL ACIDS IN THE MIXTURE

Having determined the percentage of adipic acid in the mixture, the volume of N sodium hydroxide solution equivalent to the whole of the adipic acid present in the 0·5 g of mixed acids taken for equilibration test is calculated[2, 4]. This volume, less that required in the titration of the aqueous aliquot in the equilibration test, is now added to the contents of the equilibration bottle, followed by the titrated aqueous aliquot.

In this way, the equilibration bottle now contains alkali equivalent to all the adipic acid in the bottle.

The contents of the bottle are mixed thoroughly, then transferred to a liquid-liquid extractor and extracted with ether. The ethereal solution is then evaporated to dryness and the residual sebacic acid recrystallised from water, then dried to constant weight at 120°C. The aqueous solution still in the liquid-liquid extractor is acidified with hydrochloric acid and re-extracted with ether. The ether extract is evaporated to dryness and the residual adipic acid recrystallised from water prior to drying to constant weight at 130°C.

From adipic acid–sebacic acid mixtures the above method yields acids of good quality as is shown by the following figures on recovered acids.

	Melting point	Equivalent weight
Adipic acid	147·0°C	74·8
Sebacic acid	130·5°C	103·5

BASE HYDROCHLORIDES

The base hydrochloride isolated on hydrolysis of many commercial nylons usually consists of hexamethylene diamine dihydrochloride or 5-amino caproic acid hydrochloride or mixtures of the two. There are a few cases where other hydrochlorides are produced but they are dealt with separately.

Two useful figures to determine on the hydrochlorides obtained on hydrolysis are:

1. The melting point in a capillary tube.
2. The chlorine content by potentiometric titration of an aqueous solution of a known weight of the hydrochlorides with standard silver nitrate solution.

Typical figures obtained on the hydrochlorides produced on hydrolysis of commercial nylons are given below, alongside the generally accepted figures for the pure hydrochlorides.

Hexamethylene diamine dihydrochloride produced on hydrolysis of 6 : 6 nylon

m.p. of hydrochloride	m.p. of pure hydrochloride	Chlorine content of hydrochloride	Chlorine content of pure hydrochloride
251°C	253°C	37·1 to 37·3%	37·5%

5-amino caproic acid hydrochloride produced on hydrolysis of nylon 6

| 122 to 125°C | 125°C | 20·8 to 20·9% | 21·1% |

If mixtures of hexamethylene diamine dihydrochloride and 5-amino caproic acid hydrochloride are involved then the following useful methods are available for their examination.

In the first method the 5-amino caproic acid hydrochloride in the mixture is determined by direct titration. 0·5 g of the mixed hydrochlorides

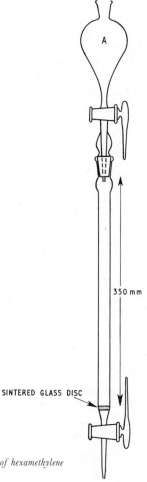

350 mm

SINTERED GLASS DISC

Fig. 5.1.   Ion exchange column for the determination of hexamethylene diamine

is dissolved in 20 ml of water and the solution is titrated with 0·1 N sodium hydroxide solution using methyl orange as indicator.

1 ml 0·1 N sodium hydroxide solution = 0·0165 g 5-amino caproic acid hydrochloride.

In the second method the hexamethylene diamine can be separated from a mixture of the hydrochlorides i.e. of hexamethylene diamine dihydrochloride and 5-amino caproic acid hydrochloride. The separated hexamethylene diamine is then determined by titration.

10 g of Amberlite I.R.A. 400, after drying at 100°C, are slurried with water, and the slurry transferred to an ion exchange column as shown in Figure 5.1.

Excess water is then run off until the resin in the column is just covered with water. Care must be taken to ensure that the liquid in the column does not drain below the head of the ion exchange resin, in order to avoid loss of ion exchange capacity of the resin.

The Amberlite I.R.A. 400 ion exchange resin is now activated by the passage of about 800 ml of N sodium hydroxide solution through the column at a rate of about 2 to 3 ml/min. The activated resin is then washed with distilled water until 50 ml of the effluent, when titrated with 0·1 N hydrochloric acid solution with methyl orange as indicator, give a titration of less than 0·1 ml. About 250 ml of water are usually sufficient to effect this. The activated ion exchange resin is now ready for immediate use.

About 0·5 g (accurately weighed) of the mixture of hexamethylene diamine dihydrochloride and 5-amino caproic acid hydrochloride are dissolved in 50 ml of water and the solution is transferred to the separating funnel A shown in Figure 5.1. This solution is now allowed to pass through the activated ion exchange column at a rate of 2 to 3 ml/min. The separating funnel and sides of the column are washed down with water and the washings passed through the column. The separating funnel is then filled with water which is then allowed to run through the column at the rate of 2 to 3 ml/min.

The first 150 ml of effluent are titrated with 0·1 N hydrochloric acid solution using methyl orange as indicator and each succeeding 50 ml of

Table 5.3

Added hexamethylene diamine dihydrochloride	5-Amino caproic acid hydrochloride	Found hexamethylene diamine dihydrochloride
g	*g*	*g*
0·5013	Nil	0·4958
0·3510	0·1480	0·3427
0·2520	0·2471	0·2445
0·1502	0·3490	0·1437
Nil	0·4997	0·0027

effluent is similarly titrated until the titration is negligible, i.e. less than 0·1 ml of 0·1 N hydrochloric acid solution. A total of 300 ml of effluent is usually required.

The total titre is calculated in terms of hexamethylene diamine dihydrochloride, on the basis that 1 ml of 0·1 N hydrochloric acid solution is equivalent to 0·009457 g of hexamethylene diamine dihydrochloride.

The method gives very satisfactory (although slightly low) results for the determination of hexamethylene diamine dihydrochloride in mixtures of hexamethylene diamine dihydrochloride and 5-amino caproic acid hydrochloride as is shown in Table 5.3.

OBSERVATIONS ON THE HYDROLYSIS OF NYLON AND RELATED POLYMERS

The observations made previously in this chapter indicate the usefulness of the method in the examination of the simple polymers e.g. nylon 6 : 6, nylon 6 : 10 and nylon 6. They also indicate quite clearly the usefulness of the procedure in the examination of simple copolymers such as nylon 6 : 6 / 6 and nylon 6 : 6 / 6 : 10; the proportions of the individual constituents being brought out quite clearly. An interesting case is that of the terpolymer, nylon 6 : 6 / 6 : 10 / 6 (40 : 30 : 30).

Cases have arisen where it was suspected that a given product did not contain the individual constituents in the right proportions. In such cases the hydrolysis method has proved to be particularly effective for it should be realised that with the right formulation:

1. A definite weight of the terpolymer should, on hydrolysis, yield a definite total weight of hexamethylene diamine dihydrochloride plus 5-amino caproic acid hydrochloride.
2. This mixed hydrochloride should, on resolution, give the right proportions of the individual hydrochlorides.
3. A definite weight of the terpolymer should give the right weight of mixed adipic and sebacic acids.
4. The mixed adipic and sebacic acids, should, on test, prove to be in the right proportions.

Then again, straight application of the hydrolysis test may prove to be particularly valuable in the examination of new or unknown polymers, copolymers and terpolymers.

Clasper, Haslam and Mooney[5] have illustrated this at some length in a paper they wrote on the examination of a terpolymer of caproamide, hexamethylene diaminoadipamide and *pp'*-diamino-dicyclo-hexylmethane adipamide (nylon 6 / nylon 6 : 6 / nylon PACM : 6). In this paper they paid particular attention to the products obtained on hydrolysis of the sample with hydrochloric acid.

In a subsequent paper[6] they went on to describe how mixtures of the three hydrochlorides obtained in the hydrolysis test viz. the hydrochlorides of hexamethylene diamine, *pp'*-diamino-dicyclo-hexylmethane and 5-amino caproic acid could be resolved.

The method involved the separation of a mixture of *pp'*-diamino-dicyclo-hexylmethane dihydrochloride and 5-amino caproic acid hydrochloride on a cellulose column with sec-butanol-formic acid-water as

299

eluting solvent. Hexamethylene diamine dihydrochloride is subsequently eluted from the column with an alcohol-water solvent. The proportions of the respective hydrochlorides are determined by analysis of the two eluates after evaporation to dryness, allowance being made for changes that take place in the composition of the individual hydrochlorides on passage through the column.

Further, the hydrolysis method quickly brings out the differences between nylon 6, nylon 7 and nylon 11. These three polymers yield different proportions of amino acid hydrochlorides of different compositions and properties.

Reference has already been made to the interesting behaviour of methoxy-methyl nylon in the hydrolysis test, but, perhaps the most interesting hydrolysis of all is that of Igamid U. Working on a rather larger scale than usual, Haslam and Clasper[3] showed that hexamethylene diamine dihydrochloride was produced in the hydrolysis of Igamid U. Moreover, they indicated that one of the first products of hydrolysis was tetrahydrofuran which was subsequently converted to 1 : 4-dichlorobutane. Further, they developed a simple and interesting test which took advantage of the fact that tetrahydrofuran was produced initially. The operation was so carried out that any volatile products of hydrolysis were brought in contact with a minimum amount of water. The appearance of an oily layer on this water was the signal to stop the hydrolysis, to withdraw the oily layer plus water, and to treat this mixture with nitric acid to yield succinic acid of high purity.

Full details of the results of application of the hydrolysis tests to many polymers, copolymers etc. are given in the papers by Clasper and Haslam[2, 3].

THE IDENTIFICATION OF NYLON AND RELATED POLYMERS BY PAPER CHROMATOGRAPHY

It has been pointed out that a great deal of very useful information may be obtained by chemical examination of the hydrolysis products of nylon and related polymers when relatively large amounts of material are available for test. Nevertheless, on occasion, quite small amounts only are available for test, and some time ago Clasper, Haslam and Mooney[7] set themselves the task of providing a simple chromatographic procedure for the examination of commercial polyamides.

Ordinarily, only 25 mg of substance are used for test, although the principle of the test may be extended to the examination of much smaller amounts of polymeric material.

The method is based on the following principles:

1. hydrolysis of the polyamide with 50% v/v hydrochloric acid, with observations of the behaviour of the polyamide in this hydrolysis,

2. after evaporation of the hydrolysis product to dryness and removal of the excess hydrochloric acid by further evaporation with water, the residue is dissolved in ethanol and observations are made on the solution process, and

3. the ethanolic solution is chromatographed in duplicate, with *n*-propanol-ammonia-water mixture as the developing solvent; one of the chromatograms is dried and examined under ultra-violet light, after which it is sprayed with ninhydrin reagent; the other chromatogram is sprayed with a methyl red-borate buffer solution.

The complete procedure incorporating the principles enumerated above has permitted the differentiation of commercial polyamides and copolymers. Full details of the tests are given below, followed by information about the behaviour of various polymers in the test (Table 5.4).

Apparatus

Aimer Universal chromatographic tank.
Filter paper, Whatman No. 1, for chromatography, 10×10 in $(250 \times 250$ mm), with corner holes.
Glass tubes, prepared from glass tubing having an internal diameter of about 6 mm; the length of each tube is approximately 100 mm.

Reagents

Developing solvent.
Mix 6 parts of *n*-propanol, 3 parts of ammonia solution (sp. gr. 0·880) and 1 part of water, by volume.
Methyl red-borate buffer reagent.
Dissolve 12·368 g of Analar boric acid together with 14·912 g of Analar potassium chloride in water in a 1 l calibrated flask, then add 35·0 ml of N sodium hydroxide solution and dilute to the mark with water. Check the pH of this solution, which should be $8·0 \pm 0·1$. Dissolve 0·03 g of methyl red indicator in 100 ml of the buffer solution, shaking frequently to ensure complete solution.
Ninhydrin reagent.
Dissolve 0·3 g of ninhydrin in a mixture of 95 parts of *n*-butanol and 5 parts of 2 N acetic acid, by volume.
Hydrochloric acid, diluted $(1+1)$.

Procedure

Weigh 25 mg of the substance into a glass tube and add 0·5 ml of diluted hydrochloric acid $(1+1)$. Seal the glass tube carefully in a gas flame,

301

Table 5.4. Examination of commercial nylon polymers and copolymers (p. 300)

	Nylon 6:6	Nylon 6:10	Nylon 6	Nylon 11
1. Observation in hydrolysis tube—				
(a) while still hot	Complete solution	Complete solution	Complete solution	Small amount of insoluble matter
(b) after cooling	Tendency for crystal formation; white precipitate on adding water	Heavy white crystalline precipitate	Complete solution	Sets solid owing to mass of fine white crystals
2. Odour on opening hydrolysis tube	No characteristic odour	No characteristic odour	No characteristic odour	No characteristic odour
3. Solubility in ethanol	Completely soluble	Completely soluble	Completely soluble	Completely soluble
4. Examination under ultra-violet light	Spot due to hexamethylene diamine dihydrochloride (R_F 0·81 to 0·86). Also slight evidence of adipic acid spot (R_F 0·28 to 0·36)	Spot due to hexamethylene diamine dihydrochloride (R_F 0·81 to 0·86). Also slight evidence of sebacic acid spot (R_F 0·59 to 0·64)	Spot due to 5-amino caproic acid hydrochloride (R_F 0·47 to 0·52)	Large spot in same place as for hexamethylene diamine dihydrochloride (R_F 0·81 to 0·86)
5. Examination after spraying with ninhydrin reagent	Purple spot due to hexamethylene diamine dihydrochloride (R_F 0·81 to 0·86). Also faint brown spot due to adipic acid (R_F 0·28 to 0·36)	Purple spot due to hexamethylene diamine dihydrochloride (R_F 0·81 to 0·86). Also faint brown spot due to sebacic acid (R_F 0·59 to 0·64)	Purple spot due to 5-amino caproic acid hydrochloride (R_F 0·47 to 0·52)	Large purple spot in same place as for hexamethylene diamine dihydrochloride (R_F 0·81 to 0·86)
6. Examination after spraying with methyl red-borate buffer reagent—				
(a) after initial 10 min; colour of paper: yellow	Pink spot due to adipic acid (R_F 0·28 to 0·36). Pink spot due to ammonium chloride (R_F 0·47 to 0·52)	Pink spot due to sebacic acid (R_F 0·59 to 0·64). Pink spot due to ammonium chloride (R_F 0·47 to 0·52)	Pink spot due to ammonium chloride (R_F 0·47 to 0·52)	Pink spot due to ammonium chloride (R_F 0·47 to 0·52)
(b) after 30 min: colour of paper: yellow turning to pink	Pink spots become more pronounced. Yellow spot develops due to hexamethylene diamine dihydrochloride (R_F 0·81 to 0·86)	Pink spots become more pronounced. Yellow spot develops due to hexamethylene diamine dihydrochloride (R_F 0·81 to 0·86)	Pink spot becomes more pronounced. No evidence of any further spots	Pink spot becomes more pronounced. No evidence of any further spots

Table 5.4.—continued

	Igamid U	Nylon 6:6:6:10 (40:60)	Nylon 6:6:6 (60:40)	Nylon 6:6:6:10:6 (40:30:30)	Nylon 6:6:6/PACM:6 (1:1:1)
1. Observation in hydrolysis tube—					
(a) while still hot	Insoluble matter present	Complete solution	Complete solution	Complete solution	Complete solution
(b) after cooling	No change. Some insoluble matter present	Heavy white crystalline precipitate	Complete solution	Complete solution	Complete solution
2. Odour on opening hydrolysis tube	Sweet odour	No characteristic odour	No characteristic odour	No characteristic odour	No characteristic odour
3. Solubility in ethanol	Insoluble matter present	Completely soluble	Completely soluble	Completely soluble	Completely soluble
4. Examination under ultra-violet light	Spot due to hexamethylene diamine dihydrochloride (R_f0·81 to 0·86)	Spot due to hexamethylene diamine dihydrochloride (R_f0·81 to 0·86). Also slight evidence of adipic acid spot (R_f0·28 to 0·36) and sebacic acid spot (R_f0·59 to 0·64)	Spots due to hexamethylene diamine dihydrochloride (R_f0·81 to 0·86) and 5-amino caproic acid hydrochloride (R_f0·47 to 0·52). Also slight evidence of adipic acid spot (R_f0·28 to 0·36)	Spots due to hexamethylene diamine dihydrochloride (R_f0·81 to 0·86) and 5-amino caproic acid hydrochloride (R_f0·47 to 0·52). Also slight evidence of adipic acid spot (R_f0·28 to 0·36)	Spots due to hexamethylene diamine dihydrochloride (R_f0·81 to 0·86) and 5-amino caproic acid hydrochloride (R_f0·47 to 0·52). Also evidence of a pp'-diaminodicyclo-hexylmethane dihydrochloride spot (R_f0·98) and an adipic acid spot (R_f0·28 to 0·36)
5. Examination after spraying with ninhydrin reagent	Purple spot due to hexamethylene diamine dihydrochloride (R_f0·81 to 0·86)	Purple spot due to hexamethylene diamine dihydrochloride (R_f0·81 to 0·86). Also faint brown spots due to adipic acid (R_f0·28 to 0·36) and sebacic acid (R_f0·59 to 0·64)	Purple spots due to hexamethylene diamine dihydrochloride (R_f0·81 to 0·86) and 5-amino caproic acid hydrochloride (R_f0·47 to 0·52). Also faint brown spot due to adipic acid (R_f0·28 to 0·36)	Purple spots due to hexamethylene diamine dihydrochloride (R_f0·81 to 0·86) and 5-amino caproic acid hydrochloride (R_f0·47 to 0·52). Also faint brown spots due to adipic acid (R_f0·28 to 0·36) and sebacic acid (R_f0·59 to 0·64)	Purple spots due to hexamethylene diamine dihydrochloride (R_f0·81 to 0·86) and 5-amino caproic acid hydrochloride (R_f0·47 to 0·52) and pp'-diaminodicyclo-hexylmethane dihydrochloride (R_f0·98). Also faint brown spot due to adipic acid (R_f0·28 to 0·36)
6. Examination after spraying with methyl red-borate buffer reagent—					
(a) after initial 10 min; colour of paper: yellow	Pink spot due to ammonium chloride (R_f0·47 to 0·52)	Pink spots due to adipic acid (R_f0·28 to 0·36) and sebacic acid (R_f0·59 to 0·64). Pink spot due to ammonium chloride (R_f0·47 to 0·52)	Pink spot due to adipic acid (R_f0·28 to 0·36). Pink spot due to ammonium chloride (R_f0·47 to 0·52)	Pink spots due to adipic acid (R_f0·28 to 0·36) and sebacic acid (R_f0·59 to 0·64). Pink spot due to ammonium chloride (R_f0·47 to 0·52)	Pink spot due to adipic acid (R_f0·28 to 0·36). Pink spot due to ammonium chloride (R_f0·47 to 0·52)
(b) after 30 min; colour of paper: yellow turning to pink	Pink spot becomes more pronounced. Yellow spot develops due to hexamethylene diamine dihydrochloride (R_f0·81 to 0·86)	Pink spots become more pronounced. Yellow spot develops due to hexamethylene diamine dihydrochloride (R_f0·81 to 0·86)	Pink spots become more pronounced. Yellow spot develops due to hexamethylene diamine dihydrochloride (R_f0·81 to 0·86)	Pink spots become more pronounced. Yellow spot develops due to hexamethylene diamine dihydrochloride (R_f0·81 to 0·86)	Pink spots become more pronounced. Yellow spot develops due to hexamethylene diamine dihydrochloride (R_f0·81 to 0·86) and pp'-diamino-dicyclo-hexyl-methane dihydrochloride (R_f0·98)

then place it in an upright position in an oven at 120°C and leave it overnight. After removing the tube from the oven, make observations on the appearance of the solution; allow it to cool and make further observations. On opening the tube, note any odour that may be evolved, and then transfer the contents of the tube to a 10 ml beaker, washing with 1 to 2 ml of water. Observe any change that may take place on the addition of water. Evaporate the contents of the beaker to dryness once or twice, with small additions of water to remove the hydrochloric acid completely, and finally dry in an oven at 105°C for 15 to 20 min. Dissolve the residue in 1 ml of ethanol and warm to effect solution. Note if there is any insoluble material. Now spot in duplicate 3 to 5 μl of the ethanolic solution on to a Whatman No. 1 10 × 10 in (250 × 250 mm) filter paper having two corner holes on a line drawn 1·5 in (40 mm) from the edge of the paper. Six spots can be placed along this line, so that at least three samples can be dealt with at the same time. After spotting, fit the paper into the frame of the Aimer Universal chromatographic outfit. Put about 200 ml of the n-propanol-ammonia-water developing solvent into the glass solvent-trough in the chromatographic tank and, after allowing the solvent to equilibrate with the atmosphere in the tank for about 1 h, place the frame in position in the tank and allow the chromatograms to develop for about 7 h. At the end of this time remove the frame and allow the paper to dry in the air. When the paper is dry, place it in an oven at 105°C for 15 min and then examine it under ultra-violet light. Now separate the duplicate chromatograms and spray the first chromatogram with the ninhydrin reagent, and, after allowing the paper to dry in the air, place it in an oven at 105°C for 5 min. Observe the position and colour of the spots obtained.

Spray the duplicate chromatogram with the methyl red-borate buffer reagent and dry it by placing it between sheets of filter paper. After 10 min make observations on the position and colour of the spots that develop, and repeat these observations after a further 20 min.

THE IDENTIFICATION OF NYLON AND RELATED POLYMERS BY RING PAPER CHROMATOGRAPHY

The chromatographic procedure developed by Clasper, Haslam and Mooney has been simplified even further in a method of ring paper chromatography devised by Haslam and Udris[8].

This method is as follows:

Procedure

A 25 mg portion of the polymer is first hydrolysed as described previously, i.e. overnight at 120°C with diluted hydrochloric acid (1 + 1). The hydrolysis product is then washed out of the tube with water, and the mixture is evaporated to dryness before extraction with 1 ml of hot ethanol.

The apparatus is then assembled as shown in Figure 5.2, the inner compartment, a small petri dish, containing approximately 20 ml of the developing solvent (a mixture of 6 volumes of *n*-propanol, 3 volumes of ammonia solution (sp. gr. 0·880) and 1 volume of water). The mixture is

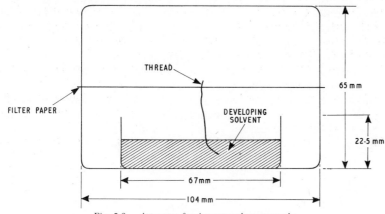

Fig. 5.2.   Apparatus for ring paper chromotography

allowed to equilibrate for a few minutes. In the test for bases and 5-amino caproic acid a 5 μl portion of the ethanolic solution of the hydrolysis products is applied to the centre of an 110 mm Whatman No. 1 filter paper by means of a micropipette and is allowed to dry. A piece of Miladi No. 16 thread* is then threaded through the centre of the filter paper, the edges of the hole being made quite smooth with the thick end of the needle, and the wick is cut so that it projects about 5 mm above and approximately 30 mm below the filter paper. The paper is then transferred to the apparatus, the wick is immersed in the solvent mixture, and the whole apparatus is closed. It is important that the upper and lower petri dishes are well ground, so that a good fit is obtained and solvent vapours cannot escape. Particular attention should be paid to the distance between the surface of the solvent and the filter paper; in our experience 22·5 mm is satisfactory.

The elution is allowed to proceed at room temperature for $1\frac{1}{2}$ to $1\frac{3}{4}$ h, although care should be taken to ensure that the developing solvent does not reach the edges of the petri dishes.

*It is possible that a commercial thread may not always contain the same proportions of natural or other lubricants and for this reason may not always behave in a standard way.

It is suggested that this difficulty may be avoided by using a scoured yarn of almost pure cotton cellulose, such as is prepared by Shirley Developments Ltd, 40 King Street West, Manchester 3.

The filter paper is then removed and dried, first in air and then at 105°C for 20 min, after which it is sprayed with a solution containing 0·3 g of ninhydrin in a mixture of 95 ml of n-butanol and 5 ml of 2 N acetic acid. The paper is then dried in air and finally at 105°C for 5 min, after which it is ready for examination.

The test for acids, such as adipic and sebacic acids, is carried out in the same way, except that three separate 5 μl portions of the ethanolic solution of the hydrolysis products are applied to the centre of the filter paper before chromatography. In this test, spraying is carried out lightly with methyl red-borate buffer reagent (see p. 301). The sprayed paper is dried between pieces of clean filter paper. It is advisable to carry out control tests on known copolymers under the same conditions.

The various bases and acids formed on hydrolysis of the polymers produce characteristic uniform rings, although the R_F values for individual constituents are not always the same as those found originally by Clasper, Haslam and Mooney and given in their paper on ascending paper chromatography[7].

Different hydrolysis products behave, in the tests, as described below.

TESTS FOR BASES

The approximate R_F values for hexamethylene diamine dihydrochloride, ω-aminoheptanoic acid hydrochloride and 5-amino caproic acid hydrochloride are 0·90, 0·78 and 0·73 respectively. pp'-Diamino-dicyclo-hexylmethane dihydrochloride has an R_F value of 1 and gives a blue ring.

TESTS FOR ACIDS

Sebacic acid produces a pink ring (R_F approximately 0·82) and the R_F value for adipic acid is approximately 0·62. In addition a pink ring (R_F approximately 0·72) due to ammonium chloride is invariably obtained in the test.

Sometimes the ring produced by ammonium chloride is not well separated from that produced by adipic acid, but in this event the combined ring is broad and when the behaviour in this test is taken in conjunction with the test for bases little difficulty in identification is experieced.

All this chemical and chromatographic work is, in day-to-day practice, supplemented by infra-red tests.

These infra-red tests are carried out on the original polymers, on separated plasticisers, on hydrolysis products and on substances separated from these products.

In our experience, there are very few polymers or polymer preparations that do not respond to this combination of chemical, chromatographic and infra-red tests.

PYROLYSIS/GAS CHROMATOGRAPHIC IDENTIFICATION OF NYLONS

This method of identification of nylons is of value in cases where only a very small amount of sample e.g. a single fibre, is available for examination. In general, nylons show a similar peak pattern but nylons 6, 6 : 6, 6 : 10 and 11 may be distinguished from one another by examination of the pyrograms obtained when the apparatus is used at high sensitivity. Differences in shape and retention time of minor peaks then become apparent. Nylon copolymers are not identified satisfactorily by this method.

MISCELLANEOUS TESTS

In addition to the identification of the type of nylon polymer or copolymer the analyst in the industry is often called upon to determine such variables as amine and carboxyl end-groups, water, solution viscosity, spin finish, stabiliser, delustrant and filler, in samples submitted for test. The principles of the methods that are employed are as follows.

AMINE END-GROUPS

Amine end-groups are most frequently determined by titration of a phenol-methanol solution of the sample with standard alcoholic hydrochloric acid solution using thymol blue as indicator. Potentiometric and conductimetric methods of amine end-group determination have been used but, in our experience, these methods do not give satisfactory results when residues from alkali catalysts are present in the samples. With such samples the method of Zahn and Rathgeber is preferred[9]. This method involves reaction of the primary amine end-groups in the sample with 2 : 4-dinitrofluorobenzene to form the yellow N-2 : 4-dinitrophenyl derivative, which, after washing free from the excess reagent, is determined colorimetrically in formic acid solution.

CARBOXYL END-GROUPS

Carboxyl end-groups are usually determined by dissolving the polymer in hot benzyl alcohol and titrating the solution with standard alkali using phenol phthalein as indicator. Various modifications of this procedure have been suggested[10], the most important being the use of a solvent mixture of 2 : 6-xylenol and chloroform in the examination of polycaprolactam samples. With this solvent, titrations may be carried out at

307

30°C. This is of great importance when coloured samples are being examined, when it is necessary to determine the end-point in the titration electrometrically.

LUBRICANTS

When the method for the determination of plasticiser and for the isolation of the plasticiser-free polymer is applied to samples which also contain organic lubricants such as stearic acid and stearyl alcohol, then the lubricant will be recovered with the plasticiser. In those cases where plasticiser is absent and lubricant is present, the lubricant may be determined by the following procedure. This has been found to be very suitable for the determination of lubricant in samples of nylon 6 : 6 and nylon 6.

Procedure

3 to 5 g of the sample are weighed into a 250 ml round-bottomed flask. 30 ml of conc. hydrochloric acid are added followed by 100 ml of diethyl ether.

The mixture is heated under reflux on the water bath at 60°C for 4 h with occasional swirling. At the end of this time, when most of the sample will be dispersed in the acid layer, the flask is removed from the water bath and allowed to cool. The ether layer is decanted off into a 250 ml beaker and the ether evaporated. The extraction is repeated with a further 100 ml of ether and finally the hydrochloric acid phase is washed by swirling with 75 ml of ether. The combined ether extracts and washings are evaporated to dryness on the water bath and the residue is cooled.

The cold residue is re-extracted with three consecutive 10 ml portions of ether and the extracts are combined in a tared 50 ml flat-bottomed flask. The ether is evaporated-off on the water bath and the residue is dried to constant weight at 105°C. From the weight of this residue the proportion of lubricant in the sample is calculated.

The identity of the recovered lubricant may be established by infra-red examination.

In cases where metal stearates have been employed it is necessary to establish the identity of the metal either by examination of the sample itself or by tests on the product obtained by extraction of the sample with methanol in a Soxhlet apparatus.

SPIN FINISH

Spin finish is present on many fibre samples and must be removed quantitatively or complications may arise in the identification of the fibre polymer. The best extractants are methanol, benzene and petroleum ether.

If identification of the spin finish itself is required, it is essential to separate and purify the finish so as to free it from such substances as mineral and fatty oils as well as ionic and non-ionic surfactants. This usually involves the use of ion exchange, column and thin layer chromatographic methods followed by examination of the separated components by chemical, n.m.r. and infra-red methods as indicated in Chapter 2. Quantitative determination of e.g. anionic surfactants and polyethylene oxide compounds may be carried out as described in Chapter 7.

A useful test for the detection of anionic surfactants on nylon yarn is based on previous work on the estimation of anionic detergents in sewage[11-13]. The test is carried out as follows.

Reagents

Chloroform. Analar.
N Sodium hydroxide solution.
Methylene blue solution.
A 0·175% w/v solution of B.P. quality material in distilled water.
2 N Sodium acetate solution.
82·04 g of anhydrous Analar sodium acetate are dissolved in water and the solution diluted to 500 ml with distilled water.
10 N Sulphuric acid solution.
Prepared from Analar sulphuric acid and adjusted exactly to strength.

Methylene blue reagent solution

The above reagents are mixed as follows:

Sodium acetate solution (2 N)	500 ml
Methylene blue solution	100 ml
Sulphuric acid solution (10 N)	A volume of this solution that is equivalent to 95·2 ml of 10 N acid

The mixture is then diluted to 1 l with water.

Procedure

5 g of the nylon, preferably in the form of filaments about 2 in (50 mm) long are placed in a test tube (6 × 1 in; 150 × 24 mm).

10 ml of distilled water are added followed by 2 drops of N sodium hydroxide solution, i.e. 0·07 ml. The tube and contents are heated over a small bunsen flame until boiling commences. This boiling is continued

for 2 min with frequent shaking. The mixture is cooled and the aqueous solution poured into a clean 100 to 150 ml glass bottle and the test tube emptied of the residual filaments. 1 ml of the methylene blue reagent solution and 5 ml of chloroform are added to the contents of the bottle and the mixture is shaken thoroughly for 30 s, then transferred back to the original test tube.

With samples of nylon on which anion active compounds are present the lower chloroform layer will show a pronounced blue tint.

With samples of nylon on which anion active compounds are not present the lower chloroform layer will be almost colourless. The test is extremely sensitive and all apparatus should be extremely clean. Moreover, the 2 drops of N sodium hydroxide solution should not be omitted from the test. Further, steps should be taken to avoid contamination of samples submitted for test. In our experience 0·00002 g of anion active agent may be detected in 10 ml of distilled water by the above test.

DETERMINATION OF WATER IN NYLON

The water content of nylon samples is determined by the method described in detail by Haslam and Clasper[14]. The sample is contained in a 8×1 in $(200 \times 20$ mm) test tube connected via a U-tube trap to a vacuum pump. The sample tube is maintained at a temperature of $260°C$ and the volatile matter given off is collected in the U-tube which is immersed in a cooling bath at $-80°C$. The condensate in the trap is washed with dry methanol into a Karl Fischer titration assembly and the water titrated electrometrically with Karl Fischer reagent.

SOLUTION VISCOSITY

The relative viscosity of a nylon polymer is normally determined at $25°C$ on a solution of the sample in 90% w/w formic acid solution containing $8·4\%$ w/w of polymer, and is an important determination in that it is a measure of the degree of polymerisation of the polymer. The strength of the formic acid must be closely controlled and allowance made, in the weight of sample taken, for the moisture content determined as described above. The density ((ρ_2) of the prepared solution is determined at $25°C$ by means of a density bottle and the dynamic viscosity at $25°C$ from the timed flow (t_2) in a calibrated B.S. No. 3 or type-E U-tube viscometer (const. K_2). The dynamic viscosity at $25°C$ of the formic acid solvent (density ρ_1) is similarly determined from a timed flow (t_1) in a B.S. No. 1 or type-A viscometer (const. K_1). The relative viscosity of the sample solution is then calculated from

$$\text{R.V.} = \frac{\rho_2 \times K_2 \times t_2}{\rho_1 \times K_1 \times t_1}$$

THE DETERMINATION OF DELUSTRANTS, FILLERS AND
STABILISING ELEMENTS

Additives such as carbon black, molybdenum disulphide and glass fibre, which are not decomposed by hydrochloric acid hydrolysis may be determined by a direct gravimetric method as follows.

A known weight of the sample is decomposed by boiling with 50% v/v hydrochloric acid solution, and the hot solution filtered through a weighed G4 crucible. The insoluble residue on the filter is washed well with hot 50% hydrochloric acid solution followed by hot ethyl alcohol in order to remove the last traces of polymer hydrolysis products. The residue in the crucible is dried at 150°C and weighed.

If an insoluble oil or wax is seen floating on the hot hydrochloric acid solution of the polymer, indicating the presence of a lubricant or plasticiser, the mixture is cooled. The lubricant or plasticiser is then removed by extraction with ether prior to reboiling and filtration. Failure to do this may result in a blocked filter. Alternatively, the test may be carried out much more simply on a prepared sample of plasticiser and lubricant-free polymer.

In the determination of stabilising elements, X-ray Fluorescence methods, where applicable, may be used with the great advantages of simplicity and speed, particularly when the stabilising elements are present only in low concentrations. The following calibration ranges serve to demonstrate the scope of the method:

Copper ($Cu_{K\alpha}$): 5–80 p.p.m.
Iodine ($I_{L\alpha}$): 50–2000 p.p.m.
Phosphorus ($P_{K\alpha}$): 10–300 p.p.m.
Titanium ($Ti_{K\alpha}$): 0·05–2·5% TiO_2
Manganese ($Mn_{K\alpha}$): 10–200 p.p.m.
Iron ($Fe_{K\alpha}$): 2–20 p.p.m.

All determinations are made on $\frac{1}{8}$ in (3 mm) thick moulded disks and ratio methods of calibration used.

THE IDENTIFICATION OF POLYIMIDE AND POLYAMIDO-IMIDE
RESINS BY HYDROLYSIS METHODS

Polyimide resins may be considered as the condensation products of diamines with aromatic compounds containing two acid anhydride groups. An example is

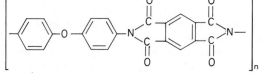

which would be derived from *pp'*-diamino diphenyl ether and pyromellitic anhydride. The related products, based on aryl diamines and substances such as trimellitic anhydride, contain both amide and imide groups in the fully condensed resin and are known as polyamido-imides.

The presence of a polyimide may be suspected if a polymer contains nitrogen and, further, if it gives evidence of the presence of carboxyl bands at 5·6 μm and 5·8 μm in the infra-red spectrum. The presence of four prominent bands in the infra-red spectrum at 5·6 μm, 5·8 μm, 6·1 μm and 6·5 μm will strongly suggest that the material is a polyamido-imide.

A polymer having the infra-red spectral features of a polyimide was received for test. This polymer, when boiled with 10% w/v sodium hydroxide solution, disintegrated and dispersed in the solution and alkaline vapours were evolved.

The hydrolysis product was extracted with ether and the ether extract dried over anhydrous sodium sulphate, then decanted and evaporated to yield the base. The aqueous layer, after extraction of the base, was acidified with concentrated hydrochloric acid, then extracted with ether. This ether layer was dried and evaporated as before to yield an acid. By the application of infra-red and n.m.r. procedures the base was identified as 4:4'-diamino diphenyl methane and the acid as pyromellitic acid.

The principle of the method has been applied to other polyimide type polymers, and bases such as 4:4'-diamino diphenyl ether and 4:4'-diamino diphenyl methane as well as acids such as pyromellitic, trimellitic and 4:4'-dihydroxy diphenyl 3:3'-dicarboxylic acid have been isolated.

The hydrolysis time seems to vary with the type of polyimide polymer and times of 4 h to 72 h have been recorded with different polymers. If, after hydrolysis and acidification, a voluminous pink precipitate not extracted by ether is produced, then incomplete hydrolysis is indicated. In such cases after extraction of any free acid as described, the aqueous solution is treated with excess 10% w/v sodium hydroxide solution and the alkali hydrolysis is completed.

Incidentally, it has been shown that the principle of the above method may be applied directly to the identification of polyimides present with polytetrafluoroethylene in coating compositions.

INFRA-RED SPECTRA OF NYLON AND RELATED POLYMERS

Nylons, as a class, are differentiated from almost all other resins by the strong and invariant secondary amide pattern shown in their spectra. The appearance of bands at 3·05 μm, 3·45 μm, 6·1 μm, 6·45 μm and 6·85 μm is as characteristic of these materials as is the well-developed band pattern which appears in *o*-phthalate esters.

Natural proteins and resins such as urea/formaldehyde condensate also contain secondary amide groups, and consequently have rather similar spectra. However, these resins contain less hydrocarbon structure than nylons, hence the 3·45 μm and 6·85 μm bands are much weaker in relation to the amide bands and in many cases the 6·85 μm band is absent. These materials also have a more complex molecular structure than the nylons; this is reflected in a broadening of the amide bands at 3·05 μm, 6·1 μm and 6·45 μm in the protein and urea/formaldehyde spectra, as opposed to the relatively sharp and well-defined bands observed in the nylons.

The urethane resins have again somewhat similar spectra, but the shift of the 6·1 μm amide band to 5·9 μm is quite obvious, and these resins have much stronger bands in the 8 μm to 13 μm region than the nylon resins.

Most of the common nylon polymers are based on linear dicarboxylic acids and linear diamines or on linear ω-amino acids; they differ chemically therefore only in the length of $-(CH_2)-$ chains separating the amide groups, and the head to head or head to tail arrangement of the amide groups. It is perhaps somewhat surprising that these resins can be distinguished by their spectra, since the chemical differences are so small. However, the useful and characteristic differences which exist between the spectra of these resins do not arise primarily from chemical features, but from molecular interactions in the crystalline regions of the material. This may be demonstrated by observing the spectra of, for example, nylon 6, nylon 6 : 6 and nylon 6 : 10 at temperatures above their crystalline melting points; under these conditions the spectra are qualitatively indistinguishable.

INFRA-RED SPECTRA OF SIMPLE NYLON POLYMERS

The spectra of the following simple nylons are reproduced:

Nylon 6—Spectra 5.1, 5.2 and 5.3
Nylon 7—Spectra 5.4(a) and 5.4(b)
Nylon 8—Spectrum 5.5
Nylon 11—Spectra 5.6 and 5.7(a) and 5.7(b)
Nylon 12—Spectra 5.8(a) and 5.8(b)
Nylon 6 : 6—Spectrum 5.9
Nylon 6 : 10—Spectrum 5.10
Methoxy-methyl modified Nylon 6 : 6—Spectrum 5.11
Polyurethane from hexamethylene di-isocyanate and 1 : 4 butane diol—Spectrum 5.12

These are the straight resins commonly encountered. Spectra of many other nylons are given in *D.M.S.*[15] (Nos. 2533 to 2569) but the majority of these are unlikely to be encountered commercially.

An additional complication in the examination of nylons is that the resins can exist in more than one crystalline form. Fortunately, as far as identification work is concerned, the different crystalline forms of nylon

6 : 6, designated α and β, appear to have identical spectra; the same is true of nylon 6 : 10. The spectra of the α and δ forms of nylon 6, however, show distinct differences; Spectrum 5.1 is mainly the α form while Spectrum 5.2 is mainly the δ form. There is also a β crystalline form of nylon 6, but, as with the other nylons mentioned, the α to β transition is apparently not accompanied by any significant effect in the spectrum. The effects due to crystallinity in nylon 6 are described by Ziabicki[16].

Although we are not aware of any work on the crystalline forms of nylon 11, differences are noted also in the spectrum of this material, and we have therefore included two Spectra (5.6 and 5.7) which appear to be recognisably different forms. Methoxy-methyl modified nylon 6 : 6 (structure p. 290) is soluble in alcohol. Because of the bulky side group, this nylon cannot crystallise; this however does not render identification difficult because, in addition to amide groups, aliphatic ether is present leading to the appearance of additional structure in the 9 μm to 10 μm region (Spectrum 5.11). This resin is decomposed by formic acid, with the production of nylon 6 : 6; thus if the recommended method of specimen preparation is followed the result will be misleading. If, therefore, a material is found to be nylon 6 : 6, and there is reason to suppose that it could be the methoxy-methyl modified material, it should be heated with alcohol, when the modified form will dissolve, and a film for identification purposes can be cast from the solution.

MEASUREMENT OF CRYSTALLINITY OF NYLON RESINS

The identification of absorption bands in the spectra of nylon polymers with either crystalline or amorphous material has been used as the basis for the measurement of crystallinity of these resins. Starkweather and Moynihan[17] have used the following bands:

Nylon 6 : 6   10·68 μm crystalline   8·78 μm amorphous
Nylon 6 : 10   11·73 μm crystalline   8·88 μm amorphous

No useful measurements seem to be possible with nylon 6, owing to the different spectra of the α and δ crystalline forms[16].

'MONOMER' IN NYLON 6

Nylon 6 may contain significant amounts of 'monomer' (free caprolactam). Caprolactam is comparatively volatile, and will be lost in the film-casting process described, but if specimens are prepared by hot pressing or by microtoming, free caprolactam, when present, will be evident from additional absorption bands at 11·2 μm and 11·5 μm (Spectrum 5.3). The absorption band at 11·5 μm forms the basis of a rapid and most satisfactory method of estimating free caprolactam in nylon 6 in amounts greater than 0·2%. A double-beam instrument is preferred for this work. Suitable specimens may be prepared by hot pressing as this process does not appear to influence the caprolactam content; the thickness required may be varied from 0·001 to 0·007 in (0·03 to 0·2 mm) as necessary to obtain a 'monomer' band of measurable intensity. Calibration is performed by placing in the spectrometer beam a sample of nylon 6 containing no free caprolactam, prepared as follows. Dissolve 1 g of nylon 6 in 5 ml of formic acid with warming. When solution is complete add 10 ml methanol to the hot solution, followed by 70 ml diethyl ether. Filter off the precipitated resin and dry in vacuo. Hot press the sample to about 0·003 in (0·08 mm) thickness.

In the sample beam of the instrument, simultaneously with the monomer free nylon sample, is placed an absorption cell of about 0·004 in (0·1 mm) thickness, and a similar cell is placed in the reference beam. The cell in the sample beam is filled with solutions of known concentrations of caprolactam in carbon tetrachloride, while that in the reference beam is filled with neat carbon tetrachloride to balance the solvent absorption. The calibration solutions are made up in the range 0·5% to 5·0% w/v.

The spectra of the combination of the monomer free nylon and each of the calibration solutions are recorded in the range 11 μm to 12 μm. As the absorption due to monomer is weak, ordinate scale expansion is helpful, particularly at the lower concentrations.

The calculation is best done by assuming that a caprolactam solution is equivalent to a layer of caprolactam of a number of mg/cm^2. Thus, a cell of d mm thickness, containing an a% w/v solution of caprolactam, is equivalent to a layer of $a.d$ mg/cm^2. The absorbance at 11·5 μm, measured by the base-line density method, is plotted against the concentration of caprolactam, expressed as mg/cm^2. This calibration curve may be used to estimate caprolactam in a particular nylon sample by measuring the absorbance at 11·5 μm by the same method, and the specimen thickness. From the specimen thickness the concentration of nylon, expressed as mg/cm^2, may be calculated, and the concentration of caprolactam, in the same units, is determined from the measured absorbance using the calibration graph. The percentage caprolactam in the sample is then given by

$$\frac{\text{Conc. of monomer in mg/cm}^2}{\text{Conc. of nylon in mg/cm}^2} \times 100\%$$

This method of calculation is useful when dealing with solid samples, as although a given thickness cannot necessarily be obtained, the thickness of the prepared sample can be measured with a dial gauge micrometer with considerable precision.

ALIPHATIC POLYURETHANES

Although only one resin of this nature is of commercial importance, the condensation product of hexamethylene di-isocyanate and 1 : 4-butane diol, the spectrum (Spectrum 5.12) shows quite clearly the chemical nature of the resin, and consequently any resin of a similar constitution could be recognised immediately.

The overall spectrum shows considerable similarity to those of the nylon resins, with an NH band near 3 μm, and a pair of strong bands in the 6 μm region. However, the carbonyl band of the urethane groups occurs at 5·9 μm, as opposed to 6·1 μm in the nylon resins (amide I band), while the C—O group, present in this resin but not in the nylons, is evident from the strong band near 8 μm. The constitution of this resin, and presumably that of any other aliphatic polyurethane, may be ascertained by hydrolysis (p. 300), followed by the recovery of the diamine equivalent to the di-isocyanate originally used, and either a diol or some recognisable decomposition product equivalent to it. Infra-red examination of diamines is discussed later in this chapter (p. 318), and the spectra of common diols are given in Spectra G.1 to G.11.

AROMATIC POLYURETHANES

The direct counterpart of the resins described above, based on simple diols and aromatic di-isocyanates, does not appear to be of any commercial interest, but the condensation product of one molecule of a trihydric alcohol with three molecules of an aromatic di-isocyanate is described by Hummel. The resulting tri-isocyanate is reacted with a hydroxyl terminated polyester. Hummel[18] gives spectra of these products but the final resins have spectra very similar to the polyester urethanes described elsewhere (p. 268) and their nature would presumably be revealed by hydrolysis in the usual way.

OTHER SYNTHETIC POLYAMIDES

The condensation of dimers and trimers of linoleic acid (which act as polycarboxylic acids) with ethylene diamine is the basis of a series of resins whose spectra are given in the *Sadtler Commercial Spectra*[19] (D1551 to D1555). The prominent secondary amide structure appears in these spectra, but the bands in the 7 μm to 14 μm region are on the whole broad and ill-defined. These resins cannot be completely characterised from their spectra, as products based on other branched long-chain acids and simple diamines would presumably be very similar. Resins of this type, however, are distinguished from natural proteins and urea/formaldehyde condensates by the relative prominence in their spectra of the C—H structure at 3·45 μm and 6·85 μm.

NYLON COPOLYMERS

The first and most important comment on the infra-red examination of nylon copolymers is to suggest the utmost caution in attempting to ascertain the chemical nature of a copolymer from its spectrum. As has been pointed out, the identification of nylon resins from their spectra depends upon the recognition of absorption bands which arise from crystalline material. The irregular nature of the individual molecular chains in a copolymer inhibits crystallisation, and, except where one amide is present in considerable excess, the characteristic patterns associated with the individual nylons cannot develop; hence there is generally no useful relation between the spectrum of a copolymer and the sum of the spectra of the individual nylons to which the copolymer may be equivalent chemically. The situation is rendered more difficult by variations due to the extent of 'blocking' of the units of individual nylons in the copolymer chains; thus two copolymers which on hydrolysis prove to have identical constitutions in terms of recovered acids and bases may show differences in their spectra.

In any event the poor crystallinity of the copolymers tends to suppress the development of sharp bands, and rather featureless spectra result. These points are borne out by Spectra 5.13, 5.14 and 5.15 which are respectively 6 : 6 / 6 : 10, 6 : 6 / 6, and 6 : 6 / 6 : 10 / 6 copolymers. Examination of the last two spectra reveals a small difference $\sim 9 \, \mu m$, while the 6 : 6 / 6 : 10 copolymer shows small differences from the other resins between 8 μm and 9 μm. In our experience however even these minor effects are not reproducible and depend upon the crystallinity of the specimen.

Even the introduction of PACM (pp'-diamino-dicyclo-hexylmethane) units in the chain produces a surprisingly small effect in the spectrum (Spectrum 5.16). We may therefore conclude our remarks on the spectra of nylon copolymers with the observation that unless, from its spectrum, a material is obviously a straight nylon resin, no firm opinion can be expressed as a result of infra-red examination other than that the unknown appears to be a nylon copolymer.

These materials are readily hydrolysed (p. 300) and the chromatographic behaviour and the infra-red spectra of the hydrolysis products give clear evidence as to the nature of the material both qualitatively and quantitatively.

INFRA-RED SPECTRA OF DICARBOXYLIC ACIDS SEPARATED FROM NYLON HYDROLYSIS PRODUCTS

The dicarboxylic acids are crystalline solids, and, in contrast to the lower monocarboxylic acids which are liquid at room temperature, show useful and characteristic infra-red spectra, whether examined as mulls in liquid paraffin or as alkali-halide disks. For this reason it is not necessary to proceed to the barium salts to obtain spectra for identification work, although in fact the barium salts also have quite satisfactory spectra, which can be used if appropriate. The spectra of the dicarboxylic acids likely to be encountered are given in Spectra A.1 to A.3. The acids most frequently recovered from commercial samples are adipic and sebacic acids.

The crystallinity of these acids does not appear to be greatly influenced if they are mixed together, and, in the case of the acid hydrolysis products of a 6 : 6 / 6 : 10 or 6 : 6 / 6 : 10 / 6 copolymer, a spectrum of the adipic/sebacic mixture obtained shows all the features of the individual acid spectra. A satisfactory quantitative analysis of these mixed acids is possible from the spectrum using the 13·6 μm band for adipic acid and the 13·3 μm or 13·8 μm band for sebacic acid. The absorbance ratio 13·6 μm/13·3 μm (or 13·8 μm) is plotted against composition ratio on a series of standard mixtures of the two acids. The plot is linear, and may be used directly to estimate the proportions of unknown mixtures of these acids, as separated from the hydrolysis products of a copolymer. For this quantitative work, the alkali-halide disk method is preferred as it is much easier to get the two acids well dispersed in this medium than in liquid paraffin.

INFRA-RED SPECTRA OF BASES SEPARATED FROM NYLON HYDROLYSIS PRODUCTS

The examination of the separated bases from nylon hydrolysis products is less straightforward than in the case of the acids. It seems more profitable to examine the base hydrochlorides, which are crystalline solids giving sharp, clear spectra. They are comparatively easy to dry. The free bases on the other hand tend to give rather broad absorption bands, and are often hygroscopic in nature. The spectra of the base hydrochlorides which are commonly encountered in nylon work are given in Spectra H.1 to H.4. Alkali chlorides can be used in the disk method for base hydrochloride spectra, but other halides must be avoided, as ion exchange occurs giving very puzzling spectra. Liquid paraffin appears to be satisfactory as a mulling agent.

Base hydrochloride spectra, as a class, are unusual in having significant structure, attributed to the NH_3^+ ion, in the 3·8 μm to 5·0 μm region. The spectra of these materials contain a large number of sharp bands, and are very satisfactory for identification by matching, but do not give much useful structural information to enable the nature of an unknown to be deduced when no comparison spectrum is available. However the base hydrochloride spectra distinguish diamines from amino acids, the latter having a strong carbonyl band in the usual acid position (5·9 μm). This observation can be useful with an unknown, as it enables the molecular weight of the base to be calculated if the nitrogen or chlorine content of the hydrochloride is known.

The spectra of mixtures of a diamine dihydrochloride with an ω-amino acid hydrochloride, such as may be obtained from the hydrolysis of copolymers containing nylon 6, appear to be straightforward, showing all the bands of both constituents. Having obtained this qualitative

318

information however there is little point in pursuing the quantitative determination, as this is straightforward by chemical methods.

The spectra of comparatively few base (amine) hydrochlorides appear in the literature, and even these may be suspect if they relate to suspensions of base hydrochloride in potassium bromide disks because there is the possibility that these are spectra of the base hydrobromide resulting from interaction of the materials in the disk. Thus, when it is not found possible to match the spectrum of a base hydrochloride with literature spectra, it will be better to release the free base about which the literature may be more useful and reliable. Again, if the base is solid, it will be important to determine its spectrum, and match with literature spectra, as a suspension in liquid paraffin, as reaction is almost certain to occur with potassium bromide or other alkali halide in a solid disk, with the formation of the base hydrohalide. Again, liquid bases are best examined as neat liquids, since many solvents, and particularly the favourites chloroform and carbon tetrachloride, may react with bases to produce products with curious and quite useless spectra for identification purposes.

In the spectra of the free bases (Spectra B.1 to B.4), those of the ω-amino acids are of interest, since they do not show the usual acidic carbonyl band at 5·9 μm. These compounds exist as zwitterions, the carbonyl band being replaced by the pair of bands for a carboxylic acid salt near 6·5 μm and 7·2 μm. The ionised ammonium structure replaces the 3·0 μm NH$_2$ bands, and gives rise to a series of bands in the 3·7 μm to 5·0 μm region. On the whole the spectra of the free bases show rather broad bands, and for identification work they are much less satisfactory than the corresponding hydrochlorides.

OTHER AMIDE RESINS

Primary amide resins show a pair of strong doublet bands in their spectra, one near 3 μm and a second near 6 μm (See Table 2.8). These features should be sufficient to enable primary amide resins to be recognised from their spectra.

The common resins of this class are polyacrylamide (Spectrum 5.17) and polymethacrylamide (Spectrum 5.18). Other than the pronounced features noted above, which are common to both spectra, the absorption bands of these resins are broad and indefinite, except for the structure in the 8 μm to 9 μm region, which is sufficiently distinctive to enable the resins to be differentiated. The only tertiary amide resin commonly encountered is polyvinyl pyrrolidone. As this resin contains no N—H groups there is no absorption in the 3 μm region, but the amide carbonyl band appears as a single strong band at 6 μm (Spectrum 5.19). Other than the band at about 7·7 μm, the remainder of the spectrum contains broad absorption bands, and there is no indication that a cyclic structure is present.

Of these primary and tertiary amide resins, polyacrylamide and poly-vinyl pyrrolidone are water soluble, and satisfactory films may be cast from aqueous solution on to a silver chloride plate. Polymethacrylamide is rather less soluble, but reasonably good spectra may be obtained by the halide-disk method.

THE INFRA-RED SPECTRA OF POLYIMIDES AND RELATED RESINS

The general structure of these resins is shown on p. 311. The most important feature in their infra-red examination is the presence of the imide ring, which gives a characteristic carbonyl pattern in the infra-red spectrum consisting of a prominent sharp band at 5·6 μm and a stronger and broader band at 5·8 μm. The latter cannot be distinguished from the carbonyl band of esters, but when it is accompanied in the spectrum of a resin by the 5·6 μm band this is strong evidence that the resin is a polyimide. The other commonly encountered chemical groups which have very short wavelength carbonyl bands, namely the acid anhydride and the peroxide group —CO—O—O—CO—, are unlikely to be present in significant concentration in synthetic resins. Even so, the result of a chemical test for nitrogen will be useful information.

The significant features of the spectra of the common polyimides are illustrated in Spectra 5.20 and 5.21. The doublet carbonyl band is evident in both spectra, but in the case of the resin based on benzophenone tetracarboxylic acid (Spectrum 5.21) an additional carbonyl band appears very near 6·0 μm due to the diaryl ketone group. The resin of Spectrum 5.20 is based on 4 : 4'-diamino diphenyl ether, and the very prominent band at about 8·0 μm arises from the diaryl ether group.

In other respects, the infra-red spectra of the polyimides consist of a very large number of rather sharp absorption bands, as might be expected from materials of a predominantly aromatic nature, although there are no other features which can be associated exclusively with the polyimide structure. These rich spectra are, however, potentially very useful for qualitative identification by comparison with reference spectra, such as those given by Hummel[18, 20].

Nevertheless, the complete identification of a polyimide resin from its infra-red spectrum alone requires considerable care, as a number of chemically modified resins exist, and it cannot be assumed that it will be possible to obtain evidence of the modification from a qualitative examination of the infra-red spectrum. Thus, in the polyesterimides (produced, for example, from trimellitic acid, a diamine and a glycol) the chemical difference is the introduction of an ester group, and this may be difficult to detect because the strongest band of the ester is that of the carbonyl group which coincides with the strong polyimide absorption at 5·8 μm. On the other hand, the polyamido-imides (produced for example from

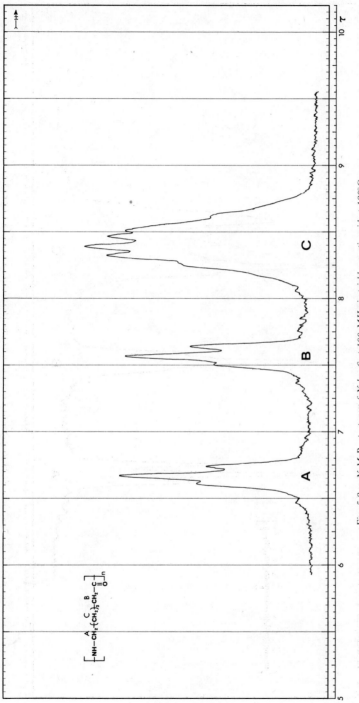

Fig. 5.3. N.M.R. spectrum of Nylon 6 at 100 MHz in trichloroacetic acid at 130°C

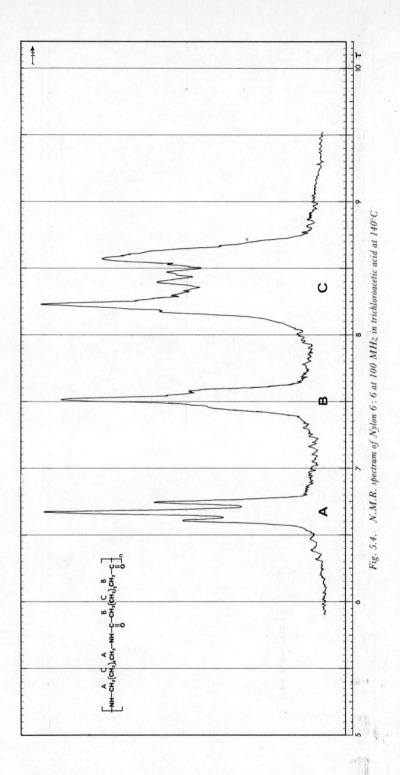

Fig. 5.4. N.M.R. spectrum of Nylon 6 : 6 at 100 MHz in trichloroacetic acid at 140°C

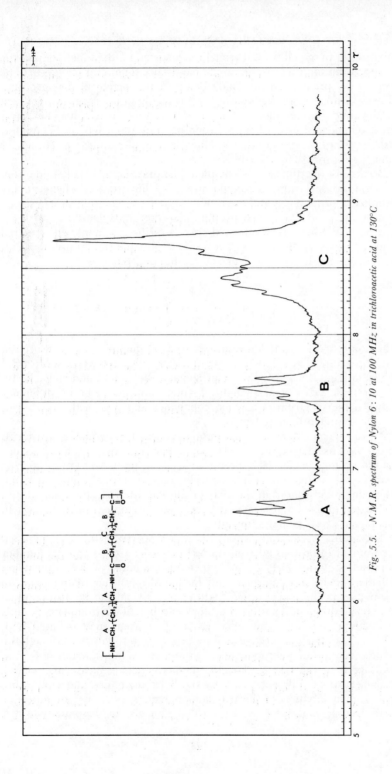

Fig. 5.5. N.M.R. spectrum of Nylon 6:10 at 100 MHz in trichloroacetic acid at 130°C

mixtures of pyromellitic and trimellitic acids and a diamine) contain the amide group, and the strong amide bands at 6·0 μm and 6·5 μm may be seen in the presence of the imide bands in this region of the spectrum. However, there is some similarity in this region of the spectrum between a polyamido-imide and a polyimide based on benzophenone tetra-carboxylic acid, as each have a strong band at about 6·0 μm. The infra-red spectra of these and other modified polyimides appear in Hummel's book[18] (Figures 1248–1262).

Because of the difficulty of complete identification of a polyimide from its infra-red spectrum, we would stress the importance, whenever any detailed information is required, of a proper examination of a resin by hydrolysis (p. 312) followed by the separation and identification of the hydrolysis products. The infra-red spectra of the commonly encountered acids are given in Spectra A.7, A.8 and A.9, while the infra-red spectra of the common bases appear in Spectra B.5 and B.6.

THE N.M.R. SPECTRA OF SIMPLE NYLON RESINS AND COPOLYMERS

The n.m.r. spectra of nylon resins are of good quality when measured in trichloroacetic acid solution at about 130°C. The 100 MHz spectra are shown in Figures 5.3, 5.4 and 5.5 for Nylon 6, Nylon 6 : 6 and Nylon 6 : 10. Under these experimental conditions there appears to be no difference between the spectrum of a nylon copolymer and that of the equivalent mixture of nylon homopolymers.

All the spectra show a triplet resonance near 6·7 τ which is attributed to the protons adjacent to the NH group. The equivalent triplet resonance for the protons on the CH_2 group adjacent to the C=O group appears near 7·6 τ but this is often not as well resolved as the lower-field band. The remaining protons in the CH_2 chains contribute to the resonance in the region of 8·5 τ and clearly the complete assignment of resonances in this region presents some difficulty.

The ability to determine from the n.m.r. spectrum the ratio between the number of protons adjacent to NH or C=O groups and the number of protons on other CH_2 groups will enable a number of the simple nylon polymers to be distinguished, and further the contour of the group of resonances in the region of 8·5 τ will enable Nylon 6 to be distinguished from 6 : 6, although the ratio of protons in different environments in these two polymers is the same. The latter observation may be used with advantage in the quantitative analysis of 6 / 6 : 6 copolymers, though this is more favourable for copolymers containing a minor amount of 6 : 6 in 6 rather than the reverse. Possible methods include measuring relative peak heights at different places in the 8·5 τ resonance and calibrating with known mixtures of the two homopolymers. A similar quantitative analysis is possible for 6 : 6 / 6 : 10 copolymers. In these analyses it is important that no constituents other than those sought are present in the sample, and re-precipitated samples are preferred.

REFERENCES

1. HASLAM, J., and SWIFT, S. D., *Analyst, Lond.*, **79,** 82 (1954)
2. GLASPER, M., and HASLAM, J., *Analyst, Lond.*, **74,** 224 (1949)
3. HASLAM, J., and CLASPER, M., *Analyst, Lond.*, **76,** 33 (1951)
4. HASLAM, J., and CLASPER, M., *Analyst, Lond.*, **75,** 688 (1950)
5. CLASPER, M., HASLAM, J., and MOONEY, E. F., *Analyst, Lond.*, **80,** 812 (1955)
6. CLASPER, M., HASLAM, J., and MOONEY, E. F., *Analyst, Lond.*, **81,** 587 (1956)
7. CLASPER, M., HASLAM, J., and MOONEY, E. F., *Analyst, Lond.*, **82,** 101 (1957)
8. HASLAM, J., and UDRIS, J., *Analyst, Lond.*, **84,** 656 (1959)
9. ZAHN, H., and RATHGEBER, P., *Melliand Text Ber.*, **34,** 749 (1953)
10. MAURICE, M. J., *Analytica chim. Acta*, **26,** 406 (1962)
11. *The Determination of Anionic Synthetic Detergents using Methylene Blue, Colorimetric Chemical Analytical Methods, Pt. 4*, 5th edn, The Tintometer Co., Salisbury, England
12. JONES, J. H., *J. Ass. off. agric. Chem.*, **28,** 398 (1945)
13. EVANS, H. C., *J. Soc. chem., Ind., Lond.*, **69,** 876 (1950)
14. HASLAM, J., and CLASPER, M., *Analyst., Lond.*, **77,** 413 (1952)
15. *Documentation of Molecular Spectroscopy*, Butterworths, London and Verlag Chemie, Weinheim/Bergstrasse
16. ZIABICKI, A., *Kolloidzeitschrift*, **167,** 132 (1959)
17. STARKWEATHER, H. W., and MOYNIHAN, R. E., *J. Polym. Sci.*, **22,** 363 (1956)
18. HUMMEL, D. O., *Infra-Red Analysis of Polymers, Resins and Additives, An Atlas*, **1,** Pt II, Wiley Interscience (1969)
19. *Sadtler Commercial Spectra Series D*, The Sadtler Research Laboratories, Philadelphia 2
20. HUMMEL, D. O., *Farbe Lacke*, **74,** No. 1, 11 (1968)

6

POLYOLEFINE AND HYDROCARBON
POLYMERS AND COMPOSITIONS

In the first part of this chapter attention is directed initially to the determination of major additives such as polyisobutene in polythene/polyisobutene compound, paraffin wax in paraffin wax/polythene blends and of carbon black in polythene preparations.

This is followed by observations on the pyrolysis/gas chromatography of hydrocarbon polymers and on the determination of minor additives such as fillers, U.V. absorbers, antioxidants, slip agents, antistatic agents, amide waxes, epoxy compounds and metal salts.

The second part of this chapter is introduced by a table which should permit the principal hydrocarbon resins to be differentiated by means of their infra-red spectra. The hydrocarbon resins discussed in this chapter are polythene, polypropylene, the aromatic hydrocarbon resins and the principal copolymers and polymer blends containing these resins as major constituents. The differentiation of different types of polythene by means of their infra-red and n.m.r. spectra is discussed, and methods are given for the quantitative estimation of the principal structural groups in these resins. Information is provided on the determination of the proportions of the constituents of ethylene copolymers by infra-red and n.m.r. spectrometry, and observations, from infra-red and n.m.r. studies, are made on both random and block propylene/ethylene copolymers.

Two important additive determinations that are carried out by infra-red methods, those of oleamide in polythene and dilaury dithiopropionate in polypropylene, are dealt with in detail.

A number of other hydrocarbon resins which are chiefly important as rubbers (polyisoprene, polybutadiene, polyisobutene, etc.) are described in Chapter 8.

INTRODUCTION AND GENERAL OBSERVATIONS

In thinking of polyhydrocarbons, one thinks inevitably of such substances as polythene, polypropylene and other poly-α-olefines. In the preparation of polythene the raw material of the process, ethylene, is obtained from oil by the cracking process in which the large molecules in the oil are broken down catalytically into ethylene and other hydrocarbons of low molecular weight.

With high-pressure polythene, pressures of the order of 1000 to 3000 atmospheres and temperatures of 100 to 300°C are used in the polymerisation. In addition, it is necessary to add catalysts such as oxygen or organic peroxides.

The polymer thus produced consists of highly branched long chain molecules, the repeating link being $-CH_2-CH_2-$. Each molecule contains many branches of various lengths.

The regularity of the individual molecules makes possible some close packing of unbranched parts of the chains, leading to highly ordered or crystalline regions known as crystallites. The branched structure of some parts of the molecules prevents a completely ordered arrangement along the whole length of the chain; the greater the amount of branching the less the degree of crystallinity.

Low-pressure polythene is made by quite a different process. Certain combinations of metal alkyl complexes and transition metal halides catalyse the polymerisation of ethylene in such a way that the reaction can be carried out at temperatures and pressures little above atmospheric. It is not surprising that the material thus produced is of a different character from high-pressure polythene.

The above process has been carried a stage further in the case of the polymerisation of propylene. Catalysts have been developed which control the way in which successive monomer units add to the growing polymer chain, and two stereoregular forms, isotactic and syndiotactic, have been prepared. A stereochemically irregular form known as the atactic form can also be obtained, and is commercially available; this is usually a rubbery solid. Stereoregular forms of polymers are discussed in Chapter 4 (p. 270); information on the stereoregularity of polypropylene may be obtained by n.m.r. examination (p. 394).

The common polypropylene of commerce consists mainly (at least 90%) of the isotactic form.

In the field of olefin polymerisation, therefore, it is apparent that many of the problems involving polymer properties depend upon polymer structure, and since such problems are on the whole more readily attacked by physical methods, the main bias of this chapter is towards infra-red examination.

The older-fashioned analytical chemist will not neglect altogether carbon and hydrogen determinations on polyhydrocarbons though, as a rule, the tests will not yield much information of value.

He may find it more profitable to have a sound knowledge of how to determine traces of titanium, aluminium, iron, molybdenum, silicon, sodium etc., left as catalyst residues.

Moreover, he will find in practice that such substances as polythene will be compounded with e.g. carbon black, butyl rubber, antioxidants and slip agents etc. and some of the methods used to determine substances such as carbon black will be given in this chapter.

Having said that, however, one must come back again to reiterate the importance of infra-red testing in this field of polyhydrocarbons. One has only to consider the questions that are asked to realise this, e.g. Is the sample a high-pressure product or not? What is the degree of crystallinity and the nature of it? Are ester, or carbonyl or aldehydic groups present in the polymer? Are methyl or ethyl or butyl groups present in the polymer and, if so, where?

Over the years, it has been the practice for many so-called analytical laboratories to be concerned with the grading of polyhydrocarbons, with the determination of molecular weights in solvents such as tetralin, with densities etc., but in this preliminary part of the chapter our bias is towards the more strictly chemical tests that are called for in the trade.

Hence it is, after this preliminary introduction, that this chapter is concerned with:

1. The determination of polyisobutene in polythene/polyisobutene compound.
2. The determination of paraffin wax in paraffin wax/polythene blends.
3. The determination of carbon black in polythene preparations.
4. Observations on the pyrolysis/gas chromatography of hydrocarbon polymers.
5. The extraction and determination of additives such as fillers, U.V. absorbers, antioxidants, slip agents, antistatic agents, amide waxes, epoxy compounds and metal salts of organic acids.

DETERMINATION OF POLYISOBUTENE IN POLYTHENE/ POLYISOBUTENE COMPOUND

In the case of well homogenised, unfilled samples, this determination may be performed directly on the sample by infra-red spectroscopy. Otherwise it is necessary to employ an indirect method, based upon solvent extraction of the polyisobutene and subsequent determination of the polyisobutene in the extract by infra-red spectroscopy.

DIRECT METHOD

Prepare the sample in the form of a film by hot pressing between poly-tetrafluoroethylene-coated stainless-steel plates. A sample of 0·01 in

(0·3 mm) thickness is suitable for polyisobutene contents between 5% and 25%, but the thickness may be varied appropriately for other compositions. A series of standard samples, prepared by blending polythene and polyisobutene by milling and creping, are required. The standards are prepared as films by the method given above.

Measure the spectra of all standards and unknowns over the ranges 2·1 µm to 2·8 µm and 9·0 µm to 12·5 µm. Measure the absorbances at the band maxima at approximately 2·4 µm and 10·5 µm by the base-line absorbance method, drawing linear backgrounds as shown in Figure 6.1. In the case of the standard samples, plot the ratio

$$\frac{\text{Absorbance at } 10\cdot53 \ \mu\text{m}}{\text{Absorbance at } 2\cdot391 \ \mu\text{m}}$$

against weight % polyisobutene in the composition. The graph should be linear and pass through the origin. Determine the absorbance ratios as

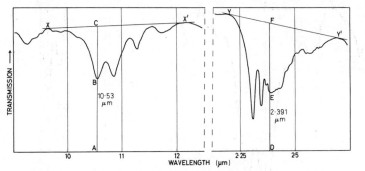

Fig. 6.1.   Determination of polythene/polyisobutene mixtures (direct method)

above for the unknown samples, and determine their composition from the calibration graph.

The band at 2·4 µm is measured as a thickness reference, but if the samples to be determined are rigid (i.e. not rubbery) it may be possible to measure their thicknesses with a micrometer gauge. In this case the measurement of the 2·4 µm band is omitted from the method, and the ratio

$$\frac{\text{Absorbance at } 10\cdot53 \ \mu\text{m}}{\text{Sample thickness}}$$

is determined in all cases, and treated in the same way.

INDIRECT METHOD

This method is applied in the case of filled samples, or samples which by the test method above are found to be inhomogeneous, as determined by the reproducibility of the result on different pressings prepared from the same original sample.

329

1 g of the sample is weighed accurately into a 500 ml stoppered conical flask. Approximately 0·5 g filter aid and 50 ml toluene are added and the mixture is heated on a hot plate to just below its boiling temperature. This temperature is maintained, with constant swirling, until solution of the sample is complete. 100 ml petroleum ether are added to the *hot* solution, by pouring the solvent rapidly down the side of the flask. The flask is stoppered immediately, shaken for about 1 min to precipitate the

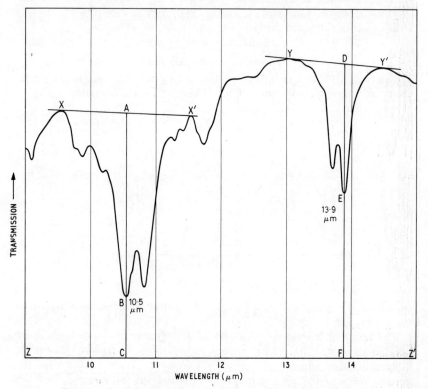

Fig. 6.2.   Determination of polythene/polyisobutene mixtures (indirect method)

polymer, then allowed to stand for 1 h. When precipitated correctly the precipitated polythene and filter aid will settle out as individual particles and the subsequent filtration will then proceed at a reasonable speed.

After the standing period, the mixture is filtered through a dry, grade-3 sintered-glass crucible and the filtrate is collected in a 400 ml beaker. The conical flask and the precipitate on the filter are washed with three 20 ml portions of petroleum ether, giving the precipitate on the sinter an occasional stir to aid filtration.

The filtrate is boiled down to a volume of about 20 ml on a boiling-water bath, then transferred quantitatively to a weighed 50 ml Pyrex evaporating basin, using a little petroleum ether to wash out the original beaker.

The extract is evaporated to dryness on the water bath, swirling the dish occasionally at the last stages of the evaporation to avoid skinning. Finally, the residue is dried to constant weight at 100°C and the percentage total toluene-petroleum ether extract in the sample is calculated.

The proportion of polyisobutene in the dried extract is then determined by the following infra-red method.

Sample preparation

Samples are required of about 0·002 in (0·05 mm) thickness. These are prepared by hot pressing using 0·002 in (0·05 mm) feeler gauges as spacers, between stainless steel plates, preferably coated with polytetrafluoroethylene.

Standard sample

A standard sample is required, prepared by mill blending polythene and polyisobutene in known proportions and pressing as above. It is advisable to prepare several samples from the blend and measure the spectra of these by the procedure below, to ensure that the blend is uniform. A blend of about 50/50 w/w is suitable.

Procedure

Using a double-beam spectrometer, record the spectrum of the standard sample between 9·5 μm and 15 μm. Similarly record the spectra of the unknown samples. The spectra obtained will be similar to Figure 6.2, but the intensity of the polyisobutene structure between 10 μm and 11 μm will vary relative to the 14 μm polythene structure according to the proportions of the blend.

Calculation

Draw straight line backgrounds on the spectrum and measure the absorbances at the 10·5 μm and 13·9 μm band maxima. Perform this calculation on all the spectra. Calculate the constant K from the expression

$$\frac{\text{Absorbance at } 10 \cdot 5 \ \mu\text{m}}{\text{Absorbance at } 13 \cdot 9 \ \mu\text{m}} = \frac{\log \text{AC/BC}}{\log \text{DF/EF}}$$

$$= \text{K} \times \frac{\text{weight \% polyisobutene in blend}}{\text{weight \% polythene in blend}}$$

331

using the absorbances determined on the spectrum of the standard blend, and using the known weight proportions of the standard blend. Using the determined value of K, and the measured absorbances on the spectrum of an unknown sample, the ratio polyisobutene to polythene is calculated from the expression

$$\frac{\text{Absorbance at } 10\cdot5 \ \mu m}{\text{Absorbance at } 13\cdot9 \ \mu m} = K \times \frac{\text{weight \% polyisobutene in unknown}}{\text{weight \% polythene in unknown}}$$

from which the percentage polyisobutene may be calculated, and hence the composition of the original sample.

THE DETERMINATION OF PARAFFIN WAX IN POLYTHENE/ PARAFFIN WAX BLENDS

The following method has been found useful for the determination of paraffin wax in polythene/paraffin wax blends.

1 g of the sample is refluxed with 100 ml of ether for 1 h. The solution is cooled and the polythene removed by filtration.

The filtrate is collected in a tared flask, the ether removed by evaporation on a water bath and the flask and contents brought to constant weight in an oven at 100°C. The percentage of paraffin wax in the original sample is calculated from the weight of the extract. The nature of the extract should be confirmed by infra-red examination if there is reason to suspect that substances other than paraffin wax may be present. The method will not extract long-chain amide waxes quantitatively; ethyl acetate is a better solvent for these waxes, but the preferred method is given on p. 350.

THE DETERMINATION OF CARBON BLACK IN POLYTHENE PREPARATIONS

This method for the determination of carbon black pigments in certain polymers such as polythene, butyl rubber and polystyrene was developed initially by the Bell Telephone Co. Cincinatti. The determination is often called for, and, in practice, the method has been found to be very successful indeed.

The test involves the pyrolysis of the polymer in a stream of nitrogen at 500°C and the test is a very reproducible one.

With polythene containing no carbon black, no residue is obtained and specimens of carbon black in ordinary use show losses of less than 5% under the conditions of the test. No appreciable inaccuracies are therefore likely to be encountered due to incomplete removal of the polythene or to significant losses from the carbon black.

On occasions, of course, the possibility of false values being obtained due to the presence of additives other than carbon black has to be considered.

Non-volatile substances would give high results, and the value of the test has always to be considered in relation to the additives present.

Apparatus

Porcelain combustion boat $75 \times 9 \times 8$ mm (or larger).
Pyrex tube 29 mm diameter $\times$ 43 cm long.
Electric furnace—capable of maintaining 500°C.
Thermometer 0 to 550°C.
Rotameter—range 1 to 10 l/min.
Two cold traps with ground-glass connecting leads and 10 mm diameter inner tubes.

Procedure

The apparatus is set up as shown in Figure 6.3. The tube from the second trap leads to a fume hood or to the outside atmosphere. The boat is

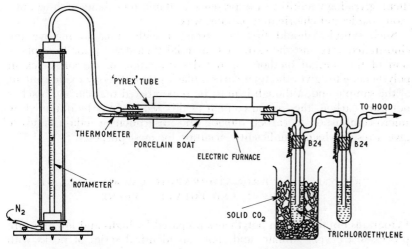

Fig. 6.3. Apparatus for the determination of carbon black in polythene preparations

ignited to red heat in a bunsen flame, cooled in a desiccator for at least 30 min and weighed.

1 g (± 0.1 g) of sample is placed in the boat and re-weighed. The nitrogen flow is adjusted to 1.7 ± 0.3 l/min. The boat and sample are then placed in the centre of the Pyrex tube, the stopper carrying the

thermometer and nitrogen inlet tube is inserted, and the thermometer is adjusted so that the bulb is touching the boat. The tube is now placed in the furnace and the cold traps are connected. The furnace is heated until the temperature reaches 300 to 350°C and the rate of heating is thereafter controlled so that a temperature of 480°C is reached in approximately 10 min, and a temperature of 500°C in a further 10 min. The temperature is maintained at 500±5°C for 10 min. The tube containing the boat is then withdrawn from the furnace and allowed to cool in the stream of nitrogen for 5 min.

The boat is finally removed from the tube via the nitrogen inlet end, placed in a desiccator, allowed to cool for 20 to 30 min, then weighed. The amount of carbon black residue is calculated as a percentage of the original sample.

The time of heating from cold to 350°C should be about 10 min. In carrying out successive runs the furnace temperature should be allowed to fall to 300°C or less between each determination.

THE DETERMINATION OF CARBON BLACK IN FLAME-RETARDANT POLYTHENE

The method given above for the determination of carbon black in polythene gives high results in the presence of flame retardants such as antimony oxide and chlorinated paraffin wax.

Such samples should first be extracted with ether to remove the chlorinated wax and the extracted material then dried prior to application of the normal method for the determination of carbon black in polythene. This gives the total carbon black plus antimony oxide content of the sample and, although it must be remembered that carbon black is not pure carbon, the carbon content of this residue may be obtained by micro-combustion to carbon dioxide and absorption in soda-asbestos to give a result which is sufficiently accurate for most purposes.

PYROLYSIS/GAS CHROMATOGRAPHIC EXAMINATION OF POLYTHENE AND POLYPROPYLENE

Pyrograms obtained from polythenes prepared by high- and low-pressure processes are very similar, and show an identical series of peaks with minor differences in peak-height ratios. The most notable characteristic of the polypropylene pyrogram is the large peak at $R_T = 0.78$ relative to styrene (see p. 61), and the presence of copolymerised ethylene may be detected by an increase in the peak height at $R_T = 0.30$ when compared with the peaks at $R_T = 0.22$ and 0.45. Although quantitative data may be obtained in some cases from calculations based on these peak-height ratios compared with those measured on standard samples, the method is not recommended for the analysis of ethylene/propylene copolymers since the pyrogram may also be modified by the degree of random, block and graft copolymerisation of the sample. The value of pyrolysis/gas chromatography in the examination of these copolymers is largely in the qualitative analysis of heavily filled samples where infra-red methods based on differences in the fine structure in the spectrum cannot be applied.

EXTRACTION AND IDENTIFICATION OF MINOR ORGANIC ADDITIVES

A polyolefine or hydrocarbon polymer composition may contain, in addition to polymer and process residues from its manufacture, additives such as U.V. absorbers, antioxidants, slip agents, epoxy compounds and metal salts of organic acids.

Some indication of the principles involved in the extraction and identi-fication of additives of this type, and some specific examples, are given below. Most of these organic additives may be extracted, either wholly or partially, with alcohol or ether, and these extracts are then available for examination directly by U.V. spectroscopy, or for separation by thin-layer chromatography, as described below. The fractions from the separation may be examined by U.V., infra-red or n.m.r. spectroscopy, or by chemical test methods.

QUALITATIVE EXAMINATION OF ORGANIC ADDITIVES IN POLYOLEFINES BY U.V. SPECTROSCOPY

This form of examination provides valuable information on those addi-tives which show strong absorption in the U.V. spectral region; these include the common substituted benzophenone U.V. absorbers and the phenolic antioxidants. The procedure is based on the work of Ruddle and Wilson[1].

Reagents

Ethanol suitable for spectroscopy.
Potassium hydroxide solution.
10 g Analar potassium hydroxide is dissolved in 10 ml distilled water and diluted to 60 ml with ethanol.
Nickel peroxide. Laboratory Reagent Grade.

Procedure

In the case of film or granular samples 1 to 2 g of the sample are boiled under reflux with 50 ml ethanol for 24 h; powder samples are boiled for

2 to 3 h. The resulting solution is cooled and filtered, and the following tests are carried out on the filtered solution:

1. The U.V. spectrum of the extract is obtained over the range 250 nm to 450 nm using 10 mm cells, and the alcohol 'blank' in the reference beam. If necessary, the alcohol extract is diluted with the alcohol blank.
2. 1 ml of distilled water and 2 ml of potassium hydroxide reagent are added to 10 ml of the extract. The U.V. spectrum of the resulting solution is now obtained over the same range as in Test 1 against a blank treated similarly.
3. 20 ml of the extract used in Test 1 are shaken with 0·5 g of nickel peroxide for 5 min. The mixture is filtered through a double Whatman filter paper and the spectrum of the filtrate over the range 250 nm to 450 nm is then obtained with the alcohol blank in the reference beam. If necessary, the filtered reaction products are diluted with the alcohol blank. If the filtered reaction product is coloured, it is examined over the range 430 nm to 800 nm.
4. 2 drops of the potassium hydroxide reagent are added to the cell containing the oxidised sample extract described in Test 3 and the spectrum obtained over the range 250 nm to 450 nm.

The four spectra, examples of which are given in Figure 6.4, are compared with those obtained from known antioxidants in order to characterise the antioxidant present in the sample.

With samples of low antioxidant content it may be necessary to repeat the extraction on a larger weight of sample.

THIN-LAYER CHROMATOGRAPHY

This procedure, which is essential for the unequivocal identification of the components of a complex mixture of additives, involves a preliminary separation of the ether-extracted additives by thin-layer chromatography and subsequent identification of the separated components by infra-red, n.m.r., or U.V. spectroscopy and other methods.

The chromatography is carried out on 1 mm thick Kieselgel G plates, using a developing solvent consisting of a 90 : 10 mixture by volume of petroleum ether (b.p. 40–60°C) and ethyl acetate. A sample loading of approximately 70 mg of the ether extract is advised and this is applied in a continuous line approximately 20 mm from the bottom of the plate. The development height of about 130 mm in the cell described on p. is satisfactory.

The plate is dried with a hot-air drier and the position of the separated bands marked whilst viewing the plate in U.V. light. The R_F values are

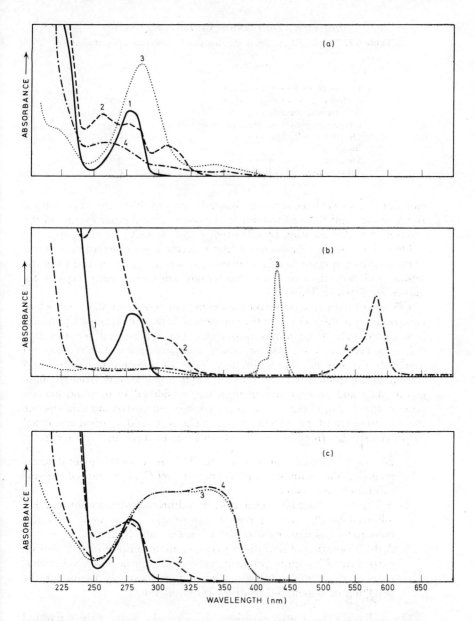

Fig. 6.4. Spectra of (a) 2 : 6-di-tert-butyl-4-methyl phenol; (b) bis (3 : 5-di-tert-butyl-4-hydroxy phenyl) methane; (c) 1 : 3 : 5-trimethyl-2 : 4 : 6-tris (3 : 5-di-tert-butyl-4-hydroxybenzyl) benzene. Curves (1), ethanolic solution; curves (2), alkaline ethanolic solution (10 ml of ethanolic solution plus 1 ml of water plus 2 ml of ethanolic potassium hydroxide solution); curves (3), ethanolic solution after reaction with nickel peroxide; curves (4), nickel peroxide reaction product made alkaline with 2 drops of ethanolic potassium hydroxide solution (After 'Analyst')

Table 6.1. Typical R_F values in the thin-layer chromatography of additives

Substance	R_F value
Low molecular weight polymer	1·0
4-Methyl-2 : 6-di-*tert*-butyl phenol	0·55
Tris(*p*-nonyl)phosphite	0·5
2-Hydroxy-4-*n*-octoxy benzophenone	0·46
Dilauryl thiodipropionate	0·42
Stearic acid	0·11
Oleamide	0·0
Alkyl-terminated polyethylene oxide	0·0

calculated and each component band is scraped from the plate into a small beaker and the additive isolated from a filtered ether extract of the fraction, for identification by infra-red, n.m.r. or U.V. spectroscopy.

One of the main advantages is the extremely good separation of the main stabiliser systems from low molecular weight polymer and from the amide and polyethylene oxide condensate additives. Some typical R_F values are listed in Table 6.1.

On occasion the separation on the 1 mm plate is not complete and before confirmation of the identity of the component is possible further separation is required to remove minor impurities. For this second separation a 0·25 mm layer is used and standard comparison substances are run alongside the sample. Approximately 50 μg of the sample fraction and standards are spotted in ether solution along a line 20 mm from the base of the plate and at least 30 mm from the outside edges to avoid uneven elution which can occur at the plate edge when petroleum ether is the main component of the solvent mixture. The plate is developed and dried as previously described and sprayed with one of three spray reagents:

1. 20% v/v sulphuric acid solution. This is a useful general spray reagent; after spraying and drying at 150°C all separated organic substances are visible.

2. A 2% w/v ethanolic solution of 2 : 6-dibromoquinonechloroimide* followed by 2% w/v aqueous borax solution. This spray produces coloured spots with most of the phenolic antioxidants.

3. A 0·1 N solution of potassium permanganate containing 5% sodium carbonate. This spray produces yellow spots on a pink background with most organic additives, and it is particularly useful for the detection of 4-methyl-2 : 6-di-*tert*-butyl phenol.

On each plate the relative positions, shapes and colours of the separated spots should be compared, but it should be noted that the particular solvent mixture used tends to form additional solvent fronts behind the main front. These subsidiary fronts are visible, however, and hence,

---

* This substance is now more accurately known as 2 : 6-dibromo-*p*-benzoquinone-4-chloroimine.

although true R_F values may be inconsistent, comparison of adjacent spots and standards is relatively simple.

CONFIRMATORY TESTS IN THE IDENTIFICATION OF MINOR ORGANIC ADDITIVES

It is sometimes necessary to support the spectroscopic identification of the separated components by confirmatory tests. For this purpose the chromatography is repeated on a 1 mm plate and the fractions are examined by the appropriate tests outlined in Chapter 2.

For example, polyethylene oxide type slip agents will respond to the heteropolyacid test given on p. 133 and epoxy compounds to Foucry's test given on p. 105. The mass spectrum of each separated component also provides very useful information.

QUANTITATIVE DETERMINATION OF MINOR ORGANIC ADDITIVES

CHEMICAL METHODS FOR ANTIOXIDANTS

In 1953 J. Haslam, S. Grossman, D. C. M. Squirrell and S. C. Loveday worked out methods for the detection and determination of antioxidants in polythene. They concerned themselves with five antioxidants.

1. NN'-di-β-naphthyl p-phenylene diamine.
2. NN'-diphenyl p-phenylene diamine.
3. Dicresylol propane.
4. Butylated hydroxy anisole.
5. Phenyl-α-naphthylamine.

The determinations of 1, 3 and 4 are very interesting ones because they illustrate the use of quite different chemical principles in the determination of antioxidants.

In their early work on the determination of NN'-di-β-naphthyl p-phenylene diamine, Haslam *et al.* developed the principle of solution of the sample in toluene followed by precipitation with alcohol to obtain a solution of the antioxidant. Initially they used vanadic acid as an oxidant of the substituted diphenylamine but later they indicated their preference for hydrogen peroxide as an oxidising agent.

The method they worked out, which is also applicable with minor modifications to the determination of NN'-diphenyl p-phenylene diamine, is given in full in BS 2782[2], together with the method for the determination of phenolic antioxidants, for example dicresylol propane and 4 : 4'-thio-bis(3-methyl 6-*tert*-butyl phenol).

The method as applied to the phenolic antioxidants depends on coupling the extracted phenol with diazotised sulphanilic acid to yield a coloured complex which may be measured at 430 nm.

On occasion, quite different types of coupling reaction may have to be employed in the determination of antioxidants in polythene, and an instance of this is the case of the determination of phenyl-β-naphthylamine.

THE DETERMINATION OF PHENYL-β-NAPHTHYLAMINE IN POLYTHENE

The level of antioxidant is 0 to 0·05%.

Preparation of sample solution

1 g sample is weighed into a clean 50 ml round-bottomed flask containing about 2 g of broken glass. 10 ml toluene (**B.D.H.** sulphur free) is added from a pipette and the flask connected to a reflux condenser and heated on a water bath with frequent swirling. When solution is complete about 20 ml of 95% ethyl alcohol is poured down the condenser, the ground-glass joint washed with alcohol and a stopper placed firmly in the flask. The flask and contents are shaken vigorously to precipitate the polymer, and the mixture is placed in a refrigerator for at least 5 min. The contents of the flask are now filtered through a 150 mm Whatman No. 31 filter paper and the filtrate collected in a 100 ml standard flask. The original flask, polymer and filter are washed thoroughly with small portions of ethyl alcohol until the volume of filtrate collected is 100 ml. The solution is shaken thoroughly and reserved as the prepared sample solution.

The colorimetric test for phenyl-β-naphthylamine at the level 0 to 0·05% antioxidant is then carried out as follows:

Reagents

Sodium nitrite solution.
A 2·5% w/v solution of sodium nitrite in water.
p-Nitroaniline solution.
1 g of p-nitroaniline is dissolved in 50 ml of 1 : 1 v/v hydrochloric acid solution and diluted to 100 ml with distilled water.
Diazotised p-nitroaniline solution.
To 5 ml of the p-nitroaniline solution is added 1 ml of the sodium nitrite solution and the solution is allowed to stand in an ice bath for 15 min. The solution is removed from the ice bath and used within 1 h.

Procedure

To 20 ml of the prepared toluene-alcohol solution of the antioxidant contained in a 6 × 1 in (150 × 24 mm) boiling tube are added 0·2 ml of the diazotised *p*-nitroaniline solution. The solution is shaken well and allowed to stand for 10 min. The optical density of the reddish-purple solution is then measured at 530 nm in a 20 mm cell with water as the reference solution.

A calibration curve is prepared by carrying out the test exactly as above on a series of standards containing 0, 0·02, 0·04, 0·06, 0·08 and 0·10 mg of phenyl-*β*-naphthylamine in 20 ml of 10% v/v toluene in ethyl alcohol solution. In our experience this calibration curve is a straight line passing through the following points:

mg phenyl-*β*-naphthyl-amine per 20 ml toluene-ethanol solution	0	0·02	0·04	0·06	0·08	0·10
Optical Density at 530 nm (20 mm cell	0	0·190	0·380	0·570	0·700	0·950

From the optical density obtained from the sample solution and the above calibration curve, the phenyl-*β*-naphthylamine content of the sample is calculated.

DETERMINATION OF BUTYLATED *p*-HYDROXY ANISOLE IN POLYTHENE

The level of the antioxidant is 0 to 40 p.p.m.

In this interesting application of the Gibbs' method, the antioxidant is extracted from the polythene in toluene-alcohol medium. In order to develop the indophenol colour, however, it is first necessary to remove the toluene by distillation with alcohol. Otherwise, low results will be obtained.

Reagents

Phenol-free toluene.
The toluene (B.D.H., sulphur free) is washed by shaking in a large separating funnel with 30% w/v sodium hydroxide solution, then distilled from sodium hydroxide sticks into a clean bottle.
Purified ethyl alcohol.
95% v/v ethyl alcohol is refluxed over sodium hydroxide for at least 1 h, then distilled into a clean bottle.
0·04% w/v solution of 2 : 6-dibromoquinone chloroimide (B.D.H.) in 95% ethyl alcohol is prepared immediately prior to use.
Borate buffer solution.

341

23·4 g sodium borate ($Na_2B_4O_7 \cdot 10H_2O$) (Analar) and 3·27 g sodium hydroxide (Analar) are dissolved in distilled water and the solution is diluted to 1 l.

Apparatus

The flasks used in the test should be 100 ml round-bottomed flasks with B24 or B19 ground-glass joints. A graduation mark is placed on the flasks at the level of the meniscus when the flask contains 25 ml ethyl alcohol, 1 g of polymer and 2 g of broken glass.

Procedure

1·00 g of the sample in the form of small chips is weighed into a 100 ml round-bottomed flask containing 2 g of broken glass and 10·0 ml of the phenol-free toluene are added. The flask is heated under reflux on a water bath, with occasional swirling to dissolve the polymer, which should be complete within 1 to $1\frac{1}{2}$ h. Whilst swirling constantly, 17 ml ethyl alcohol are poured down the condenser slowly. The flask is removed, stoppered and shaken thoroughly in order to effect complete precipitation of the polymer. Any polymer sticking to the bottom of the flask at this stage should be loosened by scraping with a glass rod.

The flask is fitted with a side arm, condenser and dropping funnel and the contents are heated gently, with dropwise addition of ethyl alcohol to maintain the volume of the mixture in the flask at the previously marked volume of 25 ml. The distillate is collected in a measuring cylinder and the distillation is continued until a volume of 100 ml of distillate has been collected. The distillation flask is cooled and the volume of the contents is adjusted to the 25 ml mark with ethyl alcohol.

50 ml distilled water and 10 ml of the borate buffer solution are added, and, maintaining agitation, 0·5 ml of the dibromoquinone chloroimide reagent are added slowly. The flask is stoppered, shaken and allowed to stand for 5 min. In a similar flask a blank solution is prepared so as to contain 25 ml alcohol, 50 ml distilled water, 10 ml of the borate buffer solution and 0·5 ml of the dibromoquinone chloroimide solution.

Both blank and sample solution are filtered separately through No. 31 filter paper and 50 ml of each filtrate is then allowed to stand for 30 min for full colour development. The optical density of the sample is then measured in a 40 mm cell at 610 nm against the blank solution in a paired cell.

The amount of butylated hydroxy anisole in the 50 ml aliquot, and hence by calculation in the original sample, is then determined by reference to a calibration graph. This graph is prepared by application of the test to known amounts of butylated hydroxy anisole in which the butylated hydroxy anisole is dissolved in 25 ml alcohol, then 50 ml of distilled water is added followed by 10 ml borate buffer solution and 0·5 ml dibromoquinone chloroimide reagent etc.

DETERMINATION OF BIS(2-HYDROXY 3-METHYL CYCLOHEXYL 5-METHYL PHENYL) METHANE IN POLYTHENE

Next, we come to the case of a well hindered phenol that may have lost a great deal of its phenolic character so far as the analytical chemist is concerned and yet this phenol may be capable of reducing phosphomolybdate in ammoniacal media.

Such a phenolic body is bis(2-hydroxy 3-methyl cyclohexyl 5-methyl phenyl) methane.

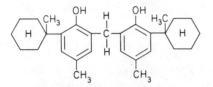

The determination of this kind of substance in polythene is as follows: The level of antioxidant is 0 to 0·1%.

Reagents

Toluene (B.D.H. laboratory reagent, sulphur free).
Ethyl alcohol (aldehyde free).
2 l of absolute alcohol are refluxed with a mixture of 20 g zinc granules and 20 g Analar potassium hydroxide for 45 min. The alcohol is distilled, rejecting the first 200 ml, the main distillate being collected in a clean bottle.
Phosphomolybdic acid solution.
A 2% w/v solution of phosphomolybdic acid in distilled water is prepared and filtered into a clean brown bottle.
Dilute ammonia solution.
One volume of conc. ammonia solution (sp. gr. 0·880) is diluted with one volume of distilled water.

Preparation of sample solution

1 g of the polythene sample in the form of small chips is weighed into a clean 50 ml Quickfit round-bottomed flask containing about 2 g of broken glass, 10 ml of toluene are added from a pipette and the flask is connected

343

to a reflux condenser and heated over a water bath. The polythene is allowed to dissolve with occasional gentle swirling of the flask to assist solution. When complete solution is obtained, i.e. after about 1 to $1\frac{1}{2}$ h, the condenser is washed with 15 to 20 ml of ethyl alcohol, the flask closed with a glass stopper, then shaken vigorously to precipitate the polymer. The mixture is cooled and then filtered through a Whatman No. 31 filter paper (150 mm) into a standard flask (100 ml). The flask and precipitate on the filter paper are washed with ethyl alcohol until the total volume of the filtrate is 100 ml. This filtrate is now mixed well and 20 ml are transferred, by means of a pipette, to a squat 100 ml beaker.

The beaker is immersed in a boiling-water bath until the last trace of solvent is just removed. The beaker is now removed from the water bath and 20 ml of ethyl alcohol are pipetted into the beaker, washing down the sides of the beaker with the solvent. The solution is swirled gently and allowed to cool to room temperature.

Colorimetric test

10 ml of the prepared sample solution are transferred, by means of a pipette, to a dry 6×1 in $(150 \times 24 \text{ mm})$ test tube. 2 ml of the phosphomolybdic acid solution are added and the whole is mixed well. 3·0 ml of the dilute ammonia solution are added from a burette with gentle swirling. *Exactly 5 minutes* after addition of the dilute ammonia solution the optical density of the blue test solution is measured in a 40 mm cell against a paired cell containing water. A blank determination is carried out on all the reagents used and the value of this blank is deducted from the figure obtained for the test solution.

A calibration curve is prepared relating mg of the antioxidant in 15 ml of the final coloured solution with optical density reading exactly as above; from the readings obtained the proportion of antioxidant in the polythene sample is readily calculated.

Notes

1. It is desirable to check two points on the calibration curve with each set of samples in order to avoid any errors due to temperature fluctuations.
2. For amounts of antioxidant less than 0·02%, greater sensitivity may be obtained by evaporating 50 ml of the initial toluene-ethanol solution to dryness and dissolving the residue in 20 ml of ethyl alcohol.

    Well hindered phenols such as the one described above may often be determined by quite a different principle, i.e. by reaction with methylene blue at high pH values to yield a coloured complex, soluble in chloroform.

3. In the methods just described an air or electrically driven stirrer may be fitted through the middle of the condenser to assist the solution process. When a stirrer is used, the addition of the broken glass is unnecessary.

SPECTROSCOPIC METHODS FOR ANTIOXIDANTS

The chemical methods so far described suffer certain disadvantages in that the procedures are lengthy, tend to be non-specific and may suffer interference by impurities in the solvents etc. The tendency has thus been to move towards the use of U.V. absorption methods.

U.V. ABSORPTION METHOD FOR READILY SOLUBLE ANTIOXIDANTS

It has been shown that many of the more soluble antioxidants such as 2:6-di-*tert*-butyl 4-methyl phenol may be extracted directly from samples by reflux or Soxhlet extraction using solvents such as chloroform; the solvent shows no interfering absorption in the U.V. and the polymer is only swollen in the solvent.

The procedure, which has been found particularly valuable for 2 : 6-di-*tert*-butyl 4-methyl phenol, is detailed below.

Apparatus

U.V. spectrophotometer.
Absorption Cells. Matched pairs of fused quartz cells. 10 mm and 20 mm path length.
Electric hotplate or heating mantle.
Reflux condenser. B 24 joint.
Extraction flasks. 150 ml. B 24 joint.
Volumetric flasks. 100 ml.
Filter funnels.
Filter papers. Whatman No. 542. 150 mm.

Reagent
Chloroform. Analar.

Extraction procedure

The appropriate weight of sample (see Table 6.2) is weighed into a 150 ml extraction flask and the requisite volume of chloroform is added from a pipette or other accurate dispenser. Two or three boiling beads are placed in the flask which is then connected to a reflux condenser and the solvent is boiled gently on a heating mantle or hot plate for 1 h. Whilst still attached to the condenser the flask is then cooled, after which it is removed, stoppered and cooled in a bath of running water. At the same time

Table 6.2. Antioxidant determination conditions

Expected antioxidant concentration p.p.m.	Weight of sample g	Volume of chloroform ml	Cell path-length mm
1000–8000	1·0	100·0	10
200–2000	2·0	50·0	10
20–250	5·0	50·0	20

a blank is carried out on the chloroform used by submitting the solvent to the same heating conditions as are described for the sample. Care is taken to avoid loss of solvent due to evaporation as no adjustment to volume is made.

Spectrophotometry

The cooled blank and sample solutions are filtered through 150 mm Whatman No. 542 filter papers supported in covered funnels. The absorbance of the filtered sample extract is measured in the appropriate cell (see Table 6.2) over the range 255 nm to 320 nm against the prepared blank in the comparison cell.

If a recording instrument is not available the absorbance measurements are made at 283 nm and at 5 nm intervals between 255 nm and 320 nm and the absorbance spectrum is plotted. The absorbance at 283 nm is calculated by the base-line method with the base line drawn from the absorbance at 255 nm tangentially to the curve at approximately 300 nm as shown in Figure 1.1(b) on p. 3.

Using the above principle, calibration curves are prepared using known concentrations of the antioxidant in chloroform.

The concentration of antioxidant in the chloroform extract of the unknown sample is read off from the appropriate calibration graph. Hence the amount in the original sample may be calculated.

U.V. ABSORPTION METHODS FOR SPARINGLY SOLUBLE
ANTIOXIDANTS

In the case of NN'-di-β-napthyl-p-phenylene diamine, bis(2-hydroxy-3-methyl cyclohexyl 5-methyl phenyl) methane and 4 : 4'-thio-bis(3-methyl 6-*tert*-butyl phenol) the rapid single-solvent extraction method fails to give complete recoveries and the toluene-alcohol extraction procedure detailed on p. 343 is used.

The substituted diamine is determined by direct measurement against a mixed-solvent blank at 320 nm, but in the case of 4 : 4′-thio-bis(3-methyl 6-*tert*-butyl phenol) it is necessary to evaporate the extract to dryness and redissolve the residue in ethanol for measurement at 280 nm.

A test of particular value in the examination of phenolic antioxidants is based on the so-called alkali shift. It may be that a phenolic antioxidant has been isolated from a polymer admixed with an impurity. When this mixture is treated with alkali the phenol will produce a phenate and, as a result, a change in U.V. absorption. More often than not the background material will remain unchanged.

If the 'difference' spectrum, i.e. between before and after addition of alkali, is examined, it is often seen to be highly characteristic of the particular phenol, and may be used in its determination.

The original paper by Harvey and Penketh[3] draws particular attention to this procedure in the case of the determination of small amounts of *o*-phenyl phenol. This principle is recommended for the spectroscopic determination of bis(2-hydroxy-3-methyl cyclohexyl 5-methyl phenyl) methane.

The toluene-alcohol extraction is carried out as previously described and 10 ml water added before the final dilution to volume with ethanol. 10 ml of the extract solution is transferred by pipette to a glass-stoppered test tube and 2 ml 90% v/v aqueous ethanol is added. The solution is mixed and the absorbance measured at 310 nm against a solvent blank similarly diluted. A separate 10 ml aliquot of extract solution is placed in another tube and 2 ml 10% w/v potassium hydroxide in 90% v/v aqueous alcohol added. After mixing the absorbance of this solution is measured at 310 nm against a blank made similarly alkaline. The difference between the two absorbance readings is referred to a prepared calibration curve and hence the antioxidant content of the sample may be calculated.

Certain phenolic antioxidants respond well to precise oxidation tests. Cook[4] showed that 4-methyl 2 : 6-di-*tert*-butyl phenol was oxidised by lead dioxide in ethanol solution and that from the reaction products, unchanged phenol, 1 : 2-bis(3 : 5-di-*tert*-butyl 4-hydroxy-phenyl) ethane and the quinone 3 : 5, 3′ : 5′-tetra-*tert*-butyl stilbene 4 : 4′-quinone could be recovered. It seemed to us that the production of a quinone of such a highly conjugated character in this oxidation process could be exploited with appropriate U.V. measurements in a method for the determination of 4-methyl 2 : 6-di-*tert*-butyl phenol in polyolefines.

Such a method was therefore devised for polythene in which the antioxidant was extracted from 1 g samples by dissolving in 10 ml re-distilled methyl cyclohexane, and reprecipitating the polymer by the addition of 35 ml ethanol. The resultant suspension was heated under reflux and vigorously stirred prior to filtration and dilution of the filtrate to 100 ml with ethanol. The extract was shaken with 1·5 g lead dioxide for 20 min and the absorbance of the filtered solution measured at 340 nm and 286 nm against a reagent blank. The difference in absorbance at the two wavelengths was referred to a calibration graph prepared by application of the oxidation procedure to known solutions of antioxidant. This method proved to be very useful for the determination of 0 to 250 p.p.m. antioxidant, but when the original supply of lead dioxide had become exhausted it was soon noticed that other batches had very different efficiencies in the oxidation process and careful selection had to be carried out. It became more and more difficult to obtain adequate supplies of satisfactory reagent and the method was thus abandoned in favour of the direct extraction procedure given on p. 345.

The principle of the oxidation procedure is still used however in the qualitative procedure outlined on p. 336, based on the work of Ruddle and Wilson[1] who found that nickel peroxide is a more efficient oxidising agent than lead dioxide for effecting spectral changes by oxidation of phenolic antioxidants.

RECOVERY OF ANTIOXIDANTS FROM POLYTHENE BY DISTILLATION IN VACUO

One other method of isolation of the antioxidant from the sample should be mentioned. This can often be of value when extremely small amounts of volatile antioxidants are present and has been used to obtain evidence

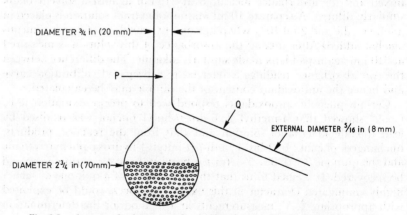

DIAMETER ¾ in (20 mm)

P

Q

EXTERNAL DIAMETER ⁵⁄₁₆ in (8 mm)

DIAMETER 2¾ in (70mm)

Fig. 6.5.   Apparatus for the recovery of antioxidant from polythene by distillation in vacuo

of the presence of less than 10 p.p.m. of 4-methyl 2 : 6-di-*tert*-butyl phenol and phenyl-α-naphthylamines in a polythene preparation.

50 g of the polymer, cube cut to approximately 40 mg cubes, are loaded into a 150 ml round-bottomed Pyrex flask fitted with a side arm just above the bulb, as shown in Figure 6.5. The flask is sealed just above the side arm

at point P and evacuated via the side arm to a pressure of less than 5 torr (0·7 kN/m^2) prior to sealing off under vacuum at point Q. The apparatus is then assembled as shown in Figure 6.6 with the flask containing the polythene immersed in a Woods' metal bath at 150°C and the side arm in a Drikold bath at −80°C. The temperature of the Woods' metal bath is raised over a period of 1 h to 300°C and the polythene sample is maintained at this temperature for 4 h, whilst the side arm is continuously cooled by repeated additions of Drikold.

At the end of this period the Woods' metal bath is allowed to cool to 180–200°C and the flask is removed. The side arm is cut at point R whilst the flask is still hot; the vacuum is broken and the side arm is removed.

In this type of work it is always desirable to carry out two distillation tests, i.e. one on the sample under test and one on a sample of polythene free from the suspected antioxidant.

The side arm receiving tubes containing the distillates which usually appear as white gel-like solids are cut again at point S and the two portions are placed in a stoppered tube (15 ml capacity) to which 10 ml

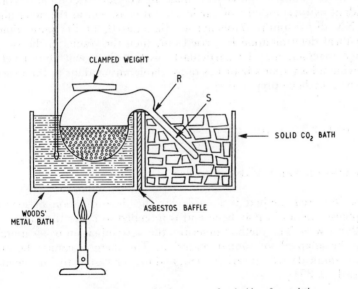

Fig. 6.6.   Assembled apparatus for the recovery of antioxidant from polythene

ethyl alcohol (spectroscopically pure) are added. After allowing the tubes to stand overnight they are shaken gently by hand until the wax-like solid is completely disintegrated into small particles.

The alcohol solutions are filtered through No. 1 Whatman filter paper into 20 mm cells and their U.V. absorptions measured spectrophotometrically in the region 250 nm to 320 nm against a paired cell containing the spectroscopically pure ethyl alcohol.

In our experience the U.V. absorption curve from pure polythene will take the form of a smooth curve exhibiting general absorption. If however, the distillate is derived from a polythene containing as little as 10 p.p.m. of 4-methyl 2 : 6-di-*tert*-butyl phenol then the U.V. absorption curve will show a marked inflection at 273 nm and a general increase in absorption over the wavelength range 260 nm to 295 nm.

On the contrary, if the distillate is derived from a polythene containing phenyl-α-naphthylamine, then the alcohol extract of the distillate will exhibit pronounced absorption in the region 315 nm to 375 nm. Moreover, it will be noted that the alcohol solution exhibits a violet fluorescence when examined under the U.V. lamp, as will the residual polymer, because of the relative, though not complete, non-volatility of the anti-oxidant.

Carried out in the above manner, the distillation test can be of very great value indeed, and particularly when information about the behaviour of known antioxidant-polythene combinations under comparable conditions of test has been made available by the analytical chemist. It has been indicated previously that direct U.V. examination of anti-oxidant extracts may be of particular value as, e.g. in the determination of NN'-di-β-naphthyl *p*-phenylene diamine. If, at the same time, the chemical determination is carried out, then the results of the two tests, taken together, may be particularly useful. They will almost certainly provide information about loss of available antioxidant in the processing of the polythene preparation.

DETERMINATION OF LONG-CHAIN AMIDE SLIP AGENTS IN POLYTHENE

INFRA-RED SPECTROPHOTOMETRIC METHOD

This infra-red method is suitable for the determination of long-chain aliphatic amides in polythene and is intended to cover the range 0·02% to 2·0% w/w. The method describes the determination of oleamide, but may be adapted for similar materials. The method cannot be applied when amine derivatives are present, and in such cases a titrimetric method is used (p. 354).

Apparatus

Absorption cells. Matched pair of stoppered infra-red quality fused silica cells 20 mm path length, e.g. 'Infrasil' type 21, from Lightpath Optical Company Ltd, Gallows Corner, Romford, Essex, England.
Beakers. 250 ml and 500 ml.
Conical flasks. 150 ml, fitted with ground-glass stoppers.

Hot plate and water bath. Both flame-proof.
Diatomite filter aid. Johns Manville 'Standard Supercel'.

Reagents

>Carbon disulphide. Analar.
>Liquid paraffin.
>Methanol.
>Methylene chloride.
>Toluene.

Procedure for the extraction of additive from polythene

According to the expected concentration of amide in the sample, an appropriate weight is taken as indicated in Table 6.3. The sample is weighed into a beaker and dissolved in 100 ml toluene by heating gently

Table 6.3. Quantities used for the extraction of additive from polythene

Expected concentration %	*Weight of sample* g	*Dilution volume* ml	*Aliquot to be evaporated* ml
0·02–0·1	10	Not diluted	Take total extract
0·1–0·2	5	500	400
0·2–0·4	5	500	200
0·4–0·8	5	500	100
0·8–2·0	5	500	50

on a flame-proof hot plate in a fume cupboard. The mixture is stirred constantly with a glass rod to aid solution.

When solution is complete 200 ml warm methanol are added, with stirring, to precipitate the polymer. The mixture is allowed to stand for 1 h.

A layer of filter aid approx. $\frac{1}{4}$ in (6 mm) deep is placed in a glass Buchner funnel. The filter aid is washed with methanol. The contents of the beaker are poured into the funnel, and filtered under reduced pressure. The beaker is washed out with four separate 30 ml portions of methanol, the washings being poured through the polymer in the Buchner funnel on each occasion. On each washing, the polymer is pressed firmly against the pad with a suitable tool. If the solution is to be diluted (see Table 6.3) it is transferred to a 500 ml flask and diluted to 500 ml with methanol. The entire solution, or the aliquot, as indicated, is transferred to a beaker and the bulk of the solvents boiled off on a flame-proof hot plate. The solution is then evaporated just to dryness on a boiling-water bath. The residue is dissolved in approx. 10 ml methylene chloride and the solution is transferred to a flask provided with a stopper. The beaker is washed out with methylene chloride. The methylene chloride is evaporated off on a water bath, and the residue is finally dried at 100°C for 1 h (this procedure is undertaken in order to ensure complete removal of methanol, which would otherwise interfere with the test). The flask is stoppered and allowed to cool.

Spectrophotometry

12·5 ml carbon disulphide are added to the residue in the flask. The flask is stoppered and the mixture is swirled gently until solution is complete.

One of the 20 mm quartz cells is filled completely with the solution and placed in the sample beam of the spectrophotometer. Similarly the second cell is filled with the carbon disulphide and placed in the comparison beam of the instrument. The spectrum is recorded between 2·6 μm and 3·2 μm. The spectrum obtained will be similar to Figure 6.7. The absorbance at the band maximum (approx. 2·93 μm) is measured against the linear background as in the figure. The absorbance is related to the concentration of oleamide by a calibration graph, prepared in the following way.

Calibration

To a series of stoppered 150 ml flasks are added 0, 2, 4, 6, 8 and 10 ml of a standard 0·1% w/v solution of oleamide in methylene chloride. To each flask are added 5 ml of a 1% v/v solution of liquid paraffin in methylene chloride and the solvent is evaporated off on a boiling water bath. The residue is finally dried at 100°C for 1 h. The procedure under the heading 'Spectrophotometry' is now followed. The graph of oleamide concentration versus absorbance is plotted for these samples.

The graph covers the range 0 to 100 mg oleamide in the test solution. If an aliquot has been taken appropriate allowance is made, otherwise the result read from the calibration graph is the oleamide content, in mg, of the sample weight originally taken.

This method may also be applied to other olefine polymers and copolymers.

DIRECT INFRA-RED METHOD FOR THE DETERMINATION OF
LONG-CHAIN AMIDES IN POLYTHENE

When 1% or more of long-chain amide is present, a direct method may be used. This method is not suitable for copolymers of ethylene with esters or acids, or compositions containing salts of carboxylic acids e.g. metallic stearates, and in these cases the extraction method given above should be used. Amine additives also interfere.

The sample is prepared in the form of a film of 0·01 in (0·3 mm) thickness by hot pressing between polytetrafluoroethylene-coated steel plates.

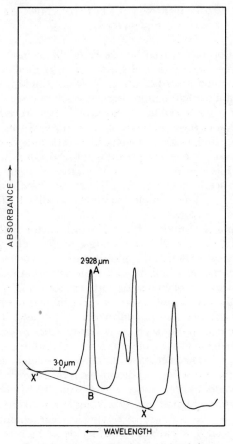

Fig. 6.7. Determination of oleamide in polythene

The pressing is carried out at the lowest possible temperature to avoid oxidation of the sample, or loss of the additive, which may otherwise bleed out from the sample.

The spectrum is measured over the range 5 μm to 7 μm. A linear background is drawn between the minima at approx. 5·5 μm and 6·5 μm. The absorbance of the short-wave component of the doublet band at approx. 6·1 μm is measured and the absorbance per mm of the sample is calculated. It is necessary to prepare a graph by measuring the absorbance per mm of a series of samples containing known amounts of oleamide, and samples determined by the solution method given above will be suitable, provided that they do not contain interfering components.

CHEMICAL METHODS FOR THE DETERMINATION OF
LONG-CHAIN AMIDES IN POLYTHENE

The determination of nitrogen by the Kjeldahl method is excellent if a sufficiently high concentration of the additive is present but, as in the infra-red test method described above, it is not applicable when amines such as long-chain diethanolamine derivatives are present. The titrimetric method of Wimer[5], as modified by Squirrell[6], has proved very useful for the determination of these additives in admixtures. In this procedure the mixture of amine and amide is dissolved in acetic anhydride and titrated, using an automatic recording potentiometric titrator, with a standard solution of perchloric acid in a 1 : 1 v/v mixture of acetic acid and acetic anhydride. An acetic acid reference electrode (Electronic Instruments Ltd, type RJ 23/4) is now preferred to the modified calomel electrode used previously.

Two end-points are produced in this titration, the first corresponding to the amine and the second corresponding to the amine plus the amide.

A preliminary extraction of the additives from the polymer is necessary. Whilst this presents no problem with material containing no inorganic additives, the presence of materials such as silica and titanium dioxide causes some difficulty. Some grades of these inorganic materials tend to adsorb the nitrogen-containing additives very strongly. This means that the normal solvents used, such as toluene, carbon tetrachloride and 1 : 2-dichloroethane, will sometimes extract only 50 to 60% of the amine or amide present. To extract these additives completely it is necessary to use very powerful extraction conditions and Taubinger[7] has shown that boiling the sample for 2 h under reflux with 0·1 M monochloroacetic acid in carbon tetrachloride will give complete recovery. Titration can be carried out in the presence of relatively large amounts of dry non-reactive solvents such as carbon tetrachloride or 1 : 2-dichloroethane, so that an aliquot of the extract with added acetic anhydride may be titrated directly with the perchloric acid titrant. As perchloric acid acts as a catalyst in the reaction of acetic anhydride with water, any solvents used must be dry: solvents containing significant amounts of water can cause relatively large errors due to the apparent titration of the water present. The tertiary long-chain N-diethanolamine derivative tends to react with the chlorinated solvents to cause minor, spurious end-points during the titration, but these end-points are easily eliminated by adding 1% v/v hydrochloric acid to the acetic anhydride, when the hydrochloride, not the free amine, is titrated.

Occasionally polymers that do not contain the above-mentioned inorganic fillers may contain salts such as calcium stearate, which would interfere in the determination of the amine or amide by being partially extracted and titrated as bases. Investigation has shown that such metal salts, although extracted by the above acidic extractant, as well as by carbon tetrachloride and isopropyl acetate, are insoluble in 1 : 2-dichloro-ethane and hence the above additives can be determined if extraction is carried out with this solvent.

DETERMINATION OF POLYETHYLENE OXIDE CONDENSATES IN POLYTHENE COMPOSITIONS

INFRA-RED ABSORPTION METHOD

This method is based on the measurement of alkyl ether groups from the absorption band at approx. 9·0 μm carried out on an extract from the polymer. A correction is necessary at this wavelength as there is some over-lapping absorption from aliphatic hydrocarbon material (e.g. low molecular weight polymer) present in the extract used.

The extraction procedure is identical to that described in the method for determination of long-chain amide in polythene (see p. 352) and the same procedure is followed also for taking aliquots according to the amount of condensate expected in the original composition. Thus, as before, the sample is submitted for infra-red examination as a residue in a 150 ml stoppered flask, to which 12·5 ml carbon disulphide are added. The flask is re-stoppered and the contents gently swirled until solution is complete.

In this case a matched pair of sodium chloride cells 3·5 mm path length are used (from Research and Industrial Instrument Co., London). One cell is filled with the carbon disulphide solution and placed in the sample beam of the spectrophotometer. The reference cell is filled with carbon disulphide and placed in the reference beam. The instrument is set to 10·5 μm wavelength, and the trimmer comb (sample-beam attenuator) is adjusted to bring the recorder pen to between 85% and 90% on the transmission scale. The spectrum is then recorded between 8 μm and 10·5 μm, and will be similar to that in Figure 6.8. The linear background is drawn as shown in the figure between the appropriate minima and the absorbance at exactly 9·0 μm is measured. Thus the uncorrected absorbance for polyethylene oxide condensate is

$$\log_{10}\frac{AC}{AB}$$

Next the absorbance due to hydrocarbon residues present in the extract is determined as follows. With the absorption cells still in the spectrometer, the wavelength is set to 2·25 μm and the trimmer comb is adjusted to bring the pen to between 85% and 90% on the transmission scale. The spectrum is recorded between 2·25 μm and 2·75 μm. The trace obtained will be similar to that in Figure 6.8. The linear background is drawn as shown between the absorption minima at approx. 2·25 μm and

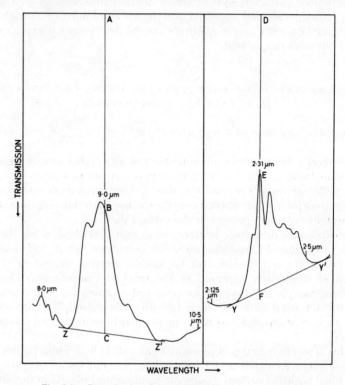

Fig. 6.8.   Determination of polyethylene oxide condensate in polythene

2·65 μm. The absorbance at the maximum at approx. 2·31 μm is measured as

$$\log_{10}\frac{DF}{DE}$$

Calibration

A calibration graph is prepared relating the absorbance of hydrocarbon oil at 2·31 μm to that at 9·0 μm so that the correction for hydrocarbon may be made. To a series of standard flasks are added 0, 2, 4, 8, 12, 16 and 20 ml of a standard 10% v/v solution of liquid paraffin in carbon disulphide. To each flask 10 ml of a standard 0·24% w/v solution of the polyethylene oxide condensate in carbon disulphide are added, and each solution is diluted to 50 ml with the same solvent. The solutions are shaken thoroughly and then each solution is treated in the same manner as has been described above, and the spectrum is recorded in the ranges quoted.

The absorbance is measured at exactly 9·0 μm and at the maximum at approximately 2·31 μm as described. The absorbance value at 9·0 μm for each solution containing hydrocarbon oil is deducted from the absorbance value of that containing no added oil. A graph is plotted of these absorbance differences (i.e. the correction to be *added* at 9·0 μm) against the corresponding absorbances at 2·31 μm.

A graph is prepared relating corrected absorbance at 9·0 μm to weight of polyethylene oxide condensate per 12·5 ml of carbon disulphide solution in the following manner:

To a series of 150 ml stoppered flasks are added 0, 2, 4, 6, 8, and 10 ml of a standard 0·1% w/v solution of polyethylene oxide condensate in methylene chloride. To each solution is added 5 ml of a 1% w/v solution of liquid paraffin in methylene chloride. The solvent is evaporated on a boiling-water bath, and the residue then dried at 100°C for 1 h after which each standard mixture is treated exactly as detailed for a sample. The correction graph as detailed above is used to determine the correction to be added to the absorbance at 9·0 μm, then the graph relating the corrected absorbance at 9·0 μm to the weight of polyethylene oxide condensate in each solution. The graph covers the range 0 to 10 mg polyethylene oxide condensate.

Calculation for samples

The correction to be added to the 9·0 μm absorbance in relation to the measured absorbance at 2·31 μm is read off from the correction graph. The correction is applied to the absorbance measured at 9·0 μm in each case. The weight of polyethylene oxide condensate corresponding to the corrected absorbance at 9·0 μm is read off from the calibration graph. The weight so obtained is then related to the weight of the original sample used, taking into account any aliquots that may have been taken during the preparation of the sample extract.

CHEMICAL METHOD FOR THE DETERMINATION OF
POLYETHYLENE OXIDE CONDENSATE IN POLYTHENE

Alternatively the polyether may be determined gravimetrically by precipitation of the complex formed by the addition of barium chloride and phosphotungstic acid solution to a filtered solution of the extract in 90% v/v methanol acidified with hydrochloric acid (see p. 418). Volumes of solvent and reagents must be carefully standardised and the method calibrated by application of the method to known weights of particular condensates. A drying temperature of 100°C is satisfactory for the complex.

THE DIRECT DETERMINATION OF DILAURYL THIODIPROPIONATE IN POLYPROPYLENE BY INFRA-RED SPECTROSCOPY

In principle it is possible to determine dilauryl thiodipropionate (DLTDP) in polypropylene by measuring the intensity of the C=O absorption band at about 5·75 μm on a film pressed from the composition. In practice there are a number of difficulties which must be considered.

1. There is broad but significant absorption of polypropylene over the region of interest for this determination. Thus the determination is best made by a difference method, placing in the reference beam of the spectrometer a sample of polypropylene which contains no DLTDP. On the other hand, this reference sample must not be oxidised, hence we prefer to use a sample of polypropylene stabilised with a phenolic stabiliser only e.g. 0·15% Topanol CA. A film of 0·018 to 0·020 in (0·46 to 0·50 mm) thickness is pressed from a stock of such material, and a fresh film is pressed monthly from this stock, the old film being rejected.

2. Many polypropylene compositions contain, in addition to DLTDP, substances such as phenolic stabilisers, U.V. absorbers, and metal salts of carboxylic acids. Interaction may occur between DLTDP and these other additives and the effect of this is that, in addition to the 'free' DLTDP band at 5·743 μm a further band due to DLTDP in a 'bonded' form occurs at 5·813 μm. In order to account quantitatively for the DLTDP, both these forms must be measured in the test.

3. Many compositions contain carboxylic acid salts and during processing these may decompose with the formation of the free acid, which has an absorption band at 5·843 μm. This may interfere with the DLTDP test, hence the method of measurement must be arranged to overcome this interference.

4. The absorption of carbonyl groups in oxidised polypropylene falls within the region of interest so it is most important that neither the test sample nor the reference polypropylene sample be oxidised. Suggestions regarding the reference sample are given above, but it is most important that the films used in the test be prepared at low temperatures (e.g. 150°C) and that the pressing be made as quickly as possible, so that the compositions are not held at elevated temperatures longer than necessary.

In spite of the difficulties that are set out above, it has been found that the method given here is both accurate and extremely rapid and reproducible in different laboratories.

Method

A film is prepared by extrusion from the material under test or by hot pressing between polytetrafluoroethylene-coated stainless-steel plates of 0·018 to 0·020 in (0·46 to 0·50 mm) thickness, taking precautions against oxidation of the sample, as suggested above.

This sample is placed in the sample beam of a double-beam spectrometer. In the reference beam is placed a reference sample, prepared from

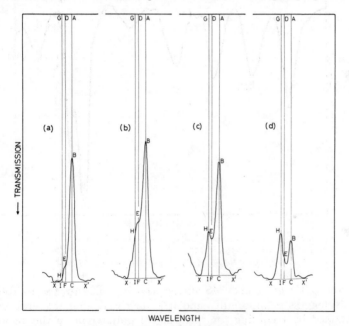

Fig. 6.9. Determination of dilauryl thiodipropionate (DLTDP) in polypropylene: $B = 5·743$ μm, $E = 5·813$ μm, $H = 5·843$ μm. Spectra with (a) 0·28% DLTDP (substantially in the free form), (b) 0·42% DLTDP (in the free and bonded forms), (c) 0·24% DLTDP (in the free and bonded forms) and stearic acid present, (d) 0·07% DLTDP (substantially in the free form) and stearic acid present

a DLTDP-free polypropylene compound, prepared as under (1) above. The spectrometer is set to 5·5 μm and the trimmer comb (sample-beam attenuator) of the spectrophotometer is adjusted to bring the pen to 80% transmission on the recorder scale. The spectrum is recorded between 5·5 μm and 6·0 μm at the slowest available recording speed. The shape of the trace recorded may vary considerably with different samples. Some examples of the recorded spectra, obtained from different compositions, are given in Figure 6.9(a) to (d).

Measurement of the spectral trace

In each case the straight line XX′ is drawn to meet the curve tangentially at the absorbance minima occurring at about 5·6 μm and 6·0 μm. The straight lines ABC, DEF and GHI are drawn parallel to the transverse lines on the recorder paper to pass through the curves at B, E and H at

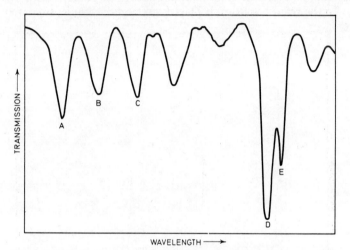

Fig. 6.10. *Reference spectrum of 0·003 in thick polystyrene: A = 5·144 μm (1944·0 cm⁻¹), B = 5·345 μm (1871·0 cm⁻¹), C = 5·551 μm (1801·6 cm⁻¹), D = 6·245 μm (1601·4 cm⁻¹), E = 6·316 μm (1583·1 cm⁻¹)*

5·743 μm, 5·813 μm and 5·843 μm respectively. The precise location of the wavelengths is important, and it is suggested that reference values be obtained by recording the spectrum of polystyrene in this region of the spectrum, and applying the values quoted in Reference 8 for the five polystyrene bands in this region (Figure 6.10). The 'free' DLTDP absorbance is given by

$$\log_{10}\frac{AC}{AB} = A_{FD}$$

The 'bonded' DLTDP absorbance is given by

$$1 \cdot 5 \log_{10}\frac{DF}{DE} - 0 \cdot 75 \log_{10}\frac{GI}{GH} = A_{BD}$$

360

From the values calculated as above, and the measured thickness of the film sample under examination, the content of DLTDP is calculated from the equation

$$\% \text{ w/v DLTDP} = \frac{(A_{FD} \times X) + (A_{BD} \times Y)}{\text{Thickness of sample}}$$

The values of the factors X and Y are determined as follows.

Calibration

The simplest method of calibration is by measurement of the spectra of known solutions of DLTDP in paraffin. Solutions of DLTDP in paraffin oil are prepared, containing 0·05, 0·10, 0·15, 0·20, 0·25 and 0·30 g per 100 ml of paraffin oil. The spectra of these solutions are measured in an absorption cell of 0·020 in (0·50 mm) thickness, in the same manner as has been used to measure the polypropylene samples, but now placing in the comparison beam of the instrument a cell of the same thickness but containing the blank paraffin oil. It is necessary to know the thickness of the cell used to contain the solutions. Backgrounds on these spectra are drawn in the way described above for sample spectra, and the absorbance at 5·743 μm, i.e. the maximum due to DLTDP, is measured. Absorbance per unit thickness versus w/v % concentration of the solutions is plotted. The plot should be linear and should pass through the origin. The slope of this graph

$$\frac{\% \text{ w/v concentration}}{\text{Absorbance per unit thickness}}$$

is the factor X in the expression given above. Although the value of Y may also be determined experimentally, it has been found that

$$Y = 1 \cdot 5 X$$

THE DETERMINATION OF ADDITIONAL ELEMENTS IN HYDROCARBON POLYMERS

Elements other than carbon and hydrogen may be present in hydrocarbon polymers. They may arise from catalyst and process residues e.g. in the cases of titanium, aluminium, sodium, iron and chlorine, or as a result of the presence of organic and inorganic additives such as stabilisers, pigments and fillers. These latter additives may contain such elements as nitrogen, chlorine, sulphur, calcium, silicon and antimony.

The general identification of the inorganic elements present is usually carried out on the ash of the sample. With appropriate precautions it may be possible, in some cases, to ash the sample quantitatively without loss of the elements concerned. In such cases the spectrographic examination of the ash may provide all the additional information that is required.

It is, however, often necessary either to make determinations of specific elements in the ash by chemical or atomic absorption methods or to determine specific elements directly in the sample by X-ray Fluorescence procedures.

Atomic absorption methods are used for the determination of aluminium and sodium and X-ray Fluorescence methods for the direct determination of titanium, iron, chlorine, calcium, and sulphur in polyolefine powders and densified compound sample disks. Percentage concentrations of antimony and chlorine containing fire retardants can also be determined. When the SbKα fluorescence is used for the antimony determination a disk greater than $\frac{1}{4}$ in (7 mm) thick must be used but not matrix absorption correction is normally required for varying amounts of chlorine. A correction is, however, required for the absorption of the chlorine Kα radiation by the antimony present.

The Kjeldahl method is used for the determination of nitrogen and when a chemical determination of chlorine is required three methods are recommended.

1. MACRO-METHOD: 0·02 TO 1% CHLORINE E.G. IN POLYPROPYLENE

A pellet (0·5 g) of the polypropylene is burnt in a Mahler–Cook type bomb calorimeter, using 25 atm (2·5 MN/m^2) oxygen pressure, and placing 10 ml of distilled water in the bottom of the bomb. After cooling, the pressure in the bomb is released by bubbling the gases through 25 ml of water. The water is transferred from the bubbler to a 250 ml beaker and the water from the bomb calorimeter and the bomb washings are added. 5 drops of 2 N nitric acid solution are added and the solution is diluted to 150 ml. A full-scale recording titrimeter similar to that described by Haslam and Squirrel[9] is adjusted to give a full-scale deflection for 200 mV and the chloride in the solution is titrated with 0·002 N silver nitrate solution. The electrode system described by these authors is used.

The silver nitrate solution is standardised by the titration of known amounts of pure sodium chloride. The salt is dissolved in 150 ml water and 5 drops of 2 N nitric acid solution are added prior to the titration.

2. MICRO-METHOD: 0·1 TO 1% CHLORINE

Apparatus

Oxygen Flask Combustion equipment. A 2000 ml oxygen combustion flask, provided with a firing head supporting a platinum gauze sample holder and appropriate remote control ignition facilities is recommended. (See p. 83)

Methyl cellulose capsules. Capacity, approx. 0·45 ml; diameter, approx. 7 mm; weight, approx. 80 mg. Size No. 1 supplied by A. H. Thomas and Co., Philadelphia, are suitable.

Cotton wool. B.P.C. super-quality.

Safety screen.

Spectrophotometer. Range 350 nm to 800 nm. The Unicam SP 600 is suitable.

Absorption cells. Matched pair of 20 mm path-length glass cells.

Reagents

Stock standard chloride solution.
Dissolve 0·8245 g Analar sodium chloride (previously dried at 270°C for 4 h) in distilled water and dilute to 1000 ml. 1 ml ≡ 0·5 mg chlorine.
Dilute standard chloride solution.
As required, dilute 10 ml stock sodium chloride solution to 100 ml with distilled water. 1 ml ≡ 0·05 mg chlorine.
Hydrogen peroxide solution. 5 volumes.
Sodium hydroxide solution. 0·1 N.
Nitric acid. Analar.
Nitric acid solution. 0·1 N.
Ammonium ferric sulphate solution.
Dissolve 12 g Analar ammonium ferric sulphate in distilled water, add 40 ml concentrated nitric acid and dilute to 100 ml with distilled water. Filter into a clean bottle.
Mercuric thiocyanate solution.
Dissolve 0·4 g mercuric thiocyanate (recrystallised from ethanol) in 100 ml absolute ethanol.
Oxygen. Cylinder.

Procedure

About 0·1 g of the sample is weighed into the larger half of a methyl cellulose capsule and covered with a small pad of cotton wool. The platinum gauze basket of the oxygen flask firing head is lined with cotton wool, the capsule is inserted and a suitable ignition wick is arranged. The basket is covered with a small piece of platinum gauze. 5 ml hydrogen peroxide reagent and 5 ml 0·1 N sodium hydroxide are added to the clean dry 2000 ml combustion flask. The flask is filled up with oxygen by blowing a stream of oxygen into it for approximately 1 min and then immediately replacing the stopper. The solution is swirled to wet the walls of the flask thoroughly, the stopper is removed and the firing head and sample holder are inserted quickly and secured into position. The joint is sealed with 1 ml of distilled water and the apparatus is placed behind the safety screen. The remote control firing system is then operated in order to initiate the combustion.

The flask is allowed to stand for 30 min after completion of the combustion. The firing head is then lifted and washed down with 5 ml 0·1 N nitric acid solution delivered from a pipette, followed by 20 ml distilled water. The washings are collected in the flask. The head is removed from the flask and, after swirling the contents of the flask to mix the solution, 2 ml ammonium ferric sulphate solution followed by 3 ml mercuric thiocyanate reagent are added, mixing well after each addition. The solution is allowed to stand for 10 min and the absorbance is measured at 460 nm in a 20 mm cell against distilled water in the comparison cell. A blank determination is carried out throughout the whole combustion and spectrophotometric procedure, omitting only the sample. The blank absorbance is deducted from that of the sample to obtain the corrected absorbance.

Calibration

A calibration curve is prepared relating mg chloride per 41 ml final coloured solution to corrected absorbance at 460 nm in a 20 mm cell by the following method. 1, 2, 4, 8, 12, 16 and 20 ml of the dilute standard chloride solution are added to a series of 100 ml beakers and each solution is diluted to 21 ml with distilled water. To each solution is added 5 ml of 5 volumes hydrogen peroxide reagent, 5 ml 0·1 N sodium hydroxide reagent and 5 ml 0·1 N nitric acid solution. Each solution is mixed in turn and the reagent addition and spectrophotometric procedure is carried out as described above. The absorbance obtained for the blank is deducted from that obtained for each of the standards and a graph of corrected absorbance against mg chloride per 41 ml final coloured solution is plotted. This graph covers the range 0 to 1·0 mg chloride.

The concentration of chloride in the sample solution is read off from the calibration graph and the amount in the original sample is calculated.

3. TRACE METHOD: 3 TO 100 P.P.M. CHLORINE

This determination is frequently required on polypropylene powders in which traces of chlorine may remain as process residues. Maltese, Clementini and Mori[10] have studied the problem in detail and have described methods which distinguish between 'fixed' chlorine present in salt form and 'labile' chlorine which may form hydrochloric acid at high temperatures. These authors found that all forms of chlorine could be fixed by a procedure the principle of which has been used by Hamilton and Ringham[11] to determine chlorine in polypropylene.

A 40 g sample is weighed into an 80 mm diameter platinum basin and 25 ml of 0·5% w/v alcoholic potassium hydroxide solution are added, together with sufficient alcohol to ensure thorough wetting of all the polymer. The basin is allowed to stand for 2 h in a chloride-free atmosphere, for example in a nitrogen-purged box, and the alcohol is then evaporated off and the basin transferred to the apparatus shown in Figure 6.11. A nitrogen flow rate of 80 ml/min is used and the rheostat on the heating mantle is adjusted to give an even volatilisation of the polymer ($\sim$400°C).

When volatilisation is complete ($\sim$30 min), leaving a small carbonaceous residue, the basin is cooled and the residue transferred as completely as possible to a 150 ml beaker using 6 ml distilled water followed

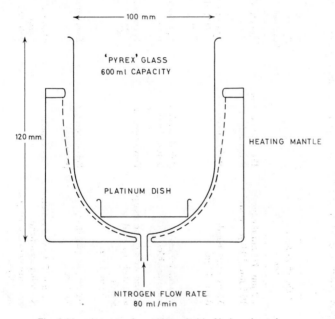

Fig. 6.11.   Apparatus for rapid 'pyrolysis' of hydrocarbon polymers

by 5 ml acetone as transfer solvents. The basin is finally washed with three 6 ml portions of distilled water and the washings added to the beaker. 3 ml 1·0 N nitric acid solution is added followed by 100 ml acetone, and the titration of the chloride is carried out with 0·01 N silver nitrate solution as described on p. 154 i.e. with automatic potentiometric recording. The titre is corrected for the small blank on the reagents and the result is calculated as p.p.m. total chlorine in the sample.

Table 6.4. Characteristic bands of hydrocarbon resins

All hydrocarbon resins show absorption bands near 3·4 μm and 6·8 μm which are usually prominent features in their spectra, but which are of only limited value in differentiating these resins.

Position and nature of prominent bands	Other characteristic bands	Resin	Ref. to Spectrum	Comments
8·1 μm Strong, medium width	7·2 μm, 7·3 μm doublet 10·5 μm, 10·8 μm doublet	Polyisobutene or Butyl rubber	Spectrum 8.18	Often encountered in blends e.g. with polythene
8·2 μm 10·8 μm 13·1 μm Medium to strong	Many weak bands 8 μm–12 μm	Polybutene-1	Spectra 6.31 and 6.32	All these poly-α-olefins in isotactic form show many sharp medium to weak bands in their spectra
8·2 μm 9·0 μm Medium	7·2 μm, 7·3 μm sharp doublet	Poly-3-methyl butene-1	Spectrum 6.33	
8·6 μm 10·0 μm 10·3 μm 11·9 μm All medium strength, sharp bands	7·3 μm strong and sharp, several sharp weak bands 7·5 μm–13 μm	Polypropylene	Spectrum 6.22	
8·6 μm Strong	7·2 μm, 7·3 μm sharp doublet	Poly-4-methyl pentene-1	Spectrum 6.34	
8·6 μm Medium	7·2 μm, 7·3 μm doublet. No other significant structure between 7·5 μm and 15 μm	Polyterpene resin	Hummel[14] Figures 436–438	Some similarity to atactic polypropylene (Spectrum 6.23)
8·6 μm 10·3 μm Medium to strong	No other significant bands	Atactic polypropylene	Spectrum 6.23	

Band position	Band description	Other bands	Polymer	Spectra reference	Remarks
10·1 μm / 11·0 μm	11·0 μm band strong, 10·1 μm band medium intensity	6·1 μm sharp and medium. Other bands weak	1:2-polybutadiene	Spectra 8.1 and 8.5	Most polybutadiene samples contain both forms, also the *cis* form (see below). Butadiene often encountered copolymerised with styrene, acrylonitrile or both
10·4 μm	Strong and broad	Other bands weak	*Trans*-1:4-polybutadiene	Spectra 8.2 and 8.3	
12·0 μm	Medium strength, broad band	6·1 μm medium strength, sharp; 7·3 μm medium strength and width	Polyisoprene	Spectra 8.14–8.17	Spectrum strongly influenced by crystallinity
12·3 μm, 12·8 μm, 13·2 μm, 13·8 μm	Prominent broad bands, variable relative intensity	6·0 μm–6·8 μm two or three sharp bands of medium intensity	Polyvinyl toluene	Hummel[14] Figures 533 and 535	Often copolymerised with butadiene, acrylonitrile, or both
13·3 μm	Strong, very broad	Many weak sharp bands between 8 μm and 12 μm	Polyindene	Hummel[14] Figure 551	Often copolymerised with coumarone (extra band near 14·3 μm)
13·3 μm, 14·3 μm	Both strong and broad	Several medium or weak sharp bands 6 μm–6·8 μm and 8 μm–12 μm	Polystyrene or Poly-α-methyl styrene	Spectra 6.37 and 6.38	Often encountered in copolymers and blends
13·6 μm	Strong, very broad	6·1 μm sharp and weak	*Cis*-1:4-polybutadiene	Spectrum 8.4	Usually contains *trans* and 1:2 isomers
13·7 μm, 13·9 μm	Strong, sharp doublet	7·3 μm medium strength and width. Very weak structure between 8 μm and 13 μm	Polythene	Spectrum 6.1	Occasionally encountered as ethylene/ester copolymers
No significant bands between 7·5 μm and 15 μm		7·3 μm medium strength and width	Cyclised rubber	Hummel[14] Figure 441	Vague, indefinite spectrum

The above authors have shown that the temperature of the pyrolysis and volatilisation of the polymer must not exceed 500°C. Above this temperature serious losses of chloride as potassium chloride can occur.

This method has the advantage over other procedures that a large weight of sample is used which minimises errors due to sample inhomogeneity and allows a lower limit of detection.

THE DETERMINATION OF WATER IN POLYOLEFINES

The remarks made in Chapter 3 on the determination of water in polyvinyl powders and compounds are also applicable to polyolefines and indeed similar methods can be used. Alternatively the 'Nylon Method' (Chapter 5) has proved valuable and a gravimetric method in which the water released by vacuum distillation is trapped on a pre-weighed absorption tube containing phosphorus pentoxide on an inorganic support is useful for samples containing more than 0·1% of water. The method now preferred, because of its sensitivity and general versatility, is the gas chromatographic procedure quoted as an example in Chapter 1 (p. 51).

INFRA-RED EXAMINATION OF HYDROCARBON POLYMERS

When it appears, following the scheme outlined in Chapter 2, or from other evidence, that a resin is a hydrocarbon, its further identification depends upon the recognition of absorption bands due to structural groups present in the molecule. The spectral features of the principal hydrocarbon resins are summarised in Table 6.4, and by the use of this table, and the reference spectra, most hydrocarbon resins, if presented in a clean condition, should be identified.

POLYTHENE

When examined as thin films (0·001 to 0·002 in or 0·025 to 0·05 mm thickness) all kinds of polythene appear very similar in their spectra (Spectrum 6.1). The strong absorption bands at 3·4 μm, 6·8 μm, 7·3 μm and 13·8 μm all arise from the —CH_2— chain. When the infra-red spectrum of a resin shows these features, and no others of significance, it may be presumed to be polythene, but there is naturally a strong similarity between the spectrum of polythene and the spectra of high molecular weight linear aliphatic hydrocarbons, including paraffin wax.

There is usually no difficulty in deciding, from its physical properties, whether a substance is polythene or paraffin wax, since the latter is soluble in ether while polythene is substantially insoluble. This test is useful when dealing for example with wax-coated paper; both poly-thene and paraffin wax are removed by boiling toluene, but only the latter is ether soluble (see p. 332).

CRYSTALLINE EFFECTS IN THE SPECTRUM OF POLYTHENE

The absorption band at 13·8 μm is split into two sharp peaks in the spectrum of crystalline polythene; a similar effect is noted in other materials which contain linear hydrocarbon chains when these materials are examined as crystalline solids. In the amorphous condition (e.g. molten polythene, or a crystalline hydrocarbon when examined in solution) this sharp doublet in the spectrum is replaced by a broader and relatively weaker band.

This doublet band is of no practical value for the quantitative measurement of crystallinity (or amorphous content) owing to the overlap of the crystalline and amorphous structure and possible anomalies in the intensity of the crystalline bands due to molecular orientation in the sample. The most satisfactory method of estimating amorphous content is by means of the absorption band at 7·65 μm; details of this method are given by Miller and Willis[12].

THE CHEMICAL FINE STRUCTURE OF POLYTHENE

OBSERVATIONS BY DIRECT INFRA-RED SPECTROSCOPY

When polythene is examined as thick layers, differences are observed in the spectra of different samples. To illustrate these effects, the spectra of a high-pressure polythene, a polythene prepared with a Ziegler catalyst, and a Phillips type polythene[13] are shown in Spectra 6.2, 6.3, and 6.4 for 0·02 in thick samples. These spectra are intended to demonstrate the nature of the differences which may be observed, and it must be emphasised that differences of this kind may be found between the spectra of polythenes made by any particular manufacturing process and may arise either from changes in the process conditions, or through the effect of modifiers such as small amounts of hydrocarbons or olefines added during the polymerisation process.

These spectra, when compared with that of high molecular weight polymethylene (Spectrum 6.5), show substantial additional structure between 10 μm and 11·5 μm. These bands arise from the presence of molecular groups other than the linear —CH_2— chain, particularly C=C unsaturated groups, details of which are given in Table 6.5. There also appears in this region a band at 11·18 μm due to the terminal methyl of a pendent alkyl group of C_4 or longer (i.e. a branch along the polymer chain). This last band is in many cases obscured by the band due to

369

Table 6.5. Unsaturated groups in aliphatic hydrocarbon polymers

Unsaturated Group	Structure	Band position		Extinction coeff[a] ε (peak height)	Factor[b] for double bonds/ 1000 C atoms
		μm	cm^{-1}		
Trans chain (trans vinylene)	$\begin{array}{c} -CH_2 \quad\quad H \\ C=C \\ H \quad\quad CH_2- \end{array}$	10·35	965	86[c]	0·18
Terminal (vinyl)	$\begin{array}{c} -CH_2 \quad\quad H \\ C=C \\ H \quad\quad H \end{array}$	10·10	990	48[c]	0·32
		11·00	910	120[c]	0·13
Pendent methylene (vinylidene)	$\begin{array}{c} -CH_2 \quad\quad H \\ C=C \\ -CH_2 \quad\quad H \end{array}$	11·27	888	133[c, d]	0·12[d]
Trisub (cis) in poly-isoprene	$\begin{array}{c} -CH_2 \quad\quad CH_2- \\ C=C \\ CH_3 \quad\quad H \end{array}$	12·05	830	55[e]	0·28[e]
Trisub (trans) in poly-isoprene	$\begin{array}{c} -CH_2 \quad\quad H \\ C=C \\ CH_3 \quad\quad CH_2- \end{array}$	12·05	830	25[e]	0·61[e]
Cis chain (cis vinylene)	$\begin{array}{c} -CH_2 \quad\quad CH_2- \\ C=C \\ H \quad\quad H \end{array}$	13·60[f]	735[f]	25[f]	0·61[f]

[a] Extinction Coefficient $\varepsilon = \dfrac{A}{cl}$

where A = peak height absorbance
  c = conc. of absorbing group in g mol/litre of sample (or solution)
  l = thickness of sample (or solution) in cm
Thus ε is the absorbance of a 1 cm thick layer of sample (or solution) containing one g mol of C=C groups/litre. 1 g mol of C=C is taken as 26 g, i.e. —CH=CH—. Values determined with rock-salt prism instrument

[b] Multiply absorbance/cm of the resin by the appropriate factor to obtain the number of C=C groups/1000 carbon atoms. Density of hydrocarbon resins is assumed to be 0·9 in obtaining this figure. These factors are calculated from the extinction coefficients in the preceding column of the table

Factor = $\dfrac{15\cdot5}{\varepsilon}$

The factor is determined as follows:
1 litre of resin, density assumed 0·9, contains 900 g
The g mol weight of the —CH$_2$— group (1 carbon atom) is 14 g

Therefore 1 litre of resin contains $\dfrac{900}{14}$ g atoms of carbon

$= 64\cdot3$ g atoms of carbon

ε is the absorbance/cm of a resin containing 1 g mol of C=C groups/litre

Therefore ε is the absorbance/cm of a resin containing $\dfrac{1000}{64\cdot3}$ g mol of C=C groups/1000 g atoms of carbon

Therefore ε is the absorbance/cm of a resin containing 15·5 C=C groups/1000 carbon atoms

If the absorbance/cm of an unknown sample be A, then the concentration of C=C groups is $A\,\dfrac{15\cdot5}{\varepsilon}$ 1000 carbon atoms,

giving the factor $\dfrac{15\cdot5}{\varepsilon}$ by which the absorbance/cm must be multiplied to obtain the result in terms of double bonds/1000 carbon atoms

[c] Figures calculated from data on monomeric hydrocarbons. The figures of Silas, Yates and Thornton[15] for trans chain and terminal (11·00 μm band) calculated for polybutadiene are 133 and 184 respectively

[d] In the case of polythene, the total measured absorbance at 11·27 μm (888 cm⁻¹) must first be corrected for underlying methyl group absorption (see text)

polyisoprene

[f] This band is very variable in position in different compounds. These figures apply to cis-polybutadiene. The coefficient determined by Silas, Yates and Thornton[15] is an area, and not a peak height absorbance, and hence is not comparable

pendent-methylene (vinylidene) unsaturation at 11·27 μm, and is only revealed by removing the unsaturation by brominating the polymer (p. 372). A method for the determination of the amounts of the different unsaturated groups present in polythene is given on p. 372.

OBSERVATIONS BY DIFFERENCE SPECTROSCOPY

Other less-obvious differences between the spectra of different polythene samples may be revealed by measuring difference spectra between a sample of the unknown and a sample of polymethylene, or a linear high molecular weight polythene, placed in the reference beam of the instrument. By this test, an absorption band due to methyl groups may be revealed at 7·26 μm; the test has been developed as a fully quantitative method (p. 374).

Absorption bands in the 13 μm region due to pendent ethyl and butyl groups may also be revealed by difference spectroscopy in the following way, with a test specimen of approximately 0·01 in (0·3 mm) thickness.

The specimen is placed in the sample beam, and a polymethylene (or linear polyethylene) wedge, prepared as described on p. 376, is placed in

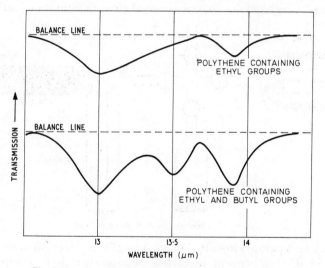

Fig. 6.12.   Absorption bands of ethyl and butyl groups in polythene

the comparison beam of a double-beam instrument. The wedge is arranged so that the thickness of the comparison sample may be altered by pushing the wedge across the beam. With the wedge thickness at approximately 0·01 in (0·3 mm), the transmission level at 12·5 μm is adjusted to about 70% with an attenuator placed in one or other of the beams of the

spectrometer, depending upon the relative transmission of the two specimens. While traversing the spectrum manually, the wedge thickness is
adjusted to obtain as far as possible the same transmission values at
12·5 μm, 13·65 μm and 14·5 μm. The spectra obtained will be somewhat
similar to those in Figure 6.12.

The absorption bands at 13·0 μm and 13·5 μm are due to ethyl and
butyl groups respectively. The latter value is rather lower than that given
in Table 2.4 for the absorption bands due to butyl side chains. Presumably
there is distortion of the spectra in Figure 6.12 because of the difficulty
in balancing the very strong band at 13·8 μm.

The application of this test in a quantitative manner has been described
by Willbourn[16].

QUANTITATIVE ESTIMATION OF UNSATURATED GROUPS IN POLYTHENE

Three types of C=C unsaturated groups occur in polythene. The type of
unsaturated group and the wavelength of the absorption band used for
quantitative estimation are given in Table 6.5.

An absorption band due to pendent alkyl groups (C_4 and longer) occurs
very close to the band due to pendent-methylene unsaturation and cannot

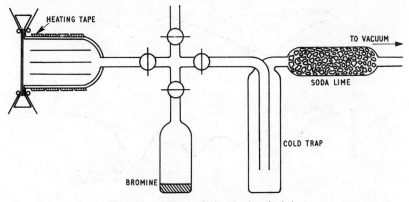

Fig. 6.13. Apparatus for bromination of polythene

normally be separated from it. The most satisfactory method of measuring
the band due to pendent methylene unsaturation is to compare the
spectrum of the sample with that of the same material from which unsaturation has been removed by bromination.

BROMINATION PROCEDURE

The apparatus used is shown in Figure 6.13. It consists of a small glass
vacuum oven which can be heated and maintained at a temperature just

above the melting point of the polythene specimen (120 to 140°C depending upon the type of polythene). The oven is connected through a tap to an oil vacuum pump and a small reservoir containing degassed liquid bromine. The oven contains a glass rack on which may be placed microscope slides to carry the specimens to be brominated.

Press the polythene sample to approx. 0·04 in (1 mm) thickness. Cut a

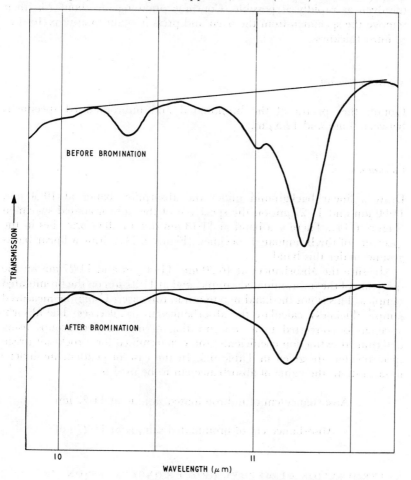

Fig. 6.14. Quantitative estimation of unsaturated groups in polythene

strip from the pressing and place it on a microscope slide, retaining the remainder of the pressing. Place the specimen and slide in the oven, and pump out the chamber to a hard vacuum. Heat the oven until the specimen is just seen to melt; maintain the oven at this temperature and open the tap to admit bromine vapour. Allow the bromine vapour to

373

remain in contact with the specimen for 5 min; then pump out the bromine vapour. During the latter process maintain a cold trap in the line to the pump to condense the bromine vapour, thus protecting the pump.

When all the bromine vapour has been pumped out, as judged by the disappearance of the brown vapour from the oven (about 5 min), cool the oven as rapidly as possible. Continue pumping for about 2 h, then remove the specimen from the oven and press it again to approx. 0·04 in (1 mm) thickness.

Spectrophotometry

Obtain the spectra of the brominated and unbrominated specimens between 9 μm and 12·5 μm.

Calculation

Draw a linear background under the absorption bands at 10·30 μm, 11·00 μm and 11·27 μm on the spectrum of the unbrominated specimen (Figure 6.14). There is a band at 11·18 μm due to alkyl branches in the spectrum of the brominated specimen (Figure 6.14); draw a linear background under this band.

Measure the absorbances at 10·30 μm, 11·00 μm and 11·27 μm on the spectrum of the unbrominated sample, and at 11·27 μm on the brominated sample. (This is not the band maximum in this case.) Using the measured sample thickness, calculate the absorbance/cm in all cases. The absorbance/cm is converted to a concentration of unsaturated groups using determined extinction coefficients; those determined for a rock-salt prism spectrometer are given in Table 6.5. In the case of pendent methylene unsaturation, the value of absorbance/cm to be used is

Absorbance/cm of unbrominated sample at 11·27 μm

− Absorbance/cm of brominated sample at 11·27 μm

DETERMINATION OF METHYL GROUP CONCENTRATION IN POLYTHENE

There has been a good deal of controversy as to the most satisfactory method of determining the methyl group content of polythene. The following method, however, developed in our laboratories, is in our opinion the most straightforward and simple method of performing this analysis.

374

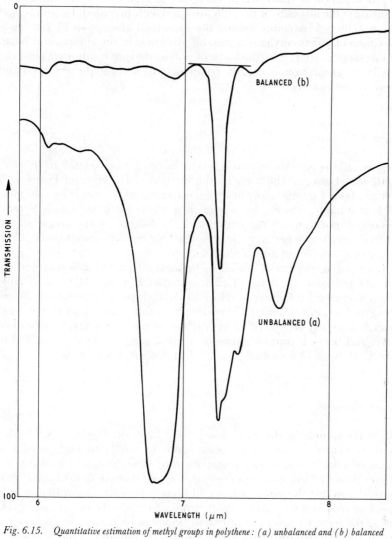

Fig. 6.15. Quantitative estimation of methyl groups in polythene : (a) unbalanced and (b) balanced spectra

The method is based upon measuring the absorption band at 7·26 μm due to methyl groups. This appears on the edge of a block of absorption with maxima at 7·32 μm and 7·40 μm, the latter arising from —CH_2— groups (wagging mode).

The situation is illustrated in curve (a) in Figure 6.15; the problem is to measure the intensity of the 7·26 μm band with precision. In our method this is achieved by compensating the unwanted absorption by placing a specimen of polymethylene of suitable thickness in the comparison beam of the spectrometer. The thickness of polymethylene is adjusted until the —CH_2— absorption is exactly compensated, leaving the —CH_3 band clear for measurement.

Polymethylene wedge

Prepare the polymethylene as a wedge varying from 0·0025 to 0·015 in (0·07 to 0·4 mm) in thickness by the method of Harvey and Peters[17]. A wedge about 4 in (100 mm) in length (tapering edge) and 1·5 in (40 mm) high is suitable. Mount the wedge in a frame for handling; it is most important not to touch the wedge with the fingers, as any grease on the surface of the wedge may lead to serious errors, particularly when measuring samples of low methyl content.

If no polymethylene is available, the best alternative is a polythene of low methyl content. It has been found that Hostalen GUR is suitable (manufactured by Farbwerke Hoechst A.G., Hoechst, Germany). When measured against polymethylene, this material is found to contain less than 0·5 methyl groups per 1000 carbon atoms on the test as described here, and hence is quite suitable for a blank sample. We are indebted to Mr T. P. Snell of Van Leer (U.K.) Ltd, for this information.

Spectrophotometry

Place the sample in the form of a pressed layer of approx. 0·005 in (0·1 mm) thickness in the sample beam, and the wedge in the comparison beam of a double-beam spectrometer arranged so that the radiation beam passes through that section of the wedge which is about 0·005 in (0·1 mm) in thickness. Set the instrument to 7·14 μm (where the absorption of both specimens is comparatively small) and adjust the beam attenuator to bring the recorder pen to an apparent transmission reading of about 70%. Turn the wavelength drum manually to 7·33 μm and adjust the wedge thickness to make the apparent transmission the same as that at 7·14 μm. Return to 7·14 μm and measure the transmission again; make a further minor adjustment to the wedge position if necessary to obtain the same transmission value at these two wavelengths. Then record the spectrum

376

between 7 μm and 7·5 μm. This should be similar to curve (b) in Figure 6.15.

Calculation

Draw a straight line background under the band (b) as in Figure 6.15. Measure the absorbance in the usual way, and calculate the absorbance/cm for methyl groups. For a rock-salt prism spectrometer the following relationship was determined:

Methyl groups/1000 carbon atoms = 0·85 × absorbance/cm at 7·26 μm

Calibration

The calibration factor above was determined from observations on a series of specially prepared samples of branched polymethylene in which the number of branches was known precisely by an independent measurement[16]. Owing to the labour involved in this method of calibration, it will generally be more convenient to select as calibration standards some of the polythene samples whose methyl group contents are known. These specimens may then be retained as permanent reference samples for the analysis.

The following grades of 'Alkathene' (Imperial Chemical Industries Ltd) may be used.

'Alkathene'	Methyl content Methyl groups per 1000 carbon atoms
XDB 58	17
XLF 28	21
WNC 18	28
WRM 19	32
WVG 23	37

The original calibration[16] assumes that the side branches are butyl groups, and it is possible that the extinction coefficient for the methyl group in other side chains, such as methyl, ethyl or very long alkyl groups may differ from that determined for the butyl group. Since, however, the band at 7·26 μm is a compound band arising from all types of methyl group present, and methods are not available for estimating independently the relative amounts of the different alkyl groups which contribute to the 7·26 μm absorption band, it seems reasonable in practice to use only one

377

extinction coefficient for this band, determined on a side chain of inter-mediate length.

OBSERVATIONS ON OXIDISED POLYTHENE

Spectra of heat- and photo-oxidised polythene were observed by Cross, Richards and Willis[18]. They examined high-pressure polythene, but the effect of oxidation on other kinds of polythene is similar.

Heat oxidation has been shown to take place through the formation and decomposition of the hydroperoxide group[19] the final products being various carbonyl and hydroxyl containing groups. Thus the most obvious change in the spectrum on oxidation is in the region of 5·8 μm where the absorption of the carbonyl group occurs. The spectrum recorded with a prism spectrometer shows a broad band in this position; under con-ditions of higher resolving power the band is seen to be irregular in shape, and evidently consists of a number of overlapping bands. The principal components are considered to arise from aldehyde (5·77 μm), ketone (5·81 μm) and carboxylic acid (5·84 μm). The firmest assignment is that of carboxylic acid since treatment of oxidised polythene with alkali causes the disappearance of the 5·84 μm maximum, and the appearance of a new band at about 6·2 μm; this is completely consistent with the forma-tion of a salt from a carboxylic acid.

There are also some features at shorter wavelengths which have been attributed to peracids and peresters[20] although this structure might also arise from the five-membered lactone ring structure.

In photo-oxidation, the most obvious spectral difference again is the development of the 5·8 μm carbonyl band. The major spectral difference between heat- and photo-oxidised specimens in which the carbonyl con-centration is the same is the development in photo-oxidised samples of prominent bands at 11·0 μm and 10·1 μm, due to introduction of the terminally unsaturated group $RCH{=}CH_2$. It is thought that this results from chain scission, the unsaturated groups being developed at the new chain ends created [18, 20].

When polythene is photo-oxidised, or thermally oxidised in the solid state, it would be expected that the surface of the sample, being in contact with the air, would be oxidised much more heavily than the interior of the sample. This suggests that, if the spectrum of the surface layer only were measured, the onset of oxidation could be observed sooner than if the transmission spectrum of the total sample were measured. The measurement of the surface layer only, by attenuated reflection spectro-scopy, to give an 'early warning' of oxidative degradation has been reported[21]. Our experience with this kind of measurement is that it is difficult to make it properly quantitative, as oxidised samples often have rough surfaces which fail to make good contact with the surface of the multiple internal reflection prism, and for this reason we prefer to use a transmission method to determine the combined oxygen content of oxidised polythene.

DETERMINATION OF COMBINED OXYGEN IN OXIDISED POLYTHENE

It is assumed in this method that the oxygen is present as ketonic carbonyl groups, since most of the carbonyl groups detected are of this type[20].

Sampling

The sample thickness to be used is not critical, and may be varied to obtain a carbonyl band of convenient intensity for measurement. Samples over 0·08 in (2 mm) thick are not suitable as the direct radiation transmission of the resin in very thick layers is poor.

The method of sample preparation depends upon the nature of the problem. If, for example, a sample of powder or granules is submitted, this may be hot pressed to a suitable thickness. On the other hand if a moulding be submitted, this will usually be preferentially oxidised on the surface, and it may be more informative to measure the spectra of sections cut from the surface of the moulding with a microtome.

Again, if it is desired to follow carbonyl formation in the oxidation of a particular sample of polythene at a temperature below its melting point, the original sample should be pressed to about 0·01 in (0·2 mm) thickness. There is no point in using a thicker sample, as the oxidised surface is then merely diluted with unoxidised polythene from the interior of the sample, since oxidation is limited by the penetration of oxygen into the material.

Spectrophotometry

Obtain the transmission spectrum of the prepared sample between 5·5 μm and 6·8 μm.

Calculation

Draw a linear background under the absorption band at 5·8 μm and measure the absorbance at the band maximum. Using the measured sample thickness calculate the absorbance/cm.

379

Calculate the percentage carbonyl oxygen from the absorbance/cm by a determined absorption coefficient.

Calibration

Calibration is best performed with a pure sample of a long-chain ketone, such as distearone. Known weights may be compounded in a known weight of polythene, alternatively known concentration solutions may be made of distearone in an aliphatic hydrocarbon medium such as *n*-heptane. The absorption coefficient E for a 10 mm path length of a solution containing $31 \cdot 7\%$ w/w distearone is determined, using the same conditions of slit width and traverse speed as for the polythene samples. The percentage oxygen as carbonyl in the unknown polythene samples is determined using the expression

$$\% \text{ Oxygen w/w} = \frac{\text{Absorbance/cm of unknown}}{E}$$

MODIFIED FORMS OF POLYTHENE

CHLORINATED AND CHLOROSULPHONATED POLYTHENE

The spectrum of chlorinated polythene is variable, and depends upon the distribution of chlorine along the polythene chains. A sample with a random distribution of chlorine (Spectrum 6.6) contains virtually no straight runs of $-CH_2-$ groups, and hence the $13 \cdot 8 \, \mu m$ $-CH_2-$ structure disappears, and the $6 \cdot 8 \, \mu m$ $-CH_2-$ band moves towards $7 \, \mu m$ (see Table 2.9).

If the chlorine atoms are bunched together along the polythene chains (Spectrum 6.7) the absorption bands of long runs of $-CH_2-$ groups still appear at $6 \cdot 8 \, \mu m$ and near $13 \cdot 8 \, \mu m$. New bands appear at $7 \, \mu m$ and $7 \cdot 9 \, \mu m$ indicative of $-CH_2-$ and C—H groups in a heavily chlorinated environment, and additional C—Cl bands appear in the $12 \, \mu m$ to $13 \, \mu m$ region. The spectra shown represent extreme conditions, and all samples of chlorinated polythene may be expected to show some combination of the features seen in these spectra.

The spectrum of chlorosulphonated polythene is given by Hummel[14] (Figure 1391). This shows a broad band near $8 \, \mu m$, rather similar to that in Spectrum 6.6, but additional strong sharp bands at $7 \cdot 3 \, \mu m$ and $8 \cdot 6 \, \mu m$ make this material easy to recognise from the spectrum.

IRRADIATED POLYTHENE

The spectrum of polythene irradiated (with high-energy radiation) in vacuo (Spectra 6.8, 6.9) shows some quantitative differences from that of the same polythene before irradiation. With high-pressure polythene, the first effect noted is a decrease of pendent-methylene unsaturation (Spectrum 6.8), which has almost disappeared at a dose of 20 mega-roentgens. With increasing dose, the chain unsaturation increases slowly (Spectrum 6.9, 60 megaroentgens irradiation). There appears to be little effect on methyl group content at these relatively small irradiation doses, but a small increase of methyl content may be observed after prolonged irradiation.

It is known from solubility measurements that cross-linking is introduced by irradiation, but there is no evidence of this from the spectrum.

When irradiation is performed in air rather than in vacuo, the changes described above are seen, but the process is accompanied by oxidation, with the appearance of the usual carbonyl structure near 5·8 μm.

Thus the effect of irradiation as observed in the infra-red spectrum is to produce quantitative rather than qualitative differences. The analyst called upon to ascertain whether a given sample of polythene has been irradiated will not usually be in possession of an unirradiated blank of the same polythene for comparison. However, if a polythene sample has a high content of chain unsaturation, as compared with either terminal- or pendent-methylene unsaturation, this can be taken to indicate that the sample may have been irradiated.

COPOLYMERS OF ETHYLENE WITH OTHER OLEFINES

Copolymers of ethylene with other unsaturated aliphatic hydrocarbons such as butene-1 and pentene-1 have been made, and may be of some importance as polythenes of controlled branching. There are usually no characteristic spectral features of the resins, and infra-red examination is usually confined to measurement of total methyl content and unsaturation by the methods above, and the detection of alkyl side chains following Willbourn's method[16].

Experience of the results of these tests on known polymers and co-polymers will often enable some deductions to be made on the nature of such a resin, and occasionally an additional clue may be provided by the presence of other characterisable hydrocarbon groups such as the iso-propyl or tertiary butyl group whose principal absorption bands are given in Table 2.4.

Copolymers of ethylene and propylene are considered on p. 391.

COPOLYMERS OF ETHYLENE WITH ESTERS

Copolymers of ethylene with small amounts of ester monomers such as acetates, acrylates and methacrylates are commercially important. All the prominent bands of polythene appear in these spectra, together with an ester carbonyl band which is relatively sharp (as opposed to the rather broad carbonyl bands due to oxidation) and prominent C—O structure in the 8 μm to 9 μm region.

In ethylene/vinyl acetate copolymers (Spectra 6.10 and 6.11) the absorption bands at 8·1 μm and 9·8 μm indicate the presence of acetate groups. Comparison of these two spectra shows the change in contour of the 13·8 μm —CH$_2$— band with change in acetate content. The sharp doublet due to crystalline packing of the —CH$_2$— groups in the low-acetate copolymer is replaced by a single broad band in the high-acetate copolymer in which the ethylene fraction is amorphous.

In the acrylate copolymers the C—O structure is near 8·6 μm and is sufficiently distinctive in the case of methyl and ethyl acrylate copolymers (Spectra 6.12 and 6.13) for these to be differentiated.

Unfortunately, as in the case of copolymers with vinyl chloride, the corresponding acrylate and maleate copolymers have very similar spectra (Spectra 6.12 to 6.15). Ethylene/methyl acrylate and ethylene/dimethyl maleate copolymers are distinguished only by the structure near 12 μm, while the corresponding ethyl acrylate and diethyl maleate copolymers differ somewhat in the shape of the structure due to C—O groups at 8·5 μm.

The spectrum of ethylene/methyl methacrylate copolymer (Spectrum 6.16) is surprisingly similar to those of the acrylate copolymers, being distinguished mainly by the appearance of a sharp band at 10·1 μm. The characteristic methacrylate C—O structure has, as usual, been greatly modified by copolymerisation.

INFRA-RED ABSORPTION METHODS FOR THE DETERMINATION OF THE COMBINED VINYL ACETATE CONTENT OF ETHYLENE/VINYL ACETATE COPOLYMERS

Method applicable to copolymers containing 3 to 30% combined vinyl acetate

For copolymers in this range the most convenient absorption band for the determination of combined vinyl acetate content is that at 16·5 μm. The sample is prepared by hot pressing between polytetrafluoroethylene-coated plates. With unknown samples the thickness of the pressing must be changed on a trial and error basis to obtain an absorption band of convenient intensity (i.e. 0·2 to 0·7 absorbance units). In practice it is not desirable to use a sample thinner than 0·0025 in (0·06 mm) or the thickness measurement is too imprecise, while samples of thickness greater than 0·05 in (1 mm) transmit so little energy that the spectroscopic measurement is unsatisfactory. Thus the practical limits are between 3 and 30% combined vinyl acetate.

The spectrum is recorded on a double-beam instrument over the range 15 μm to 18 μm and will be similar to Figure 6.16. The linear background is drawn as shown in the figure and the absorbance is measured at 16·5 μm against this background.

$$\frac{\text{Absorbance at } 16\cdot5\ \mu\text{m}}{\text{Thickness of sample}} \times \text{K} = \%\ \text{Combined vinyl acetate in sample}$$

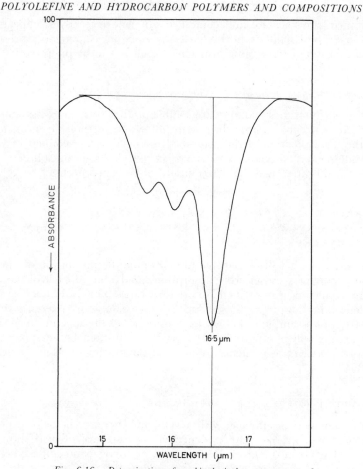

Fig. 6.16. Determination of combined vinyl acetate content of ethylene/vinyl acetate copolymer (0·00275 in thick sample containing approx. 28% acetate)

The value of the constant K must be determined either by measuring the spectra of copolymers the acetate contents of which are known by n.m.r. measurements (see p. 394) or by measuring the spectra of laminates prepared from films of polythene and polyvinyl acetate. The latter method is tedious as it is difficult to prepare films of sufficient area and uniformity, but in our experience both methods give the same result.

Method for samples containing below 3% combined vinyl acetate

Such samples are best dealt with by recording the complete spectrum over the range 2·5 μm to 15 μm on a thin film and measuring the absorbance ratio of the bands at 9·8 μm and 13·8 μm, presumed to be due to combined vinyl acetate and combined ethylene respectively. Spectrum 6.10 may be useful as a standard for such measurements, although standardisation on the same instrument is clearly to be preferred.

Method for samples containing over 30% combined vinyl acetate

The majority of these samples are soluble, and in such cases the simplest and most satisfactory procedure is to follow the method given in detail for the determination of the combined methyl acrylate content of vinyl chloride/methyl acrylate copolymers in tetrahydrofuran solution (see p. 197) calibrating with known solutions of polyvinyl acetate.

INFRA-RED SPECTROSCOPIC METHODS FOR THE QUANTITATIVE ANALYSIS OF OTHER ESTER COPOLYMERS OF ETHYLENE

In many cases it will be possible to determine the proportion of ester in these copolymers by selecting absorption bands due to each of the constituents and measuring their absorbance ratio. This ratio may be used to determine the proportions of an unknown sample by reference to the spectrum of a sample of known composition e.g. the spectra given here. The results will be approximate only, but will often be adequate, and if more precise measurements are necessary, n.m.r. methods are preferred (p. 394).

ETHYLENE COPOLYMERS WITH ACIDS

The copolymers of ethylene with acrylic and methacrylic acids, Spectra 6.17 and 6.18, show clear evidence of the presence of acid from the carbonyl band at 5·9 μm and the broad hydroxyl band underlying the 3·4 μm C—H structure together with the broad band at about 10·8 μm. These two spectra are rather similar, although a sharp methyl group band appears near 7·3 μm in the case of the methacrylic acid copolymer. These acid copolymers are often encountered in a partially neutralised state. Examples are Spectra 6.19 and 6.20, which are those of a 10·6% methacrylic acid copolymer neutralised with sodium to different extents. The weakening of the 5·9 μm carbonyl band of the acid and the increased intensity of the carboxylic acid salt band at 6·5 μm with increasing neutralisation are clearly shown in these spectra.

BLENDS CONTAINING POLYTHENE

POLYTHENE/POLYISOBUTENE BLEND

The blend of polythene and polyisobutene (Spectrum 6.21) shows the characteristic bands of both these resins. The prominent bands of polyisobutene are evident at 8·15 μm, 10·5 μm and 10·7 μm; the single methyl group band of polythene at 7·3 μm is obscured by the strong methyl doublet of polyisobutene, but the presence of polythene is clear from the strong 13·8 μm band.

It is possible to determine the proportion of polyisobutene in these blends by the method given on p. 328. It is not possible by infra-red spectroscopy to distinguish between polyisobutene and butyl rubber.

POLYTHENE/PARAFFIN WAX BLENDS

It is not possible to distinguish polythene/paraffin wax blends from straight polythene by means of their infra-red spectra. However, if such a blend is suspected, the method given on p. 332 may be applied.

POLYPROPYLENE

The spectrum of isotactic polypropylene (Spectrum 6.22) is somewhat surprising, since, in addition to the bands normally attributable to —CH_2— and —CH_3 groups at 3·4 μm and near 7 μm, a number of sharp bands of medium intensity occur in the 8 μm to 12 μm region.

In the spectrum of molten isotactic polypropylene, which is presumably amorphous, and so-called atactic polypropylene which is amorphous at room temperature (Spectrum 6.23) all the prominent structure between 8 μm and 12 μm disappears, except for bands at 8·7 μm and 10·3 μm. The only important difference in the spectra of the molten polymer and atactic polymer is the band in the latter at 11·3 μm. This is thought to be due to the chain-terminating group $R·C·(CH_3)=CH_2$ and its prominence in this spectrum of the atactic resin is presumably due to the relatively low molecular weight of the sample examined.

It is reasonable to presume therefore that the additional bands which appear in the spectrum of the isotactic resin are not due directly to the chemical structure of the resin. Nor are they due to the interaction of molecules when packed together in the crystal, since different crystalline forms of polypropylene, in which the molecules are packed in different ways, have virtually identical spectra. It is therefore concluded that the features of the spectrum of isotactic polypropylene are due to the helical arrangement of the chain of the individual polymer molecule.

Thus, although a number of papers have appeared giving methods of measurement of the crystallinity of polypropylene from the infra-red spectrum, it is doubtful whether such a measurement is possible in practice. On the other hand, measurement of the proportion of polypropylene chains present in the helical arrangement seems to be a reasonable proposition[22].

385

Infra-red spectra of syndiotactic polypropylene have been published by many authors, but some care is required here, as this material exists in different crystalline forms with different chain conformations and consequently different infra-red spectra[23].

A very large number of papers have been published on the many aspects of the spectroscopic examination of polypropylene, and Hummel[14] gives a list of the more important literature references.

UNSATURATION IN POLYPROPYLENE

Examination of polypropylene as comparatively thick specimens (~ 0.015 in or 0.4 mm thickness) reveals a shoulder (which varies in intensity in different specimens) at approx. 11.25 μm on the side of the polypropylene band at 11.1 μm. This shoulder, which is characteristic of pendent methylene unsaturation, arises from the end group on the polypropylene chain:

$$R-CH_2-C=CH_2$$
$$|$$
$$CH_3$$

DETERMINATION OF TERMINAL (PENDENT METHYLENE) UNSATURATION IN POLYPROPYLENE

Since the absorption band due to terminal unsaturation in polypropylene at 11.25 μm is a shoulder on a relatively strong polypropylene band, the measurement is made by difference, cancelling the polypropylene absorption by means of a polypropylene wedge containing no unsaturation, placed in the comparison beam of the spectrometer. The unsaturation from a sample of polypropylene is removed by bromination.

The following conditions have been found satisfactory.

Preparation of polypropylene wedge

Use the apparatus previously described (p. 372) for bromination of polythene, but in this case the vacuum oven is not required, and is replaced by a glass bulb into which is placed about 5 g of finely powdered polypropylene.

Degas the bromine, and degas the polypropylene powder. Open the tap between the bromine container and the bulb containing the polypropylene. Leave the bromine vapour in contact with the powder for about 3 d. Then pump off the bromine vapour, and continue pumping until the powder appears colourless (1 to 2 h). Remove the polypropylene powder and using the method of Harvey and Peters[17] prepare a wedge about 4 in (100 mm) in length and 1·5 in (40 mm) high, tapering from 0·005 to 0·02 in (0·12 to 0·5 mm) in thickness.

Spectrophotometry

A double-beam spectrometer, either rock-salt prism or grating, to operate in the 10 μm to 12 μm region is required. A grating instrument is pre-

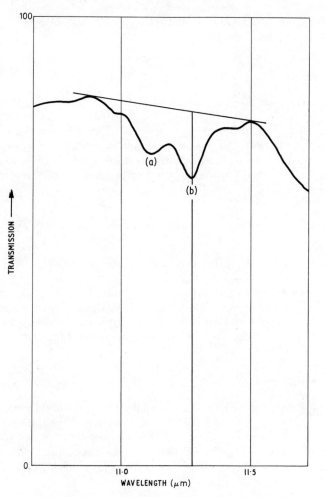

Fig. 6.17. Determination of unsaturation in polypropylene

ferred because the unsaturation band is very close to a strong polypropylene band, and the balancing operation is much simpler on an instrument of good resolving power.

Press the unknown specimen to 0·015 in (0·4 mm) thickness and mount it in the sample beam of the instrument. Place the prepared wedge in the comparison beam and adjust it so that its effective thickness is ∼0·015 in (0·4 mm). Set the spectrometer to 10·9 μm. Adjust the apparent transmission to about 70% by means of an attenuator in one of the spectrometer beams. Rotate the wavelength drum slowly by hand; the polypropylene band at 11·1 μm ((a) in Figure 6.17) must be slightly positive (about 65% peak transmission). Adjust the comparison wedge as necessary to achieve this. Then record the spectrum between 10·5 μm and 11·5 μm at a speed of 1 μm in 10 min.

Calculation

The spectrum will appear similar to Figure 6.17. Draw a straight line under the pair of bands. Measure the absorbance at the band peak at approx. 11·25 μm ((b) in Figure 6.17). Calculate the absorbance/cm. Convert this to the concentration of pendent methylene groups in g mol/l using an extinction coefficient determined on a pure hydrocarbon containing pendent methylene unsaturation, or use the coefficient given in Table 6.5.

DETERMINATION OF OXYGEN AS CARBONYL GROUPS IN OXIDISED POLYPROPYLENE

The determination of carbonyl groups in oxidised polypropylene is best done by a difference method as there is significant absorption of polypropylene near to the carbonyl absorption band. In view of this additional absorption, it is not desirable to use a thickness greater than 0·02 in (0·5 mm) of sample, and the remarks on sample preparation for carbonyl estimation in polythene (p. 379) are equally applicable here.

Preparation of a comparison wedge

The wedge is prepared by the usual method[17], and should vary between 0·005 and 0·02 in (0·12 and 0·5 mm) in thickness. If oxidation of a particular sample is to be studied, it will be convenient to make the wedge from the polypropylene to be used for the experiment. Otherwise a sample should be selected which is known, from its history, not to be oxidised.

Spectrophotometry

Measure the thickness of the prepared sample and place this in the sample beam. Place the wedge in the comparison beam and adjust the wedge

thickness to be equal to that of the sample. Record the spectrum from
5·5 μm to 6·5 μm at a speed of about 1 μm in 5 min.

Calculation

Proceed as for carbonyl content of polythene, using the same coefficient.

Many samples of polypropylene contain dilauryl thiodipropionate, or similar aliphatic esters, and show a weak but sharp absorption band at 5·75 μm; it is important to distinguish between this and the band due to oxidation which is broader and centred on 5·80 μm.

BLENDS AND MIXTURES CONTAINING POLYPROPYLENE

We have encountered the following blends containing polypropylene.

POLYPROPYLENE/POLYTHENE BLENDS

The spectrum of such a blend is shown in Spectrum 6.24. The complete spectrum of helical polypropylene is evident, while the presence of crystalline polythene is shown by the doublet near 14 μm. As in the case of polythene/paraffin wax blends described earlier, the 14 μm band could be due to paraffin wax, and again polythene and paraffin wax must be differentiated by ascertaining whether the component responsible for the 14 μm absorption can be removed by treatment of the material with ether.

QUANTITATIVE DETERMINATION OF THE PROPORTIONS OF A POLYPROPYLENE/POLYTHENE BLEND

Prepare a number of standard blends of polythene in polypropylene by milling the two polymers together. The range 0 to 50% polythene should be covered with about five blends. Press at least two samples from each blend to test homogeneity, to a thickness of 0·002 to 0·003 in (0·05 to 0·08 mm).

Press the unknown samples to the same thickness.

Spectrophotometry

A single- or double-beam spectrometer, either prism or grating, may be used. A single-beam spectrometer is preferred, as commercial samples may contain additional components which scatter energy, and under such conditions a result on a single-beam instrument is more reliable. Record the spectrum at about 1 μm/min traverse speed on all the prepared samples over the ranges 8 μm to 9·5 μm and 13 μm to 14·5 μm (Figure 6.18). The straight line backgrounds XX′ and YY′ are drawn under the bands at 8·7 μm and 13·9 μm due to polypropylene and polythene respectively,

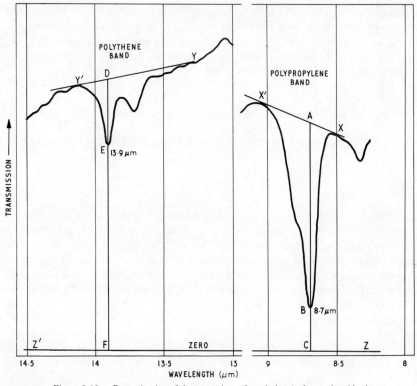

Figure 6.18.   Determination of the proportions of a polythene/polypropylene blend

and the zero line ZZ′ is drawn. The polypropylene absorbance is given by

$$\log_{10}\frac{AC}{BC}$$

The polythene absorbance is given by

$$\log_{10}\frac{DF}{EF}$$

Calculation

Measure the absorbances of the polypropylene band at 8·7 μm and the long-wave maximum of the polythene doublet at 13·9 μm. On all standard samples calculate

$$\frac{\text{Absorbance at } 13 \cdot 9 \ \mu\text{m}}{\text{Absorbance at } 8 \cdot 7 \ \mu\text{m}}$$

Plot this against

$$\frac{\text{Concentration w/w polythene in the sample}}{\text{Concentration w/w polypropylene in the sample}}$$

This graph should be linear.

Calculate the same absorbance ratio for the unknown samples, and determine the proportions of polythene and polypropylene from the calibration graph.

POLYPROPYLENE/POLYISOBUTENE BLENDS

Again in the spectra of these blends (Spectrum 6.25) all the polypropylene helical bands are seen; polyisobutene (or butyl rubber) even at low concentrations is evident by the absorption band at $8 \cdot 1$ μm and at higher concentrations the polyisobutene band at $10 \cdot 8$ μm may also be seen.

The proportions of a blend may be determined using standard blends, the method being precisely similar to that given for polythene/polypropylene blends, but here measuring at $8 \cdot 1$ μm for polyisobutene and $12 \cdot 3$ μm for polypropylene.

PROPYLENE/ETHYLENE COPOLYMERS

The spectra of these copolymers vary considerably and depend principally upon the way in which the ethylene units are distributed in the copolymer chains.

In one extreme case, in which ethylene units are present entirely as blocks, the spectrum is almost indistinguishable from that of a blend of polythene and polypropylene, and appears similar to Spectrum 6.24. Where there is a substantial amount of random copolymer, however, the propylene units do not assume the helical chain form, and the ethylene fraction of the copolymer is unable to crystallise, so that the spectrum has the appearance of a mixture of atactic polypropylene and amorphous polythene (Spectra 6.26 and 6.27).

Some further effects are noted in copolymers of low combined-ethylene content, and copolymers of this composition are those which are very frequently encountered as commercial resins. If there are no adjacent ethylene units, runs of three $-CH_2-$ groups are present, and the $-CH_2-$ band which occurs in polythene at $13 \cdot 8$ μm becomes a fairly sharp single band at about 13·6 μm (Spectrum 6.28). If pairs of ethylene units are present, the major absorption band in this region appears at about 13·7 μm and thus random copolymers of low combined-ethylene content often have the appearance of Spectrum 6.29, where the maxima at 13·85 μm and 13·7 μm arise from short runs and pairs of ethylene units respectively, and the shoulder at 13·6 μm arises from single ethylene units. This may be compared with Spectrum 6.30 where, although the combined-ethylene content is lower, the ethylene units are present as blocks, and the doublet characteristic of crystalline ethylene may be seen in the region of 13·8 μm. A more detailed study of this problem has been made by Willis[24]. In Spectrum 6.27 a shoulder is evident at about 13·3 μm; Van Schooten, Duck and Berkenbosch[25] suggest that this band arises from the group —CH_2—CH_2— which they believe arises from some head-to-head polymerisation of propylene units.

The determination of the combined-ethylene content of these copolymers is a matter of considerable difficulty, and a number of methods have been suggested[26-28]. The only simple method appears to be by comparing the absorbance at 13·9 μm (or some other maximum in this region of the spectrum) with that at 8·7 μm; and choosing as a standard a copolymer of similar contour in the 13·9 μm region, the composition of which is known. The various spectra compiled here should be adequate to obtain reasonable results on commercially available resins. The band of polypropylene at 8·7 μm appears to be the most satisfactory, being that least influenced by the degree of randomness of the copolymer.

OTHER ISOTACTIC α-OLEFINE POLYMERS

Certain other isotactic polymers made from α-olefines also take up a helical chain form in the crystalline state, and the spectra of these resins have a character similar to that of isotactic polypropylene, with a number of sharp bands in the 8 μm to 12 μm region. Thus if an unknown hydrocarbon resin has a spectrum of this nature, an isotactic α-olefine polymer is indicated.

Examples of such resins are polybutene-1 (Spectra 6.31 and 6.32), poly-3-methyl butene-1 (Spectrum 6.33), and poly-4-methyl pentene-1 (Spectrum 6.34). These resins are often encountered as copolymers with minor amounts of other olefines.

When this minor component introduces a relatively long side chain into the polymer, the corresponding —CH_2— rocking mode will be observed in the 13 μm to 14 μm region. Examples are Spectra 6.35 and 6.36, which are copolymers of 4-methyl pentene with hexene-1 and decene-1 respectively. The bands due to the butyl and octyl side chains appear at 13·72 μm and 13·88 μm, although, as shown in Table 2.4, side chains over —$(CH_2)_6CH_3$ produce a band so close to 13·88 μm that it is not possible to distinguish them on this test. Furthermore, reference to Table 2.4 shows that runs of $-CH_2-$ groups along the polymer backbone, such as would occur in copolymers of higher olefines with ethylene, would have absorption bands in their spectra very similar to those arising from the alkyl side chains described above.

Another effect of copolymerisation may be to influence the conformation of the chain of the major constituent, thus for example the three-fold helix may open up to a four-fold helix, with some modification of the spectrum of the major constituent. Such a change may also be induced by heat treatment of the higher olefine homopolymers, of which perhaps the best example is polybutene-1, which may be encountered in either a three-fold or four-fold helix form (Spectra 6.31 and 6.32 respectively). Fortunately the change of spectrum which accompanies this change of physical form is unlikely to lead to any ambiguity in the identification of these resins; on the other hand it is not sufficient evidence that a sample is a copolymer with a particular monomer, or even that it is a copolymer.

AROMATIC HYDROCARBON RESINS

The infra-red spectrum of polystyrene (Spectrum 6.37) is so familiar to spectroscopists that no explanation is called for. The spectrum of poly-α-methyl styrene (Spectrum 6.38) is very similar since this resin also contains the mono-substituted benzene nucleus, but differences occur in the 6 μm to 10 μm region which are sufficient to differentiate the resins if they are in a moderately pure condition.

The spectrum of polyvinyl toluene is much more distinctive. This may be encountered as the pure para isomer (Hummel[14] Figure 533). Para substitution is indicated by the very strong band at 12·3 μm. This band is still prominent in the spectrum of the polymer from mixed isomers (Hummel Figure 535) but the presence of ortho and meta isomers is shown by the bands at 13·8 μm and 13·2 μm respectively. Hummel also gives the spectra of a number of copolymers of vinyl toluene.

Copolymers of styrene are extremely common. Those with butadiene are described in Chapter 8, and with methyl methacrylate in Chapter 4. In the spectrum of styrene/acrylonitrile copolymer (Spectrum 6.39) the major features arise from styrene, but acrylonitrile is evident at 4·4 μm. Measurement of the absorbance ratio of the bands at 4·4 μm and 6·25 μm, due to acrylonitrile and styrene respectively, appears to be a satisfactory method of estimating the proportions of this copolymer, using reference samples of known composition for calibration. Spectra of several more styrene copolymers are given by Hummel and in Ref. 29.

Polymers based on indene are of interest as coating resins. The spectrum of polyindene (Hummel Figure 551) has a strong and distinctive band at 13·4 μm. The indene/coumarone copolymer is however more common.

Spectra of several of these copolymers appear in Ref. 29. The additional band near 14·25 μm presumably arises from the copolymerised coumarone.

N.M.R. SPECTRA OF ALIPHATIC HYDROCARBON POLYMERS

POLYTHENE

The proton n.m.r. spectrum of polythene as measured in the normal way is not informative as almost all the protons are equivalent, and give rise to a single resonance at about 8·8 τ. However, it has been shown[30] that a signal due to methyl group protons at about 9·1 τ and resonances in the 4·5 τ to 5·2 τ region due to protons on C=C groups can be revealed by spectrum accumulation (up to 400 scans). However this does not constitute a practical method of differentiating between different types of polythene. In the field of ethylene polymers, n.m.r. measurements have found their principal application in the quantitative analysis of copolymers with polar monomers, particularly esters (see below).

POLYPROPYLENE

The resonance of the α-methyl protons near 9·1 τ is split into a doublet because there is also a single proton on the α-carbon atom. In spite of this, the resonance shows clearly the effects of stereochemical arrangement, and may be used to determine the proportions of isotactic, syndiotactic and heterotactic triads in a sample of the resin[31]. In other respects the spectrum is extremely complex, as the resonances are split due to spin-spin coupling (Figure 6.19).

HIGHER α-OLEFINE POLYMERS

The n.m.r. spectra of these materials are of little practical interest as these spectra consist of complex overlapping resonances in the 8 τ to 9 τ region.

COPOLYMERS OF ETHYLENE

The n.m.r. spectrum is of outstanding value in the determination of the proportions of copolymers of ethylene, especially the ester copolymers, many of which are of considerable commercial importance. The following examples will illustrate the general principle.

Ethylene/vinyl acetate copolymers

In the n.m.r. spectrum of these resins, the resonance at 8·1 τ is due to the methyl protons of the acetate group (Figure 6.20).

Let the area of this resonance be M
and the area of all other resonances be R

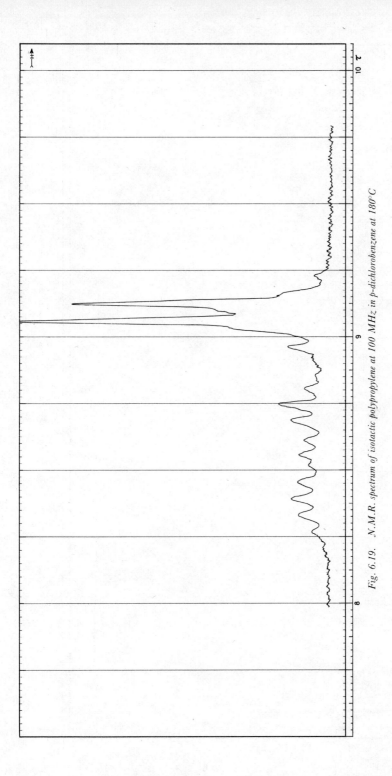

Fig. 6.19. N.M.R. spectrum of isotactic polypropylene at 100 MHz in p-dichlorobenzene at 180°C

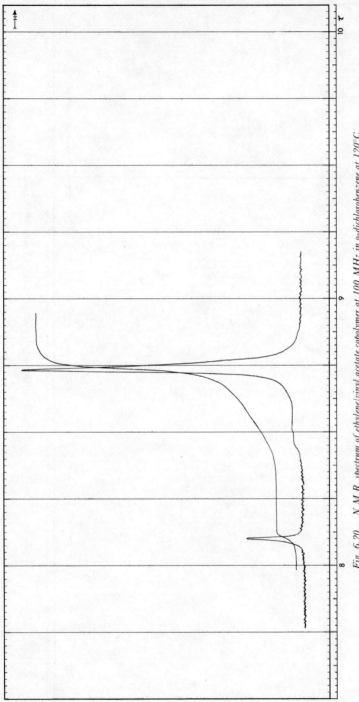

Fig. 6.20. N.M.R. spectrum of ethylene/vinyl acetate copolymer at 100 MHz in o-dichlorobenzene at 120°C

Then the mole fraction of combined vinyl acetate is proportional to $M/3$ because there are 3 protons in the methyl group. However, for each methyl group proton in the vinyl acetate monomer unit, there is one proton in the rest of this molecular unit. Hence in the remaining area R, M arises from combined vinyl acetate, thus leaving $R-M$ arising from combined ethylene units in the copolymer. Because there are 4 protons in the ethylene monomer unit, the quantity

$$\frac{R-M}{4}$$

is proportional to the mole fraction of combined ethylene present. Thus

$$\frac{M}{R-M} \times \frac{4}{3}$$

gives the molar proportion of vinyl acetate to ethylene, from which the weight proportion may be obtained.

The analysis of the structure arising from protons other than those in the acetate methyl group is difficult. However, if the spectrum is recorded in solution in benzene, structure may be observed on the methyl resonance, and this structure can give evidence on the distribution of the vinyl acetate units in the copolymer[32].

ETHYLENE/ETHYL ACRYLATE COPOLYMERS

In this case, the distinct resonance is that due to the two protons adjacent to the oxygen atom in the group $-O-CH_2-$. This occurs at about 5·9 τ (Figure 6.21). A calculation similar to that in the example above is used, taking the area of the 5·9 τ resonance, and the area of all other resonances in the spectrum, but in this case

If the area of the 5·9 τ resonance be E

and the remaining area be R

Then the molar proportion of ethyl acrylate to ethylene in the copolymer is

$$\frac{E}{R-3E} \times \frac{4}{2}$$

because there is a total of 8 protons in the ethyl acrylate molecule, of which 2 are in the $-O-CH_2-$ group. Thus the additional number of protons to be accounted for in the ethyl acrylate molecule is three times the number of $-O-CH_2-$ protons present. The area due to the remaining protons, $R-3E$, arises from ethylene, and must be divided by 4 because there are 4 protons in the ethylene molecule.

ETHYLENE/METHYL METHACRYLATE COPOLYMERS

Here the methyl resonance of the $-O-CH_3$ group appears well clear of other resonances at about 6·5 τ (Figure 6.22). The area of the 6·5 τ

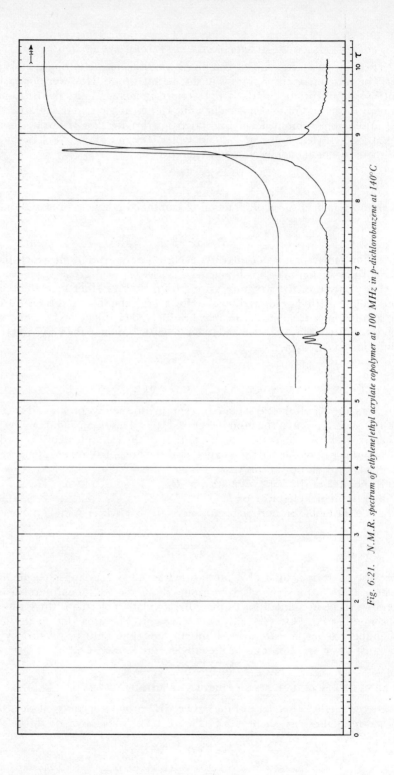

Fig. 6.21. N.M.R. spectrum of ethylene/ethyl acrylate copolymer at 100 MHz in p-dichlorobenzene at 140°C

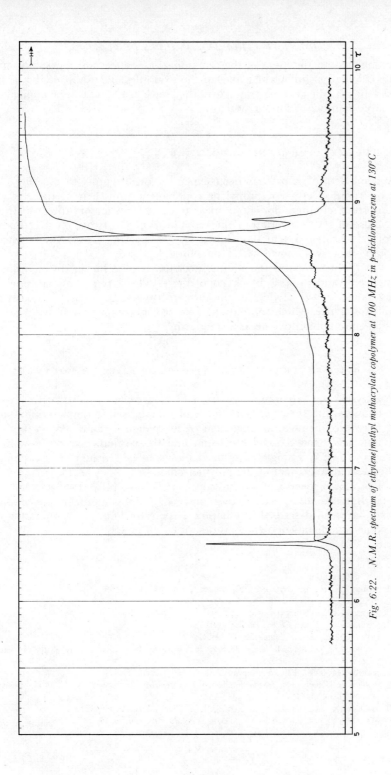

Fig. 6.22. N.M.R. spectrum of ethylene/methyl methacrylate copolymer at 100 MHz in p-dichlorobenzene at 130°C

resonance and the areas of the remaining resonances are measured, and the calculation follows the lines of the examples given above. It will be evident that the general principle of the above examples can be applied to a large number of similar systems.

ETHYLENE/PROPYLENE COPOLYMER

A good deal of effort has been expended on attempts to measure the proportions of ethylene/propylene copolymers from their n.m.r. spectra. Unfortunately there does not appear to be any satisfactory direct method because of severe overlap of the resonances of the two monomer units; Porter[33] proposed a method based on the spectra of copolymers whose composition was known by C^{14} labelling.

On the other hand the n.m.r. spectrum is of value in distinguishing between random and block copolymers. The latter show a 'singlet' resonance at 8·78 τ while in the former ethylene gives rise to a heavily split resonance, which can only be detected by comparison of band areas using pure polypropylene as a standard.

N.M.R. SPECTRA OF AROMATIC HYDROCARBON POLYMERS

Polystyrene has an n.m.r. spectrum of rather poor quality with broad resonances; this spectrum does not readily lend itself to the determination of the stereochemical arrangement of the polymer chain. Poly-α-methyl styrene has a more useful spectrum, and the resonance of the α-methyl group, which is a singlet, has been shown to be sensitive to the stereochemical arrangement of the polymer chain[34].

The n.m.r. spectra of styrene copolymers have proved satisfactory for quantitative analysis. The analysis of styrene/methyl methacrylate copolymers has been described in Chapter 4 (p. 278), and the examination of styrene/butadiene copolymers is described on p. 456.

REFERENCES

1. RUDDLE, L. H., and WILSON, J. R., *Analyst, Lond.,* **94,** 105 (1969)
2. BS 2782, Pt 4/405 (1970)
3. HARVEY, D., and PENKETH, G. E., *Analyst, Lond.,* **82,** 498 (1957)
4. COOK, G. D., *J. org. Chem.,* **18,** 260 (1953)
5. WIMER, D. C., *Analyt. Chem.,* **30,** 77 (1958)
6. SQUIRRELL, D. C. M., *Automatic Methods in Volumetric Analysis,* Hilger & Watts, London (1964)
7. TAUBINGER, R. P., Private Communication
8. *Tables of Wavenumbers for the Calibration of Infra-Red Spectrometers,* I.U.P.A.C., Butterworths, London (1961)
9. HASLAM, J. and SQUIRRELL, D. C. M., *J. appl. Chem., Lond.,* **9,** 65 (1959)
10. MALTESE, P., CLEMENTI, L., and MORI, A., *Chimica Ind., Milano,* **49,** 1070 (1967)

11. HAMILTON, J. B., and RINGHAM, M., Private Communication
12. MILLER, R. G. J., and WILLIS, H. A., *J. Polym. Sci.*, **19,** 485 (1956)
13. KRESSER, T. O. J., *Polyolefin Plastics*, Van Nostrand and Reinhold, New York (1969)
14. HUMMEL, D. O., *Infra-Red Analysis of Polymers, Resins and Additives, An Atlas*, **1,** Pt. II, Wiley Interscience (1969)
15. SILAS, R. S., YATES, J., and THORNTON, V., *Analyt. Chem.*, **31,** 529 (1959)
16. WILLBOURN, A. H., *J. Polym. Sci.*, **34,** 569 (1959)
17. HARVEY, M. C., and PETERS, L. L., *Analyt. Chem.*, **32,** 1725 (1960)
18. CROSS, L. H., RICHARDS, R. B., and WILLIS, H. A., *Discuss. Faraday Soc.*, **9,** 235 (1950)
19. BURNETT, J. D., MILLER, R. G. J., and WILLIS, H. A., *J. Polym. Sci.*, **15,** 591 (1955)
20. WOOD, D. L., and LUONGO, J. P., *Mod. Plast.*, 132, Mar. (1961)
21. CHAN, M. G., and HAWKINS, W. L., *Polymer Preprints*, **9,** (2) 1638 (1968)
22. BRADER, J. J., *J. appl. Polym. Sci.*, **3,** 370 (1960)
23. PERALDO, M., and CAMBINI, M., *Spectrochim. Acta*, **21,** 1509 (1965)
24. WILLIS, H. A., *Chemy Ind.*, 742, June (1969)
25. VAN SHOOTEN, J., DUCK, E. W., and BERKENBOSCH, R., *Polymer*, **2,** 357 (1961)
26. DRUSHEL, H. V., and IDDINGS, F. A., *Analyt. Chem.*, **35,** 28 (1963)
27. LOMONTE, J. N., *Jnl Polym. Sci.*, Pt. B, **1,** 645 (1963)
28. BLY, R. M., KIENER, P. E., and FRIES, B. A., *Analyt. Chem.*, **38,** 217 (1966)
29. *Infra-Red Spectroscopy. Its use in the Coatings Industry*, Federation of Societies for Paint Technology, Philadelphia (1969)
30. KATO, Y., and NISHIOKA, A., *Bull. chem. Soc. Japan*, **39,** 1426 (1966)
31. WOODBREY, J. C., and TREMENTOZZI, Q. A., *Jnl Polym. Sci.*, Pt. C, No. 8, 113 (1965)
32. WU, T. K., *Macromolec*, **2,** 520 (1969)
33. PORTER, R. S., *Jnl Polym. Sci.*, Pt. A1, **4,** 189 (1966)
34. RAMEY, K. C., and STATTON, G. L., *Makromolek. Chem.*, **85,** 287 (1965)

7

FLUOROCARBON POLYMERS AND POLYMER EMULSIONS

The first part of this chapter is concerned with the determination of fluorine in large quantities e.g. in polytetrafluoroethylene and also in small amounts e.g. in herbage and in volatile products. A relatively new method is described for the determination of fluorine in fluorocarbon polymers, based on Oxygen Flask Combustion followed by photo-fluorimetric titration of the combustion products.

Observations are made on the pyrolysis/gas chromatography of fluorocarbon polymers.

The determination of additives in fluorocarbon polymers is discussed and particular attention is paid to the examination of polytetrafluoro-ethylene dispersions, paints and paint primers.

The chapter includes a qualitative description of the infra-red spectra of the more important fluorinated resins and observations are made on the proton and fluorine magnetic resonance spectra of these materials.

POLYTETRAFLUOROETHYLENE

The monomer tetrafluoroethylene, which is a gas at atmospheric pressure, is produced as follows:

Fluorspar reacts with sulphuric acid to yield hydrofluoric acid. This acid then reacts with chloroform to yield chlorodifluoromethane.

$$2HF + CHCl_3 = 2HCl + CHClF_2$$

This chlorodifluoromethane, on pyrolysis at 800°C, yields the monomer tetrafluoroethylene.

$$2CHClF_2 = 2HCl + C_2F_4$$

The tetrafluoroethylene is scrubbed and distilled. The purified mono-mer is polymerised under pressure in the presence of excess water which acts as a heat-transfer medium for the removal of the large amount of heat evolved in the polymerisation.

Polymerisation is initiated by a water-soluble catalyst capable of generating free radicals e.g. a peroxide.

$$C_2F_4 + C_2F_4 \rightarrow \quad \begin{matrix} F & F & F & F \\ | & | & | & | \\ -C{-}C{-}C{-}C{-} \\ | & | & | & | \\ F & F & F & F \end{matrix}$$

Depending on the detailed conditions, the polymer may be obtained in either of two basic forms—as an ivory white granular powder or as an aqueous dispersion.

The product from the polymerisation vessel may be washed, dried and sold as polytetrafluoroethylene granular powder.

Alternatively the product from the polymerisation vessel may be coagulated, washed, dried and sold as a coagulated dispersion polymer.

Finally, the polymerisation product may be transferred to a stabilisation vessel, stabilised with surface active agents, then concentrated by electrodecantation to yield a polytetrafluoroethylene dispersion.

Whether engaged in the analytical examination of powders, dried dispersions or other products, the analytical chemist will find again and again that one of the most useful tests will be that of the very accurate determination of fluorine; this is particularly useful, as is often the case, when large amounts of fluorine are involved.

The method was worked out originally for the determination of fluorine in polytetrafluoroethylene but the principle of the test may be readily extended to other circumstances[1].

THE DETERMINATION OF FLUORINE IN POLYTETRAFLUOROETHYLENE

Reagents

Alizarin Red S solution.
A 0·1% w/v aqueous solution of Alizarin Red S (sodium alizarin sulphonate).
Thorium nitrate solution 0·05 N.
6·9 g of $Th(NO_3)_4 \cdot 4H_2O$ are dissolved in water and the solution is diluted to 1 l.
60% w/w Perchloric acid solution. Analar.
Sodium fluoride.
Reagent sodium fluoride is recrystallised from water. The crystals are dried and heated to fusion in a platinum dish. The product is then

ground to a powder in an agate mortar. Sufficient of the powder is dried for 2 h at 105°C before use.

Sodium fluoride solution 0·05 N.

2·1 g of sodium fluoride are dissolved in water and the solution is diluted to 1 l.

Cobalt sulphate solution 1·0% w/v.

10·0 g of Analar $CoSO_4 \cdot 7H_2O$ are dissolved in water and the solution is diluted to 1 l.

Potassium dichromate solution 0·1% w/v.

1 g of Analar $K_2Cr_2O_7$ is dissolved in water and the solution is diluted to 1 l.

Sodium hydroxide solution N.

Sodium hydroxide solution 0·1 N.

Hydrochloric acid solution 0·1 N.

Sodium peroxide.

Dried starch (Analar soluble starch).

Quartz chips.

Apparatus

Laboratory mill. Christy Norris or similar type.

Metal peroxide bomb; electrically fired.

Steam distillation apparatus Figure 7.1.

pH meter with glass and calomel electrodes. The meter is standardised against a pH 4·0 buffer solution before use.

Alternatively and more conveniently an instrument such as an automatic titrimeter may be used to maintain pH during the titration.

Two mechanical stirrers, preferably the magnetic type.

Preparation of the sample

Sufficient of the material for test is ground in the laboratory mill and the material that passes a 30 mesh sieve is collected and dried in an air oven at 100°C for 1 h.

Procedure

0·7 g of the dried starch is weighed into the cup of the metal peroxide bomb. One scoopful (15 g) of the sodium peroxide is added, the lid is placed on the cup and the contents are mixed well. About 0·12 to 0·14 g of the sample of polytetrafluoroethylene are weighed accurately into a weighing bottle. About two thirds of the bomb charge is now transferred from the bomb cup to a dry 50 ml beaker and after emptying the sample from the weighing bottle into the bomb cup, the portion of the charge in the beaker is poured back into the cup. The weighing bottle is re-weighed to obtain the exact weight of sample. The total charge in the cup is

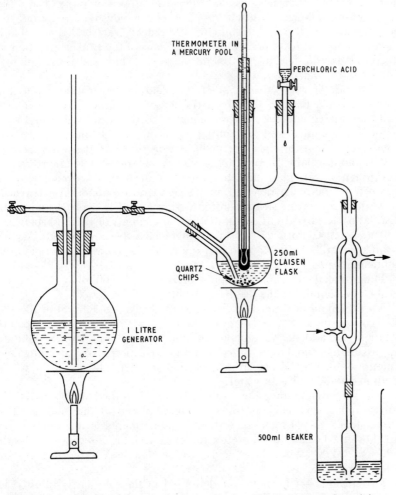

Fig. 7.1. Steam distillation apparatus for the determination of fluorine in polytetrafluoroethylene

mixed with a stout piece of nickel wire, any material adhering to the wire being brushed back into the cup.

After attaching the fuse wire to the bomb head, the bomb is assembled and the charge is fired electrically. After allowing 5 min for the bomb to cool in the cooling water bath surrounding it, the bomb is opened and the bomb head is washed with water into 50 ml of boiling water contained in a 500 ml beaker. The cup is placed upright in the beaker and a boiling rod is inserted so that its bottom tip rests in the cup. The beaker is then covered with a watch glass and by means of a wash bottle a few drops of water are allowed to run down into the fused mass. When the first vigorous reaction has subsided, the cup is tipped on one side and the fused mass is allowed to leach out. The cup is now removed from the beaker, rinsed well, and the contents of the beaker are boiled for 10 min to destroy the excess of peroxide. The contents of the beaker are allowed to cool, then transferred to a 100 ml measuring flask and the solution is diluted to the mark with water.

By means of a pipette 50 ml of the solution are transferred to the 250 ml Claisen flask of the steam distillation apparatus. A few quartz chips are added and the flask is connected to the steam boiler and condenser. A 500 ml beaker containing 50 ml of water is placed under the condenser so that the tip of the delivery tube dips into the water so as to collect the distillate. 50 ml of 60% w/w perchloric acid solution are run in through the tap funnel and the contents of the flask are heated, with the steam from the boiler diverted until a temperature of 130°C is reached.

At this temperature steam is introduced into the Claisen flask and steam distillation is carried out, maintaining the contents of the flask between 130 and 140°C until 250 ml of distillate have been collected. At this point the distillation is stopped and the beaker is removed, the top of the condenser being rinsed with water.

The artificial end-point solution is prepared in a 500 ml beaker similar to the one containing the distillate as follows:

About 150 ml of water are placed in the beaker and 15 ml of the 1% w/v cobalt sulphate and 1·0 ml of the 0·1% w/v potassium dichromate solution are added. An amount of standard sodium fluoride solution approximately equal to the amount of fluorine expected and an equivalent amount of 0·05 N thorium nitrate solution are then added and the whole is diluted to 300 ml with water.

The beaker containing the distillate is placed on the base of a magnetic stirrer and a glass enclosed motor rotor is placed in the solution. The electrodes of the pH meter or titrimeter are immersed in the solution and up above the beaker three burettes are set, containing N sodium hydroxide solution, 0·1 N hydrochloric acid solution and 0·05 N thorium nitrate solution respectively.

The stirrer is set in motion and the pH of the solution is adjusted to 6·0 with the N sodium hydroxide solution. The pH is then adjusted to 3·3 using 0·1 N hydrochloric acid.

The beaker containing the artificially prepared end-point solution is placed on another stirrer adjacent to the sample solution and the stirrer is set in motion at the same speed as that of the sample solution.

1·0 ml of Alizarin Red S indicator is added to the sample solution which is then titrated slowly with the 0·05 N thorium nitrate solution. The pH of the solution is allowed to fall to 3·0 and is then maintained at this level during the remainder of the titration by the addition of 0·1 N sodium hydroxide solution. When nearing the end-point the thorium nitrate solution is added 2 drops at a time, 15 sec being allowed to elapse between the additions.

The end-point is taken as the point at which the colour of the sample solution matches the colour of the artificially prepared end-point solution.

The number of ml of thorium nitrate solution required is noted.

The thorium nitrate is standardised against the recrystallised sodium fluoride which has been through the same process as the sample of polytetrafluoroethylene. Where titrations of the order of 50 ml are expected, about 0·2 g of sodium fluoride is weighed accurately and fused in the metal peroxide bomb with the customary amount of fusion mixture. The fusion products are leached out and made up to 100 ml with water. 50 ml of this solution are distilled with perchloric acid solution and quartz chips and the distillate is titrated with 0·05 N thorium nitrate solution of pH 3·0 as before.

The thorium nitrate titre of the sodium fluoride is calculated to g fluorine per 1·0 ml of thorium nitrate solution. Using this factor the percentage of fluorine in the sample under test is calculated.

As has been shown by Haslam and Whettem[1] the method gives excellent results in the determination of fluorine in polytetrafluoroethylene e.g. 76·07 and 75·98% against a theoretical figure of 76·0%.

This original paper should be consulted for details of the amounts of acid, alkali etc. required at various stages of the final titration.

The analytical chemist engaged in work on fluorine-containing polymers should always be on the lookout for alternative methods of test.

Not all the problems can be solved by sodium peroxide fusion.

We have had the opportunity of making a thorough examination of Wickbold's method[2], which, in certain circumstances, can be used with great advantage.

WICKBOLD'S METHOD

In this method, which is described in great detail in the original paper, the vapourised organic fluorine compound or its partial combustion products are passed through an oxy-hydrogen flame which, in the case of polytetrafluoroethylene, decomposes the material completely to carbon dioxide and hydrogen fluoride. The hydrogen fluoride is absorbed in dilute sodium hydroxide solution and the fluoride is in our case titrated with standard thorium nitrate solution at a controlled pH of 3·0.

We have the following observations to make on the method.

1. The method suffers from the limitation that it cannot be applied to the determination of fluorine in samples containing inorganic

material since these would leave an ash of inorganic fluoride. Solution of this ash and addition of this solution to the absorber solution is hardly a practical proposition.

2. In our hands the method tends to give rather low figures for the fluorine contents of polymers i.e. about 0·5% low for polytetrafluoroethylene.

3. Organic sulphur compounds interfere and tend to produce high values for fluorine contents.

4. However, the method is an excellent one to use when questions are asked, e.g. how much fluorine is given off when polytetrafluoroethylene is held at elevated temperatures, e.g. 250 to 450°C, with a stream of air passing over the sample at a definite rate. The air can be passed directly to the oxy-hydrogen flame and the method for fluorine applied directly; the point to bear in mind is that a lot of nitrite is produced at the same time as the fluoride arising from the polytetrafluoroethylene.

THE DETERMINATION OF SMALL AMOUNTS OF FLUORINE

The analytical chemist in the plastics industry may encounter problems concerned with the determination of quite small amounts of fluorine compounds in the presence of quite large amounts of organic matter e.g. traces of fluorine in herbage.

In such cases we have disintegrated the moist grass in a homogeniser in the presence of fluorine-free lime. The slurry is then evaporated to dryness, dried and ignited at 600°C in a muffle furnace.

The ignited residue is treated with perchloric acid solution and silica and distilled in steam to yield a distillate containing the fluorine.

An aliquot of this distillate is brought to a definite degree of acidity, alongside a similar volume of distilled water brought to the same degree of acidity. A known volume of Alizarin Red S solution is now added to sample and comparison solutions. Standard thorium nitrate is now added to the sample solution until a pink colour is obtained. This same amount of thorium solution is now added to the comparison solution after which standard fluoride solution is added to the comparison solution until sample and comparison solution match. The fluorine content of the original grass or sample is then readily calculated.

More recent work on the determination of small amounts of fluoride in aqueous solutions has shown the fluoride ion selective electrode to be of value. The electrode manufactured by Orion Research Inc., Cambridge, Mass. is used. Over a useful range this electrode gives a change in potential of approx. 59 mV at 25°C for each tenfold change in fluoride ion activity. The pH of the sample solutions is adjusted to between 6 and 8 to allow complete dissociation of HF and HF_2^- which are present in acid solution: fluoride complexes with polyvalent cations such as Si^{4+}, Al^{3+} and Fe^{3+}

are destroyed by the addition of a polyvalent anion such as citrate to release the bound fluoride.

Rather than take direct measurements on the prepared sample solution and refer these to a calibration graph prepared under identical conditions of pH and total ionic strength, relating fluoride concentration to electrode response, we prefer to use the 'Known Increment Method'[3].

This method, which can also be used with other ion selective electrodes, involves taking two potential difference readings, E_1 and E_2, before and after the addition of a known concentration D of the ion being determined. From the equation

$$E_2 - E_1 = 2.303 \frac{RT}{ZF} \log \frac{C+D}{C}$$

the value of C, the concentration of the ion sought, may be calculated or derived from a nomogram. Z is the valence of the ion sought and $2.303 RT/F$ is the Nernst constant = 59.16 mV at $25°C$.

The optimum addition is such that $C = D$, which gives a change of about 20 mV for the monovalent fluoride ion.

In applications of this method to acid sample solutions containing aluminium and fluoride at concentrations between 10^{-5} and 10^{-3} M, sodium citrate was first added and the pH adjusted to 7 ± 0.5. It was shown that as long as the fluoride addition made produced a change of between 15 and 25 mV then results accurate to $\pm 5\%$ relative were obtained.

THE DETERMINATION OF FLUORINE IN ORGANIC COMPOUNDS
BY OXYGEN FLASK COMBUSTION AND PHOTOFLUORIMETRIC
TITRATION OF FLUORIDE

On many occasions the analyst in the plastics industry may be asked to determine fluorine in organic compounds and plastics materials in which the amount of fluorine is below 20%. For such materials, which contain sufficient carbon and hydrogen to permit ready ionisation of the fluorine by combustion in oxygen, the following method based on the work of Willard and Horton[4] and Gel'man and Kiparenko[5] has proved particularly useful.

10 to 20 mg sample together with about 25 mg polyethylene foam are wrapped in approx. 75 mg filter paper impregnated with potassium nitrate and combusted in oxygen in the apparatus described on p. 83. 5 ml distilled water is used as absorption solution and when absorption of the combustion products is complete (30 min) the contents of the flask are washed into a 100 ml square-section titration cell with 20 ml distilled water. The pH of the solution is adjusted to 3 by the addition of 2 N trichloroacetic acid solution and 5 ml of a buffer solution (94.5 g mono-chloroacetic acid and 30 g sodium hydroxide per 1) are added. The

solution is now diluted with 30 ml ethanol and 5 ml of quercetin indicator (0·02% in 95% v/v ethanol) added, prior to photofluorimetric titration of the fluoride with 0·02 N thorium nitrate solution. We have successfully used automatic recording titrimetry both with the apparatus shown in Figure 7.2 assembled in the laboratory and with the Baird Atomic Fluorispec SF 100 E spectrofluorimeter as the measuring instruments. A typical titration curve is shown in Figure 7.3.

In our experience the method gives accurate results with organic compounds containing less than 20% fluorine. With polytetrafluoroethylene, however, low results are obtained due to incomplete breakdown of C—F bonds and absorption of fluoride on the glass combustion vessel. The figures obtained vary from 66 to 72% (theory 76·0%). Light and Mannion[6] however claim to obtain satisfactory results in this determination. They prefer a polycarbonate combustion flask and use dodecyl alcohol as a combustion aid prior to titration of the ionised fluoride with thorium nitrate solution using a fluoride ion selective electrode to monitor the end-point.

THE DETERMINATION OF CARBON, HYDROGEN AND CHLORINE IN FLUOROCARBON POLYMERS AND COPOLYMERS

In our experience, in the micro-determination of carbon and hydrogen in fluorine-containing hydrocarbons and polymers; the lead peroxide section at the beak end of the combustion tube should be modified with granular sodium fluoride (14–20 mesh). This absorbs any silicon tetrafluoride formed by the reaction of hydrogen fluoride with the silica of the combustion tube. Any elementary fluorine present is absorbed by the silver wool packing. The sodium fluoride packing is maintained at 180°C by means of a thermostatic heater although it is noted that some workers prefer temperatures of 270°C±20°C. For most compounds the normal combustion temperature of 750°C±20°C is satisfactory, but very resistant fluorocarbons have to be burnt at a temperature of 950°C. The method gives excellent results on such substances as trifluorochloroethylene and polytetrafluoroethylene.

Small amounts of chlorine as e.g. from polyvinyl chloride may be determined with great accuracy in polytetrafluoroethylene by applying the method worked out by Haslam, Hamilton and Squirrell[7] for the determination of chlorine by the Oxygen Flask Combustion method, i.e. combustion of the polymer in oxygen, collection of the products of combustion in alkaline sulphite, followed by addition of hydrogen peroxide and nitric acid and titration with standard silver nitrate solution.

PYROLYSIS/GAS CHROMATOGRAPHY OF FLUOROCARBON POLYMERS

Under normal pyrolysis conditions, at 550°C, the distinctive feature of tetrafluoroethylene polymers is high thermal stability. The pyrograms

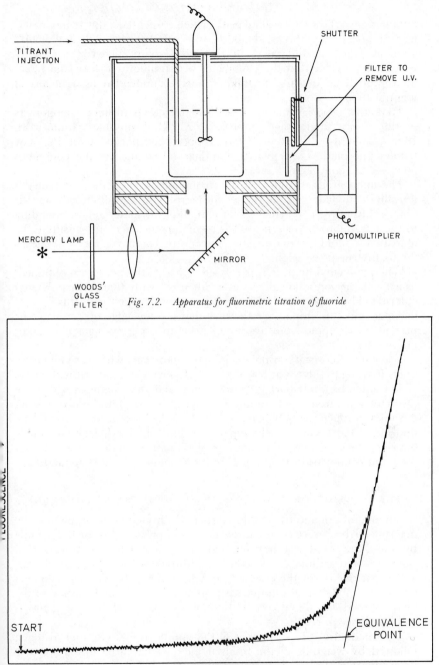

TITRANT
INJECTION

SHUTTER

FILTER TO
REMOVE U.V.

MERCURY LAMP

WOODS'
GLASS
FILTER

MIRROR

PHOTOMULTIPLIER

Fig. 7.2.   Apparatus for fluorimetric titration of fluoride

START

EQUIVALENCE
POINT

FLUORESCENCE

TITRANT ADDITION ⟶

Fig. 7.3.   Photofluorimetric titration curve of fluoride with thorium nitrate solution (Quercetin indicator)

obtained each show only a small peak, which is essentially due to monomer. Careful stepwise pyrolysis at 100°C intervals between 550 and 850°C gives evidence of differences in thermal stability between e.g. polytetrafluoroethylene and the copolymer of tetrafluoroethylene and hexafluoropropylene, but this method is not recommended as a means of identification.

To do this a special column is used capable of separating the monomers produced in a single stage pyrolysis at 750°C. A suitable column is an 18 in (450 mm) $\times \frac{1}{4}$ in (6 mm) o.d. copper tube packed with 1% w/w tritolyl phosphate on activated alumina (50–80 mesh); the column is operated at room temperature.

The main product from both the homopolymer and the copolymer is tetrafluoroethylene, but a higher proportion of hexafluoroethylene is obtained from the copolymer. After elution of the material corresponding to the tetrafluoroethylene peak it is recommended that the sensitivity be increased tenfold to permit more accurate comparison of the hexafluoroethylene peak heights.

When polyvinyl fluoride is pyrolysed at 550°C the pyrogram obtained, using a flame ionisation detector, is identical with that from polyvinyl chloride. This is because HF and HCl are not detected with this form of detector and this emphasises the fact that a knowledge of the elements present in a sample is always useful prior to a pyrolysis/gas chromatographic test.

The chlorofluorocarbon polymers and copolymers with e.g. vinylidene fluoride when pyrolysed at 550°C give a characteristic pattern with peaks of low intensity and short retention times and this enables them to be distinguished from polytetrafluoroethylene and tetrafluoroethylene/hexafluoropropene copolymers. Viton A (hexafluoropropene/vinylidene fluoride) and Kel F elastomer (trifluorochloroethylene/vinylidene fluoride), however, give the same pyrolysis pattern under the conditions used and cannot be distinguished from one another by this method.

DETERMINATION OF ADDITIVES IN FLUOROCARBON POLYMERS

Small amounts of additives evenly dispersed throughout a sample may be determined by direct examination of a cold-pressed disk of the sample by the X-ray Fluorescence method. It is important, however, that standards and sample should contain additives of the same particle size and when this is not the case, as e.g. with samples containing unknown grades of molybdenum sulphide or glass fibre, then a modification of the method used for the determination of carbon black in polythene is employed (see p. 332).

This method involves pyrolysis of the sample at 700°C under nitrogen followed by weighing of the residual filler. It has been applied to the determination of glass, asbestos, graphite, molybdenum sulphide, bronze and lead, in amounts not greater than 2%, and could be applied to the determination of other inorganic fillers. Certain pyrolysis products of polytetrafluoroethylene are toxic and it is absolutely essential that the apparatus should be placed in an efficient fume chamber or that the vapours from the sample should be led into an efficient extraction system.

Apparatus

Combustion boat, approx. $80 \times 10 \times 10$ mm.
Silica combustion tube, approx. 25 mm bore and at least 500 mm long.
Electric furnace to accommodate the combustion tube and maintain a temperature of 700°C.
Thermocouple to measure temperatures in the range 500 to 750°C.
Flow-meter to measure nitrogen flow rate in the region of 1 to 3 l/min.
Nitrogen. Oxygen-free, B.O.C. white-spot nitrogen is suitable.

The apparatus must be assembled and situated so as to eliminate any risk of escape of toxic pyrolysis products into the laboratory atmosphere.

Procedure

The combustion boat is heated to red heat, cooled in a desiccator for at least 30 min and weighed. $1 \cdot 0 \pm 0 \cdot 1$ g of the sample is transferred to the combustion boat and the boat and sample are re-weighed. The combustion boat is placed in the middle of the combustion tube and a stopper, carrying the thermocouple and an inlet tube for the nitrogen, is fitted so that the thermocouple is adjacent to the boat.

The nitrogen flow is adjusted to $1 \cdot 7 \pm 0 \cdot 3$ l/min, the combustion boat placed in position in the furnace, and a stopper carrying an outlet tube is fitted. This outlet tube enables the pyrolysis products to be conducted away to an efficient extraction system.

The temperature of the furnace is raised to 700°C and this temperature is maintained for 60 min. The combustion boat is then withdrawn from the furnace and the boat transferred to a desiccator, cooled for 30 min and weighed. If the filler is carbon black, the combustion tube and boat are cooled for several minutes with the stream of nitrogen still passing prior to transfer of the boat to the desiccator.

$$\text{Filler } \% = \frac{\text{Weight of residue}}{\text{Weight of sample}} \times 100$$

POLYTETRAFLUOROETHYLENE DISPERSIONS, PAINTS AND
PAINT PRIMERS

In addition to polymer and emulsifying agents many of these preparations contain additional components such as pigments, chromic acid, phosphoric acid and hydrocarbon solvents.

Over many years the following general scheme of analysis has proved to be of particular value.

1. TOTAL SOLIDS

Weigh accurately 5 g sample into a previously ignited tared platinum dish. Evaporate on a water bath and dry the residue to constant weight at 100°C.

2. ASH

Ignite the residue obtained above over a small bunsen flame in the fume cupboard and finally complete the ashing process in a muffle furnace at 700°C. Cool and re-weigh the dish containing the ash. Examine the ash spectrographically for major metallic constitutents.

3. EMULSIFYING AGENTS

Weigh accurately an amount of sample (approx 5 g) into a beaker (250 ml) and add 40 ml methanol to precipitate the polymer. Boil the suspension gently on a hot plate until any chromium present is completely reduced to the green chromic hydroxide. This usually takes about 5 min.

Cool, filter the mixture through a lined Soxhlet thimble supported in the siphon tube and collect the filtrate in a weighed fat-extraction flask. Wash the beaker with methanol until all the polymer has been transferred to the thimble.

Connect the flask to the Soxhlet jacket and continue extraction for 6 h using the methanol filtrate as the extraction medium.

Disconnect the flask, and evaporate the methanol on the water bath. Dry the residue to constant weight at 100°C.

This residue contains the total emulsifying agents.

Selective solubility methods are used to separate the components as follows:

(a) Polyethylene oxide type emulsifying agents

To the total methanol extract in the extraction flask add approximately 20 ml dry ether, then allow to stand in the cold for 1 h with occasional stirring. Decant the ether solution through a No. 2 porosity glass sinter and collect the filtrate in a weighed glass evaporating dish. Wash the residue in the flask and the filter with dry ether, evaporate the ether and dry the residue to constant weight at 100°C.

414

This ether-soluble emulsifier may be identified as a polyethylene oxide type either by the heteropoly acid test or by infra-red examination.

(b) Sodium alkyl sulphate or fluorinated emulsifier

To the ether-insoluble residue add approximately 10 ml cold water and stir well, prior to filtering through the same No. 2 porosity glass sinter as in (a) above. Collect the filtrate in a second weighed glass evaporating basin. Wash any residue, flask and filter with water and evaporate the combined filtrates to dryness on a water bath.

Dry the extract to constant weight at 100°C.

This ether-insoluble but water-soluble emulsifier is identified by chemical and infra-red tests.

E.g. sodium lauryl sulphate gives positive tests for sulphur and for anion active compounds in the methylene blue test and gives reactions characteristic of alkyl sulphates.

Fluorinated emulsifiers give strong positive tests for fluorine, are anion active and fume strongly on warming with mineral acid i.e. liberating the free fluorinated acid.

4. HYDROCARBONS

Hydrocarbon solvents are often added to polytetrafluoroethylene dispersions in an amount approximately 5% of the total sample. This hydrocarbon can be isolated by continuous steam distillation in the return type Dean and Starke apparatus, as follows:

Weigh approximately 20 ml of sample into a 250 ml round-bottomed flask and add 100 ml distilled water and 2 drops silicone anti-foam agent M.430. Connect the flask to the 5 ml capacity return type Dean and Stark apparatus and reflux condenser.

Reflux gently for 1 h using a heating mantle and, after cooling to room temperature, read off the volume of hydrocarbon which has collected as the upper layer in the side limb of the apparatus.

This hydrocarbon, after measurement, is drawn off by means of a pipette, dried over anhydrous sodium sulphate and identified by such tests as refractive index, b.p., gas liquid chromatographic and infra-red tests.

5. PIGMENT AND POLYMER

Weigh accurately a clean dry 50 ml centrifuge tube and into it weigh 5 to 8 g sample. Precipitate the polymer by the addition of 25 ml acetone followed by 20 ml water. Stir the suspension well and then centrifuge at 2500 rev/min for 30 min. Decant the clear supernatant liquid through a Whatman No. 541 filter paper and collect the filtrate in a 250 ml volumetric flask. Wash the polymer and pigment in the centrifuge tube by stirring with water (about 40 ml) and recentrifuge. Repeat this washing at least 4 times and collect all the filtered washings in the 250 ml flask.

Dry the centrifuge tube containing the isolated polymer plus pigment to constant weight at 100°C. Some measure of the ratio of pigment to polymer in this mixture may be obtained by a quantitative ash determination in a platinum crucible.

A spectrographic examination of the ash is always desirable.

6. CHROMIUM

Dilute the filtrate in the 250 ml volumetric flask, from (5) above, to the mark with distilled water and mix well.

(a) Determination of chromium present as dichromate

Transfer a 50 ml aliquot of the solution to a conical flask and add 100 ml of 2 N sulphuric acid and 50 ml 0·1 N ferrous ammonium sulphate solution.

Titrate the excess ferrous ammonium sulphate with standard 0·1 N potassium permanganate solution. Carry out a blank titration on 50 ml of the 0·1 N ferrous ammonium sulphate solution.

(b) Determination of total chromium

Transfer a 50 ml aliquot of the solution to a conical flask, add 5 ml 0·1 N silver nitrate solution followed by 20 ml 10% w/v ammonium persulphate solution.

Boil the solution for 30 min to decompose the persulphate and cool the solution. Add 100 ml of 2 N sulphuric acid solution followed by 50 ml 0·1 N ferrous ammonium sulphate solution and from this point proceed as in (a) above.

7. PHOSPHATE

Transfer 100 ml of the sample solution to a 800 ml conical flask and add 10 ml 0·1 N silver nitrate solution and 40 ml 10% w/v ammonium persulphate solution to convert the chromium to dichromate. Boil for 30 min to decompose the excess persulphate and add 15 ml conc. nitric acid. Cool to 70°C and add 50 ml of Johnson's molybdate solution. Shake vigorously for 2 min and allow the precipitate of ammonium phospho-molybdate to cool and settle. Filter through paper pulp using suction and wash the flask and precipitate with 2% w/v potassium nitrate solution until the washings are neutral.

Transfer the pulp pad and precipitate to the original flask. Add 0·5 N sodium hydroxide solution from a burette until the yellow precipitate is completely dissolved and back titrate the excess with 0·5 N nitric acid solution using phenolphthalein as indicator.

Using a previously determined factor for the 0·5 N sodium hydroxide solution, calculate the total phosphorus to the proportion of phosphoric acid in the sample.

As an additional test that the chromium and phosphorus are present as CrO_3 and H_3PO_4 a potentiometric titration may be carried out on about 1 g of the original sample after dilution with water.

Titrating with 0·1 N sodium hydroxide solution two end-points will be noted:

(1) at pH 4·7
(2) at pH 9·3

The usage of 0·1 N sodium hydroxide solution at pH 9·3 corresponds to the titration of the CrO_3 to Na_2CrO_4 and the H_3PO_4 to Na_2HPO_4 and should correspond to that calculated from (6) and (7) above.

8. PH

This is determined directly on the sample using the glass electrode/calomel electrode system.

9. FREE AND COMBINED AMMONIA

15 g of sample, after dilution and addition of excess 30% w/v sodium hydroxide solution are distilled in a Kjeldahl distillation apparatus.

The distillate is collected in 0·1 N hydrochloric acid solution and the excess acid finally back titrated with 0·1 N sodium hydroxide solution to obtain a measure of the ammonia in the distillate. If free ammonia is required then this may be determined by direct distillation without addition of alkali.

QUANTITATIVE DETERMINATIONS OF EMULSIFIER COMPONENTS

Having determined the nature of the emulsifier system present it is often necessary to make quantitative determinations of the components present.

THE DETERMINATION OF POLYETHYLENE OXIDE CONDENSATE
IN POLYTETRAFLUOROETHYLENE DISPERSION CONTAINING
AMMONIA

This method has been shown to be very useful for preparations containing approximately 60% polytetrafluoroethylene, 3 to 4% polyethylene oxide condensate and ammonia sufficient to give a pH of 9·5 to 10·0.

Reagents

Methyl alcohol.
Barium chloride solution.
A 10% w/v solution of barium chloride ($BaCl_2 \cdot 2H_2O$).
Phosphotungstic acid solution.
A 10% w/v solution of the Analar acid in water.
Hydrochloric acid solution.
25 ml conc. HCl are diluted to 100 ml with distilled water.
Standard solution of the polyethylene oxide condensate.
5 g of the condensate are dissolved in water and the solution diluted to 100 ml with water. 10 ml of this strong solution are diluted to 250 ml with distilled water, 1 ml of the dilute solution = 0·002 g polyethylene oxide condensate.

Procedure

2 g of the dispersion are weighed into a stoppered conical flask. 3 ml of the hydrochloric acid solution are added followed by 3 ml barium chloride solution and 25 ml of methyl alcohol. The mixture is shaken vigorously, then boiled gently for a few minutes to coagulate the polymer. The mixture is filtered hot through a 150 mm Whatman 31 filter paper into a beaker (250 ml) and the polymer and filter paper are washed with 100 ml methyl alcohol.

80 ml of distilled water are added to the filtrate, and after placing a boiling rod in the beaker and covering with a watch glass, the solution is boiled down to a volume of 30 ml.

The solution is cooled, when any turbidity present in the hot solution will clear, then washed into a beaker (100 ml) and diluted to approx. 50 ml. 3 ml of phosphotungstic acid solution are added with constant stirring, and the solution brought to the boil to coagulate the precipitate, then allowed to stand overnight.

The precipitated complex is filtered off on a weighed Gooch crucible (Porosity 4), then washed with 40 to 50 ml of cold water, and dried at 100°C for 2 h to yield the weight of dry complex.

A calibration graph is prepared by submitting to the above test known volumes of the dilute standard polyethylene oxide condensate diluted to 50 ml with water; from the data obtained the proportion of polyethylene oxide condensate in the dispersion is readily calculated.

THE DETERMINATION OF POLYETHYLENE OXIDE CONDENSATE AND SODIUM LAURYL SULPHATE IN POLYTETRAFLUOROETHYLENE DISPERSIONS

Here the procedure is somewhat different. 10 g of sample are weighed into a 150 ml stopped conical flask. 50 ml methanol are added and the mixture is shaken well and boiled to coagulate the polymer. The hot solution is filtered through a 150 mm Whatman 31 filter paper into a beaker and the paper and polymer are washed with methanol until the total volume of filtrate is approx. 100 ml. 70 to 80 ml of water are added to the filtrate, and, after placing a boiling rod in the beaker and covering with a watch glass, the solution is boiled down to a volume of 35 ml.

After cooling, 3 g Amberlite IRA 400 resin (in the OH form) are added and the beaker and contents allowed to stand with occasional swirling for 1 h. The resin is filtered off on a small Whatman 31 filter paper and the original beaker and resin washed with distilled water until the volume of filtrate collected in a 100 ml beaker is 50 ml.

3 ml barium chloride ($BaCl_2 \cdot 2H_2O$) 10% w/v solution, 3 ml 25% v/v hydrochloric acid solution and 3 ml phosphotungstic acid solution (10% w/v) are added and the complex is now precipitated and weighed as in the determination of polyethylene oxide condensate just described. The calibration is also carried out in a similar manner.

The method gives excellent results for preparations containing approx. 16% polytetrafluoroethylene, 0·6% sodium lauryl sulphate, 0·1 to 0·2% polyethylene oxide and 4% toluene.

The determination of sodium lauryl sulphate in this type of preparation may be accomplished by coagulation of the polymer with sodium chloride and 50% v/v aqueous methanol. After filtration and boiling off of excess methanol the anion active sodium lauryl sulphate is titrated with standard cetyl ammonium bromide solution in the presence of ethylene dichloride and bromophenol blue.

INFRA-RED SPECTRA OF FLUOROCARBON RESINS

The outstanding feature of the spectra of fluorocarbon resins is the high intensity of the absorption bands of the C—F group. This is in marked contrast with hydrocarbon spectra, and indeed the difference in intensity is so great that in a compound such as polyvinylidene fluoride, with a 1 : 1 ratio of CH_2 and CF_2 groups, a spectrum which shows the C—F stretching bands at high intensity shows such weak C—H stretching bands at 3·4 μm that the material may be thought to contain no hydrogen.

Since comparatively few fluorocarbon polymers exist, there is little call for elaborate structural diagnosis from the spectrum. The presence of fluorine and chlorine should be evident from the elements test, but the presence of fluorine should also be apparent from the spectrum because of the exceptionally high intensity of the bands in the 8 μm to 11 μm region, which are the prominent feature of the spectra of these resins.

Resins containing the CF_2 group have a pair of intense bands between 8 μm and 9 μm, although an exceedingly thin sample ($\sim$0·0001 in or 0·002 mm) is required to observe these separately, and often only a single broad band between 8 μm and 9 μm will be seen. Resins containing the single CF group, whether as CHF, CClF or $C(CF_3)F$, show strong bands in the 9 μm to 11 μm region. Observation of the presence of C—H groups is useful, but, as noted above, care is required here because the great intensity of the C—F bands makes the C—H structure at 3·4 μm appear deceptively weak. However, in a specimen of about 0·001 in (0·03 mm) thickness there is no difficulty in ascertaining whether the resin contains C—H groups.

The principal straight fluorocarbon resins are:

Polytetrafluoroethylene	Spectrum 7·1
Polytrifluorochloroethylene	Spectra 7.2
Polyvinylidene fluoride—'Kynar'	Spectra 7.3
Polyvinyl fluoride	Spectra 7.4

The following copolymers may also be encountered:

Tetrafluoroethylene/hexafluoropropene—'Teflon FEP'	Spectrum 7.5
Hexafluoropropene/vinylidene fluoride—'Viton A'	Spectra 7.6
Trifluorochloroethylene/vinylidene fluoride— 'Kel F elastomer'	Spectra 7.7

Polytetrafluoroethylene (Spectrum 7.1) shows very little structure in the 2·5 μm to 15 μm region other than the strong 8 μm to 9 μm absorption block; further bands of interest occur in the long wave infra-red region, and are discussed by Krimm[8]. The absorption band at 4·3 μm is the overtone of the strong CF_2 absorption band. Polytetrafluoroethylene as made is highly crystalline, but the weak structure in the 12 μm to 14 μm region arises from amorphous material, and these bands have been used to measure amorphous content[9].

Polyvinyl fluoride (Spectra 7.4) containing only CF groups has a strong band centred on 9·2 μm; the 3·4 μm and 7 μm structures arise from the CH and CH_2 groups. Polyvinylidene fluoride (Spectra 7.3) contains CF_2 groups, and the main C—F band returns to the 8 μm to 9 μm region. The C—H stretching band at 3·4 μm is surprisingly weak in this spectrum, but the 7·2 μm band, presumably the CH_2 deformation band, is prominent. The spectrum of this substance shows marked differences when different methods of sample preparation are used (e.g. Spectra 7.3(a) and 7.3(b)); these are presumably due to changes in the degree of crystallinity.

The spectrum of tetrafluoroethylene/hexafluoropropene copolymer (Spectrum 7.5) is very similar to that of polytetrafluoroethylene itself, but the additional band at 10·2 μm is evident on samples of over 0·0005 in (0·01 mm) thickness. This band is thought to arise from the single C—F group, rather than the CF_3 group, since a band in almost the same position (10·3 μm) occurs in polytrifluorochloroethylene (Spectra 7.2). The greater relative intensity of this band in polytrifluorochloroethylene as opposed to the hexafluoropropene copolymer is presumably due to the much higher concentration of C—F groups in the former resin.

The other two copolymers (Spectra 7.6 and 7.7) have rather similar spectra. Both contain substantial concentrations of CF_2 groups with the band maxima between 8 μm and 9 μm, while the CH_2 stretching and deformation bands appear at 3·4 μm and 7·2 μm as in polyvinylidene fluoride. The band pair at 11·3 μm and 11·9 μm from this constituent is also evident in both copolymer spectra.

Spectra of these copolymers are also given in Hummel's collection, and these show minor differences in some cases from the spectra reproduced here. It is presumed that these differences arise from crystallinity differences, due to different methods of sample preparation. It is unlikely that these spectral differences will prevent the qualitative identification of a fluorocarbon copolymer, but, on the other hand, without a careful investigation of a particular material it would be unwise to attempt quantitative analysis of the proportions of a copolymer. In many cases, however, the qualitative identification, combined with quantitative elementary analysis, will give a completely satisfactory result.

N.M.R. SPECTRA OF FLUOROCARBON RESINS

The study of fluorocarbon resins by n.m.r. spectroscopy has revealed some interesting structural features which do not occur in the corresponding chlorocarbon resins. Thus the proton n.m.r. spectrum of polyvinylidene chloride has only one resonance (at about 5·1 τ), from which it is deduced that all protons in the polymer chains are equivalent, and that the polymerisation is entirely head to tail:

$$-CH_2-CCl_2-CH_2-CCl_2-CH_2-CCl_2-$$

By analogy, the F^{19} n.m.r. spectrum of polyvinylidene fluoride should consist of one resonance (which may be split by proton-fluorine heteronuclear coupling), but in practice is found to contain four resonances. By comparison with model compounds, it has been possible to identify these with the CF_2 group (marked CF_2^*) in the following structures[10]:

$$-CH_2-CF_2-CH_2-CF_2^*-CH_2-CF_2-$$

$$+CF_2-CH_2+CH_2-CF_2^*-CH_2-CF_2-$$

$$-CH_2-CF_2-CH_2-CF_2^*+CF_2-CH_2+$$

$$-CF_2-CH_2-CF_2+CF_2^*-CH_2+CH_2-$$

It is therefore concluded that polyvinylidene fluoride has an irregular structure, which arises because some head-to-head polymerisation occurs. Examination of the structures detected shows that these are produced by inserting an occasional monomer unit 'backwards' into the polymer chain. These units are shown as $+CF_2-CH_2+$ in the structures drawn above. The proton n.m.r. spectrum of this resin (decoupled from F^{19} nuclei) is also complex, but it can be interpreted satisfactorily on the same basis.

Although the proton and F^{19} resonance spectra of polyvinyl fluoride are more complex than those of polyvinylidene fluoride, the analysis of these spectra[10] leads to the conclusion that this polymer also contains 'backwards' units. It is possible that this is a property common to a number of fluorocarbon resins, since the presence of 'backwards' vinylidene fluoride units has also been demonstrated by n.m.r. examination of vinylidene fluoride/hexafluoropropylene copolymer.

This surprisingly detailed analysis of fluorocarbon resins from their F^{19} n.m.r. spectra is possible because chemical shifts between F^{19} nuclei in different environments are very large compared with the chemical shifts observed in the proton n.m.r. spectra. This means also that, provided suitable solvents can be found and the resonances can be assigned, quantitative analysis of copolymers of fluorocarbon monomers is comparatively straightforward.

REFERENCES

1. HASLAM, J., and WHETTEM, S. M. A., *J. appl. Chem., Lond.*, **2**, 339 (1952)
2. SWEETSER, P., *Analyt. Chem.*, **28**, 1766 (1956)
3. FRANT, M. S., ROSS, J. W., and RISEMAN, J. M., *Am. Lab.*, 14, Jan. (1969)
4. WILLARD, H. H., and HORTON, C. A., *Analyt. Chem.*, **22**, 1194 (1950)
5. GEL'MAN, N. E., and KIPARENKO, L. M., *Zh. analit. Khim.*, **20**, 229 (1965)
6. LIGHT, T. S., and MANNION, R. F., *Analyt. Chem.*, **41**, 107 (1969)
7. HASLAM, J., HAMILTON, J. B., and SQUIRRELL, D. C. M., *Analyst, Lond.*, **85,** 556 (1960)
8. KRIMM, S., *Adv. Polym. Sci.*, **2**, 51 (1960).
9. MILLER, R. G. J., and WILLIS, H. A., *J. Polym. Sci.*, **19**, 485 (1956)
10. WILSON, C. W., and SANTEE, E. R., *Jnl Polym. Sci., Pt. C*, No. 8, 97 (1965)

8

RUBBER-LIKE RESINS

The first part of this chapter describes the identification of rubbers by chemical tests and by pyrolysis/gas chromatography. After a description of methods for the determination of elements in rubbers, observations are offered on the examination of latices in respect of total solids content, the nature and amount of emulsifier and the determination, by chemical and gas chromatographic methods, of the free monomer content of these compositions. The extraction of antioxidants from rubber latices and solid compositions for subsequent identification and determination is then described.

The second part of the chapter is concerned with the infra-red and n.m.r. spectra of commercially important rubbers. The spectra may be used for identification, and in many cases may be applied also to determine the proportions of the constituents of copolymers. The relative amounts of different structural isomers may be determined in those polymers based on diene monomers.

A very large number of substances could be dealt with under this heading, e.g. natural rubber, gutta percha, butadiene/acrylonitrile copolymers, butadiene/styrene copolymers, butadiene/methyl methacrylate copolymers, butadiene/styrene/acrylonitrile terpolymers, polybutadienes, chloroprenes, chlorinated rubber and rubber hydrochloride, cyclised rubber, polyether and polyester rubbers, polyisobutene and polyisoprene.

The analytical chemist in the plastics field will, for the present at any rate, hardly be expected to be equally proficient in all aspects of rubber analysis, whether natural or synthetic.

He will be familiar with books and papers dealing particularly with the analysis of rubber, such as Wake's textbook[1], and Memorandum No. U9A on '*The Identification and Estimation of Natural and Synthetic Rubber*'[2].

Although he may have some doubts about the efficiency of qualitative schemes for the separation of mixtures of natural and synthetic rubbers put forward in various publications, he should be familiar with some of the tests that, in favourable circumstances, the rubber chemist employs, e.g.:

1. Heating and burning tests.
2. Behaviour in the Weber test for natural rubber.
3. Oxidation with chromic acid to yield acetic acid as in tests for rubber hydrocarbon.
4. Behaviour in oxidation tests with nitric acid, i.e. yielding p-nitrobenzoic acid from styrene-containing substances. This acid may be reduced, diazotised and coupled.
5. Behaviour in acid disintegration tests and freezing tests.
6. Behaviour with organic solvents in solubility and swelling tests.
7. Behaviour in pyrolysis tests.

In our experience, the one test that the analyst in the plastics industry will make frequent use of is the Modified Weber Test for Natural Rubber.

THE MODIFIED WEBER TEST FOR NATURAL RUBBER

The directions for this test are taken from the above mentioned Memorandum No. U9A and are as follows:

Place approximately 0·05 g of the dried acetone-extracted sample in a clean dry test tube and add about 5 ml bromine solution containing 10% v/v bromine in carbon tetrachloride.

Place the test tube in a beaker of water and raise slowly to boiling point. Continue heating until not more than a trace of bromine remains, add 5 to 6 ml phenol solution (containing 10% w/v phenol dissolved in carbon tetrachloride) and continue heating until the colour develops.

A violet colour is given by natural rubber in a few minutes.

If the heating is continued for 15 min, small amounts of natural rubber in the presence of other rubbers may be readily detected. In certain special circumstances Neoprene FR may give a faint violet colour.

Oil-type reclaims give a positive result by the above method but no coloration is given by alkali reclaims. However, the presence of the latter in admixture with natural rubber does not inhibit the development of the violet colour.

Butadiene/styrene copolymers do not cause interference in mixtures of these materials with natural rubber.

PYROLYSIS/GAS CHROMATOGRAPHY

The rubber-like polymers give very characteristic pyrograms which provide a useful means of identification particularly for small samples

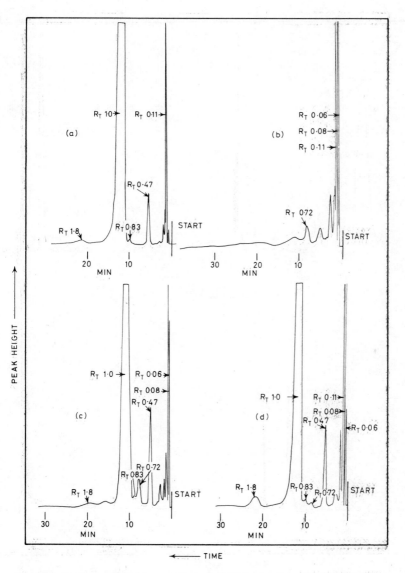

Fig. 8.1. Pyrograms of copolymers: (a) acrylonitrile/styrene; (b) acrylonitrile/butadiene; (c) butadiene/styrene; (d) acrylonitrile/butadiene/styrene

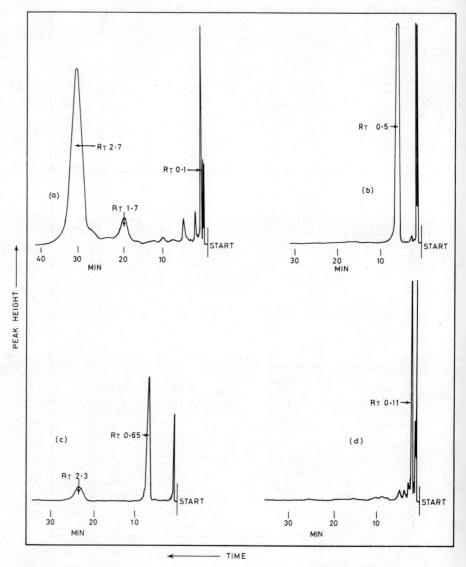

Fig. 8.2. Pyrograms of rubbers: (a) unvulcanized natural rubber; (b) polyester rubber; (c) silicone rubber; (d) polychloroprene

and samples which prove difficult to prepare for infra-red examination. The copolymers of butadiene break down to give essentially the monomers, the presence of which is quickly established using the retention times of the various peaks (Figure 8.1), relative to that of styrene monomer which is the main peak obtained in the pyrolysis of polystyrene (see p. 61).

Styrene $R_T = 0.47$ 0.83 1.0 1.8
Butadiene $R_T = 0.06$ 0.08 0.72
Acrylonitrile $R_T = 0.11$

Other rubbers may also be characterised by their pyrograms (Figure 8.2), the characteristic peaks being:

Natural rubber $R_T = 0.1$ 1.7 2.7
Polyester rubber $R_T = 0.5$
Polychloroprene $R_T = 0.11$
Silicone rubber $R_T = 0.65$ 2.3

DETERMINATION OF ELEMENTS

In ordinary circumstances chlorine and sulphur in natural and synthetic rubbers may readily be determined by fusion with sodium peroxide in a Parr bomb. One point to bear in mind in the sulphur determination is to ensure that the final precipitation of barium sulphate is carried out at a sufficiently high dilution. Large amounts of sodium salts are present in the products of a sodium peroxide fusion and, if the precipitation of barium sulphate is carried out in a highly concentrated solution of sodium salts, then the results obtained will be low. Of course, in innumerable cases modern methods of Oxygen Flask Combustion will give entirely reliable results in these determinations of chlorine and sulphur.

NITROGEN

It is generally agreed that nitrogen in natural and synthetic rubbers is best determined by the Kjeldahl method, although there are many forms of the Kjeldahl test.

The following method has been found to be very suitable for the determination of nitrogen in butadiene/acrylonitrile copolymers and may obviously be applied to nitrogen determinations in other polymers.

Apparatus

800 ml Kjeldahl flask.
Ammonia distillation apparatus. This consists of a bulb splash-head, a

condenser and a delivery tube leading to a 650 ml Phillips beaker. Kjeldahl stand.

Reagents

B.D.H. sodium sulphate tablets containing 1 g sodium sulphate.
B.D.H. copper catalyst tablets containing 1 g sodium sulphate and 0·1 g copper sulphate.
30% w/v sodium hydroxide solution.
Hydrochloric acid solution 0·1 N.
Sodium hydroxide solution 0·1 N.
Methyl red indicator.

Procedure

About 0·5 g (accurately weighed) of the comminuted sample of the butadiene/acrylonitrile copolymer is weighed into a 800 ml Kjeldahl flask. 6 B.D.H. sodium sulphate tablets, 3 B.D.H. copper catalyst tablets, 10 ml of distilled water and 20 ml of conc. sulphuric acid are added and a glass 'pear' is placed in the neck of the flask. The flask is placed on the Kjeldahl stand and the flask and contents heated fairly strongly with occasional swirling until the solution is a pale bluish green. The digestion is now continued for a further 15 min. The contents of the flask are cooled thoroughly and sufficient water is added to give a total volume of about 480 to 500 ml. A few drops of methyl red indicator are added and the flask is connected to the ammonia distillation apparatus.

The contents of the flask are heated to boiling and 30% w/v sodium hydroxide solution is added through the tap funnel until the liquid is alkaline, after which a further 25 ml of the sodium hydroxide solution are added.

The ammonia liberated is distilled into 50 ml of 0·1 N hydrochloric acid solution in the Phillips beaker and the distillation is continued at such a rate that 250 to 300 ml of distillate is collected in about 40 min. After boiling out carbon dioxide the excess hydrochloric acid is titrated with 0·1 N sodium hydroxide solution using methyl red as indicator.

A blank test is carried out at the same time using the same conditions of digestion and distillation.

The nitrogen content of the copolymer sample is calculated from the amount of acid used up in the test.

With some butadiene/acrylonitrile copolymers the nitrogen determined by the Kjeldahl procedure may be slightly low by an amount equivalent to up to 1% acrylonitrile. The reason for this is not fully understood and in some cases an improved result is obtained by the following method which is also useful when a rapid analysis is required. Since the weight of sample taken is small, however, complete homogeneity is essential.

RAPID DETERMINATION OF NITROGEN

In cases where a very rapid determination of nitrogen in butadiene/acrylonitrile rubbers is required the following method has proved useful.

Reagents

Digestion mixture.
1 volume of nitrogen-free sulphuric acid is diluted with 1 volume of water, cooled and 8 g of sodium sulphate and 1 g of copper sulphate crystals are added per 100 ml of acid and dissolved by warming.
Sodium sulphate solution.
20 g of anhydrous sodium sulphate are dissolved in water and the solution diluted to 100 ml.
Neutralising solution.
30·9 g of boric acid crystals are dissolved in 1 l of 4 N sodium hydroxide solution.
Hypobromite solution.
2 volumes of 'A' solution are mixed with 5 volumes of 'B' solution as required.
'A' bromine solution.
8 g of potassium bromide are dissolved in water, 1 ml of bromine is added and the solution is diluted to 400 ml with water. The solution deteriorates with storage and a fresh solution should be prepared when the titre of 5 ml of the hypobromite solution falls below 12 ml of 0·01 N sodium thiosulphate.
'B' borate solution.
24·7 g of boric acid crystals are dissolved in 100 ml of 4 N sodium hydroxide solution and the solution is diluted to 1 l.
Potassium iodide solution.
25 g of potassium iodide are dissolved in water, the solution diluted to 50 ml and stored in an amber bottle in a dark cupboard.
Sodium thiosulphate solution 0·01 N.
This is checked from time to time with 0·01 N potassium bromate solution.
Starch solution.
1% w/v starch in 20% w/v NaCl solution.

Procedure

About 0·05 g of rubber (accurately weighed) is digested in a small Kjeldahl flask (100 ml) with 10 ml of digestion mixture and 10 ml of the sodium sulphate solution. Heating is continued until white fumes of sulphuric acid are evolved and the solution turns a clear greenish blue. Heating is continued for a further 10 min, the solution cooled and 10 ml of distilled water added. The solution is brought just to the boil again and then allowed to cool prior to transferring to a 100 ml graduated flask and diluting to the mark with distilled water.

25 ml of the diluted solution is transferred to a stoppered bottle (8 oz or 250 ml narrow necked) and neutralising solution is added from a burette until a blue precipitate forms, after which a further 0·1 ml is added.

After cooling, 25 ml of the sodium hypobromite solution are added and the contents mixed and set aside for 2min. 10 drops of potassium iodide solution are now added followed by 3·0 ml of 3 M H_2SO_4 and, after mixing, the liberated iodine is titrated with standard 0·01 N sodium thiosulphate solution, using starch as indicator.

A blank test is carried out on all the reagents omitting only the rubber.

The equations for the essential reactions of the ammonia determination are as follows:

$$3NaOBr + 2NH_3 = N_2 + 3NaBr + 3H_2O$$
$$NaOBr + 2KI + H_2SO_4 = NaBr + I_2 + K_2SO_4 + H_2O$$

and 1 ml 0·01 N sodium thiosulphate solution $= \dfrac{0·00014008}{3}$

$$= 0·00004669 \text{ g nitrogen}$$

The nitrogen content of the sample is calculated from the final iodine titration, taking, of course, the blank titration into account.

Excellent results have also been obtained in trials of the automatic Dumas and C, H and N analysers based on a gas chromatographic finish for this analysis.

EXAMINATION OF LATICES

In the examination of synthetic rubber or resin latices the questions to be answered by the analyst usually concern the total solids content of the latex, the nature of the resin, the nature and amount of the emulsifiers and antioxidants present, and the amount of free monomers in solution. The nature of the resin may be determined by the infra-red examination described later, after preparation of the sample as was described on p. 22. If it is necessary to obtain sufficient resin for further tests then the product coagulated during the isolation of the emulsifier or antioxidant system may be used.

DETERMINATION OF TOTAL SOLIDS

For the determination of the total solids content the most commonly used method involves drying a known weight of latex on sand or on a glass filter disk contained in a glass or aluminium weighing vessel. Disposable aluminium patty dishes may be used for this purpose and drying temperatures between 100 and 150°C are satisfactory. Alternatively the 'hot-wire' method described by Meardon[3] has been shown to give reliable results, although it is essential that the sample is homogeneous since only a small amount is taken for test. In this method a small wire heating element is weighed before and after dipping in the sample under test. A current is passed through the element to give the required drying temperature and after a predetermined time the element is reweighed to give the weight of solids. The coil is then brought to red heat to burn off the solids and is then ready for the next analysis. Test conditions must be evaluated for each type of latex since decomposition of the sample during test must be avoided. The total solids content can also be obtained during the gas chromatographic procedure for the determination of free monomers (p. 433).

EMULSIFYING SYSTEM

The emulsifying system is isolated from rubber and resin latices by breaking the emulsion by the addition of alcohol, filtration of the coagulated resin and evaporation of the clear filtrate. In some cases the emulsion is difficult to break and it is advisable to make preliminary tests on small quantities of each sample to ascertain the best way to break the emulsion without the addition of large quantities of salts, i.e. electrolytes. Alternative procedures that may be useful are as follows:

1. Shake the emulsion with a little ether before the addition of methanol.

2. Add the emulsion dropwise to a large volume of methanol, with constant stirring during the addition.

3. Proceed as in procedure 1 or 2 above but using hot methanol.

4. Add a small quantity of saturated potassium chloride solution immediately prior to the addition of the methanol.

Should these methods fail or succeed only partially then it may be necessary to evaporate the emulsion to dryness, to break up the gel produced and to extract this with hot water.

The aqueous extract is filtered and evaporated to dryness. This process of extraction and evaporation with hot water is repeated four or five times using smaller volumes of water for each successive extraction.

Using this method it is usually possible to recover the surface active agent in a reasonably pure form suitable for examination.

431

Identification and determination of the emulsifier is carried out using the principles described in Chapters 2 and 7.

CHEMICAL DETERMINATION OF FREE MONOMERS

ACRYLONITRILE

In general work on synthetic polymers the analytical chemist has to become acquainted with the determination of many monomers in air, e.g. butadiene, styrene, methyl methacrylate and acrylonitrile, because of problems concerned with toxic concentrations of these monomers.

For similar reasons he may find it desirable to be acquainted with some of the principles of the determination of free monomers in synthetic rubber latices for the concentration of these monomers must obviously be very low indeed. Over the years we have had particular experience of this and, in the past, two methods were commonly used for the determination of free acrylonitrile in butadiene/acrylonitrile latices. In the first of these methods the sample is allowed to react with a standard volume of a 2·5% w/v solution of lauryl mercaptan in isopropanol in the presence of 1 ml 5% w/v alcoholic potassium hydroxide solution. After acidification and dilution with more isopropanol the excess mercaptan is titrated with 0·1 N iodine solution. A blank titration is carried out and the difference in titres calculated as percentage acrylonitrile. An alternative method also often used involves steam distillation of the free monomer from the latex into tetramethyl ammonium iodide base solution and polarographic examination over the applied voltage range $-1·5$ to $-2·5$ V.

Recent work by Taubinger[4], however, has largely displaced these methods in our laboratories. He showed that the sulphite addition method originally described by Critchfield and Johnson[5] could be applied under alkaline conditions and that with the addition of more emulsifier the reaction and subsequent titration of the liberated sodium hydroxide could be carried out without causing coagulation of the rubber. Thus, with a suitable combination of automatic dispensers, a timer and an automatic pre-set end-point titrator, the whole analysis could be mechanised.

In the method, 2 ml 0·5 N sodium hydroxide solution is dispensed into the reaction vessel together with 2 ml 15% v/v Teepol emulsifier solution and 40 ml 1·6 M sodium sulphite in 1% v/v Teepol solution (a combined reagent may be prepared if required). An aliquot of the latex sample (10 or 50 ml) is now added and after 5 min reaction time the total alkali titrated automatically to the pre-set pH of the end-point with 0·25 N hydrochloric acid solution. The difference between this titre and that of a blank on the reagents is calculated in terms of free acrylonitrile monomer. The exact pH at the end-point (normally between 10·8 and 11·2) is determined for each new batch of reagents by a preliminary manual titration and standardisation of the method is carried out with a standard solution of acrylonitrile. The method can be used for acrylonitrile concentions down to $0 \cdot 05\%$.

The main reaction in this method is as follows:

$$H_2O + Na_2SO_3 + CH_2 = CHCN \rightarrow NaSO_3CH_2 - CH_2 - CN + NaOH$$

and the reaction is exothermic. Taubinger[6] has also designed an enthalpimetric analyser in which a continuous flow of sample diluted with additional emulsifier is mixed with excess sulphite reagent in a specially designed mixer and the heat produced measured by a thermistor circuit and recorded. This method is of great value when many analyses have to be carried out.

STYRENE AND METHYL METHACRYLATE

Free methyl methacrylate in butadiene/methyl methacrylate latices and free styrene in butadiene/styrene latices are readily determined by steam distillation in a modified Dean and Starke apparatus. For this purpose the design described by Kennedy and Taubinger[7] is recommended. For small amounts of free styrene an ultra-violet absorption method is described in BS 3397, measurements being taken at the absorption maximum occurring near 282 nm.

GAS CHROMATOGRAPHIC DETERMINATION OF FREE MONOMERS

ACRYLONITRILE

In addition to the methods previously described, the apparatus detailed in Chapter 1 (Figure 1.8) for the gas chromatographic determination of volatiles may be used for the examination of butadiene/acrylonitrile latices, and in addition to the percentage free acrylonitrile, the total solids may be obtained in the same test[8].

A small piece of quartz wool is placed inside a 2 in (50 mm) length of $\frac{1}{4}$ in (6 mm) o.d. aluminium tube. 4 to 7 drops of latex (60 to 100 mg) are weighed on to the quartz wool plug inside the aluminium tube. The aluminium tube is then placed in the centre of a glass sample tube. The glass tube is inserted into the sample heater between the two $\frac{3}{8}$ in couplings with the heater assembly isolated from the chromatographic column. An interval timer is set for 5 min and then started. After the 5 min heating period the pneumatic valve is switched, purging carrier gas over the sample and sweeping all volatile material on to the chromatographic column. The chromatographic peak for the acrylonitrile is then measured and the

433

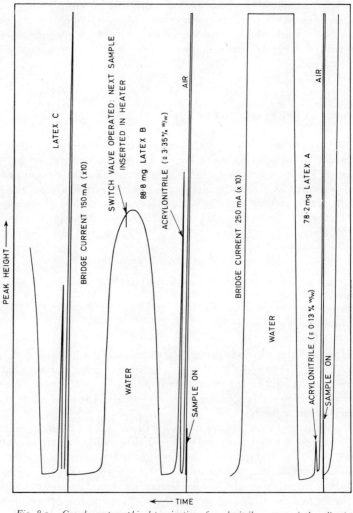

Fig. 8.3. *Gas chromatographic determination of acrylonitrile monomer in butadiene/acrylonitrile latices*

amount of acrylonitrile obtained read off from a calibration graph relating weight of monomer to peak height (or area).

If the aluminium tube is removed from the apparatus and cooled in a desiccator for 10 min, then reweighed, the total solids content of the latex can be calculated.

The calibration should cover the range 0·5 to 5·0 μl of acrylonitrile and is carried out by application of the method to glass sample tubes packed with quartz or glass wool, on to each of which has been injected 40 μl of water and a known volume of acrylonitrile monomer. The smaller volumes of monomer are best added in the form of a 5% aqueous solution. Typical chromatograms obtained for a butadiene/acrylonitrile latex are shown in Figure 8.3.

Suitable gas chromatographic conditions are as follows.

Column

10 in (250 mm) of 30% w/w glycerol + 1 ft 11 in (600 mm) of 30% w/w di-*n*-decyl phthalate, both on Gas Chrom. CLH (60–80 mesh) packed into a 2 ft 9 in (850 mm) length of $\frac{3}{16}$ in (5 mm) o.d., $\frac{1}{8}$ in (3 mm) i.d. copper tubing. The glycerol end of the column should be connected to the inlet side and the phthalate end of the column to the detector, to minimise bleed of the glycerol stationary phase. Gas Chrom. CLH is a hexamethyl-disilazane-treated Celite manufactured by Applied Science Laboratories Inc., State College, Pennsylvania.

Column and katharometer temperature: 85°C.
Carrier gas: Helium (inlet pressure 10 lbf/in^2 or 70 kN/m^2).
Sample heater temperature: 125°C (± 5°C).

Under these conditions the retention time for acrylonitrile is $1\frac{1}{2}$ min. Water is eluted completely after 8 min and the method, therefore, is capable of giving a result every 13 min (5 min heating time + 8 min development time). In practice, however, the time can be shortened by inserting a second sample for its 5 min heating period whilst the water from the first sample is being eluted (Figure 8.3). A sample throughput of 6 per hour can thus be obtained.

METHYL METHACRYLATE

The same conditions may also be used for the examination of butadiene/methyl methacrylate latices, the ester having a retention time of about $2\frac{1}{2}$ min. Methyl methacrylate monomers tend to polymerise during the pre-heat cycle however and care must be taken to calibrate under identical conditions to those used for the sample or results low by an amount up to

10% *relative* can be obtained. For example, if 2% is the true figure, the figure that is obtained may be as low as 1·8%

STYRENE

For the examination of butadiene/styrene latices a different column and conditions have been found necessary.

Column

2 ft 5 in (750 mm) of 30% w/w glycerol + 10 in (250 mm) of 30% w/w di-nonyl phthalate, both on Gas Chrom. CLH (60–80 mesh) packed into a 3 ft 3 in (1 m) length of $\frac{1}{4}$ in (6 mm) o.d., $\frac{3}{16}$ in (5 mm) i.d. copper tubing.

Column and detector temperature: 95°C.

Carrier gas: Helium (inlet pressure 8 lbf/in² or 55 kN/m²).

Bridge current: 240 mA.

Attenuation: × 10 for styrene contents of 2·5 to 10%.
   × 2 for styrene contents of 0 to 2·5%.

Sample heater temperature: 165°C (± 5°C).

For these conditions the retention time for styrene is $3\frac{1}{4}$ min and the water is completely clear in $9\frac{1}{2}$ min. Again by inserting the second sample into the heater while the water from the first sample is being eluted a throughput of 5 samples per hour can easily be maintained. Typical chromatograms are shown in Figure 8.4.

SAMPLING AND ADDITIONAL DETERMINATIONS

Methods for sampling rubber latices and for some additional determinations including pH, the Determination of Coagulum, the Determination of Surface Tension, the Determination of Dry Polymer Content of Styrene/Batadiene Rubber Latices and the Alkali Reserve of Polychloroprene Latices, are given in BS 3397.

ANTIOXIDANTS

The principles of the methods described for the identification and determination of antioxidants in polyolefines (Chapter 6) may also be applied

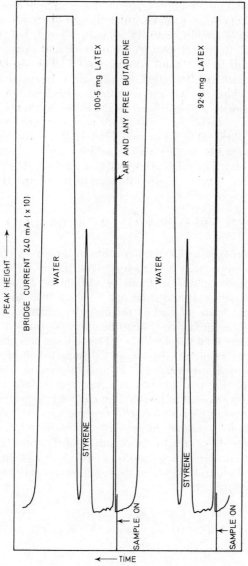

Fig. 8.4. Gas chromatographic determination of styrene monomer in a butadiene/styrene latex containing 5% free styrene

to rubber latices and resins after extraction of the antioxidant by an appropriate treatment with alcohol.

In the case of a latex, 50 to 60 ml of ethanol are transferred to a 100 ml beaker and stirred on a magnetic stirrer. 1 g of sample is added dropwise from a weight pipette and the coagulated resin filtered from the solution prior to dilution of the filtrate to 100 ml with ethanol. For rubber samples the specimen is cut into strips 1 to 2 mm thick and extracted by boiling under reflux with 80 to 100 ml ethanol for 1 h. The alcohol is decanted into a 250 ml standard flask and the extraction repeated with a further 80 ml ethanol. The extracts are combined with washings from the extraction flask and the whole diluted to 250 ml with ethanol. If the diluted extract solution is turbid, a 50 to 60 ml portion should be shaken with about 0·2 g of Celite Filter-Cel and filtered through a porosity No. 3 sintered glass crucible to give a clear solution for colorimetric or ultraviolet examination. The solids recovered from the solution by evaporation may be examined if necessary by infra-red or n.m.r. spectroscopy.

INFRA-RED SPECTRA OF RUBBER-LIKE RESINS

Rubbers are distinguished from other resins in their infra-red examination not by any special peculiarity of their spectra, but by the difficulty of preparing samples. Suggested methods of sample preparation have been considered in Chapter 1, and from among these it should be possible to select some appropriate method for the majority of materials presented for examination. The qualitative identification of rubbers by infra-red examination of the total pyrolysate is not considered here, but an excellent collection of spectra of such pyrolysates is included in Hummel's book[9] (p. 79–83). Our own preference for the identification of unknown rubbers by infra-red spectroscopy is to examine the sample as a solid film, by transmission spectroscopy, or if a suitable film cannot be prepared, to measure the spectrum by multiple internal reflection spectroscopy. Where pyrolysis is necessary, as, for example, in the case of heavily filled samples which will not yield useful spectra by examination of the solid resin, we would employ pyrolysis/gas chromatography, or pyrolysis/gas chromatography/ infra-red examination, rather than the examination of the total pyrolysate by infra-red spectroscopy.

IDENTIFICATION OF RUBBERS FROM THEIR INFRA-RED SPECTRA

As a very wide variety of resins may be encountered in the form of rubbers, it is better in the first instance to consider the spectrum alongside the results of a qualitative elementary analysis. Separation of extraneous material such as plasticisers and other additives by solvent extraction, or solution and precipitation of the resin in the case of a soluble product, are preferred where possible, to avoid confusion both with the elements test and the infra-red spectrum.

NO ADDITIONAL ELEMENTS DETECTED

The spectrum is examined for strong bands indicative of the presence of oxygen-containing groups.

1. STRONG BAND OR BANDS FROM 5·6 μM TO 6·0 μM

These show that the unknown contains carbonyl groups; ester or acid groups are most probable since rubbers containing ketone, lactone, and similar structures are unusual. Weak structure in the 5·8 μm to 6·0 μm region may often arise from minor additives or oxidation, particularly with hydrocarbon rubbers. Although this is of no consequence in identification work, it may be important for example in studies of oxidation rate in ageing experiments. A number of the polyacrylates are rubbery in nature, and another important group of ester rubbers are the polyester urethanes. These resins have been described in Chapter 4. The polyester urethanes, of course, contain nitrogen, but often only in minor amounts. Some ester copolymers of butadiene may be encountered; in such cases the unsaturated groups in the butadiene units are evident in the spectrum, as described for polybutadiene itself, in addition to the ester structure. A common resin of this type is butadiene/methyl methacrylate copolymer, also described in Chapter 4.

Although no commercial rubbers are straight polymers of carboxylic acids, a number of copolymers containing carboxylic acids are of importance, such as butadiene/styrene/maleic acid. In such cases the acid carbonyl band near 5·9 μm is usually the only evidence of an acid constituent, the rest of the spectrum arising from the other components of the copolymer and, since the latter will usually be the major components, the identification of these can be pursued neglecting the presence of the acidic structure.

In addition to those resins containing carbonyl groups, many rubber-like compositions exist which contain plasticisers with carbonyl groups. Evidence of this will usually be obtained by examining the products extracted from the material with ether.

2. STRONG BANDS NEAR 9 μM

This will suggest the presence of ether groups in the resin. Many rubbers, such as butadiene/styrene copolymer, may be made in the presence of substantial amounts of polyether surface active agents; evidence of this may be obtained by examining the products of methanol extraction. There are, however, a number of rubber-like polymers which are ethers, and in the spectra of such substances the band near 9 μm will usually be the strongest in the spectrum. The vinyl ethers, many of which are rubber-like in character, have been described (Chapter 3). The other important class of ether rubbers are the 'urethane' ethers (p. 452); these contain nitrogen, but in many cases the amount is small. Some polysulphide rubbers also contain ether groups (p. 452).

Silicone rubbers have strong absorption bands in the 9 μm to 10 μm region, and if no preliminary test is applied for the presence of silicon, these must be considered. Fortunately these substances have highly characteristic spectra. They are described in Chapter 10. Confusion may also arise with heavily filled compositions, since silica, diatomaceous earth, and some other common mineral fillers have broad and intense absorption bands near 9 μm wavelength. Evidence of the presence of such components will usually be forthcoming from examination of the ash from the composition by chemical and spectrographic methods, and the spectrum of pyrolysis residues, as described in the discussion of the pyrolysis method (p. 26) may prove helpful in elucidating the spectrum of a filled composition.

3. NO IMPORTANT BANDS FROM 5·6 μM TO 6·0 μM OR NEAR 9·0 μM

When there is no evidence of carbonyl or ether groups, a material containing no additional elements will probably be of a hydrocarbon nature. The absorption band wavelengths, used with the tables given in Chapter 2, are particularly useful, since most rubbers contain aliphatic unsaturated groups, aromatic nuclei or other recognisable features. The characteristics of the commonly encountered hydrocarbon rubbers are summarised in Table 6.4; there may be additional features in the spectrum if the material is a copolymer.

SULPHUR DETECTED

Many rubber-like compositions contain minor amounts of sulphur, as this is a constituent of many of the additives used. There is, however, a group of rubbers containing substantial amounts of sulphur (30% or more). The spectra of these products are described in this chapter.

CHLORINE DETECTED

A number of heavily plasticised polyvinyl chloride compositions have a rubber-like appearance; if the unknown material is of this type the identification should present no difficulty by the methods given in Chapter 3. In addition there are a number of chlorine-containing rubbers which can be identified from their spectra, described in this chapter.

FLUORINE DETECTED

Fluorine-containing rubbers are of importance due to their high temperature stability. The spectra of these materials are described with those of the other fluorocarbons in Chapter 7.

NITROGEN DETECTED

The only important rubbers containing large amounts of nitrogen are the butadiene/acrylonitrile copolymers and butadiene/styrene/acrylonitrile compositions (p. 446). The 'urethane' rubbers contain smaller amounts of nitrogen and exhibit in their spectra the characteristics of either aliphatic esters (ester urethanes, p. 268) or ethers (ether urethanes p. 452).

POLYBUTADIENE

Butadiene monomer can be polymerised to form a number of structurally different materials.

1 : 4-addition of the monomer can give units in the polymer chains which have either the *trans* or *cis* configuration (Figure 8.5).

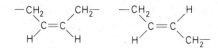

Fig. 8.5.   *Polybutadiene chain units for 1 : 4-addition of the monomer: cis (left) and trans (right)*

Butadiene units may also polymerise by addition across the 1 : 2-double bond, leaving pendent vinyl groups along the chain; with suitable choice of catalyst the successive units may have an isotactic or syndiotactic arrangement (Figure 8.6).

If it is substantially one of these forms throughout, the polymer is, in principle, able to crystallise. In such cases the spectrum shows the features associated with the chemical structure, and additional bands if crystalline material is present.

By the use of conventional catalysts a mixed product is obtained in which 1 : 2- and *trans*- and *cis*-1 : 4-structures are randomly distributed in the polymer chains. Hence the spectra of these products are comparatively straightforward and no crystallinity effects are involved. The sodium polymerised product (Spectrum 8.1) is of this type, although it is very rich in the 1 : 2-isomer. The prominent features of the spectrum at 10·1 μm,

11·0 μm and 6·1 μm arise from the —CH=CH₂ group. The absorption band at 10·3 μm and the broad band near 14·0 μm due to the *trans* and *cis* RCH=CHR′ groups respectively, are relatively weak. On the other hand, the product obtained by emulsion polymerisation with a peroxide

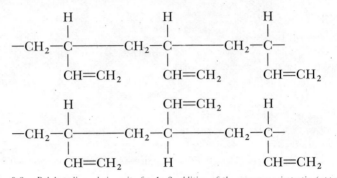

Fig. 8.6. Polybutadiene chain units for 1 : 2-addition of the monomer: isotactic (upper) and syndiotactic (lower)

catalyst (Spectrum 8.2) shows a predominance of the *trans*-1 : 4-polymer, with weaker bands from the *cis*-1 : 4- and 1 : 2-structures.

The newer stereospecific catalysts enable materials to be made which are essentially the four pure forms drawn out above[10]. The *trans*-1 : 4-resin crystallises readily, and will usually be encountered in the crystalline state (Spectrum 8.3). The spectrum shows the usual C—H structure, a prominent band due to the chain unsaturated group at 10·3 μm and additional sharp bands at 8·1 μm and 12·9 μm. The latter arise from crystalline material, and disappear if the sample is taken above the crystalline melting point.

By contrast, *cis*-1 : 4-polybutadiene (Spectrum 8.4) even if pure, crystallises only with difficulty, hence it is likely to be found in the completely amorphous state. The spectrum contains a strong broad band in the 13 μm to 14 μm region due to the *cis*-chain unsaturation. The structure at 10·3 μm is thought to arise from the presence of the *trans*-1 : 4-structure as an impurity.

Spectrum 8.5 is of syndiotactic 1 : 2-polybutadiene. The expected CH bands at 3·4 μm and 7 μm are present, with the —CH=CH₂ group showing strong absorption at 10·1 μm, 11 μm and 6·1 μm. The spectrum of isotactic 1 : 2-polybutadiene in Natta's paper[10] is rather similar in its main spectral features, but there are pronounced differences in the 14 μm to 15 μm region.

Many papers have appeared giving infra-red methods for the estimation of the concentration of *trans*- and *cis*-1 : 4-units and 1 : 2-units in polybutadiene. Silas, Yates and Thornton[11] mention the main work in this field, and give a detailed method which appears from the results given in the paper to be extremely satisfactory. Using solutions of the resins in carbon bisulphide, they have measured *trans*-1 : 4-content from the 10·3 μm band, and 1 : 2-content at 11·0 μm. The estimation of *cis*-1 : 4-content proved more difficult, as the maximum of the broad band near 14·0 μm was found to vary in wavelength with different samples. The method adopted was to integrate the absorbance between 12·0 μm and 15·75 μm. This procedure for the estimation of *cis*-1 : 4-units is likely to be difficult with resins of high 1 : 2-content, as 1 : 2-polybutadiene, in the isotactic and syndiotactic forms, and presumably also as isolated units, shows considerable absorption in the 14 μm to 15 μm region. It might be more satisfactory to restrict the range to 13·0 μm to 14·0 μm. The method of Silas, Yates and Thornton is laborious, particularly if only a small number of samples is to be examined. We have used a method which appears quite adequate for comparative work, taking measurements on samples in the form of solvent cast or pressed film. The thickness required is 0·0005 to 0·001 in (0·01 to 0·025 mm), being varied as necessary to keep all measured absorbances within the range 0·1 to 0·8. It will be noted from Silas, Yates and Thornton's paper that the overlap of absorbances is small except for the *cis*-1 : 4-band; the latter overlap is also small if the peak height absorbance at 13·6 μm is measured, rather than the integral absorbance over a wavelength range. Thus no overlap of absorbance is considered in our method. The measured absorbances at 10·3 μm, 11·0 μm and 13·6 μm are converted to concentrations using the extinction coefficients given in Table 6.5.

TYPICAL CALCULATIONS OF 1 : 2- AND *trans*- AND *cis*-1 : 4-CONTENTS

In Figure 8.7 draw the straight line backgrounds XX′ and YY′ and the zero energy line. Draw the lines ABC and DEF through the band maxima

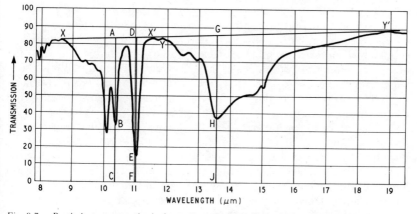

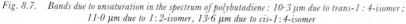

Fig. 8.7. Bands due to unsaturation in the spectrum of polybutadiene: 10·3 μm due to trans-1 : 4-isomer; 11·0 μm due to 1 : 2-isomer, 13·6 μm due to cis-1 : 4-isomer

B and E at 10·3 μm and 11·0 μm respectively, cutting XX' at A and D and the zero line at C and F. Draw GHJ through the band maximum H at 13·6 μm cutting YY' at G and the zero line at J.

Calculation

$Trans$-1 : 4-isomer. Absorbance $= \log_{10}\dfrac{AC}{BC} = A_1$

1 : 2-isomer. Absorbance $= \log_{10}\dfrac{DF}{EF} = A_2$

Cis-1 : 4-isomer. Absorbance $= \log_{10}\dfrac{GJ}{HJ} = A_3$

Weight % $trans$-1 : 4-isomer $= \dfrac{A_1/86}{A_1/86 + A_2/120 + A_3/25} \times 100\%$

Weight % 1 : 2-isomer $= \dfrac{A_2/120}{A_1/86 + A_2/120 + A_3/25} \times 100\%$

Weight % cis-1 : 4-isomer $= \dfrac{A_3/25}{A_1/86 + A_2/120 + A_3/25} \times 100\%$

Since the film thickness is not measured, the result is calculated by assuming that the total of all the unsaturated groups measured is 100%.

COPOLYMERS OF BUTADIENE

BUTADIENE/STYRENE COPOLYMERS

These copolymers are encountered over a wide composition range, but the spectra are quite straightforward and characteristic. Spectrum 8.6 of a copolymer of approx. 60/40 proportions shows, almost unmodified, all the absorption bands of both polystyrene (Spectrum 6.37) and polybutadiene (Spectrum 8.2). The 60/40 copolymer is rubber-like in nature; with an increasing proportion of polystyrene the copolymer becomes a rigid solid similar in appearance to straight polystyrene, and in the spectrum of an 85/15 copolymer (Spectrum 8.7) the bands of polystyrene become relatively much more prominent. As the proportion of copolymerised butadiene is decreased, the bands at 10·1 μm, 10·3 μm and 11·0 μm due to this constituent decrease in intensity until at the 5% level (Spectrum 8.8) only the 10·3 μm band of copolymerised butadiene can be identified with certainty. However, since there are no other common resins in which the 10·3 μm band is so prominent, even 1% of butadiene copolymerised with styrene can be recognised from the spectrum, although polystyrene itself has a weak band close to this wavelength. The aromatic phosphate plasticisers, such as tritolyl phosphate, have a very prominent absorption band at 10·3 μm, and polystyrene containing a small amount of such a plasticiser has an infra-red spectrum which is remarkably similar to that of a butadiene/styrene copolymer containing a minor amount of combined butadiene. This difficulty may be overcome by comparing the spectra of the resin before and after ether extraction, and measuring the spectrum of the ether extract, since any phosphate plasticiser will be concentrated in the extract.

'Carboxylated' butadiene/styrene copolymers are strictly terpolymers of butadiene, styrene, and an unsaturated acid such as maleic acid. The spectrum of such a composition (Spectrum 8.9) shows the characteristics of a butadiene/styrene copolymer, as previously described, and in addition an absorption band near 5·9 μm due to the acidic carbonyl group. Since the concentration of carboxylic acid present in such copolymers is small in most cases it will not be possible to ascertain from the spectrum which acid has been used. When only a weak carbonyl band is observed there is danger of confusing a 'carboxylated' copolymer with an oxidised specimen of butadiene/styrene copolymer. However, many of the acids used readily form anhydrides, and a small amount of anhydride is often evident in the spectra of true 'carboxylated' copolymers as weak carbonyl bands at about 5·5 μm and 5·65 μm.

If the infra-red spectrum is to be used as a test to decide whether a resin is 'carboxylated' it will be important to prepare a sample for examination under conditions under which oxidation will not occur, e.g. to prepare a film, either from latex or solution, at room temperature, and preferably in the absence of oxygen, to avoid oxidation. Under some circumstances the acid may be present as carboxylate ion. This will occur if the film examined is prepared from an alkaline latex, or in some cases if a film is prepared on the surface of a potassium bromide or sodium chloride plate. Silver chloride appears to be a better substrate to use for supported films. If the acid is present as carboxylate ion, the 5·9 μm band disappears, and is replaced by a broader band near 6·4 μm. Whether the resin contains free carboxylic acid or carboxylate ion, it is necessary to ensure that this is combined in the polymer rather than present as an additive, by solvent extraction of the sample prior to infra-red examination.

Quantitative estimation of the concentration of carboxyl groups in 'carboxylated' butadiene/styrene copolymer is straightforward, using the 5·9 μm carbonyl band. The unknown sample is hot pressed into a film of such a thickness that the spectrum shows a band of convenient intensity (absorbance between 0·1 and 0·5) at 5·9 μm. The absorbance at 5·9 μm may not be entirely due to carbonyl groups as polystyrene has some weak bands in this region. The correction to be applied is ascertained by measuring the spectrum of a straight butadiene/styrene copolymer of approximately the same composition, and of a similar and known thickness.

445

The molar extinction coefficient for a carboxylic acid at 5·9 μm in an aliphatic hydrocarbon solvent is then measured under the same conditions. Having measured the thickness of the unknown sample with a micrometer, the molar concentration of carboxyl groups may be calculated. If the actual acid used to make the sample is known, the weight percentage present may be calculated.

QUANTITATIVE ESTIMATION OF THE PROPORTIONS OF BUTADIENE AND STYRENE IN BUTADIENE/STYRENE COPOLYMERS

The method of Miller and Willis[12] is based upon the measurement of absorption bands at approx. 2·2 μm and 2·4 μm arising from aromatic and aliphatic C—H groups respectively. Calibration samples of known compositions are specially prepared taking polymerisation to completion.

Alternatively, soluble polymers may be examined by n.m.r. spectroscopy (p. 456) to determine their composition, and these samples may be used as calibration standards for the infra-red measurements. Since the infra-red measurements are made on solid films, they are equally applicable to soluble and insoluble (cross-linked) resins.

The infra-red method suggested above is applicable only to resins containing more than about 10% combined butadiene. For lower butadiene contents, measurement of the ratio of the absorbance of the bands at 10·3 μm to that at 6·3 μm, based on Spectrum 8.8 as a standard, will suffice for most purposes.

BUTADIENE/ACRYLONITRILE COPOLYMERS

The absorption spectrum of polyacrylonitrile (Spectrum 3.16) is weak compared with that of polybutadiene; on the other hand the nitrile absorption band near 4·4 μm lies in a region in which polybutadiene is relatively transparent, and there is therefore no difficulty in ascertaining the presence of combined acrylonitrile in these copolymers from the infra-red spectrum (Spectrum 8.10). Some wavelength modification of the strong bands of copolymerised butadiene is also evident, the 11·0 μm band appearing near 10·85 μm. Since the identification of combined acrylonitrile depends only upon the appearance of the nitrile absorption band, it is to be presumed that copolymers of butadiene with other nitriles, for example methacrylonitrile, would be difficult to distinguish from butadiene/acrylonitrile copolymers on the basis of the infra-red spectrum.

To obtain an indication of the combined acrylonitrile content of a resin from its infra-red spectrum, the ratio of the absorbances at 10·3 μm and 4·4 μm may be measured since

$$\frac{\text{Absorbance at } 10\cdot3 \ \mu m}{\text{Absorbance at } 4\cdot4 \ \mu m} = k \times \frac{\text{w\% combined butadiene}}{\text{w\% combined acrylonitrile}}$$

The value of k is determined by measuring a known copolymer (e.g. Spectrum 8.10). It must be emphasised, however, that the result so obtained is an indication only, and it may well be preferable to rely upon the chemically determined nitrogen content. It appears that the infra-red measurement is more reliable in solution than on a solid film, and a quantitative method for the analysis of these copolymers in solution from the infra-red spectrum has been reported[13].

As with the butadiene/styrene system, 'carboxylated' butadiene/acrylonitrile copolymers may be encountered. The observations on 'carboxylated' butadiene/styrene (p. 445) are equally applicable to these resins.

TERPOLYMERS AND BLENDS OF BUTADIENE, STYRENE AND ACRYLONITRILE

The spectra of these materials show the characteristics of polybutadiene, polystyrene and polyacrylonitrile. Spectrum 8.11, for example, is the spectrum of a blend of butadiene/styrene and butadiene/acrylonitrile copolymers. However, the spectrum is virtually indistinguishable from that of a terpolymer, or a blend of corresponding proportions of styrene/acrylonitrile copolymer with butadiene/styrene copolymer, and it is not possible to distinguish the many possible blends of copolymers from the terpolymers of butadiene, styrene and acrylonitrile purely from the infra-red spectrum of the complete composition.

Further evidence may be obtained in many cases by solvent extraction of the resin with, for example, acetone, followed by infra-red examination of the soluble and insoluble material. While it is unreasonable to expect a complete separation of such a polymer mixture, some indication will usually be obtained as to the nature of the composition.

Quantitative estimation of the amount of butadiene, styrene and acrylonitrile present in a blend or terpolymer seems to be reasonably satisfactory, measuring the absorbance at $4\cdot4 \ \mu m$ for acrylonitrile, $6\cdot25 \ \mu m$ for styrene, and $10\cdot3 \ \mu m$ for butadiene. The thickness of the film specimen used for the measurement is about $0\cdot002$ in ($0\cdot05$ mm), but does not need to be known accurately:

$$\frac{\text{Absorbance at } 4\cdot4 \ \mu m}{\text{Absorbance at } 6\cdot25 \ \mu m} = k' \times \frac{\text{w\% acrylonitrile}}{\text{w\% styrene}}$$

$$\frac{\text{Absorbance at } 10\cdot3 \ \mu m}{\text{Absorbance at } 6\cdot25 \ \mu m} = k'' \times \frac{\text{w\% butadiene}}{\text{w\% styrene}}$$

The values of k' and k'' are determined by measuring the absorbances on known copolymers (e.g. Spectra 6.39, 8.7 and 8.11).

OTHER COPOLYMERS CONTAINING BUTADIENE

The infra-red spectrum of butadiene/vinyl toluene copolymer (Ref. 14, Figure 297) is similar in general features to that of a butadiene/styrene copolymer, except in the 12 μm to 13 μm region, where the strong band at 13·3 μm due to the mono-substituted benzene nucleus is replaced by a strong pair of bands at about 12·3 μm and 12·8 μm. These bands arise from *para*- and *meta*-substituted aromatic nuclei. It is, on the other hand, very difficult to see any evidence of the presence of combined α-methyl styrene in the spectrum of the terpolymer styrene/α-methyl styrene/butadiene, although the band at 10·3 μm shows the presence of combined butadiene (Hummel[9] Figure 507).

The copolymer of butadiene with 2-methyl-5-vinyl pyridine (Hummel Figure 469) shows the expected bands in its spectrum due to copolymerised butadiene, together with clear and prominent bands at 12·1 μm and 13·6 μm, attributable to copolymerised 2-methyl-5-vinyl pyridine. The terpolymer butadiene/acrylonitrile/2-methyl-5-vinyl pyridine (*Sadtler Commercial Spectrum*[15] D2006) has similar characteristics, but shows in addition the nitrile band at about 4·4 μm.

Butadiene/methacrylate copolymers are considered in Chapter 4.

POLYMER BLENDS CONTAINING BUTADIENE RUBBERS

Butadiene/styrene copolymer blended with butadiene/methyl methacrylate copolymer is sometimes encountered (Spectrum 8.12). This spectrum shows the characteristics of polybutadiene and polystyrene, but a modified polymethyl methacrylate pattern is seen, as described in the discussion on this material (p. 256). This enables such materials to be distinguished by means of their infra-red spectra from blends of butadiene/styrene copolymer with polymethyl methacrylate, in the spectra of which the polymethyl methacrylate pattern appears unmodified. The latter compositions may be encountered as impact-modified acrylics. Similarly, impact-modified P.V.C. may consist of blends of P.V.C. with butadiene rubbers. Such compositions have been discussed in Chapter 3.

Blends of butadiene rubbers, particularly butadiene/acrylonitrile copolymers, with thermosetting resins, especially phenol-formaldehyde condensate, may be encountered as mouldings, and in coating compositions and adhesives. The infra-red spectrum of this blend (Spectrum 8.13) shows the expected polybutadiene and polyacrylonitrile structure, together with bands of moderate intensity near 3·0 μm and 8·0 μm arising from phenol-formaldehyde condensate. Bentley and Rappaport show[16] the spectra of pyrolysates of such blends, and suggest a method of determining the blend proportions from the spectra. This procedure could be particularly useful in the examination of heavily filled compositions, but the quantitative analysis of such blends by transmission spectroscopy is straightforward, provided suitable reference samples are available. The bands at 3·0 μm, 4·4 μm and 10·3 μm can be measured for hydroxyl (phenol-formaldehyde), nitrile (copolymerised acrylonitrile) and chain unsaturation (copolymerised butadiene) respectively.

POLYISOPRENE (POLY-2-METHYL 1 : 3-BUTADIENE)

Structural isomers of polyisoprene can exist which are analogous to those of polybutadiene, but one more form, that made by addition across the 3 : 4-double bond, is possible (see Figure 8.8).

By the use of stereospecific catalysts, the *cis*-1 : 4-polymer may be prepared in a state sufficiently pure to be crystallisable. This form occurs

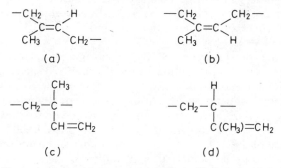

Fig. 8.8.   Structural units in polyisoprene: (a) trans-1 : 4; (b) cis-1 : 4; (c) 1 : 2; (d) 3 : 4

naturally as hevea rubber, or natural rubber. The *trans*-1 : 4-polymer also occurs naturally, as gutta percha, or balata, and this can also be crystalline.

In the spectrum of polybutadiene the bands due to the motions of the residual hydrogen atoms on the C=C group are a very prominent feature, and allow the *trans*- and *cis*-1 : 4-polymers to be distinguished. However, the corresponding forms of polyisoprene, because they contain the tri-substituted double bond, cannot be distinguished in this way, both showing absorption at 12·0 μm. Thus, in the amorphous state, the spectra of *cis*- and *trans*-1 : 4-polyisoprene are almost identical (Spectra 8.14 and 8.15) showing the usual C—H structure, including a prominent methyl group band at 7·3 μm, the tri-substituted unsaturation bands at 12·0 μm and 6·1 μm and many rather weak bands in the rest of the spectrum. In fact, although the 12·0 μm band is not modified in wavelength by the *trans/cis* isomerism, the band is considerably weaker in the *trans*- than in the *cis*-isomer, but this effect may pass unnoticed in a

449

qualitative run, particularly if the sample is impure. There is also a small spectral difference between 8 μm and 9 μm; the *cis*-isomer absorbs at 8·85 μm, while the corresponding *trans*-band occurs at 8·58 μm.

In practice, the two forms can usually be distinguished if the sample is reasonably pure, because the *trans*-isomer crystallises readily at room temperature, while the *cis*-isomer crystallises only on stretching at this temperature. Thus, as normally encountered, gutta percha and balata are crystalline, introducing additional sharp bands in the spectra of these substances. The situation here is complicated by the existence of two crystalline forms, designated α and β, which have some different features in their spectra (Spectra 8.16 and 8.17), although the basic —CH— saturated and unsaturated structure is present in all cases. The additional bands in the α form occur at 11·3 μm, 11·6 μm and 12·5 μm; these in the β form are at 11·4 μm, 12·6 μm and 13·3 μm.

The spectra of other forms of synthetic polyisoprene are given by Hummel[9] (Figure 479) and Richardson and Sacher[17]. The form of the polymer produced by sodium catalysis, and that produced under high pressure, contain significant amounts of the 3 : 4-isomer, as evidenced by a strong band in the spectrum at 11·3 μm. Emulsion polymerisation produces a resin containing significant amounts of 1 : 2-, 3 : 4-, and 1 : 4-addition polymer.

Quantitative determination of the proportion of these different unsaturated structures in polyisoprene from the infra-red spectrum has been the subject of a good deal of controversy, although the method of Binder and Ranshaw[18] appears to be satisfactory except in respect of the *cis*- and *trans*-1 : 4-ratio. The preferred infra-red method here is that of Corish[19] who measures *cis*-1 : 4 content from the absorption band at 2·46 μm. Measurement of the *cis/trans* ratio from the n.m.r. spectrum would appear to be much more satisfactory (p. 459) although the spectrum can only be recorded if the polymer is soluble.

POLYISOBUTENE

This hydrocarbon rubber is unusual in containing no significant concentration of unsaturated groups. The spectrum (Spectrum 8.18) is nevertheless highly distinctive, showing a strong sharp band at 8·2 μm thought to be due to the group

$$H_3C-\overset{|}{\underset{|}{C}}-CH_3$$

The geminal dimethyl group also produces the characteristic split methyl group band at 7·22 μm and 7·32 μm.

Butyl rubber, a copolymer of isobutene with a small amount of diene (butadiene or isoprene), has a spectrum which cannot be distinguished from that of polyisobutene itself.

Blends of polyisobutene (or butyl rubber) with polythene and with polypropylene are of commercial importance. These are described in Chapter 6.

RUBBERS CONTAINING CHLORINE

As stated in Chapter 2, the C—Cl group produces bands of reasonably high intensity, but very variable in position; hence it is necessary to approach the spectra of chlorine-containing compounds in terms of matching, rather than chemical group analysis. There are comparatively few chlorine-containing rubbers, and all have spectra which enable them to be identified without difficulty. However, confusion may arise because other chlorinated resins, notably polyvinyl chloride, may be encountered as compositions which have a rubber-like appearance.

Poly-2-chloro-butadiene (polychloroprene) (Spectrum 8.19) is distinctive in having a sharp and strong absorption band in the spectrum near 6 μm. This presumably is the C=C stretching vibration of the group. Apart from the usual C—H bands the rest of this spectrum has rather broad bands of complex shape, and there are minor variations in the spectra of different samples. These arise because 2-chloro-butadiene, as with isoprene, can polymerise by 1 : 2-, 3 : 4- and 1 : 4-addition, and in the case of 1 : 4-addition *trans*- and *cis*-isomers can be formed. Furthermore, Ferguson[20] has shown from n.m.r. measurements that the resin made with a free-radical catalyst contains also head-to-head and tail-to-tail monomer sequences. In fact the resin as commercially available is predominantly the *trans*-1 : 4-isomer, but this introduces an additional difficulty because this isomer can crystallise leading again to some differences in the spectrum. A full account of these effects is given by Maynard and Mochel[21] and Mochel and Hall[22]. The former paper is concerned with crystalline effects, and the latter with the estimation of the proportions of 1 : 2-, 3 : 4- and also *cis*-1 : 4-isomers in various samples of poly-2-chloro-butadiene prepared under different conditions. In practice, owing to the predominance of the *trans*-1 : 4-form, there is no difficulty in identifying this resin from its spectrum. The spectra of the substantially pure *cis*- and *trans*-1 : 4 forms have been given by Ferguson[20].

A further group of chlorine-containing rubbers is based on natural rubber. In our experience chlorinated rubber and rubber hydrochloride (Spectra 8.20 and 8.21) are commonly encountered.

In both cases the spectrum shows some similarity to that of natural rubber with prominent C—H bands at 3·4 μm, 6·9 μm and 7·3 μm; chlorinated rubber shows a pronounced doubling of the 6·9 μm band, and a similar but less marked effect is evident in the rubber hydrochloride spectrum. This feature presumably arises, as in the polyvinyl chloride

451

spectrum, from the CH_2 group adjacent to the CHCl group. The broad bands of chlorinated rubber at 8 μm and of rubber hydrochloride at 8·3 μm presumably arise from the CHCl group. The absorption band of rubber at 12 μm which is due to the unsaturated group largely disappears in the spectra of these products.

Other than the absorption bands mentioned above, the spectra of these polymers are rather broad and indefinite in character, and small spectral differences occur between different samples. This is probably due to the complex nature of the products which, by their method of manufacture, may be expected to have an irregular and not necessarily reproducible structure. Checkland and Davison[23] have examined the infra-red spectrum of rubber hydrochloride and have shown that minor changes may arise through crystallisation, while Salomon *et al.*[24] have examined the changes occurring in the spectrum of natural rubber with varying degrees of chlorination.

RUBBERS CONTAINING SULPHUR

There is a small group of rubbers which contain relatively large amounts of sulphur. Although the S=O group, which shows extremely strong and highly characteristic bands in the infra-red spectrum, is not present in the sulphur-containing rubbers considered in this section, nevertheless these materials have infra-red spectra which are easily recognised and therefore very useful in qualitative identification. These rubbers are of two general kinds (Hummel[9] p. 175).

Thiokol A is an example of a poly(ethylene polysulphide) rubber. The infra-red spectrum (Spectrum 8.22) is useful and distinctive, with prominent sharp bands between 7·0 μm and 8·5 μm. These bands evidently arise from vibrations of the CH_2 group adjacent to sulphur.

The other group is poly(diethyl formal polysulphide) rubbers. The group —O—CH_2—O— leads to a complex C—O pattern in the spectrum (Spectra 8.23) where the most prominent feature is a series of strong bands between 8 μm and 10 μm similar to those observed in polyvinyl formal (Spectrum 3.12).

The spectra of a number of these polysulphide rubbers appear in the literature, e.g. Hummel Figures 1066 to 1074, but all the spectra are similar to the two given here.

POLYETHER URETHANES

These resins, which at one time were almost always encountered as foamed rubbers (e.g. in upholstery applications) are now found also as paints and coatings. They are similar to the polyester urethanes in that both materials are made by the chain lengthening, and often also cross-linking, of a low molecular weight hydroxyl-terminated resin (i.e. the polyether or polyester) with a di- or poly-isocyanate. The polyethers used are usually aliphatic, but are often copolymers (e.g. ethylene/propylene oxide copolymer). According to Dono and Shelton[25], the reaction may be described as follows:

1. *Chain-lengthening steps*

$$NCO—R'—NCO+HO—R''—OH+NCO—R'—NCO$$

$$→NCO—R'—\underset{\underset{H}{|}\,\,\underset{O}{\|}}{N—C}—O—R''—O—\underset{\underset{O}{\|}\,\,\underset{H}{|}}{C—N}—R'—NCO$$

The terminal —NCO groups may react with further molecules of the type HO—R''—OH to produce a long linear polymer, or may alternatively be broken down by water to produce terminal amino groups.

$$—NCO+H_2O→ —\underset{\underset{H}{|}\,\,\underset{O}{\|}}{N—C}—OH→ —NH_2+CO_2$$

The carbon dioxide evolved helps to form the foam in a foamed composition, silicone foam stabilisers being added. The terminal amino groups, produced as above, react with further di-isocyanate with more chain extension and the production of urea bridges.

$$—NH_2+NCO—R'—NCO+NH_2—$$

$$→ —NH—\underset{\underset{O}{\|}}{C}—NH—R'—NH—\underset{\underset{O}{\|}}{C}—NH—$$

2. *Cross-linking reaction*

The hydrogen atoms in the urea bridges can react with the isocyanate group, thus cross-linking the resin (see formulae on p. 454).

The catalysts used in the process are amines and various stannous salts of organic acids. The spectra of all polyether urethanes are rather similar and, since the process of formation is complex with many possible reactions, the spectra of different samples of the same ether/isocyanate resin may show some differences, particularly as, under some conditions, the polyether chains can crystallise, introducing further minor differences in the spectra. However, the spectrum of this class of compound is easily recognised. The strongest band in the spectrum is invariably that due to the aliphatic ether group at 9 μm. One or more bands in the 12 μm to 15 μm region arise from the aromatic nucleus of the di-isocyanate used; further bands appear from the urethane groups, with weak bands from the small number of urea groups present.

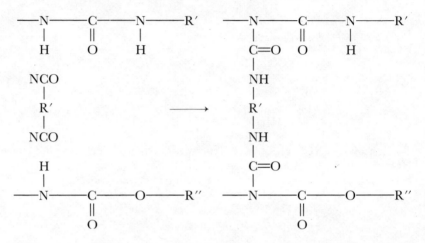

Spectrum 8.24, that of polyethylene oxide/1 : 5-naphthalene di-isocyanate, is typical of these compounds. The ether band at 9 μm dominates the spectrum; the band at 8·2 μm also originates from the polyethylene oxide structure. Naphthalene di-isocyanate residues are evident from the usual substitution band at 12·3 μm. Bands from the urethane group are the N—H stretching band (∼3·0 μm), the C=O (or Amide I) band at 5·9 μm and the N—H (or Amide II) band at 6·5 μm. The carbonyl band at 5·8 μm may also be a urethane carbonyl band[26] although it seems possible that it is due to ester groups formed by oxidation. The urea structure is shown by a weak C=O band at 6·1 μm.

It is not unusual to observe a band near 4·4 μm in the infra-red spectra of these resins due to unreacted isocyanate groups, and indeed this band is so strong that this is a very sensitive test for the isocyanate group.

In a paper giving infra-red spectroscopic data on a number of polyether urethanes, Corish[26] points out that it is possible in many cases to identify, from the infra-red spectrum of the polymer, the isocyanate on which the resin is based. On the other hand it is unwise to attempt to identify the polyether from the infra-red spectrum as the spectra of aliphatic ethers are all very similar in their features. It is preferred, whenever the resin is soluble, to identify the polyether from the n.m.r. spectrum. When this is not possible, it will be better to hydrolyse the resin, following the method suggested by Corish. From the hydrolysis product the diamine corresponding to the original di-isocyanate may be recovered for identification, together with a sample of the polyether, which is more readily identified as a pure substance than in the combined form in the original compound.

N.M.R. SPECTRA OF RUBBER-LIKE RESINS

The application of n.m.r. spectroscopy to commercial samples of rubber-like resins is limited because many are not soluble enough for examination. However, when samples of any natural or synthetic rubber are sufficiently soluble to allow useful n.m.r. spectra to be measured, it will be possible to obtain information on both their structure and composition.

A number of common rubbers are homopolymers of aliphatic diene monomers, in which the chief structural interest is in the arrangement of groups about the $C=C$ bond. A number of these structural isomers can be distinguished from the n.m.r. spectrum and, if this evidence is considered alongside that from the infra-red spectrum, it may be possible in many cases to determine quantitatively the proportions of the different $C=C$ structures present. Similar observations can be made on a number of the copolymers of diene monomers, and in addition it is possible with the majority of such copolymers to determine the proportions of the combined monomers.

The other important class of rubbers which has been extensively studied by n.m.r. is the polyurethanes, and it may well be possible to determine from the spectrum the nature of the polyether or polyester on which the resin is based. This is of particular interest when the base is a copolymer, since its composition can usually be determined, both qualitatively and quantitatively, from the spectrum. Observations can be offered in many polyurethanes on the nature of the isocyanate used. On the other hand, if a polyurethane is made from a high molecular weight base polymer, the resonances of protons associated with the urethane groups may be so weak that they are not detected in the normal spectrum (i.e. without spectrum accumulation) especially if the solvent used obscures the resonances of the aryl protons. This problem is more acute with polyester urethanes because, while the resonance of the CH_2 protons of the urethane group $-N-CO-O-CH_2-$ at about $6 \cdot 0\ \tau$ lies well clear of any resonance of a polyether, it falls in the same range as the resonance of the CH_2 protons of the group $R-CO-O-CH_2-$ which is present in polyesters. Thus there may well be difficulty in distinguishing the n.m.r. spectrum of a polyester urethane from that of the polyester resin on which it is based.

POLYBUTADIENE

The most interesting resonances in the n.m.r. spectrum of polybutadiene are those in the $5\ \tau$ region associated with protons around the $C=C$ groups.

Nature of Polymer   Structure

1:4-addition. *Cis* and *trans* chain unsaturation

$$-CH_2-CH_\alpha=CH_\alpha-CH_2-$$

1:2-addition. Terminal or vinyl unsaturation $-CH_2-CH-$

$$\begin{array}{c} | \\ C=C(H_\beta)_2 \\ | \\ H_\alpha \end{array}$$

$$H_\alpha = 4 \cdot 5 \ \tau \text{ to } 4 \cdot 7 \ \tau; \ H_\beta = 4 \cdot 9 \ \tau \text{ to } 5 \cdot 2 \ \tau$$

Durbetaki and Miles[27] give the n.m.r. spectra of polybutadienes prepared by different methods, and it is satisfactory to note that there is qualitative agreement between their conclusions and the conclusions from the infra-red spectra on the compositions of the products made by sodium catalysis and by free-radical catalysis of butadiene. The infra-red spectra (Spectra 8.1 and 8.2) show that the 11·0 μm band due to vinyl (terminal) unsaturation is very strong in the spectrum of the sodium-catalysed product, and comparatively weak in the spectrum of the product prepared by free-radical catalysis. Correspondingly, Durbetaki and Miles[27] show that the resonance at 5·0 τ is prominent in the n.m.r. spectrum of the sodium-catalysed product, while that at 4·7 τ is the stronger in the spectrum of the product made by free-radical catalysis.

Senn[28] shows how the relative concentrations of chain and terminal (vinyl) unsaturation may be determined from the n.m.r. spectrum, although *trans*- and *cis*-chain unsaturation cannot be distinguished. As we have seen above (p. 441), it is a relatively simple matter to determine

$$\frac{\text{Concentration of } trans \text{ chain unsaturation}}{\text{Concentration of terminal (vinyl) unsaturation}}$$

from the infra-red spectrum, and if this measurement is combined with the ratio

$$\frac{\text{Concentration of total chain unsaturation}}{\text{Concentration of terminal (vinyl) unsaturation}}$$

from the n.m.r. spectrum it is evident that it is possible to give a most satisfactory description of the composition of a sample of polybutadiene.

BUTADIENE/STYRENE COPOLYMERS

A number of workers[28, 29] have examined the n.m.r. spectra of butadiene/styrene copolymers. In general the spectra are similar to Figure 8.9. The resonances near 5 τ originate from protons around the C=C group of the combined butadiene in the copolymer chain, as detailed above for

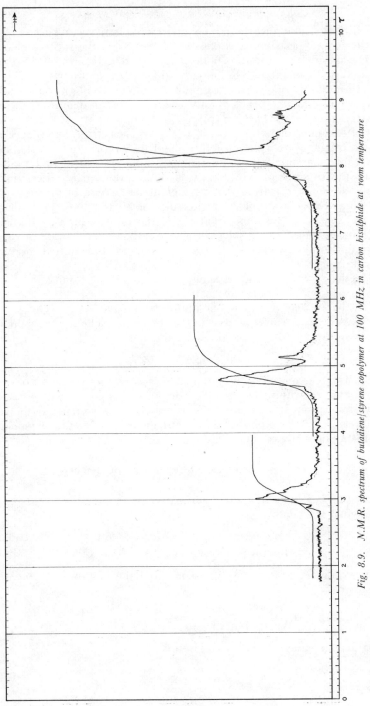

Fig. 8.9. N.M.R. spectrum of butadiene/styrene copolymer at 100 MHz in carbon bisulphide at room temperature

polybutadiene. Because there is no interference from styrene in this region of the n.m.r. spectrum, it is possible to determine the ratio of chain to terminal (vinyl) unsaturation of the combined butadiene in the copolymer from these resonances. This is very difficult to do from the infra-red spectrum of the copolymer because of overlap of the absorption bands of polystyrene with the bands of polybutadiene used to determine the concentration of unsaturated groups.

Furthermore, the resonances in the $3\,\tau$ region of the n.m.r. spectrum which originate from the protons around the aryl ring of the styrene in the copolymer are free of interference from resonances of butadiene. Thus the situation is particularly favourable for quantitative analysis from the n.m.r. spectrum, having measured the area of the resonances due to aryl protons in the $3\,\tau$ region and the area of all other resonances in the spectrum (assuming that the spectrum has been measured in solution in a proton-free solvent such as carbon tetrachloride or carbon bisulphide, otherwise solvent resonances must be omitted from the integral).

Let the area of the aryl resonances ($3\,\tau$ region) be A units

Let the area of all other resonances be R units

Let the area equivalent to one proton be a units

Then there are A/a aryl protons in the copolymer. The styrene monomer unit requires 5 aryl protons, there are therefore

$A/5a$ styrene units in the copolymer

Each styrene unit also contains 3 alkyl protons, the resonances of which are contained within the area R. This area R is equivalent in total to R/a protons and it contains $3A/5a$ protons due to styrene. It therefore contains $(R-3A/5)/a$ protons due to butadiene. The butadiene monomer unit contains 6 protons. Hence there must be

$$\frac{5R-3A}{30a} \text{ butadiene units in the copolymer}$$

From this it follows that the ratio of styrene to butadiene units in the copolymer is given by

$$\frac{\text{Styrene}}{\text{Butadiene}} = \frac{A}{5a} \times \frac{30a}{5R-3A} = \frac{6A}{5R-3A}$$

The gram molecular weight of styrene is 104 and that of butadiene is 54. Therefore

$$\frac{\text{Weight styrene}}{\text{Weight butadiene}} = \frac{104 \times 6A}{54(5R-3A)} = X$$

And the percentage of styrene is given by

$$\% \text{ w/w styrene} = \frac{100X}{1+X}$$

The aryl resonance at 3 τ as shown in Figure 8.9 is that seen in random copolymers which do not contain extended runs of styrene units. According to Matsuo *et al.*[29] when runs of ten or more styrene units are present, a further resonance is seen at 3·5 τ, and this effect is illustrated in a series of spectra in their paper. There are evidently differences also in the appearances of the resonances near 8 τ in the spectra of random and block copolymers of butadiene and styrene. However, the spectra of block copolymers cannot be distinguished from those of blends of the two polymers.

POLYISOPRENE

Although infra-red spectroscopy has been very useful in the study of polybutadiene (p. 441) it is of less value in the examination of polyisoprene because it does not offer a simple means of differentiating the *trans*- and *cis*-1 : 4-structures which are the predominant structures in the naturally occurring forms of this polymer. However Chen[30] has shown that in the n.m.r. spectrum the resonances of the methyl protons in the *cis*- and *trans*-1 : 4-structures are well separated, especially if the spectrum of a benzene solution is recorded, when the resonances occur at 8·21 τ and 8·35 τ respectively. Unfortunately measurement of the *trans/cis* ratio by measuring the relative areas of these two resonances is complicated by the resonance of the methyl group of the 3 : 4-structure which is coincident with that of the *trans*-1 : 4-group. This difficulty is resolved in practice by the analysis of the resonances in the 5 τ region, which arise from the protons directly attached to the C=C group. The following structures may be distinguished (see Table 8.1).

Table 8.1. Polyisoprene unit structures

Nature of unit	Structure	Position of resonance
1 : 2-addition	—CH$_2$—C(CH$_3$)— | C$\underline{H}$=CH$_2$	4·5 τ
	—CH$_2$—C(CH$_3$)— | CH=C$\underline{H}_2$	5·1 τ
cis- and *trans*-1 : 4-addition	—CH$_2$—C$\underline{H}$=C(CH$_3$)—CH$_2$—	5·0 τ
3 : 4 addition	—CH$_2$—CH— | C=C$\underline{H}_2$ | CH$_3$	5·3 τ

The resonances at $5\cdot0\,\tau$ and $5\cdot1\,\tau$ are not usually resolved, but the resonance at $5\cdot3\,\tau$, due to the 3:4-unit, is well separated from the other structure and is therefore well suited to the determination of the 3:4-isomer content of polyisoprene[31]. When the 3:4-content has been determined in this way, it is then possible to determine the *trans*-1:4-content from the methyl resonance at $8\cdot35\,\tau$, which, as explained above, arises from both the *trans*-1:4- and the 3:4-units.

The most satisfactory resonance for the measurement of the 1:2-content of polyisoprene is that at $9\cdot1\,\tau$ due to the methyl group, and Chen[30] achieved satisfactory analyses of the 1:2-contents of known mixtures in this way. In many samples, however, the 1:2-content is very low, and for this reason determination of the 1:2-content from the infra-red spectrum is often preferred, because of the very high intensity of the absorption band at $11\cdot0\,\mu$m due to the vinyl group.

Copolymers of polyisoprene are not commonly encountered. Chen[30] proposes a method for the analysis of copolymers of butadiene and isoprene and no doubt other copolymers could be examined quantitatively in a similar manner.

N.M.R. SPECTRA OF POLYURETHANE RUBBERS

The general chemical nature of these materials has been described else-where (p. 452). The more interesting of the polyurethanes for n.m.r. examination are those based on polyethers. Whereas the protons on the carbon atom α to the oxygen atom in the ether and urethane are reliably distinguished in the n.m.r. spectrum, this is not the case with the ester and urethane groups. Thus the n.m.r. spectrum gives more direct evidence of the structure of the polyether urethanes. This is fortunate, because while there is an effective alternative procedure for the identification of polyester urethanes following the method suggested by Corish[26], there is no corresponding method available for polyether urethanes.

Figure 8.10 is representative of the n.m.r. spectra of polyether urethanes, and is that of the product based on a polymer of tetramethylene glycol, chain lengthened with toluene di-isocyanate.

Protons on the carbon atom α to the ether oxygen atom show a resonance at about $6\cdot5\,\tau$, while those on a β carbon show up near $8\cdot3\,\tau$. The resonance near $5\cdot8\,\tau$ arises from protons on the carbon atom α to the oxygen of the urethane group; this resonance is well separated from that of the α-methylene group of the ether. The di-isocyanate residue contains a methyl group on the aromatic ring, and the resonance of these protons appears near $7\cdot8\,\tau$. The multiple structure in this position presumably arises because the toluene di-isocyanate used to prepare the resin is a mixture of isomers, and the τ-value of the methyl resonance is somewhat different in the different structures. When *o*-dichlorobenzene is used as solvent it is not possible to observe the resonances of the aryl protons of

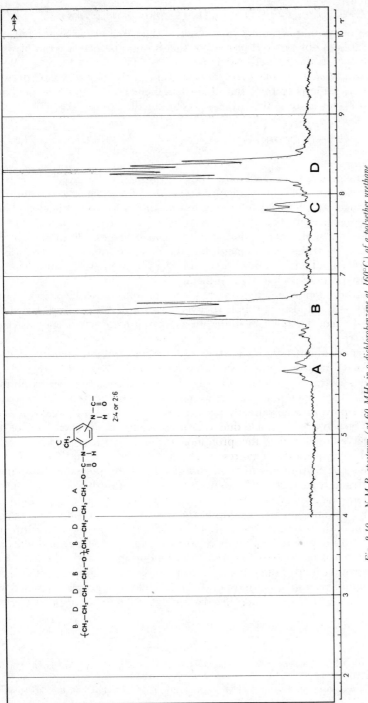

Fig. 8.10. *N.M.R. spectrum (at 60 MHz in o-dichlorobenzene at 160°C) of a polyether urethane based on a polymer of tetramethylene glycol, chain lengthened with toluene di-isocyanate*

the di-isocyanate. Brame *et al.*[32] used arsenic trichloride as solvent and were able to observe aryl protons of the di-isocyanate as a series of weak resonances in the 2 τ to 3 τ region.

Brame *et al.* have observed the spectrum of the polyurethane derived from the same polyether, but chain lengthened with diphenyl methane 4 : 4 -di-isocyanate so that the resin contains the group

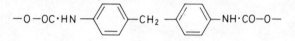

The resonance of the protons of the central methylene group appears at 6·1 τ, which may well be difficult to observe as it lies between the multiple resonances at 5·8 τ and 6·5 τ referred to above. On the other hand the aryl resonance at about 2·8 τ is relatively prominent. In order to avoid the use of arsenic trichloride as solvent it may be possible to observe the spectrum in trifluoroacetic acid, but the material appears to degrade in this solvent. Thus the detection and identification of the different di-isocyanates may present some difficulty.

On the other hand, the n.m.r. spectrum appears to be a most satisfactory way of distinguishing between the different polyethers which may be used. Thus, for example, in a polyurethane based on propylene glycol (1 : 2) the methyl protons give rise to a prominent doublet resonance at about 9·0 τ[33]. Further observations on the identification of polyethers from their n.m.r. spectra appear on p. 528.

The spectra of a number of polyester urethanes are given by Brame *et al.*[32] The superficial similarity of these spectra to those of the constituent polyesters has already been noted. Apart from the weak (or very weak) structure at low field due to protons around the benzene ring of the di-isocryanate residue, the prominent resonance near 6 τ arises from —CH$_2$O— protons. The resonance near 7·8 τ due to CH$_2$—C=O protons is a feature absent from the spectra of the polyether urethanes. The structure at high field (above 8 τ) arises from other alkyl groups of either the acid or glycol residues in the polyester. By making an appropriate allowance for the overlap of the resonances near 4·0 τ, and taking account of those resonances above 6 τ which arise from alkyl substitutents on the di-isocyanate residue, it is possible from the shape, position and relative areas of the resonances to determine in many cases the nature of the polyester used. Brame *et al.*[32] give a number of examples of this form of qualitative and quantitative analysis, including polyesters based on diethylene glycol which are, in effect, polyurethanes based on polyester/ethers.

REFERENCES

1. WAKE, W. C., *Analysis of Rubbers and Rubber-like Polymers,* 2nd edn, Maclaren, London (1969)
2. *The Services Rubber Investigations,* Memo No. U9A, H.M.S.O., London, 285 (1954)

3. MEARDON, J. I., *RAPRA Rep.* No. 143 (1965)
4. TAUBINGER, R. P., *Analyst, Lond.*, **94,** 628 (1969)
5. CRITCHFIELD, F. E., and JOHNSON, J. B., *Analyt. Chem.*, **28,** 73 (1956)
6. TAUBINGER, R. P., *Analyst, Lond.*, **94,** 634 (1969)
7. KENNEDY, T., and TAUBINGER, R. P., *Analyst, Lond.*, **94,** 625 (1969)
8. JEFFS, A. R., *Analyst, Lond.*, **94,** 249 (1969)
9. HUMMEL, D. O., *Infra-Red Analysis of Polymers, Resins and Additives, An Atlas,* **I,** Pt II, Wiley Interscience (1969)
10. NATTA, G., *Rubb. Plast. Age,* **38,** 495 (1957)
11. SILAS, R. S., YATES, J., and THORNTON, V., *Analyt. Chem.*, **31,** 529 (1959)
12. MILLER, R. G. J., and WILLIS, H. A., *J. appl. Chem., Lond.*, **6,** 385 (1956)
13. ADAMS, M. L., and SERANN, M. H., *Off. Dig. Fed. Socs Paint Technol.*, **30,** 646 (1958)
14. *Infra-Red Spectroscopy. Its Use in the Coatings Industry.* Federation of Societies for Paint Technology, Philadelphia (1969)
15. *Sadtler Commercial Spectra Series D,* The Sadtler Research Laboratories, Philadelphia 2
16. BENTLEY, F. F., and RAPPAPORT, G., *Analyt. Chem.*, **26,** 1980 (1954)
17. RICHARDSON, W. S., and SACHER, A. J., *J. Polym. Sci.*, **10,** 353 (1953)
18. BINDER, J. L., and RANSHAW, H. C., *Analyt. Chem.*, **29,** 503 (1957)
19. CORISH, P. J., *Spectrochim. Acta,* **15,** 598 (1959)
20. FERGUSON, R. C., *Jnl Polym. Sci., Pt. A,* **2,** 4735 (1964)
21. MAYNARD, J. T., and MOCHEL, W. E., *J. Polym. Sci.*, **13,** 251 (1954)
22. MOCHEL, W. E., and HALL, M. B., *J. Am. chem. Soc.*, **71,** 4082 (1949)
23. CHECKLAND, P. B., and DAVISON, W. H. T., *Trans. Faraday Soc.*, **52,** 151 (1956)
24. SALOMON, G., VAN DER SCHEE, A. C., KETELAAR, J. A. A., and VAN EYK, B. J., *Discuss Faraday Soc.*, **9,** 291 (1950)
25. DONO, G. K., and SHELTON, G. G., *Chemy Ind.*, **26,** 1158 (1962)
26. CORISH, P. J., *Analyt. Chem.*, **31,** 1298 (1959)
27. DURBETAKI, A. J., and MILES, C. N., *Analyt. Chem.*, **37,** 1231 (1965)
28. SENN, W. L., *Analytica chim. Acta,* **29,** 505 (1963)
29. MATSUO, M., UENO, T., HORINO, H., CHUJYO, S., and ASAI, H., *Polymer,* **9,** 425 (1968)
30. CHEN, H. Y., *Analyt. Chem.*, **34,** 1793 (1962)
31. CHEN, H. Y., *Jnl Polym. Sci., Pt. B,* **4,** 891 (1966)
32. BRAME, E. G., FERGUSON, R. C., and THOMAS, G. J., *Analyt. Chem.*, **39,** 517 (1967)
33. MATHIAS, A., and MELLOR, N., *Analyt. Chem.*, **38,** 472 (1966)

9

THERMOSETTING RESINS

In the first part of this chapter attention is directed to the determination of free formaldehyde in phenol- (or cresol-)formaldehyde syrups. Conventional methods of detection and determination carried out on phenol-formaldehyde moulding powders are described e.g. those of moisture, nitrogen content, acetone extract, free formaldehyde, amount and composition of the free phenols, ash content, hexamine, aniline content, naphthalene, resorcinol and boric acid. Information is given on the microscopic examination of phenol-formaldehyde moulding powders for the presence of fillers, as well as the isolation of such fillers from mouldings, and comment is made on methods of determination of the acetone-soluble content of mouldings. In particular, attention is drawn to the gas chromatographic determination of free phenol and cresols in phenol-formaldehyde resins.

Urea-formaldehyde resins and moulding powders are dealt with on similar lines to the above, but the main concern is with the determinations of urea and formaldehyde. In addition, urea-formaldehyde syrup starch preparations as well as bonded wood wastes are dealt with. Information is provided about urea-formaldehyde resins modified with furfuryl alcohol.

The reaction products of melamine and thiourea with formaldehyde are discussed and the first part of the chapter concludes with some observations on hardeners and miscellaneous plywood glues.

The pyrolysis/gas chromatography of phenol-formaldehyde, urea-formaldehyde and other resins is discussed.

In the second half of this chapter the infra-red spectra of the common thermosetting resins are described. Attention is drawn particularly to the changes which occur in these spectra due to cross-linking, since these appear to be a major source of difficulty in identifying thermosetting resins from their infra-red spectra.

The chapter concludes with some observations on the structure and composition of phenol-formaldehyde resins as determined by n.m.r. spectroscopy.

FORMALDEHYDE CONDENSATION PRODUCTS

Here we are mainly, though not wholly, concerned with reaction products of formaldehyde with such substances as phenol, urea, melamine and thiourea.

Phenol-formaldehyde, cresol-formaldehyde and phenol-cresol-formaldehyde condensation products submitted for test may include syrups, resins, moulding powders and moulded articles.

The phenol-formaldehyde resins produced with the aid of basic catalysts, and usually known as resole resins, are synthesised as follows:

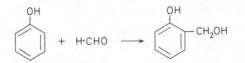

This reaction product may then react with another molecule of phenol and then with another molecule of formaldehyde in this way:

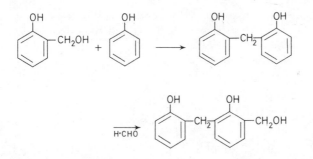

Such resinous products have reactive —CH_2OH groups which, under the influence of heat, can condense with other molecules to give large cross-linked and insoluble molecules.

On the other hand, resins produced by acid catalysts are known as Novolak or non-heat hardening types. These Novolaks are characterised by the absence of the reactive —CH_2OH group and they are incapable of condensing with other Novolak molecules on heating. They require 'hardening agents' to effect condensation.

The samples submitted for test may include:

1. Phenol- (or cresol-) formaldehyde syrups.

   These are intermediate syrupy reaction products, and tests carried

465

out on them e.g. the determination of free formaldehyde[1] are intended to throw some light on the degree of condensation.

In the determination of free formaldehyde in phenol-formaldehyde syrups 5 ml of the sample, accurately weighed, are treated with 100 ml of 90% v/v ethyl alcohol and 4 drops of bromophenol blue indicator.

The solution is now titrated with N HCl, slightly overstepping the end-point, then back titrated with N NaOH to the greenish-yellow colour of the indicator. 25 ml of 5% w/v hydroxylamine hydrochloride solution are now added and the mixture is allowed to stand for a minimum of 30 min.

The hydrochloric acid liberated, which is equivalent to the free formaldehyde in the syrup, is now titrated with N NaOH solution, rapidly to pH 2·75, then dropwise to pH 3·4.

2. Resins.

3. Moulding powders and moulded articles.

These are the principal samples with which the analytical chemist has to deal.

These products may contain:

(a) Resins.
(b) Dyes such as Nigrosine, Monolite Fast Scarlet, Rhodamine and Methyl Violet.
(c) Pigments such as Iron Oxide and Titanium Dioxide.
(d) Additives such as Lime, Magnesia and Naphthalene.
(e) Accelerators such as Hexamine.
(f) Lubricants such as Stearine.

TESTS ON MOULDING POWDERS

1. *Moisture*

In this test 5 g of the sample are dried for 24 h in vacuo over P_2O_5.

2. *Total nitrogen content*

This is determined by the Kjeldahl method. 1 g of the sample is digested for 5 h with 7·0 g of sodium sulphate-selenium mixture (35 : 1) and 25 ml conc. sulphuric acid. The digestion product is diluted, made alkaline, and the ammonia that is liberated is distilled into standard acid. The excess acid is back titrated.

3. *Acetone extract*

The powder is first dried for 24 h over P_2O_5.

2 to 2·5 g of the dry powder are weighed into a previously dried and

weighed Soxhlet extraction thimble. The powder is then extracted with acetone in a hot Soxhlet extraction apparatus. After extraction, the thimble and contents are transferred to a weighing bottle, then dried at 100°C and weighed. The loss in weight is calculated as a percentage of acetone-soluble matter.

4. *Nitrogen content of the acetone-extracted material*

This is determined by the Kjeldahl method given in 2.

5. *Free formaldehyde*

This is determined by the dimedone method. Dimedone is 5 : 5-dimethyl cyclohexane 1 : 3-dione.

10 to 15 g of the sample are mixed with 250 ml of water. The mixture is distilled in steam, 200 ml of steam distillate being collected. This steam distillate is diluted to 500 ml. To 100 ml of this solution are added 10 ml of a 5% w/v solution of dimedone in alcohol, and a few drops of

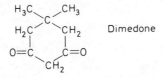

Dimedone

bromophenol blue indicator. Just sufficient dilute hydrochloric acid solution (20% v/v) is added to produce a yellow colour, 10% w/v sodium hydroxide solution is then added to give the purple tint of the indicator, after which 50 ml of a mixture of 2 volumes N sodium acetate solution and 1 volume N hydrochloric acid solution are added.

After standing for 12 h the precipitate is filtered off on a 1G3 crucible and dried at 100°C. The weight of precipitate multiplied by 0·103 gives the equivalent weight of formaldehyde.

The reaction proceeds as follows:

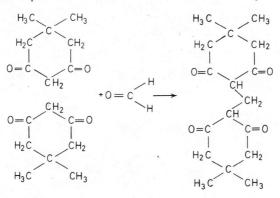

467

THE DETERMINATION OF THE AMOUNT AND COMPOSITION OF
FREE PHENOLS IN PHENOL-FORMALDEHYDE AND CRESOL-
FORMALDEHYDE RESINS AND MOULDING POWDERS:
CHEMICAL METHOD

This subject has been dealt with at great length by Haslam, Whettem and Newlands[2] and they have developed methods for the determinations.

The first principle involved in all the determinations is that of the isolation of the free phenols or cresols or mixtures of phenols and cresols. This is accomplished by dissolving the sample in aqueous sodium hydroxide, precipitating the resin by neutralisation of the solution to pH 4·5, filtering off the resin (and fillers where moulding powders are concerned) and steam distilling the phenols or cresols or mixtures of phenols and cresols in the filtrate.

Having isolated this phenolic matter there are three cases.

1. *Phenol-formaldehyde resin*

In that case the free phenolic matter will be phenol. This is brominated by a standard bromination procedure, and calculated to free phenol on the basis that, in the final titration,

1 ml of 0·1 N sodium thiosulphate ≡ 0·00157 g of phenol.

This corresponds to 2·98 (theoretical 3·0) bromine molecules combining with 1 molecule of phenol to give tribromophenol and hydrobromic acid.

2. *Cresol-formaldehyde resin*

Here we base our reasoning on the knowledge that, when a resin is made from commercial cresylic acid containing all three cresol isomers, then, because of the very high reactivity of *m*-cresol with formaldehyde, no free *m*-cresol is likely to be present in the free cresylic matter present in the resin. These free cresols will therefore consist of *p*-cresol (the least reactive) and possibly a small amount of *o*-cresol.

This *p*-cresol is brominated by a standard bromination procedure and calculated to free cresol on the basis that, in the final titration,

1 ml of 0·1 N sodium thiosulphate ≡ 0·00236 g of cresol.

This corresponds to 2·27 (theoretical 2·0) bromine molecules combining with 1 molecule of cresol to give a bromocresol and hydrobromic acid.

3. *Mixture of phenol- and cresol-formaldehyde resin*

Here the free phenolic matter is assumed to be a mixture of phenol and *p*-cresol. In deducing the proportions of phenol and cresol two factors are taken into account.

(a) The behaviour of phenol and *p*-cresol on bromination as given in 1 and 2.

(b) The application of Chapin's method[3] for the determination of phenol in the presence of certain other phenols to an aliquot of the steam distillate, the bromination value of which is known. Chapin's method is based on the knowledge that, under certain conditions, Millon's reagent gives a red colour with phenol but not with certain other phenols.

Based on the above considerations, the following methods have been worked out.

DETERMINATION OF TOTAL FREE PHENOLS

Sample

The size of the sample to be taken depends on the amount of free phenol expected to be present. For resins containing from 10 to 20% of free phenols, 2 g is sufficient. For moulding powders, 5 g is generally sufficient, but sometimes it may be necessary to take as much as 10 g of sample.

Procedure

Place the weighed sample in a 500 ml beaker and add 50 to 200 ml of 10% w/v sodium hydroxide (according to the size of sample). Boil the contents of the beaker for about 10 min to bring everything except the fillers, if present, into solution. Cool the solution and dilute to about 200 ml.

Adjust the pH of the solution to 4·5 with 20% v/v sulphuric acid at first, and then with N sulphuric acid, using B.D.H. 4·5 indicator externally.

Filter off the precipitated resin and wash it well with cold water. Transfer the filtrate and washings, which contain the free phenols, to a 500 ml flask, dilute to the mark and shake thoroughly.

Transfer 250 ml of this solution to a steam distillation apparatus. Steam distil until 350 to 400 ml of distillate have been collected in a 500 ml calibrated flask. Dilute the distillate to the mark with water and mix well. Transfer 250 ml to a 500 ml 'Iodine' flask for bromination.

Add 50 ml of an approximately 0·1 N solution of potassium bromide-potassium bromate mixture and 10 ml of 20% v/v sulphuric acid. Stopper the flask and put a few ml of 10% w/v aqueous potassium iodide solution into the trough formed by the stopper and the mouth of the flask so that no bromine escapes. Carry out a blank determination simultaneously and allow both flasks to stand in the dark for 1 h. Ease the stoppers and allow the potassium iodide to run into the flasks; add a further 15 ml of 10% w/v potassium iodide solution to each, taking the usual precautions to prevent loss of bromine. Titrate the liberated iodine with 0·1 N sodium thiosulphate solution until the iodine colour has been discharged to a

469

pale straw colour, add a few ml of chloroform to dissolve the very voluminous precipitate of tribromophenol, as this absorbs appreciable quantities of iodine, and complete the titration after adding starch as indicator. The amount of phenol (or cresol) is calculated from the difference between the blank and sample titres.

$$1 \text{ ml of } 0 \cdot 1 \text{ N sodium thiosulphate} \equiv 0 \cdot 00157 \text{ g of phenol}$$
$$1 \text{ ml of } 0 \cdot 1 \text{ N sodium thiosulphate} \equiv 0 \cdot 00236 \text{ g of cresol}$$

METHOD FOR DETERMINING THE RATIO OF FREE PHENOL TO FREE CRESOL

Sample

From the steam distillate prepared as described above, an amount is taken that would give a titre of 3·0 ml of 0·1 N potassium bromide-potassium bromate. This amount of distillate is measured from a burette into a 100 ml calibrated flask, diluted to the mark with water and mixed thoroughly. This constitutes the sample solution for this method.

Reagents

Nitric acid, diluted.
Prepare this by bubbling air through pure conc. nitric acid until the acid is colourless and then diluting 1 volume of the acid with 4 volumes of water.
Millon's reagent.
This reagent is prepared as follows: Measure 2 ml of mercury from a burette into a 100 ml conical flask and add 20 ml of pure conc. nitric acid. Hasten the resulting reaction by occasional shaking. When all action has ceased (after about 15 min), add 35 ml of water. If any separation of basic mercury nitrate occurs, add diluted nitric acid reagent drop by drop until the solution clears. Add 10% w/v aqueous sodium hydroxide drop by drop with thorough mixing until the curdy precipitate that forms no longer redissolves but just produces a slight permanent turbidity. Add 5 ml of dilute nitric acid reagent and thoroughly mix the contents of the flask. Millon's reagent so prepared will remain stable for 2 d.
Standard phenol solution.
Prepare a stock 1% w/v solution of phenol in water. Prepare the standard phenol solution from this on the day it is used by taking 2·5 ml of stock solution and making up to 100 ml with water, so that the solution contains exactly 0·025 g of phenol per 100 ml.
Formaldehyde solution.
Dilute 2 ml of 40% v/v formaldehyde solution with water to make 100 ml of solution.

Procedure

Measure 5 ml of the sample solution into each of two 6×1 in (150×24 mm) test tubes and mark them A and B, respectively. Into each of two similar tubes measure 5 ml of the standard phenol solution and mark them C and D. Measure from a burette 5 ml of Millon's reagent into each of the four tubes and mix thoroughly. Place the tubes in a water-bath maintained at 100°C for 30 min. Remove the tubes and cool them immediately in a bath of cold water for 10 min. Remove them from the cold water and add 5 ml of dilute nitric acid reagent to each tube and gently shake to mix the contents. To tubes A and C add 3 ml of the formaldehyde solution. Dilute the contents of the four tubes to 25 ml with water (the tubes should be previously marked with a file scratch at the 25 ml level), shake them thoroughly and set them aside overnight.

The contents of tubes A and C will then be found to be yellow because of the bleaching action of the formaldehyde; these are used as blanks in the colorimetric procedure that follows.

Measure 20 ml of the contents of tubes C and D by means of a pipette into separate 100 ml calibrated flasks and mark these C and D, respectively. Add 5 ml of diluted nitric acid reagent to each and make up to the 100 ml mark with water. Fill two burettes, C and D, from the flasks C and D, respectively, so that C contains the yellow 'phenol blank' solution and D the 'phenol standard' solution. The strength of the latter is now such that 1 ml contains 0·00001 g of phenol.

Place 10 ml of the contents of the test tubes A and B (i.e. 'sample blank' and 'sample') into 50 ml Nessler cylinders, marking these A and B, respectively.

Run, from burette D, a small quantity of the 'phenol standard' solution into Nessler cylinder A containing the 'sample blank' and run an equal volume from burette C of the 'phenol blank' solution into Nessler cylinder B containing the 'sample' solution.

Continue adding successive equal volumes of the 'phenol standard' and 'phenol blank' solutions to the appropriate Nessler cylinders with intermediate shaking until the contents of the cylinders are equal in colour. It is essential that these colour matching operations are performed speedily. Note the volume in ml of 'phenol standard' solution required for colour matching.

Calculation

Strength of 'phenol standard' solution in burette D =

$$\frac{5}{100} \times 0{\cdot}025 \times \frac{20}{25} = 0{\cdot}001 \text{ g/100 ml}$$

or 1 ml of solution $\equiv 0{\cdot}00001$ g of phenol

Volume of sample solution matched in Nessler cylinder =

$$5 \times \frac{10}{25} = 2 \text{ ml}$$

$\therefore$ (Volume of 'phenol standard' solution required to match sample solution) $\times 0{\cdot}00001 \times 50$

$$= \text{(weight in g of phenol/100 ml of sample solution)} = P$$

Then $\dfrac{P}{0{\cdot}00157} = \left(\begin{array}{c}\text{bromine absorbed by phenol in terms of} \\ 0{\cdot}1 \text{ N potassium bromide-bromate}\end{array}\right) = X$

Then $\left(\begin{array}{c}\text{Total bromine absorption of 100 ml of sample solution} \\ \text{in terms of } 0{\cdot}1 \text{ N potassium bromide-bromate}\end{array}\right) - X$

and $\quad = \text{(bromine absorbed by cresol)} = Y$

$$Y \times 0{\cdot}00236 = \text{(weight in g of cresol/100 ml of sample solution)} = C$$

With P and C found, the percentage composition of free phenols is easily calculated.

Example

Volume of 'phenol standard' solution required to match sample = 8·0 ml.

$\therefore P = 8{\cdot}0 \times 0{\cdot}00001 \times 50$

$$= 0{\cdot}004 \text{ g of phenol/100 ml of sample solution}$$

$\therefore X = \dfrac{0{\cdot}004}{0{\cdot}00157} = 2{\cdot}55 \text{ ml}$

and $Y = 3{\cdot}00 - 2{\cdot}55 = 0{\cdot}45 \text{ ml}$

$\therefore C = 0{\cdot}45 \times 0{\cdot}00236$

$$= 0{\cdot}001 \text{ g of cresol/100 ml of sample solution}$$

$\therefore$ Ratio of free phenol to free cresol = 4 to 1

$\therefore$ Percentage composition of free phenols

$$= 80\% \text{ phenol, } 20\% \text{ cresol.}$$

In their original paper, Haslam, Whettem and Newlands[2] incidentally draw attention to the problems connected with the special class of resins prepared by formaldehyde condensation of *m*-cresol and *p*-cresol only.

Here the free phenolic matter consists essentially of *m*-cresol and *p*-cresol only. The authors show that it is possible to analyse an aqueous solution of *m*- and *p*-cresols by a combination of two methods.

(a) Amperometric bromometric titration of the two cresols.

(b) Application of the colorimetric nitrosamine method to the solution of the two cresols.

GAS CHROMATOGRAPHIC DETERMINATION OF FREE PHENOL AND CRESOLS

Phenol and the three cresols are readily separated by gas chromatography and advantage has been taken of this in the method detailed below which is particularly useful in the examination of resin samples. The principle could be applied to the examination of alcohol extracts from moulding powder samples.

A gas chromatograph fitted with a flame ionisation detector is required for this determination.

Chromatographic conditions

Column	2 m ($\frac{1}{8}$ in or 3 mm o.d. stainless steel) 10% w/w tricresyl phosphate + 1% w/w phosphoric acid on Celite (72–80 mesh)
Carrier gas	Nitrogen
Nitrogen pressure	10 lbf/in^2 (70 kN/m^2)
Hydrogen pressure	25 lbf/in^2 (170 kN/m^2)
Air pressure	30 lbf/in^2 (200 kN/m^2)
Column temperature	125°C

Method

About 1 g of the sample is weighed into a 5 ml graduated flask. The resin is dissolved in absolute ethyl alcohol and the solution diluted to the mark with the solvent. If the resin is not completely soluble the mixture is allowed to stand for 4 h with frequent shaking. 1 μl of the solution thus prepared is injected on to the chromatograph by means of a Hamilton syringe; the chromatograph is set to the conditions outlined above.

Note: After each injection the syringe is washed and cleaned immediately.

The chromatogram is developed, attenuating as necessary in order to keep all peaks on scale. The chromatogram is repeated using attenuator settings found from a previous run. The result will be similar to Figure 9.1.

The peak areas are calculated from the formula

$$\text{Area} = \text{Height} \times \text{Width at Half Height}$$

Calibration

A standard solution is prepared containing 15 mg phenol, 12 mg o-cresol, 10 mg m-cresol and 10 mg p-cresol in 5 ml absolute ethyl alcohol.

1 μl of this solution is chromatographed under the same conditions and the peak areas of the components are calculated from the resulting chromatogram. The following formula is used to calculate the w/w

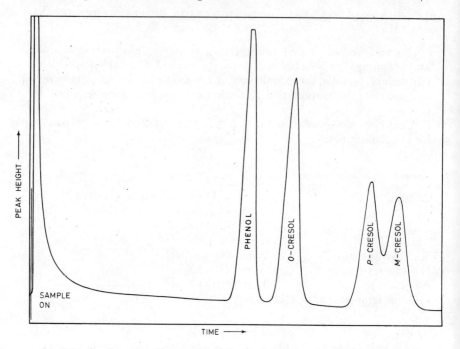

Fig. 9.1. Gas chromatographic separation of phenol and the three cresols (1 μl in ethyl alcohol)

percentage of each component in the resin:

$$\% \ w/w = \frac{A_S}{A_{ST}} \times \frac{W_{ST}}{W_S} \times 100$$

where A_S = Area for component in sample
A_{ST} = Area for component in standard
W_{ST} = Weight of component in standard
W_S = Weight of resin taken

ASH CONTENT

This is determined by igniting 5 g of the powder in a silica basin, firstly over a low bunsen flame and finally to constant weight in a muffle furnace at 750°C. It is useful to carry out a spectrographic examination of this ash and to supplement the information obtained by qualitative and quantitative inorganic analysis, if necessary.

HEXAMINE

This may be detected in phenol-formaldehyde moulding powders by homogenising, e.g. in a M.S.E. homogeniser or Waring blender or similar apparatus, 5 g of the sample with 100 ml of water. After filtration of the aqueous extract through a No. 1 Whatman filter paper, 10 ml of the filtrate are treated with 2 ml of 5% w/v aqueous mercuric chloride solution. In the presence of hexamine an immediate white precipitate is obtained.

To a further 10 ml of the aqueous extract are added 1 ml of a saturated aqueous solution of potassium ferricyanide and 1 ml of a saturated aqueous solution of magnesium sulphate. The mixture is boiled for 1 to 2 min, then filtered hot through a Whatman No. 1 110 mm filter paper. In the presence of hexamine, the filtrate, on cooling, yields a precipitate.

A semi-quantitative measure of the proportion of water-soluble hexamine may be obtained as follows: 5 g of the sample are extracted with 100 ml of water as above and the extract filtered through a Whatman No. 1 150 mm filter paper into a measuring cylinder (100 ml) and the volume of the filtrate is noted.

This filtrate is transferred to a separating funnel and extracted twice with 50 ml of ether in order to remove phenolic matter. The aqueous phase is then transferred to a 250 ml beaker and heated on a water bath to remove the ether. 2 g of potassium bromide (Analar) are added, followed by 2 or 3 drops of 20% v/v hydrochloric acid solution to render the solution acid.

50 ml of saturated bromine water are now added and the mixture is stirred well. The precipitate is filtered off immediately on a No. 1 Whatman 150 mm filter paper, and the beaker and precipitate are washed twice with a bromine-water wash liquor. This wash liquor is prepared by dissolving 0·5 g hexamine and 5 g potassium bromide in 200 ml of water, then adding 50 ml of bromine water and filtering to yield the wash liquor.

The precipitate is then washed with water (to which a few drops of saturated sodium sulphite solution have been added) into the beaker originally used for precipitation. This decomposes the hexamine tetrabromide. The solution obtained is transferred to a beaker (500 ml) and diluted to 200 ml with water. 10 ml of 20% v/v sulphuric acid solution are added and the solution is boiled on a hot plate until the odour of formaldehyde is no longer perceptible, the volume of solution being maintained at 200 ml during the boiling by adding distilled water. The solution is now transferred to an ammonia distillation apparatus, made alkaline with 30% w/v sodium hydroxide solution and the ammonia liberated distilled into 50 ml of 0·1 N hydrochloric acid solution.

The excess acid is back titrated preferably with an automatic titrimeter. The ammonia found is calculated in terms of hexamine.

Incidentally, Busfield and Chipperfield[4] have drawn attention to another useful method of determination of added hexamine in hexamine first-stage phenol-formaldehyde mixtures. The weak base hexamine, in the aqueous extract of the mixture, is titrated conductimetrically with standard strong acid.

In the absence of other bases it is possible to carry out a direct automatic recording titration of the hexamine using 0·05 N hydrochloric acid as titrant. A glass/calomel electrode pair is recommended and a pH meter/recorder sensitivity of 400 mV full-scale is satisfactory. Pure hexamine is used in the standardisation of the acid.

ANILINE

Certain phenol-formaldehyde resins may contain aniline reaction products and it may be desirable to obtain qualitative and semi-quantitative evidence of the presence of the aniline.

Two tests may be used. The aniline may first be released by reaction with alkali, followed by distillation in steam. The aniline in the steam distillate may then be detected and determined either,

(a) by coupling of the diazotised base with alkaline R-salt solution

i.e. with 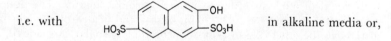 in alkaline media or,

(b) by application of the colorimetric phenol hypochlorite test for aniline.

R-SALT METHOD

2 g of the moulding powder, 20 ml of 10% w/v sodium hydroxide solution and 100 ml of water are taken and the mixture is distilled in steam, the distillate being collected in 10 ml of 20% v/v sulphuric acid solution. 500 to 600 ml of steam distillate are collected and the solution is then diluted to 1 l with distilled water.

Reagents

Approx. 0·05 N R-salt solution.
The equivalent of 3·75 to 4·0 g of 100% R-salt (mol. wt 304) is dissolved in 100 ml of boiling water and the solution is made slightly alkaline to brilliant yellow paper with 2 N sodium carbonate solution. The solution is cooled to room temperature, filtered into a volumetric flask (250 ml), then diluted to volume. This solution is stored in an amber-coloured bottle.
2 N sodium carbonate solution.

0·5 N sodium nitrite solution freshly prepared.
N hydrochloric acid solution.
Analar potassium bromide solid.

Procedure

2 ml of the diluted steam distillate are diluted to 10 ml with water and 2 N sodium carbonate solution is added until the solution is only faintly acid to Congo Red paper. The solution is cooled in ice to 5°C after which 2 ml of N hydrochloric acid solution are added followed by 2 g potassium bromide. After the bromide has dissolved, the solution is allowed to stand for 15 min. This solution is then poured into a solution of 10 ml of 2 N sodium carbonate and 0·5 ml of the 0·05 N R-salt solution which is maintained at 10 to 15°C. In the presence of aniline a characteristic orange coupling product is obtained that may be matched, if necessary, against standards prepared by the above procedure from a solution of a known amount of pure aniline in dilute sulphuric acid.

PHENOL HYPOCHLORITE TEST

Though not entirely specific for aniline, this test is not affected by toluidines and certain other primary amines.

The preliminary steam distillation is carried out exactly as for the R-salt test, except that the steam-distilled aniline is absorbed in 25 ml conc. hydrochloric acid instead of 10 ml of 20% v/v sulphuric acid solution.

Reagents

0·25 N hydrochloric acid solution.
Bleaching powder solution.
5 g of bleaching powder, containing not less than 25% available chlorine, are diluted to 100 ml with water. The mixture is warmed to 50 to 60°C with constant shaking, then filtered hot. It is desirable that the reagent should be prepared afresh each week.
Phenol reagent.
5 g of Analar grade phenol are dissolved in 100 ml of a solution prepared by diluting 50 ml of ammonium hydroxide (sp. gr. 0·88) to 1 l with distilled water. This reagent is prepared afresh each day, immediately prior to use.

Procedure

To 2 ml of the diluted steam distillate are added 8·0 ml of 0·25 N hydrochloric acid solution followed by 2 drops of bleaching powder solution; the solution is mixed, and, after standing for 5 min, the solution is immersed in a boiling water bath for $2\frac{1}{2}$ min.

5 ml of the phenol reagent are now added and colour development

allowed to proceed at room temperature for 30 min.

The blue colour produced by aniline in this test is now ready for optical density measurement and, in semi-quantitative work, the figure obtained may be compared with those obtained when the test is applied to known amounts of aniline in dilute hydrochloric acid solution.

In extreme cases, where it is desirable to establish the identity of the aniline beyond all question by the preparation of a crystalline derivative of agreed m.p., it may be desirable to work on much larger amounts of moulding powder. The impure aniline may then be isolated and converted:

(a) to tribromoaniline by bromination in glacial acetic acid medium and

(b) to benzanilide by reaction with benzoyl chloride in sodium hydroxide medium.

If, as is sometimes the case, furfuraldehyde has been used in conjunction with the aniline, then the crude base, after solution in glacial acetic acid in the tribromoaniline preparation, will yield a brilliant red coloration.

NAPHTHALENE

Naphthalene is sometimes present as an additive in phenol-formaldehyde moulding powders.

To obtain evidence of naphthalene the moulding powder is distilled in steam and the steam distillate is extracted with ether. The ethereal solution is extracted with alkali, then dried and evaporated carefully. The residual naphthalene is treated with saturated picric acid solution to yield naphthalene picrate, which may, if required, be titrated with standard alkali using 2 : 4-dinitrophenyl acetate as indicator.

Furthermore, the identity of the naphthalene may be confirmed by means of its U.V. and infra-red spectra.

FREE RESORCINOL

This may be detected by extraction of the moulding powder with ether. The ether solution is extracted with alkali, the alkaline solution made slightly acid, filtered from precipitated resin and the filtrate extracted with ether. The presence of resorcinol in the evaporated ethereal extract is confirmed by fluorescein and resorufin tests and by determination of the U.V. and infra-red spectra.

BORIC ACID

Boric acid or sodium borate may be present in phenol-formaldehyde moulding powders. Clues to the presence of boron compounds will be given in the preliminary spectroscopic examination of the powder.

If a determination is required, the ground sample is first evaporated to dryness with saturated calcium hydroxide solution. The residue is ashed and the ash is dissolved in dilute sulphuric acid. The free mineral acid is then neutralised potentiometrically with standard alkali, after which mannitol is added and the resulting mannitoboric acid determined by titration with standard alkali.

If questions arise as to whether the boric acid is present in the powder as such or as sodium borate, it is desirable to carry out aqueous extraction of the powder and to evaporate known aliquots of this aqueous extract with hydrochloric acid and methyl alcohol to ensure that all borates are removed. The sodium in the residue is then determined by the zinc uranyl acetate method or by flame photometric or atomic absorption methods.

NATURE OF THE FILLERS PRESENT IN PHENOL-FORMALDEHYDE MOULDING POWDERS

It will probably be useful to draw attention to a principle, originally devised by Wallis[5], that has been used e.g. to estimate the proportion of coconut-shell flour in a phenol-formaldehyde moulding powder containing coconut-shell flour and wood flour as fillers.

The method is based on the fact that, when examined under the microscope, coconut-shell flour is shown to contain characteristic stone cells that can be distinguished readily and counted. The standard against which these stone cells are counted consists of a definite weight of lycopodium, which consists of a definite number of lycopodium spores, of great uniformity and of easy identification.

An example will clarify the procedure.

A phenol-formaldehyde moulding powder was first extracted with acetone to remove any resin; the acetone-insoluble matter amounted to 60% of the original powder.

Then 0·2 g of the acetone-insoluble matter, 0·1 g lycopodium and 0·24 g gum tragacanth were mixed intimately in a beaker and 1 ml ethanol was added. When all the material was wetted, 3 ml of glycerol was added, followed by 10 ml of water, and the mixture was stirred carefully for several minutes.

1 drop of the mixture was placed on a clean glass slide and a cover slip was placed on top and pressed down gently.

The slide was examined under a microscope; a Vickers Projection Microscope with magnification $\times 200$ (16 mm objective) was found to be very convenient.

50 fields were counted for lycopodium spores, and 200 fields for coconut-flour stone cells, giving the following results:

$$\frac{\text{Lycopodium spores in 100 fields}}{\text{Coconut stone cells in 100 fields}} = \frac{100}{19}$$

The test was calibrated on a 50% wood flour — 50% coconut shell flour using a preparation of: 0·1 g wood flour, 0·1 g coconut shell flour,

0·1 g lycopodium and 0·24 g gum tragacanth.
In this case

$$\frac{\text{Lycopodium spores in 100 fields}}{\text{Coconut stone cells in 100 fields}} = \frac{100}{14}$$

Hence the proportion of coconut-shell flour in the acetone-extracted material was

$$\frac{50 \times 100/14}{100/19} = 67\cdot9\%$$

and the proportion of coconut-shell flour in the original moulding powder was

$$\frac{67\cdot9 \times 60}{100} = 40\cdot7\%$$

IDENTIFICATION OF THE FILLER

A more difficult problem concerns the isolation and identification of the filler in a phenol-formaldehyde moulding. The following method, devised by Haslam and Hill[6], and based on earlier work of Ersch and Nietsche, has often proved to be very useful.

The moulding is ground in a mortar (a Torrington mortar has been found suitable) and the ground material sieved. The product passing 30 B.S.S. and retained on 60 B.S.S. is suitable for the test.

A 0·1 g portion of the ground sample and 5 g of 2-naphthol are placed in a Pyrex glass tube 300 mm long and 13 mm in internal diameter. The tube is sealed and then placed in the continuous-shaking apparatus devised by Haslam and Swift[7] in their study of the hydrolysis of nylon samples. The tube and its contents are then maintained in an oven at 160° to 180°C for 4 h with constant rotation of the tube and hence intimate mixing of sample and molten 2-naphthol.

The tube and contents are cooled, the tube is opened and the resulting product treated in a beaker with 50 ml of ethanol until all the 2-naphthol appears to be dissolved.

The mixture is centrifuged and the ethanol decanted off; further washings with ethanol are carried out until the washings are colourless. The ethanol is again decanted off and a final washing with ether is made, after which the insoluble filler is allowed to dry.

Microscopical examination of this residue has given excellent results in the identification of the following fillers: wood flour, coconut-shell flour, glass, asbestos and cotton. With nylon-filled mouldings, 6 : 6 nylon may be present in the residue as smooth fibres. If this material is suspected, solution of the residue in 90% w/v formic acid with subsequent production of a film of the nylon and infra-red examination will settle the point. Chloroprene-filled preparations yield a rubber-like residue containing chlorine. In such cases, infra-red examination will yield conclusive evidence of the presence of chloroprene.

ACETONE-SOLUBLE MATERIAL IN PHENOL-FORMALDEHYDE MOULDINGS

One of the most contentious tests in the synthetic-resin field concerns the determination of the acetone-soluble matter content of phenolic mouldings.

A committee of analysts nominated by the Moulding Powder Committee of the British Plastics Federation made important observations on this test. They were as follows:

1. A definite method of preparation of the sample should be employed, i.e. with a definite milling machine working in a precise way.
2. The milled material should be sieved so that the samples for analysis are of a definite sieve size.
3. The acetone extraction should be carried out in a definite way with a precise rate of syphoning.
4. The method used by the A.S.T.M. is preferred for the final weighing of the acetone extract. This involves the evaporation of the acetone extract at 55 to 60°C on a water bath; final drying is carried out in an oven at 50°C±2°C.

The current standard method (Method 401 A) in Great Britain is that given in BS 2782 Part 4/401:1970 and embodies the principles outlined above.

UREA-FORMALDEHYDE RESINS AND MOULDING POWDERS

The types of sample normally received for analytical examination are resins and moulding powders. For plant control the tests generally carried out are those of colour, sp. gr., stoved-solids content, pH, heat stability, storage life and viscosity. It is also generally important to obtain some idea of the nature of the condensation product and this is usually ascertained by a determination of the urea and formaldehyde contents followed by calculation of the molar formaldehyde to urea ratio.

Moulding powders are, of course, much more complex, as they usually contain various other ingredients such as fillers, lubricants, accelerators, pigments, stabilisers and plasticisers.

Direct determination of the constituents is not always feasible and a more empirical approach is to some extent necessary. In our experience the following tests have proved to be useful in day to day practice. They are given in principle only. The most important tests are those for urea and formaldehyde and these are given with full working details.

1. Moisture. The determination is carried out in a vacuum oven at 60°C.

2. Water-extractable matter. This is determined under definite conditions; the residue obtained on evaporation of the aqueous extract is dried in an oven at 80°C.

3. Total nitrogen. This is determined by the Kjeldahl method.

4. Ash content. Determined in a silver crucible using a muffle at 650°C.

5. Nitrogen content of the water-extractable matter. A determination of nitrogen is carried out on the water-insoluble matter by the Kjeldahl method. The figure obtained, subtracted from the total nitrogen, gives the nitrogen content of the water-extractable matter.

6. Fillers. In urea-formaldehyde moulding powders the fillers are usually the same as those found in phenol-formaldehyde moulding powders, e.g. wood flour, cellulose, walnut-shell flour and asbestos, and microscopic identification is often of great value.

   Inorganic additives in urea-formaldehyde moulding powders are usually detected by a spectrographic examination of the sample or its ash. The metallic elements may be determined quantitatively. Separation of the filler from the resin may often be achieved by breakdown with 20% v/v acetic acid solution and filtration of the resulting suspension. The filler when recovered in this way is ideal for infra-red spectroscopic examination and indeed both organic and inorganic fillers may often be identified in this way.

   If the particle size of the filler is small, this method may be used quantitatively. When the filler is of larger particle size (6–12 mesh), however, high results may be obtained due to the presence of unchanged resin in the insoluble filler that is isolated. In such cases, once the filler has been identified, it may be better to calculate the proportion of filler by difference, after determination of the resin content of the sample from the combined nitrogen and formaldehyde figures.

   In the case of fully cured resins it is sometimes necessary to boil with phosphoric acid as described in the determination of the formaldehyde content (see p. 485).

7. Hexamine. This determination is carried out on the aqueous extract by the hexamine tetrabromide method.

   In the absence of other bases hexamine may be determined by potentiometric titration with hydrochloric acid.

8. Water-soluble zinc. In this determination an aliquot of the aqueous extract is wet digested to remove organic matter and the zinc then titrated amperometrically with the sodium salt of ethylene diamine tetra-acetic acid according to Washbrook's method[8].

   Alternatively, the zinc may be determined directly by atomic absorption spectrophotometry without prior removal of the dissolved organic matter.

9. Water-soluble sulphate. This is determined by barium sulphate precipitation.

10. Plasticisers. Plasticisers such as monocresyl glyceryl ether and

dibutyl phthalate have been encountered. These have been removed by extraction with ether or chloroform, identified by U.V., infra-red and n.m.r. spectroscopy, and determined quantitatively in a mixture of the two plasticisers by measurement of the U.V. absorbance in alcohol solution at 262 and 245 nm. The concentrations are calculated by solving simultaneous equations derived from the absorbances of standard solutions of the individual compounds.

The two most important determinations are those of total urea and formaldehyde and these tests are carried out as follows.

DETERMINATION OF UREA CONTENT

The method is based on the work of Kappelmeier[9]. The urea reacts with benzylamine to yield dibenzyl urea which is separated and weighed in the case of urea-formaldehyde resins.

With moulding powders the dibenzyl urea obtained is contaminated with filler. Hence it is necessary to determine the nitrogen content of the impure product to obtain a measure of the urea.

Apparatus

A test tube (8 × 1 in or 200 × 24 mm) blown out in the middle to give a tube of capacity about 80 ml similar to that shown in Figure 9.2.
Electric hotplate with thermostatic control.
Mechanical stirrer.
Sintered glass crucible, porosity Grade 3.

Reagents

Benzylamine, pure.
Hydrochloric acid solution.
1 volume of conc. hydrochloric acid is diluted with 3 volumes of water.
Congo Red paper.

Procedure

0·5 g of the sample is weighed into the test tube and 15 ml of benzylamine are added.

The tube is then placed on the electric hotplate and the mechanical stirrer is started. The benzylamine is brought to the boil and the water that is liberated is boiled off. The heating is now adjusted so that the benzylamine condenses about half way up the test tube.

Boiling is now continued for 2 to 4 h in the case of liquid and spray dried resins and for 4 to 8 h in the case of cured resins.

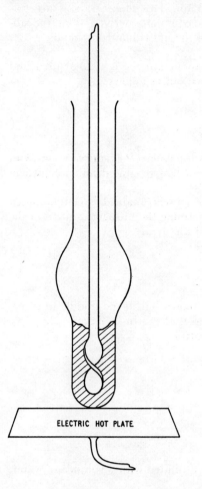

ELECTRIC HOT PLATE

Fig. 9.2.   Apparatus for the determination of urea

The reaction mixture is then allowed to cool to about 40°C and hydrochloric acid solution is added until the mixture is distinctly acid to Congo Red paper.

If an oily product is obtained, the tube and contents are placed in an ice bath and stirred until crystals are formed. If an oily product still remains then it is desirable to repeat the determination but reducing the period of refluxing with benzylamine by 1 h.

When crystallisation is achieved, the crystals are collected on a tared sintered glass crucible porosity 3, washed well with water and dried to constant weight at 105°C.

The m.p. of the crystals of dibenzyl urea obtained from a resin is 165 to 167°C. The urea content of the resin is calculated from the weight of dibenzyl urea that is obtained.

In the case of a moulding powder containing a nitrogen-free filler, the nitrogen content of the impure material collected in the sintered glass

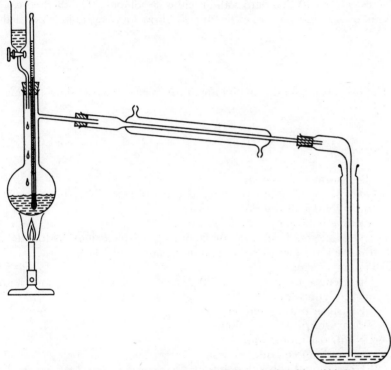

Fig. 9.3.   *Hydrolysis and distillation apparatus for the determination of formaldehyde*

crucible is determined by the Kjeldahl method, and this nitrogen is calculated to urea.

Considerable information may be obtained about the degree of condensation of a resin if the benzylamine test is carried out for two different boiling periods e.g. for 2 h and 4 h.

TOTAL FORMALDEHYDE CONTENT

The method used is based on the work of Grad and Dunn[10].

In our application of the method the condensation product is distilled

485

with 1 : 1 v/v phosphoric acid solution and the distillate is collected in alkaline potassium cyanide solution, forming the cyanohydrin. The liquid in the distillation flask is maintained at 110°C throughout by controlling the addition of fresh phosphoric acid solution.

At the conclusion of the distillation an aliquot of the distillate is added to an acid solution of silver nitrate, when the cyanide not used up by the formaldehyde is precipitated as silver cyanide. The acid solution is now adjusted to the correct acidity prior to potentiometric back titration of the excess silver with standard sodium chloride solution. The method is very satisfactory, and the rather lengthy finish as silver cyanide is avoided.

Apparatus

The hydrolysis and distillation apparatus is assembled as shown in Figure 9.3.

Reagents

Phosphoric acid solution.
One volume of syrupy phosphoric acid (sp. gr. 1·75) is diluted with an equal volume of water.
Alkaline potassium cyanide solution.
24·8 g of potassium cyanide and 10 g of potassium hydroxide are dissolved in water and the solution is diluted to 1 l.
Porous tile chips.
These chips are boiled in a 20% v/v phosphoric acid solution, then washed with water.
0·1 N silver nitrate solution.
50% v/v nitric acid solution.
2 N nitric acid solution.
0·1 N sodium chloride solution.
0·1 N sodium hydroxide solution.
Methyl red indicator.

Procedure

0·5 g of the sample is weighed into the distillation flask and 30 ml of the phosphoric acid solution are placed in the dropping funnel. 50 ml of the alkaline potassium cyanide solution are then measured by means of a pipette into a 500 ml standard flask and the delivery tube from the condenser is inserted in the flask so that the tip of the tube is covered completely by the potassium cyanide solution. The phosphoric acid in the dropping funnel is now allowed to run dropwise into the distillation flask. When empty, the dropping funnel is filled with water. The phosphoric acid solution is now brought to the boil, and the flow of water from the dropping funnel is adjusted so that the temperature of the phosphoric acid solution in the distillation flask is maintained at 110°C. Distillation is continued until about 450 ml of distillate have been collected. The apparatus is then disconnected and the condenser and delivery tube washed with water, the washings being collected in the standard flask. The solution in this flask is then diluted to 500 ml with water.

50 ml of 0·1 N silver nitrate solution are now placed in a 250 ml beaker and made acid with 2 ml of the 50% v/v nitric acid solution. Whilst stirring the solution, 100 ml of the distillate from the standard flask are measured into it by means of a pipette.

The resulting mixture containing the silver cyanide precipitate is then neutralised with 0·1 N sodium hydroxide solution using 2 drops of methyl red as indicator, and then made just acid by the addition of 0·25 ml of 2 N nitric acid solution. The excess silver nitrate is now titrated electrometrically with the 0·1 N sodium chloride solution.

The 0·1 N sodium chloride is standardised against the 0·1 N silver nitrate solution by a similar titration procedure.

50 ml of the alkaline potassium cyanide solution are pipetted into a 500 ml standard flask and diluted to the mark with water. 100 ml of this solution are now treated in exactly the same way as the sample.

From the cyanide used up by the formaldehyde in the sample calculated in terms of the equivalent silver nitrate, the formaldehyde content of the sample is calculated

$$1 \text{ ml } 0\cdot1 \text{ N silver nitrate solution} \equiv 0\cdot003003 \text{ g } H\cdot CHO.$$

UREA-FORMALDEHYDE RESINS MODIFIED BY FURFURYL ALCOHOL

The properties of urea-formaldehyde plastics may be altered by modification of the resin itself or by the inclusion of additives such as fillers, plasticisers etc.

The determination of total furfuryl alcohol in resins modified by furfuryl alcohol may be carried out using the principle of the method described by Angell[11] for the bromination of furan and its derivatives. The reagent, pyridine sulphate bromide, effects complete bromination of the furan ring and hence the percentage total furfuryl alcohol (free and combined) may be calculated from the bromine value of the resin.

A gas chromatographic method is preferred for the determination of free furfuryl alcohol. This method, which may also be applied to phenol formaldehyde resins, is given below in detail.

A gas chromatograph fitted with a flame ionisation detector is required.

Chromatographic conditions

Column	2 m ($\frac{1}{8}$ in or 3 mm o.d. stainless steel) 20% w/w silicone gum SE 30 on celite (60–80 mesh)
Carrier gas	Nitrogen
Nitrogen pressure	10 lbf/in^2 (70 kN/m^2)
Hydrogen pressure	25 lbf/in^2 (170 kN/m^2)
Air pressure	30 lbf/in^2 (200 kN/m^2)
Column temperature	90°C

Method

About 1 g of the resin is weighed into a 5 ml graduated flask. The resin is dissolved in absolute alcohol and the solution diluted to the mark with the solvent. If the resin is not completely soluble, the mixture is allowed to stand for 4 h with frequent shaking. 0·05 ml of *n*-heptanol are now added to the mixture by means of an Agla micrometer syringe. After mixing thoroughly, 1 μl of the solution is injected on to the chromatograph.

The chromatogram is developed, attenuating as necessary in order to keep both *n*-heptanol and furfuryl alcohol peaks on scale. The peak areas for both *n*-heptanol and furfuryl alcohol are calculated from the formula

$$\text{Area} = \text{Height} \times \text{Width at Half Height}$$

Calibration

10 to 15 mg of furfuryl alcohol are weighed accurately into a 5 ml graduated flask and diluted to the mark with absolute alcohol. 0·05 ml *n*-heptanol are now added to the solution and the whole mixed thoroughly. 1 μl of the solution is chromatographed under the conditions outlined above and the peak areas are calculated as in the Method. All peak areas are reduced to the same attentuation.

Calculation

The % w/w free furfuryl alcohol is calculated from the formula.

$$\%\text{w/w} = \frac{A_{FS}}{A_{FST}} \times \frac{H_{ST}}{H_S} \times \frac{W_F}{W} \times 100$$

where A_{FS} = Peak area for furfuryl alcohol in sample
$\quad\quad A_{FST}$ = Peak area for furfuryl alcohol in standard
$\quad\quad H_{ST}$ = Peak area for *n*-heptanol in standard
$\quad\quad H_S$ = Peak area for *n*-heptanol in sample
$\quad\quad W_F$ = Weight of furfuryl alcohol in standard
$\quad\quad W$ = Weight of sample taken

UREA-FORMALDEHYDE SYRUP-STARCH PREPARATIONS

Certain problems may arise in the examination of urea-formaldehyde syrups. Commercial samples, such as plywood adhesives, may contain of the order of 50% of resin solids and about 5% of starch may be added to increase the viscosity of the syrup.

The addition of the starch helps to prevent the penetration of the syrup into wood veneers.

Determinations of starch may be called for in this type of preparation. Haslam and Jeffs[12] have worked out a test in which the starch is hydrolysed with acid to glucose and this glucose exerts a catalytic effect on the production of formose (a mixture of C_6 sugars) from formaldehyde. This formose reacts with flavianic acid to produce a reaction product that may be measured colorimetrically.

BOARD PREPARED FROM PRESSED WOOD CHIPS AND UREA-FORMALDEHYDE SYRUP

These preparations may require the determination of free formaldehyde.

Haslam and Jeffs have worked out a test based on the following principles.

Air is passed for 30 min at the rate of 150 ml/min over a 1 in (25 mm) square piece of the board, maintained at the temperature of boiling methanol.

The air, containing any formaldehyde, is passed through phenyl hydrazine solution. The reaction product is then treated by the principle of the Schryver test, i.e. by reaction with potassium ferricyanide prior to acidification and colour development with a definite strength of hydrochloric acid. The red coloured solution obtained is suitable for absorptiometric measurement.

REACTION PRODUCTS OF MELAMINE AND THIOUREA WITH FORMALDEHYDE

In Chapter 2 a considerable amount of information is given about tests based on Widmer's work that distinguish between urea- and melamine-formaldehyde preparations, and information is given on tests for thiourea resins based on the work of Fearon.

The work of Hirt, King and Schmitt[13] has been found to be invaluable in distinguishing between urea-formaldehyde and melamine-formaldehyde resins. Carried out in the following way the results are quite striking:

0·1 g of the resin is refluxed with 100 ml 0·1 N hydrochloric acid solution for 1 h. The solution is filtered through a sinter and a 25 ml aliquot of the filtrate is diluted to 50 ml with water.

To a second 25 ml of the filtered solution are added 3 ml of N sodium hydroxide solution and the solution is diluted to 50 ml with water.

Both acid and alkaline solutions are now examined in the U.V. over the wavelength range 225 to 255 nm in a 5 mm or appropriate cell.

The extract from the melamine-formaldehyde shows pronounced

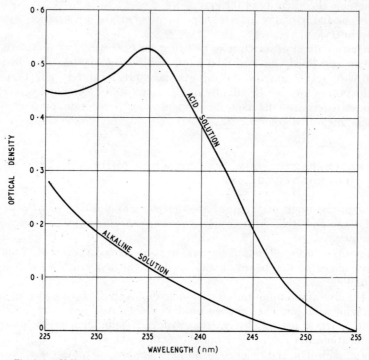

Fig. 9.4.   U.V. absorption of melamine-formaldehyde resin in acid and alkaline solutions

absorption on the acid side which is very much diminished on the alkaline side as in Figure 9.4.

The extract from the urea-formaldehyde resin exhibits negligible absorption on either the acid or alkaline side.

PLYWOOD GLUES

The analytical chemist may be called upon to identify the adhesive present in a sample of bonded plywood. It has been found that the procedure developed by Rendle and Franklin[14] is very useful.

In this method the sample is first boiled in water until the wood is waterlogged. Sections of the wood (10 to 20 μm thick) are then cut by means of a microtome[15].

Unstained sections of the plywood are then. examined under the microscope along with sections stained with:

1. Eosin.
2. Eosin with methyl blue.
3. Martius yellow.
4. Neutral red chloride.

The staining reactions of common plywood glues are summarised in Table 9.1 as well as their behaviour under U.V. light.

Table 9.1. Staining reactions and behaviour under U.V. light of common plywood glues

Glue	Resistance to water	Unstained	Eosin	Eosin and Methyl Blue	Martius Yellow	Neutral Red Chloride	U.V. Light
Urea-formal-dehyde	Poor	Clear, Colourless	Bright Pink	Pale Pink	Bright Yellow		Faint Purple
Phenol-formal-dehyde	Very Good	Light Brown		Green or Blue		Red Scarlet	
Animal glue	Very Good	Grey to Colourless					Brilliant White
Casein	Poor	Distinctly Grey or White					Faint Purple
Blood albumen	Very Good	Dark red or Black	Dark Red	Brown			Grey

HARDENERS

In the work on resins and syrups etc. the analytical chemist is bound to encounter hardeners of varying degrees of complexity. An example of the kind of procedure employed is given below; i.e. for a sample shown to contain

	% w/w
Hexamine	16·3
Urea	44·0
Ammonium chloride	10·8
Ammonia	17·7

These notes are likely to be of value in the examination of other hardeners.

1. *Total nitrogen*

1·0 ml of the sample is pipetted into a mixture of 7 g sodium sulphate-selenium catalyst mixture and 25 ml conc. sulphuric acid. The subsequent procedure is that adopted in the usual Kjeldahl determination.

2. *Free ammonia*

This is determined by titration of a known volume of the sample with standard acid.

3. *Total ammonia*

30 ml of distilled water are added to 2 ml of the sample and the solution is neutralised with 0·5 N sulphuric acid using phenolphthalein as indicator; 2·0 ml of 0·5 N sulphuric acid are added in excess. 15 ml of formaldehyde solution (Analar, 37% w/w) previously neutralised to phenolphthalein with 0·5 N sodium hydroxide solution are now added and the mixture brought to the boil. The hot solution is now titrated with 0·5 N sodium hydroxide solution, an excess of 0·5 ml being added. The final titration to the end-point is carried out with 0·5 N sulphuric acid solution. In the final calculation allowance is made for the 2·0 ml of 0·5 N sulphuric acid purposely added in the initial part of the test.

4. *Hexamine (dimedone method)*

1·0 ml of the sample is diluted to 100 ml with water; 10 ml of this solution are now diluted to 100 ml with distilled water and 10 ml dimedone solution added (10% w/v in alcohol). The solution is then neutralised to the purple tint of bromophenol blue indicator using N NaOH and N HCl solutions. 50 ml of buffer solution are then added (2 parts 62 g/l sodium acetate solution and 1 part N HCl). The mixture is digested on a water bath for 6 h, then cooled and allowed to stand overnight. The precipitate is then filtered off on a porosity 3 sintered glass crucible, washed free from chloride with the minimum amount of water and dried in an oven at 100°C.

$$\text{wt of precipitate} \times 0·08011 = \text{wt of hexamine}$$

5. *Urea (xanthydrol method)*

5·0 ml of the sample are diluted to 1 l and 5·0 ml of this solution are taken for the test. 5·0 ml of this dilute solution containing between 10 and 12 mg urea are treated with 44 ml glacial acetic acid and 6 ml xanthydrol solution (10% w/v in methyl alcohol) are added. The mixture is stirred well and then allowed to stand overnight. The precipitate is filtered off through a porosity 4 crucible, washed with a little methyl alcohol and then dried at 100°C.

$$\text{1 g urea} \equiv \text{7 g xanthydrol precipitate}$$

6. *Ionisable chlorine*

1 ml sample is diluted with distilled water, acidified with nitric acid and the chloride titrated potentiometrically with standard silver nitrate solution.

7. *Total solids*

These are determined by evaporation of a known volume of the sample to dryness on a water bath, followed by drying at 100°C in an oven or alternatively by drying over P_2O_5 in a vacuum desiccator.

PYROLYSIS/GAS CHROMATOGRAPHY OF THERMOSETTING RESINS

In our experience, pyrolysis of thermosetting resins is not a reproducible process and the method is not recommended for the detailed identification of this type of resin.

The pyrolysis method, however, is useful in the general classification of samples that are heavily filled, particularly when only very small amounts of sample are available. The pyrograms of phenol-cresol-formaldehyde resins invariably show peaks due to phenol, cresols and xylenols and small peaks due to benzene and toluene.

The retention times relative to styrene of these components under the chromatographic conditions defined in Chapter 1 (see p. 57) are as follows:

Benzene $R_T = 0.23$; Toluene $R_T = 0.45$; Phenol $R_T = 1.7$; *o*-Cresol $R_T = 2.8$; Mixed *m*- and *p*-Cresols $R_T = 3.2$, and Mixed Xylenols $R_T = 5.2$ and 7.2.

Urea-, thiourea- and melamine-formaldehyde resins yield only small proportions of volatile products on pyrolysis.

INFRA-RED SPECTRA OF THERMOSETTING RESINS

Thermosetting resins have received comparatively little attention from infra-red spectroscopists, no doubt because of the unpromising physical nature of many of the samples submitted for examination and the differences noted when comparing the spectra of various samples of what is nominally the same material, or even when comparing the spectra of different specimens prepared from the same sample of resin. Furthermore, the commonly encountered thermosetting resins such as phenol-, urea-, thiourea- and melamine-formaldehyde condensates, are readily identified

493

by straightforward chemical tests as set out in Chapter 2. These observations, together with the many specific tests described earlier in the present chapter, will often lead to a most satisfactory qualititative and quantitative analysis of a sample while the spectroscopist is still endeavouring to understand its spectrum.

As we have discussed earlier, comparatively simple reproducible infrared spectra are obtained from the majority of thermoplastic resins because of their extremely regular structures. However, when a monomer can polymerise to produce chemically different structures, as for example with isoprene or chloroprene which can polymerise by 1 : 2-, 1 : 4- or 3 : 4-addition, the spectra of products containing all possible chemical species are complex, the weaker bands in the spectra become broad and indistinct, and the relative intensities of the absorption bands may vary considerably in different samples of the resin. These effects become even more evident in the spectra of thermosetting resins, where many different reactions may occur during polymerisation, leading to a product which is chemically a complex mixture.

Thermoplastic resins are in most cases very little affected chemically by the normal methods employed in fabrication (such as compounding and moulding). These processes may cause modification of the spectrum of the resin through physical changes (crystallinity or orientation), but these need not cause difficulty in identification work since they are reversible and exactly the same spectrum can, in principle, be obtained from any sample of a pure thermoplastic resin by a correct choice of preparation method. Any chemical changes which do occur, e.g. thermal degradation or cross-linking, are insufficient to interfere with the qualitative identification of the resin.

However, the extent of chemical reaction and cross-linking which may occur during the moulding of thermosetting materials, or even as the result of heat treatment during sample preparation, is so great that, unless the effect of these structural changes on the infra-red spectrum can be recognised, there is bound to be difficulty in identifying thermosetting resins from their infra-red spectra. It is probably these effects which have most impeded the progress of infra-red spectroscopy in the interesting and important field of thermosetting resins, since materials which will not yield reproducible spectra are naturally regarded with some suspicion.

When the significance of these effects is appreciated, it becomes reasonably simple to recognise the principal types of thermosetting resins from their spectra, since prominent bands arise from significant groups present in each case which are not influenced to any great extent by cross-linking. Thus phenol-formaldehyde condensates show absorption bands due to phenolic hydroxyl groups, and the amide group in urea-formaldehyde and the triazine ring in melamine-formaldehyde condensates both give rise to useful characteristic bands. Sometimes it is impossible to understand other features of the spectrum, but where a systematic study has been made, as for example in the case of phenol-formaldehyde resins, the spectrum gives useful information on the chemical structures present, and in principle the same sort of detailed structural information as may be deduced from the spectra of olefine and diene polymers is available from the spectra of thermosetting resins.

In this description of the infra-red examination of these products we shall draw attention to that information which we feel can be reliably deduced from the spectra to supplement the results of other tests.

FILLED THERMOSETTING COMPOSITIONS

Thermosetting resins are often encountered in the form of heavily filled mouldings. The spectra of the thermosetting resins, although character-istic, usually consist of a small number of strong bands. Thus the spectrum of the composition will usually reveal absorption bands of the filler, giving evidence both as to its presence and nature. This observation is of the greatest practical value when the filler is a synthetic resin, although useful evidence may be obtained on inorganic fillers such as talc or calcium carbonate.

However, rather than to attempt the identification of a filler from the infra-red spectrum of the complete composition, it is preferable whenever possible to effect a separation. The thermosetting resins, before cross-linking, are usually extremely soluble, and in such cases a simple solvent separation may prove possible with methanol or acetone, leaving the added resin unaffected. After cross-linking, thermosetting resins are practically insoluble and this offers opportunities to remove thermoplastic resins by solvent extraction. Cross-linked phenol-formaldehyde resins may be 'dissolved' in 2-naphthol, and this often leaves the filler in a clean state for identification (p. 480). Similarly, the filler present in a urea-formaldehyde composition may be recovered by decomposition of the resin with acetic acid (p. 482) leaving the filler for examination by infra-red spectroscopy and other methods.

Samples containing relatively large amounts of inorganic fillers are often too opaque to radiation to permit useful spectra to be measured, and it is under these conditions that pyrolysis followed by infra-red examination of the complete pyrolysate, which has frequently been recommended for the identification of insoluble resins[16], is probably of the greatest value, particularly if the results are considered alongside data from pyrolysis/gas chromatography examination. Furthermore, the residue from pyrolysis may sometimes be a reasonably pure sample of the filler.

PHENOL-FORMALDEHYDE CONDENSATES

The major features of the spectrum of the condensation product of a phenol with formaldehyde are largely independent of the phenol used or the reaction conditions and it is therefore reasonably simple to identify these resins, as a class, from their infra-red spectra. The strong bands at

3 μm (O—H group) and near 8 μm (aromatic C—O group) are accompanied by bands of moderate intensity in the 6 μm to 7 μm region and between 11 μm and 15 μm due to substituted benzene nuclei. These spectral features are sufficient to prevent confusion with most other resins, although the spectra of epoxy resins (e.g. Spectrum 10.9) and of some phenol-terminated polyethylene oxide resins (e.g. Spectrum 10.5) are rather similar. These resins also show a strong band at 8 μm and a band of medium to high intensity at 3 μm, with the whole spectrum showing 'aromatic' character. However epoxy resins show a pronounced band near 9·6 μm due to the alkyl C—O group, while the polyethylene oxide resins have an even stronger band (due to simple aliphatic ether) at 9·0 μm. Some phenolic resins show bands in this region, but at different wavelengths, namely at 9·9 μm (resoles before cross-linking and novolaks containing free hexamine) or at 9·5 μm (cross-linked resoles). Absorption bands in other regions of the spectrum arise principally from substituted aromatic nuclei and do not therefore necessarily distinguish phenol-formaldehyde from other resins containing these groups.

NOVOLAKS BASED ON PHENOL

The spectrum of a phenol-based novolak resin, before cross-linking, is given in Spectrum 9.1. The major features have been referred to above; many of the other absorption bands are found to vary in intensity in different specimens.

Bender[17] has shown that the absorbance ratio 13·3 μm/12·2 μm can be related to the rate of cure of the resin with hexamine. He suggests the following assignments:

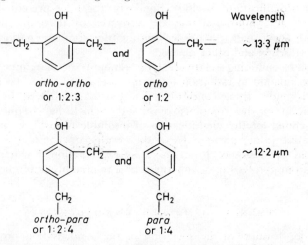

He concludes that the 'ortho-ortho' substituted structure is to be preferred for a rapid-curing resin since the regular chemical structure facilitates cross-linking.

496

The spectrum of a phenolic novolak in which the phenolic nuclei are linked by methylene bridges in the *ortho-ortho* positions (Hummel[18] Figures 275 and 276) shows a strong band at 13·3 μm and the corresponding *ortho-para* linked resin (Hummel Figure 277) shows bands of almost equal intensity at about 13·3 μm and 12·2 μm. These observations are entirely in agreement with Bender's assignments.

Further absorption bands suggested by a number of authors[17, 19, 20] might be indicative of other forms of ring substitution in phenolic novolaks, but the bands are weak and of doubtful value. Grisenthwaite and Hunter[21] consider that the weak 11·0 μm band arises from internuclear methylene bridges. However, most authorities[22] agree and our n.m.r. measurements show (p. 507) that resole resins also contain methylene bridges and since no band of significance occurs in these resins at 11·0 μm the observation would not seem to be well founded.

Phenolic novolaks are usually cross-linked by the addition of hexamine, and moulding powders with substantial amounts of free hexamine may be encountered. Hexamine is evident in the spectrum of the mixture (Spectrum 9.2) from the strong sharp doublet band at 10 μm. This extra band must be carefully distinguished from the methylol absorption which occurs in this region in resole resins, described later (p. 498). Spectrum 9.2 also shows a band at 14·4 μm. This is almost certainly due to residual free phenol in the resin, and could presumably be used to estimate free-phenol content, although we can find no reference to work on this subject.

This phenolic novolak-hexamine mixture was cross-linked at 150°C for 1 h, and alcohol extracted to remove any residual hexamine or soluble resin. The spectrum of the cross-linked product (Spectrum 9.3) is not greatly different from that of the original soluble novolak, and no new structure in the spectrum can be ascribed to the cross-linking bridges; since these are usually thought to be methylene groups which have a fundamentally weak absorption, this is perhaps not surprising. However, the 14·4 μm band disappears due to the removal of free phenol. With further cross-linking the 12·2 μm and 13·2 μm bands both weaken relative to the 8 μm C—O band. This reduced intensity of the 'substitution pattern' denotes the removal of substituent hydrogens from the benzene nuclei, presumably because these are removed in the cross-linking process. With prolonged heating, leading to heavy cross-linking, all the 12 μm to 15 μm 'substitution pattern' bands become substantially reduced in intensity, and this may cause difficulty in recognising the resins in this condition from their spectra. However under all circumstances the 8 μm and 3 μm bands appear to be very little affected in the cross-linking process.

NOVOLAKS FROM SUBSTITUTED PHENOLS

Phenolic novolaks may also be based on substituted phenols. The most common is the cresylic novolak made from mixed cresols (Spectrum

9.4). The general features are the same as those in the spectrum of the phenol-based resin, but the presence of additional alkyl groups is evident from the enhanced intensity of the —CH$_2$— structure at 3·4 μm and 6·8 μm in the cresylic novolak spectrum compared with that of the phenol novolak. Other differences are evident, for example the relatively greater intensity of the 12·3 μm band in the cresylic novolak spectrum, but this is not a reliable means of differentiating these materials since, as noted above, variations in the polymerisation conditions may appreciably modify those spectral features arising from the substituted benzene ring.

Some alkyl substituents show a characteristic absorption pattern which enables them to be recognised in the novolak spectrum. An example is the spectrum of tertiary butyl phenol novolak (Hummel[18] Figure 346). Here the split methyl band at 7·3 μm shows the presence of tertiary butyl groups but this would not differentiate between different alkyl chains with a terminal tertiary butyl group. In many cases, however, the only evidence of the presence of an alkylated phenol will be the prominence of the 3·4 μm and 6·8 μm alkyl bands and unfortunately these features could well arise from additives rather than the resin itself.

Phenols with aromatic rather than alkyl substitution produce a more obvious modification of the novolak spectrum, e.g. *p*-phenyl phenol novolak (Hummel[18] Figure 360). In this case a strong monosubstituted benzene ring pattern is superimposed on the spectrum. However, a qualitatively similar effect is noted in a phenol novolak containing a relatively large amount of free phenol, and some care is therefore required in the interpretation of these spectra.

In general it must be said that direct infra-red examination is not a satisfactory method of ascertaining the nature of the phenol used to prepare a novolak resin. In the case of soluble materials the n.m.r. spectra are particularly valuable, but when the sample is insoluble the best approach is by pyrolysis followed by G.L.C. or infra-red examination of the pyrolysis products. Such work must be done on an empirical basis, with known samples for comparison.

As in the case of a novolak based on unsubstituted phenol, the alkyl and aryl phenol novolaks show only minor changes in their spectra on cross-linking. These changes occur in the 11 μm to 15 μm 'substitution pattern' region but since with substituted phenol novolaks the band pattern here is complex before cross-linking, the effects are difficult to interpret.

PHENOLIC 'RESOLE' RESINS

The spectra of the self-hardening 'resole' resins show distinct differences from those of the corresponding novolak resins. If a resole resin solution is carefully dried at room temperature (Spectrum 9.5), the spectrum shows all the features associated with the novolaks, but in addition a strong band near 9·9 μm is evident, and the 3 μm hydroxyl band is very intense. These features, which serve to distinguish a resole from a novolak, arise from the methylol group, —CH$_2$OH, which is present in resoles and not in novolaks. Unfortunately these extra bands are easily confused with those due to the hexamine which may be mixed with a soluble novolak resin as cross-linking agent (Spectrum 9.2), but careful examination shows that the band near 9·9 μm is very sharp and doubled in the novolak-hexamine mixture, while it is smooth and much broader in the resole.

Marked changes occur in the phenolic resole spectrum when the specimen is cross-linked by heating to 150°C for 2 h (Spectrum 9.6). In the cross-linking process, most of the methylol groups disappear, as shown by the virtual disappearance of the 9·9 μm band and the much reduced intensity of the 3 μm hydroxyl band. The residual 3 μm band is now due to the phenolic hydroxyl groups, which play little or no part in the cross-linking process. Similarly, the 6·9 μm absorption band (alcoholic —OH deformation mode) largely disappears.

Intercondensation of the methylol groups in the cross-linking reaction produces the 'methylene ether' bridge

$$-CH_2-O-CH_2-$$

between the aromatic nuclei, and the corresponding ether band appears in the spectrum at 9·5 μm. A decrease of intensity of the 12·1 μm and 13·2 μm bands is noted, while the 11·3 μm and 12·7 μm bands become stronger. These changes denote increased substitution of the benzene nuclei due to condensation between the —CH$_2$OH groups and ring hydrogen atoms producing internuclear methylene bridges. Thus the marked changes in the spectrum which accompany cross-linking may be satisfactorily explained, and measurements of the intensity changes at 9·5 μm and 11·3 μm may be used to study the chemistry of the cross-linking process.

A very similar effect is noted in the corresponding spectra of a cresylic resole based on mixed cresols. In the material obtained by drying the resole solution at room temperature (Spectrum 9.7) the strong alcoholic hydroxyl structure at 3 μm, 6·9 μm and 9·9 μm is observed. After heat treatment (Spectrum 9.8) this structure virtually disappears, the shift of the 'substitution pattern' to shorter wavelengths is seen, and the 'methylene ether' band at 9·5 μm appears.

Comparison of the spectra of the phenolic and cresylic resoles in the soluble state (before cross-linking) (Spectra 9.5 and 9.7) shows that these are extremely similar, although the difference of relative intensity of the bands in the 11 μm to 14 μm region suggests a predominance of more heavily substituted nuclei in the cresylic resole. However, in view of the variation of intensity noted in these bands in the cross-linking process, it would be unwise to use them to differentiate phenolic and cresylic resoles. Even the 3·4 μm and 6·8 μm bands which can be used to differentiate phenolic and cresylic novolaks (p. 498) are not useful in this case owing to interference in each case from hydroxyl absorption bands.

Differentiation of phenolic and cresylic resoles is more certain on cross-linked samples, where there is comparatively little interference from alcoholic hydroxyl structure. In general, however, the remarks made on the identification of phenols in novolaks (p. 498) are applicable here, and in our opinion it is unwise in general to attempt to ascertain from the infra-red spectrum the nature of the phenol used in the preparation of a resole. As in the case of novolaks, n.m.r. spectroscopy is well suited to the identification of soluble resins, but for insoluble materials it is usually necessary to resort to pyrolysis followed by G.L.C. and infra-red examination of the pyrolysate on an empirical basis in relation to the behaviour of known samples.

MODIFIED-RESOLE RESINS

The alcoholic hydroxyl groups of a resole may be alkylated (etherified) to produce a resole ether. The most common example of such resins is the butylated resole (Hummel[18] Figures 323 and 324). The spectra of these materials are generally similar to those of phenol-formaldehyde resins, but the alkyl ether group is evident from the appearance of a strong and rather broad band at 9·2 μm, at a shorter wavelength than the band due to methylene ether in resole resins. Modification with fatty acids produces ester absorption bands superimposed on those of the simple resole, while additional bands due to both ester and ether appear in fatty-acid-modified etherified resoles (Hummel Figures 325 and 326). It is difficult to see how these materials would be distinguished from mixtures of etherified resoles and aliphatic esters, except by hydrolysis and identification of the products of hydrolysis. Reaction of the phenolic hydroxyl groups produces a much more dramatic change in the infra-red spectrum, as the strong hydroxyl absorption at 3·0 μm disappears. The most commonly encountered material of this type is the epoxydised phenol-formaldehyde resin; the spectra of these resins are very similar to those of epoxy resins based on bisphenol A (p. 525).

SIMPLE UREA-FORMALDEHYDE RESINS

The strong bands at 3 μm, 6·1 μm and 6·5 μm shown by all urea-formaldehyde condensates also appear in the spectra of nylons, natural proteins, and other resins containing the secondary amide group. However, all urea-formaldehyde condensates contain alcohol or ether groups, and the additional strong broad absorption bands found in their spectra in the 9 μm to 10 μm region are features which distinguish urea-formaldehyde condensates from the other secondary amide resins mentioned above. There is the possibility of confusion with methoxy-methyl modified Nylon 6 : 6 (Spectrum 5.11) since this contains side-chain ether groups, giving a prominent band at $9 \cdot 3$ μm, but the physical character of this resin is so different that confusion is unlikely. Furthermore, the methoxy-methyl modified Nylon $6:6$ resin is readily converted by solution in formic acid to Nylon $6:6$, and no doubt will then arise as to the nature of the resin.

The reaction between urea and formaldehyde is complex and, although the spectra of the products always possess the same general character, there is some variation in the spectrum of the resin before cross-linking which appears to be related to the ratio of urea to formaldehyde used in the preparation. This is illustrated by Spectrum 9.9 (U/F molar ratio $1:2 \cdot 2$) and Spectrum 9.10 (U/F molar ratio $1:1 \cdot 6$). The only difference in these spectra which can be easily interpreted in terms of a chemical difference between the resins is that the $9 \cdot 9$ μm band, due to methylol ($-CH_2OH$) groups, is more intense, relative to the 6 μm amide structure, in the resin made with a higher proportion of formaldehyde. This might be explained by assuming that the following reactions can occur in the condensation (formaldehyde is assumed to react as methylene glycol, $CH_2(OH)_2$):

$$NH_2.CO.NH_2 + n(HO.CH_2.OH + NH_2.CO.NH_2) + HO.CH_2.OH$$

$$\rightarrow NH_2.CO.NH - \left[CH_2.NH.CO.NH \right]_n - CH_2OH + (2n+1)H_2O$$

1. Formation of linear secondary polyamide.

$$\begin{array}{ccc}
NH_2 + HO.CH_2.OH & & NH.CH_2.OH \\
| & & | \\
O{=}C & \rightarrow & O{=}C \qquad +2H_2O \\
| & & | \\
NH_2 + HO.CH_2.OH & & NH.CH_2.OH
\end{array}$$

2(a). Formation of dimethylol urea.

$$n \left[\begin{array}{c} NH.CH_2.OH \\ | \\ O{=}C \\ | \\ NH.CH_2.OH \end{array} \right] \rightarrow$$

$$\begin{array}{ccccc}
NH{-}CH_2{-} & \left[N{-}CH_2{-\!-\!-\!-}N{-}CH_2 \right. & & \left. \right] & {-}N{-}CH_2.OH \\
| & | \qquad\qquad | & & & | \\
C{=}O & C{=}O \qquad\quad C{=}O & & & C{=}O \\
| & | \qquad\qquad | & & & | \\
NH.CH_2.OH & \left[NH.CH_2.OH \quad NH.CH_2.OH \right. & & \left. \right] & NH.CH_2.OH
\end{array}$$

$$_{n-2}$$

$$+(n-1) H_2O$$

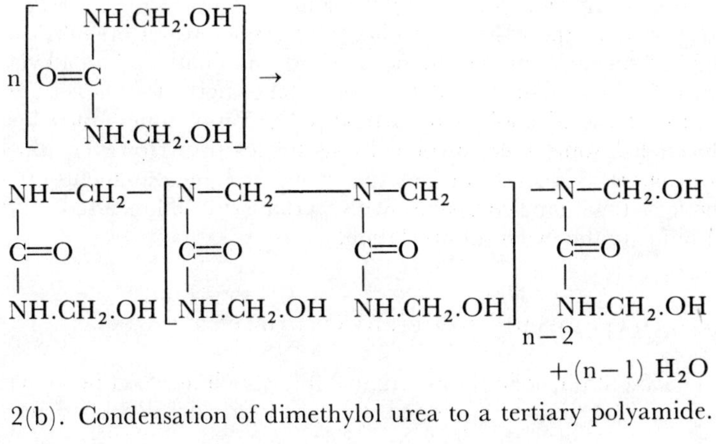

2(b). Condensation of dimethylol urea to a tertiary polyamide.

501

While the first structure is of a 'nylon' type containing only terminal hydroxyl groups, the second is very rich in hydroxyl; the latter requires relatively more formaldehyde and would presumably be favoured by a high F/U ratio.

This is in agreement with the relatively greater hydroxyl content of the high formaldehyde resin (Spectrum 9.9).

The spectrum of a simple urea-formaldehyde resin is not greatly changed when the resin is cross-linked (Spectrum 9.11). However, the decreased intensity of the hydroxyl group bands at 9·9 μm and 3·0 μm is evident, and the appearance of a shoulder at about 9·5 μm suggests that the hydroxyl groups which disappear on cross-linking are partially replaced by methylene-ether bridges, $-CH_2-O-CH_2-$, as in the phenolic resole resins. This additional feature is however poorly defined, which suggests that the methylene-ether bridges may further decompose (with loss of formaldehyde) to form methylene bridges.

ALKYLATED UREA-FORMALDEHYDE RESINS

The partial alkylation of a urea-formaldehyde condensate in which some of the free methylol groups are replaced by ether groups produces a well-marked change in the infra-red spectrum. The 'butylated' resin (Spectrum 9.12) shows a prominent band at about 9·2 μm with much reduced hydroxyl absorption at 9·9 μm and 3·0 μm in comparison with the simple urea-formaldehyde resin spectrum. Fortunately the ether band lies at a shorter wavelength than that due to the ether bridges which appear in the spectrum of urea-formaldehyde resin after cross-linking, the latter falling at about 9·5 μm.

As with the simple urea-formaldehyde resin, there are only minor differences in the spectrum on cross-linking, but any band due to ether bridges would be difficult to observe owing to the relatively strong bands already present in the 9 μm to 10 μm region.

As discussed in the case of novolaks and resoles based on substituted phenol, it seems unlikely that direct infra-red examination of an alkylated urea-formaldehyde resin would be a promising method for ascertaining the nature of the alcohol used to modify the resin, unless the alcohol introduced has some spectroscopically useful feature. However, alcohol modification is evident from the spectrum, and measurements of the absorbance at 9·2 μm, compared with perhaps the 6 μm amide band, should indicate the extent of alkylation.

UREA-FORMALDEHYDE MODIFIED WITH OTHER RESINS

As in the case of phenolic resins, organic fillers such as wood flour, paper and cotton are frequently added to urea-formaldehyde condensates.

Since the urea-formaldehyde spectrum is comparatively simple, such additives are usually evident from the spectrum, but complete identification will depend in most cases upon effecting a separation.

THIOUREA-FORMALDEHYDE CONDENSATES

The spectrum of a thiourea-formaldehyde resin is given by Hummel (Figure 1393). The amide absorption near 6·5 μm remains but the 6 μm amide band is absent and a band near 7·5 μm appears, presumably due to C=S groups.

The co-condensate, urea-thiourea-formaldehyde, is encountered more frequently than the simple thiourea resin. Spectrum 9.13 is similar to that of simple urea-formaldehyde, but the 6 μm amide band is seen to be weaker relative to the 6·5 μm band, and the band near 7·4 μm in this resin, not found in urea-formaldehyde, is similar to that seen in straight thiourea resin.

Comparison of the spectrum of the sample after cross-linking (Spectrum 9.14) with the original (Spectrum 9.13) shows changes which are generally similar to those observed in the cross-linking of urea-formaldehyde resin (Spectra 9.10 and 9.11); both were subjected to 2 h heating at 160°C. In Spectrum 9.14 the hydroxyl band at 9·9 μm has now virtually disappeared and is replaced by an ether band of considerable intensity at 9·6 μm. The 3 μm band has also decreased considerably in intensity and presumably the residual 3 μm absorption is due to NH groups.

MELAMINE-FORMALDEHYDE RESINS

Spectrum 9.15 of a melamine-formaldehyde resin, before cross-linking, shows clear differences from that of urea-formaldehyde. The amide band at 6 μm (due principally to the amide carbonyl group) is absent, the 6·5 μm band, here due to the triazine ring, becomes extremely intense, and the 12·2 μm band, also characteristic of the triazine ring, is seen. These two bands appear to be little influenced by substituents on the pendent amine groups and the appearance of both bands in a spectrum can be taken as good evidence for the presence of the triazine ring.

Since the melamine resins (like the urea condensates and the resoles) contain CH_2OH groups before cross-linking, the band at 9·9 μm and the intense 3 μm band are very prominent in the spectra.

Cross-linking of a simple melamine-formaldehyde resin (heated for 1 h at 160°C) produces an effect (Spectrum 9.16) very similar to that noted in both the resoles and urea-thiourea-formaldehyde condensates. The hydroxyl bands at 9·9 μm and 3 μm are reduced in intensity, the latter band being now much sharper, indicating that it originates mainly from NH groups. The expected ether band appears at 9·5 μm. Other minor changes in the spectrum occur, but the reason for these is not clear.

ALKYLATED MELAMINE-FORMALDEHYDE RESINS

Alkylation of a proportion of the free alcohol groups in a melamine condensate to produce an 'alcohol modified' resin has an effect on the spectrum which is remarkably similar to that observed on alkylation of a simple urea-formaldehyde resin. An example is butylated melamine-formaldehyde (Spectrum 9.17). Comparing this with Spectrum 9.15, a considerable decrease is noted in the alcoholic absorption bands at 9·9 μm and 3 μm while the ether band appears at about 9·2 μm. This band is well separated from the 9·5 μm ether band which occurs in un-modified melamine-formaldehyde resins after cross-linking. For the identification of the modifying alcohol, the infra-red spectrum of the resin does not give unambiguous information, and chemical methods are to be preferred.

MELAMINE-FORMALDEHYDE MODIFIED WITH OTHER RESINS

Melamine-formaldehyde, either as the simple or alcohol-modified resin, may be mixed with fillers such as cotton or wood flour, or 'modified' with synthetic resins such as alkyl resins. In most cases the spectrum will reveal the presence of the additive and give some evidence of its chemical nature. Where possible, separation of the resins is to be preferred for identification, although the spectra of a number of these 'modified' melamine resins appear in the *Sadtler Commercial Spectra Series D*[23], but these are of somewhat limited value as the constituents of the resins are not always quoted.

N.M.R. SPECTRA OF PHENOL-FORMALDEHYDE RESINS

N.M.R. measurements of phenol-formaldehyde resins are restricted to soluble samples, but in these cases a considerable amount of information is available from the n.m.r. spectrum. The interpretation given here is based on the extensive work of Woodbrey *et al.*[24] and Hirst *et al.*[25]

Figure 9.5 is the n.m.r. spectrum of a phenol-formaldehyde resole in acetone-d$_6$ solution. The resonance of the aryl protons is the broad feature near 3 τ, while the hydroxyl (phenolic) proton resonance appears near 3·3 τ. The latter resonance is very variable in position, depending upon the sample temperature, solvent, and the concentration of the

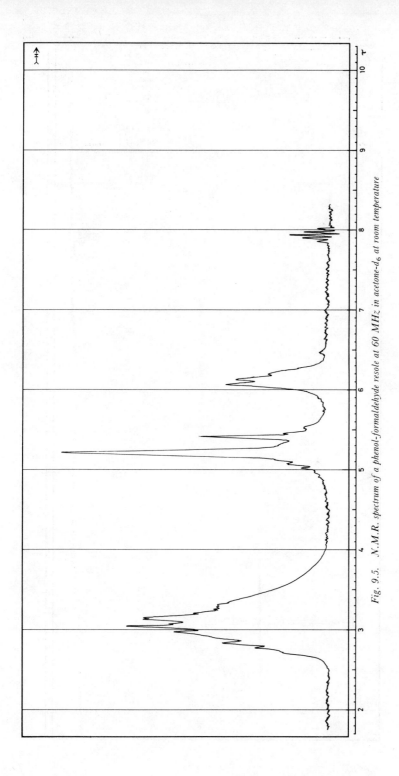

Fig. 9.5.   N.M.R. spectrum of a phenol-formaldehyde resole at 60 MHz in acetone-d₆ at room temperature

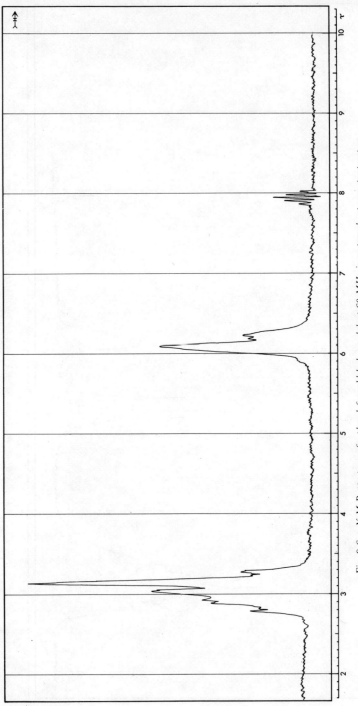

Fig. 9.6. N.M.R. spectrum of a phenol-formaldehyde novolak at 60 MHz in acetone-d₆ at room temperature

solution. Under some circumstances it may not be observed because it falls outside the $0\,\tau$ to $10\,\tau$ region.

ALCOHOL AND ETHER GROUPS IN RESOLES

In Figure 9.5 the resonances in the $5\cdot0\,\tau$ to $5\cdot7\,\tau$ region are ascribed to the protons of the methylene group next to an oxygen atom, as e.g. in the methylol group and the methylene ether bridge. It is generally accepted that these groups are present in resoles but do not occur in novolaks, and indeed there are no resonances in this region in the spectra of novolaks, e.g. Figures 9.6 and 9.7 which are spectra of phenolic novolaks. Thus there is a clear and easily interpreted difference between the n.m.r. spectra of resoles and novolaks.

The assignment of the resonances between $5\cdot0\,\tau$ and $5\cdot7\,\tau$ presents some difficulty, as a number of resonances are observed in this region which differ in position and relative intensity in different samples. Woodbrey et al.[24] have studied the n.m.r. spectra of a number of model compounds and have also observed the spectra of resoles before and after acetylation of the alcoholic hydroxyl groups; this enables resonances due to $-OCH_2O-$ groups to be distinguished from those of $-CH_2OH$ groups. They consider that, in addition to the methylol and methylene ether groups, the structure $Ar-OCH_2O-CH_2OH$ is present in some resoles and they propose a scheme for determination of the proportions of the different $-OCH_2-$ groups in these resins.

METHYLENE BRIDGES

Resonances are present in the spectra of both resoles and novolaks in the region $6\cdot0\,\tau$ to $6\cdot5\,\tau$. These are ascribed to the protons in the methylene bridge $Ar-CH_2-Ar$.

The pattern of resonances in this region differs considerably in different resins, as is evident from Figures 9.6 and 9.7. Woodbrey et al.[24] suggest

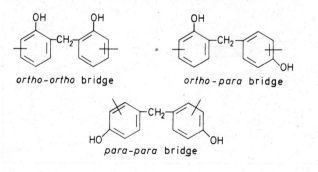

ortho-ortho bridge   *ortho-para* bridge

para-para bridge

507

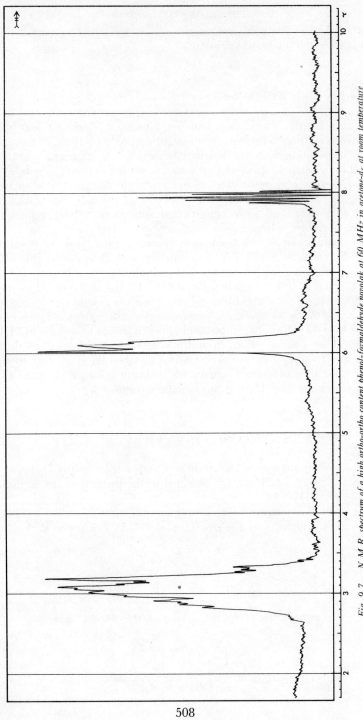

Fig. 9.7. N.M.R. spectrum of a high ortho-ortho content phenol-formaldehyde novolak at 60 MHz in acetone-d_6 at room temperature

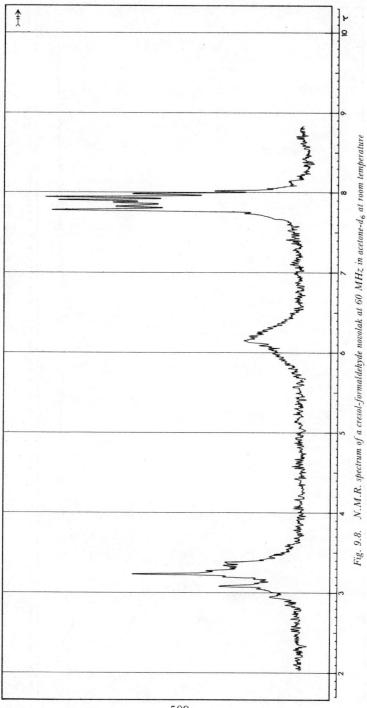

Fig. 9.8. N.M.R. spectrum of a cresol-formaldehyde novolak at 60 MHz in acetone-d_6 at room temperature

509

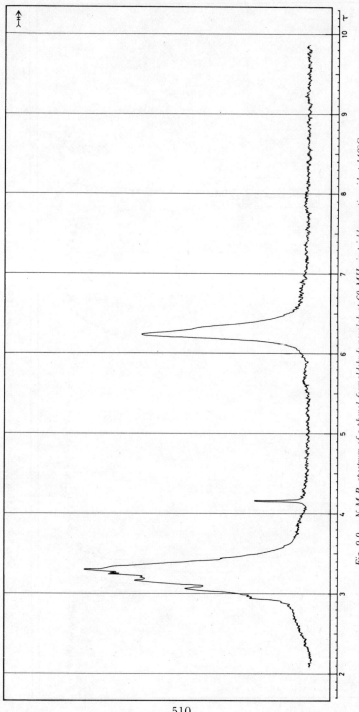

Fig. 9.9. N.M.R. spectrum of a phenol-formaldehyde novolak at 60 MHz in trichloroacetic acid at 140°C

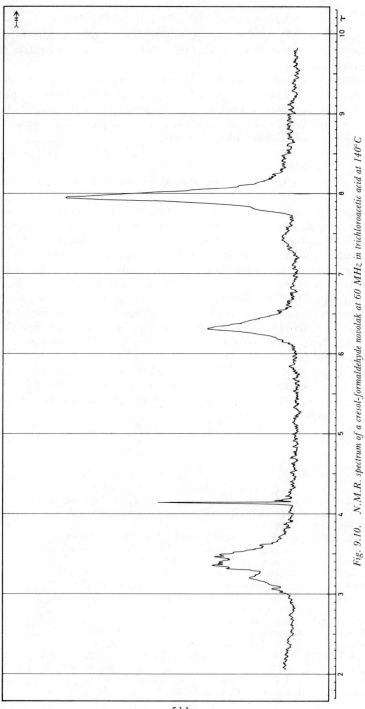

Fig. 9.10. N.M.R. spectrum of a cresol-formaldehyde novolak at 60 MHz in trichloroacetic acid at 140°C

that by an analysis (resolution) of this resonance pattern, the ratio of *ortho-ortho* plus *ortho-para* bridges to the number of *para-para* bridges may be determined.

This measurement is related to the infra-red measurement of bands at 13·3 μm and 12·2 μm (see p. 496), which gives information on the reactivity of the resin. Unfortunately, the structure in the 6·0 τ to 6·5 τ region is often very broad and indefinite, as e.g. in Figure 9.8 which is the spectrum of a cresol-formaldehyde novolak.

DIFFERENTIATION OF RESINS BASED ON DIFFERENT PHENOLS

The resonances of alkyl substituents on benzene rings (as opposed to those associated with methylene bridges between rings) fall well to high field of all the features so far described, and hence in the case of either resoles or novolaks which may be dissolved in a solvent which contains no alkyl protons it should be possible to identify alkyl substitutents on the benzene rings from the n.m.r. spectrum. A very simple example is shown in Figures 9.9 and 9.10 which are the spectra of a phenolic and a cresylic novolak dissolved in trichloroacetic acid. The very prominent resonance at 8 τ is due to the methyl group on the benzene ring in the cresylic novolak; as expected there is no resonance in this region in the spectrum of the phenolic novolak.

Measurement of the area of this methyl resonance compared with the area of the resonance due to aryl protons is a satisfactory method of determination of the proportions of cresol and phenol in a mixed cresol-phenol-formaldehyde resin.

REFERENCES

1. HASLAM, J., and SOPPET, W. W., *J. appl. Chem., Lond.*, **3,** 328 (1953)
2. HASLAM, J., WHETTEM, S. M. A., and NEWLANDS, G., *Analyst, Lond.*, **78,** 340 (1953)
3. CHAPIN, *Standard Methods for Testing Tar and its Products*, Standardisation of Tar Products Test Committee, 3rd edn, London, 217 (1950)
4. BUSFIELD, H., and CHIPPERFIELD, E. H., *Analyst, Lond.*, **78,** 617 (1953)
5. WALLIS, T. E., *Practical Pharmacognosy*, J. & A. Churchill, London, 6th edn (1953)
6. HASLAM, J., and HILL, R. V., *Analyst, Lond.*, **80,** 317 (1955)
7. HASLAM, J., and SWIFT, S. D., *Analyst, Lond.*, **79,** 82 (1954)
8. PICKLES, D., and WASHBROOK, C. G., *Analyst, Lond.*, **78,** 304 (1953)
9. KAPPELMEIER, C. P. A., *Paint Oil Chem. Rev.*, **14,** 115 (1952)
10. GRAD, P., and DUNN, R. J., *Analyt. Chem.*, **25,** 1211 (1953)
11. ANGELL, F. G., *Analyst, Lond.*, **72,** 178 (1947)
12. HASLAM, J., and JEFFS, A. R., *J. appl. Chem., Lond.*, **4,** 653 (1954)
13. HIRT, R. C., KING, F. T. and SCHMITT, R. G., *Analyt. Chem.*, **26,** 1273 (1954)
14. RENDLE, B. J., and FRANKLIN, G. L., *J. Soc. Chem. Ind., Lond.*, **62,** 11 (1943)
15. *The Preparation of Wood for Microscopic Examination*, Leaflet No. 407, D.S.I.R. Forest Products Research Laboratory
16. HARMS, D. L., *Analyt. Chem.*, **25,** 1140 (1953)

17. BENDER, H. L., *Mod. Plast.*, **30,** 136 (1953)

18. HUMMEL, D. O., *Infra-Red Analysis of Polymers, Resins and Additives, An Atlas,* **1,** Pt II, Wiley Interscience (1969)

19. RICHARDS, R. E., and THOMPSON, H. W., *J. chem. Soc.*, 1260 (1947 Pt 2)

20. RANDLE, R. R., and WHIFFEN, D. H. (Ed. SELL, G.), *Molecular Spectroscopy*, Institute of Petroleum, London, 111 (1955)

21. GRISENTHWAITE, R. J., and HUNTER, R. F., *J. appl. Chem., Lond.*, **6,** 324 (1956)

22. CARSWELL, T. S., *High Polymers Volume vii, Phenoplasts*, Interscience, New York, (1947)

23. *Sadtler Commercial Spectra Series D*, The Sadtler Research Laboratories, Philadelphia 2

24. WOODBREY, J. C., HIGGINBOTTOM, H. P., and CULBERTSON, H. M., *Jnl Polym. Sci., Pt. A*, **3,** 1079 (1965)

25. HIRST, R. C., GRANT, D. M., ROFF, R. E., and BURKE, W. J., *Jnl. Polym. Sci., Pt. A*, **3,** 2091 (1965)

10

NATURAL, CELLULOSE, EPOXY, POLYETHER AND SILICONE RESINS

This chapter deals with miscellaneous resins. The first section is concerned with chemical tests on naturally occurring resins, materials such as casein formaldehyde, and with cellulose acetate and nitrate, where attention is principally directed to methods of assay.

The next section describes the infra-red examination of polyethers, including polyethylene oxide, polyformaldehyde and the epoxy resins. The infra-red spectra of cellulose and its derivatives are then discussed.

Finally the chapter contains a section devoted to the examination of silicone resins, in which both chemical and spectroscopic methods are discussed.

Where applicable, information is provided on the pyrolysis/gas chromatographic examination and the n.m.r. spectra of the various resins.

The authors appreciate that they have included very little information on natural resins and resins derived from natural products. This is mainly because their experience of these is limited, but these resins are already well covered in existing textbooks, and their infra-red spectra are well described in Hummel's book[1].

There are of course many synthetic resins to which no specific reference has been made. The far-seeing analyst should be aware of the importance of the information, both chemical and spectroscopic, supplied by Hummel[2] on a number of interesting and unusual synthetic resins.

NATURAL RESINS, E.G. SHELLAC, COLOPHONY ETC.

The analyst in the plastics industry is occasionally called upon to identify natural resins such as shellac, copal and colophony.

The Liebermann–Storch reaction of the last is well known.

Nevertheless, the fact that most natural resins are optically active whereas the synthetic resins are not, can be of value, e.g. a sample of colophony when examined in alcohol solution in our laboratory had an optical rotation of $+12°$.

One finds in practice, and here Kline[3] is of value, that the natural resins are distinguished from one another by such values as:

Appearance
Softening point
Refractive index
Solubility
Acid number
Saponification number
Iodine value
Acetyl value
Carbonyl value
Methoxy number

CELLULOSE ACETATE

It is useful to know something about the chemical determination of the proportion of acetate. In the old days the sample was left overnight with sulphuric acid solution (1 : 1 by volume of sulphuric acid and water). On the following day the acetic acid was distilled in steam and titrated in the steam distillate with standard alkali using phenolphthalein as indicator.

Nowadays quite different methods are in use and it is probable that the most generally applicable of all is that given for the determination of acetic acid yield in BS 2880 : 1957.

METHODS OF TESTING CELLULOSE ACETATE FLAKE

This method has been found suitable for cellulose acetates of acetic acid yields ranging from about 40% to a nominal 62·5%. The term acetic acid yield is preferred for describing what has erroneously for decades been known as 'acetyl value'. With cellulose acetates of acetic acid yields of more restricted range, other methods give good results and may be easier to apply, but the method described has the advantage of being applicable to all the types of cellulose acetate likely to be met in present-day commercial manufacture. The method consists essentially in allowing finely divided cellulose acetate to stand in contact with a mixture of acetone and aqueous sodium hydroxide and then measuring by titration the amount of alkali consumed in hydrolysing the cellulose acetate.

Reagents

All reagents shall be of recognised analytical reagent quality. Distilled water shall be used whenever water is specified.

Approx. N sodium hydroxide, carbonate free.

0·5 N sodium hydroxide, carbonate free.

Approx. N sulphuric acid.

In order to ensure that there will be a positive back-titration value in the blank, the normality of the sulphuric acid should be greater than that of the sodium hydroxide.

Acetone complying with BS 509.

0·5% w/v phenolphthalein solution in alcohol.

Procedure

Grind (a beater cross mill has been found suitable) at least 2 g of the sample so that all will pass a 72 mesh BS test sieve. The addition of solid carbon dioxide to assist grinding is permissible.

Place a test portion of the ground material equivalent to about 1·5 g of oven-dry material in a weighing bottle, dry for 3 h at a temperature of $110 \pm 2°C$ and cool in a desiccator. Stopper the bottle and weigh. Transfer the contents of the bottle to a 300 ml stoppered conical flask* of borosilicate glass. Reweigh the stoppered bottle and determine the weight taken by difference. Shake the test portion evenly over the base of the flask and, without lifting the flask from the bench, carefully run 15 ml of water round the sides of the flask to ensure even distribution over the base; then add 65 ml of acetone. In order to prevent the formation of lumps, the first 10 ml are added very slowly by being poured carefully round the sides of the flask while the flask is turned gently without its base leaving the bench. Allow the flask and contents to stand for 30 min; then shake for 3 h or allow to stand overnight.

Add 25 ml N sodium hydroxide slowly with continual swirling. Shake for 3 h. Wash down the stopper with water, adding approximately 50 ml of water to the contents of the flask. Add 25 ml of N sulphuric acid and about 0·5 ml of the phenolphthalein solution. Allow to stand, shaking if necessary, until any signs of pink coloration have disappeared from the insoluble matter. Titrate with 0·5 N sodium hydroxide. Carry out a blank determination on the reagents used.

Calculation

$$\text{Acetic acid yield} = \frac{3 \cdot 002 \times (A - B)}{W}\%$$

where W = weight of oven-dry test portion in g

* If rubber stoppers are used they must previously be shaken in a mixture of acetone, water and N sodium hydroxide for 3 h and then thoroughly washed with water before use.

A = volume of 0·5 N sodium hydroxide solution required for the test portion in ml

B = volume of 0·5 N sodium hydroxide solution required for the blank in ml

The above method is that described in BS 2880 : 1957; it is probable that the precision of the test may be improved still further by the application of full-scale titrimetric procedures to the final titration.

CASEIN, GELATIN, GLUE AND BLOOD ALBUMIN

It is desirable that the analyst in the industry should have a working knowledge of the properties of casein, gelatin, blood albumin and glue (for plywood glues see p. 490). In this particular field most of the problems that are encountered are concerned with casein formaldehyde.

The following notes about casein formaldehyde, used for the manufacture of buttons, may be of value.

CASEIN FORMALDEHYDE

Specific gravity   1·33 to 1·35 or higher depending on the pigment and its concentration

Refractive Index   1·47

Formaldehyde   3 to 4%

Casein formaldehyde plastics fluoresce in U.V. light with a bluish-white visible light. Casein formaldehyde is unaffected by dilute acids and alkalis.

Casein formaldehyde plastic is resistant to concentrated hydrochloric acid at the boil for at least 15 min and for some 24 h at room temperature. Unformalised casein is readily attacked.

Casein formaldehyde plastic ignites in a flame but is self-extinguishing when withdrawn. On ignition it discolours brown, swells and bubbles, and chars with the production of alkaline gases and an odour of burning protein (hair, feather). The formaldehyde given off may be detected.

Formaldehyde in casein formaldehyde plastics may be detected and determined quantitatively by the distillation (to whitish fumes) of shavings in 40% w/w phosphoric acid solution, the end of the condenser being immersed in water or excess sodium sulphite solution.

The formaldehyde in the distillate may be detected by three methods

(a) with chromotropic acid
(b) by the Schryver phenyl hydrazine test
(c) by the phloroglucinol test

and determined by a volumetric method e.g. the sodium sulphite test.

Protein in casein formaldehyde plastics may be detected by the Biuret reaction. Nitrogen, sulphur and phosphorus may be detected by the usual sodium fusion tests or by Oxygen Flask Combustion procedures. Nitrogen may be determined by the Kjeldahl method and phosphorus may be determined after wet digestion.

Casein formaldehyde plastics are usually made from rennet casein which really consists of a casein-calcium phosphate complex.

Rennet casein contains from 0·70 to 0·74% phosphorus in organic combination and approximately 1% inorganic phosphorus. The P/N ratio for commercial casein is almost constant at 0·053 provided that it is determined by dissolving the plastics shavings in 0·05 N NaOH, precipitating with trichloroacetic acid, wet oxidising the washed precipitate by the Kjeldahl method and determining the nitrogen and the phosphorus in separate aliquots of the digestion product. For pure rennet casein, nitrogen is about 15·6% and phosphorus about 0·70 to 0·74% after allowing for ash, fat and moisture. This P/N ratio may be regarded as constant for casein which has not been fractionated.

The sulphur content of casein is of the same order as phosphorus i.e. about 0·7%. This distinguishes casein from keratinous proteins which, unlike casein, have a sulphur content of about 3·3% owing to their high cystine content.

Another test that distinguishes keratinous proteins from casein plastics involves titration of the sulphuric acid extract of the sample with standard potassium permanganate solution. The usage of potassium permanganate is much higher with keratinous proteins than with casein.

NITROCELLULOSE

Most of the tests encountered in the literature and in textbooks on the subject are tests devised for specification purposes, e.g. tests on: moisture in water-wet and dehydrated samples, total volatile matter, spirit strength, matter insoluble in acetone, matter insoluble in ether-alcohol, solubility in butyl alcohol, mineral matter, total iron, acidity, viscosity, gelation, film test, flash point of solution and water compatibility.

Nevertheless, workers in the plastics field will meet nitrocellulose and may, on occasion, require methods that will give some guidance in its quantitative assessment.

The authors draw attention to three useful methods of assay.

1. A nitrometer method.
2. A Devarda's-alloy reduction method.
3. The Schultz–Tieman method as modified by Rubens[4].

NITROMETER METHOD

The nitrometer method is an old one; indeed it is so old as to be omitted from the 1932 B.P. but, nevertheless, is of ready applicability in plastics analysis.

Prior to 1932 the test was described in the B.P. in terms of Pyroxylin as follows:

'Pyroxylin is a nitrated cellulose obtained by the action of a mixture of nitric and sulphuric acids on cotton wool, which has been freed from fatty matter, and by subsequent purification. It contains, when dry, not less than 11·5 and not more than 12·3% of nitrogen. Assay. Rub a portion through a sieve, place in a weighing bottle and dry at 100°C for 1 h. Transfer to a mercury filled nitrometer about 0·5 g (accurately weighed), with 5 ml of nitrogen free sulphuric acid, and wash it in with four 2·5 ml portions of the same acid. Shake the nitrometer well for 1 min, or until the evolution of gas ceases and after allowing it to stand for 2 min, again shake for 1 min, and allow it to stand for a further 15 min; measure the volume of liberated nitric oxide. Each ml of nitric oxide, at normal temperature and pressure, is equivalent to 0·000625 g of nitrogen.'

DEVARDA'S ALLOY REDUCTION METHOD

This method, as applied to industrial nitrocellulose, is as follows.

If necessary, remove excess volatile matter from a portion of the sample as received by pressing the material between sheets of filter paper. Mix the sample thoroughly by rubbing it gently between the hands and transfer the material to a bottle. Dry a portion of the sample on an aluminium tray in an oven at 50 to 55°C for 16 h or alternatively at 60 to 70°C for at least 2 h. The material thus prepared is referred to subsequently as the 'dried sample'.

Transfer 0·4 g of the dried sample to a small weighing bottle. Dry the bottle and contents in a boiling-water oven for 1 h. The nitrogen determination is carried out as follows: 0·4 g of the dried sample, 20 ml of water, 20 ml of hydrogen peroxide solution (20 volumes), 50 ml of potassium hydroxide solution (30% by weight), and 5 ml of alcohol (95% v/v), are allowed to stand in a distillation flask until the sample is dissolved. This solution process usually takes about 2 to 3 h.

3 g of Devarda's alloy (passing 72 BS mesh) are added. The distillation flask is fitted with a spray trap condenser etc. and any ammonia that is volatilised is collected in 50 ml of 0·1 N hydrochloric acid solution containing 3 drops of methyl red indicator.

This initial reaction is allowed to proceed without the application of heat. The mixture is then boiled and distillation is carried out until only about 30 ml of liquor remain in the distillation flask.

The excess acid remaining in the collection flask is back titrated with 0·1 N sodium hydroxide solution using methyl red as indicator. A blank determination is carried out on the whole process. The nitrogen content of the dried sample is calculated from the amount of acid used up in the test.

SCHULTZ–TIEMAN METHOD

This method is very useful for the determination of nitrogen in fibrous nitrocellulose samples which present a large surface area to the reagents. In this test the sample is decomposed with ferric chloride and hydrochloric acid to yield nitric oxide quantitatively. The gas is collected and measured over 30% w/v sodium hydroxide solution.

PYROLYSIS/GAS CHROMATOGRAPHY OF CELLULOSIC MATERIALS

Our limited experience in the examination of cellulosic materials by pyrolysis/gas chromatography indicates that under the conditions used (p. 57) the various forms of cellulose show a general pattern of main peaks with retention times relative to styrene of R_T 0·06, 0·08 and 0·10

Table 10.1. Retention times relative to styrene of main additional peaks from cellulose esters and ethers

Cellulose acetate	0·18 and 0·4
Cellulose propionate	0·33 with small peak at 0·7
Cellulose butyrate	0·6
Cellulose nitrate	4·9 with small peaks at 3·8 and 6·5
Methyl cellulose	0·15 (much larger than from cellulose)
Ethyl cellulose	0·22
Benzyl cellulose	0·45 and 1·5 with small peaks at 0·7, 1·0 and 2·5

with minor peaks at R_T 0·15, 0·25, 0·45 and 0·65. Within this general pattern the intensities of the peaks show variations depending upon the nature of the cellulose (i.e. between cotton, linen and rayon).

When the various cellulose esters or ethers are pyrolysed, the pyrograms show additional strong peaks which characterise the material. The retention times of these peaks relative to styrene are listed in Table 10.1.

INFRA-RED SPECTRA OF POLYETHERS

The ether group gives rise to one or more very strong bands in the region 8 μm to 11 μm, and if a resin with these spectral features contains no elements other than carbon, hydrogen and oxygen, and shows no carbonyl or hydroxyl band (or only weak bands in these positions) the material is likely to be an ether polymer.

The position and nature of the absorption bands due to the ether group vary considerably with different ether structures, and this forms a useful method of preliminary classification of these substances.

1. A STRONG BAND NEAR 9·0 μM

This is shown by all simple dialkyl ethers, except those containing the geminal diether with two ether oxygens on a common carbon i.e.

$$-O-CH_2-O-, \quad -O-CHR-O- \text{ and } -O-CR_2-O-$$

which have a quite different spectral character. The simple dialkyl ethers in the liquid or solid amorphous state typically show a broad 9 μm band, often with poorly developed shoulders and which is usually strong enough to completely dominate the spectrum. The band is so constant in position and appearance that the infra-red spectra of polymers containing the simple ether group are superficially similar to one another.

Some ether polymers with a regular structure, e.g. ethylene oxide polymers, can crystallise, and in this condition the spectrum still shows the very strong 9 μm band, but it is greatly reduced in width and sharp satellite peaks appear on either side of the main band. When the sample is changed to the amorphous condition, for example by melting, the ether band reverts to its normal shape described above.

The band in amorphous ethers is somewhat similar in nature to that appearing at about this wavelength in some metallic oxides, particularly silica, and in inorganic sulphates; this may cause confusion in samples containing these as inorganic fillers.

This important class of compounds includes the vinyl ethers, described in Chapter 3, the cellulose ethers described later in this chapter and the polyalkylene oxides (other than polymethylene oxide) which are described in this section.

A modification of this pattern occurs in ethers with a cyclic (alkyl or aryl) substituent or a nitrogen atom on the α-carbon atom. In the spectra of such compounds the band remains single and broad, as in the other alkyl ethers described above, but moves to longer wavelengths (9·5 μm). The presence of this band has been noted as arising from the methylene ether bridges in the spectra of urea, melamine and phenolic resole resins (Chapter 9) and this band is a useful feature in the spectra of cellulose derivatives, which have a strong band near 9·5 μm, irrespective of the way in which the cellulose is modified. Infra-red examination of cellulose and its derivatives is described later in this chapter.

2. TWO OR MORE STRONG BANDS BETWEEN 8 μM AND 11 μM

Strong interaction of the group vibrations occurs in the C—O—C—O—C linkages of the geminal diethers and here the single ether band is replaced by two and sometimes four bands of high intensity in the 8 μm to 11 μm region. The precise pattern is much less well-defined than in the simple alkyl ethers but generally the considerable intensity of the bands combined with the absence of significant carbonyl or hydroxyl structure will suggest a compound of this type. Several resins fall into this group, including the cyclic vinyl ethers described in Chapter 3 and the 'polysulphide' rubbers described in Chapter 8. The latter will be easily recognised by elementary analysis, since they are very rich in sulphur. The other important resin is polyformaldehyde, discussed in this section.

3. ONE STRONG BAND BETWEEN 8·0 μM AND 8·6 μM

The diaryl ethers have a strong band in the region 8·0 μm to 8·6 μm, often the strongest band in the spectrum. Some esters and phenols have a band similar in position and appearance but the absence of bands due to carbonyl and hydroxyl groups should exclude these possibilities.

The polyphenylene oxide resins contain the diaryl ether linkage, and some of the polysulphone resins contain this functional group. Both these series of resins are described later in this chapter.

4. STRONG BANDS NEAR 8·0 μM AND 9·5 μM

In the aryl alkyl ethers two C—O modes may be distinguished. Thus the C—O aryl band occurs near 8·0 μm, as in the diaryl ethers, while the C—O alkyl band is displaced, as in some of the other structures discussed above, to about 9·6 μm. The important compounds in this group are the phenol-based epoxy resins and the infra-red spectra of these materials are discussed later in this chapter.

INFRA-RED SPECTRA OF DIALKYL ETHER POLYMERS

Apart from the vinyl ethers described in Chapter 3 and the cellulose ethers described later in this chapter, the polyethers usually encountered are the polyalkylene oxides (or polyalkylene glycols). These materials are of comparatively low molecular weight and find application as synthetic resins only after chain lengthening as in polyether rubbers described in Chapter 8. However, both as simple polyethers and modified in various ways, they may be encountered as additives to synthetic polymers, and some examples of the many possible products are therefore included here.

Polyethylene oxide may be either amorphous or crystalline at room temperature depending upon its molecular weight (Spectra 10.1(a) and (b)). The major spectral difference between the crystalline and amorphous material is at 9 μm, where the single band in the amorphous state splits into three sharper bands in the crystalline state; many other bands decrease in intensity and become broader in the amorphous spectrum.

In unmodified polyethylene oxide the molecular chains are terminated by alcoholic hydroxyl groups causing the appearance of the band at 3 μm. Although this band may in principle be used to estimate the molecular weight of the resin, assuming that chain branching does not occur, in fact it is extremely difficult to dry these resins completely and, since water has strong absorption at 3 μm, the method is not to be recommended. In such cases we prefer an alternative method[5] which takes advantage of the fact that the 2·1 μm hydroxyl band is not influenced by the presence of even relatively large amounts of water in the sample.

The other common polyalkylene oxide, polypropylene oxide (Spectrum 10.2) is usually a liquid. The general spectrum is very similar to that of amorphous polyethylene oxide but bands due to methyl groups are evident near 3·4 μm and at 7·3 μm.

Although higher ether polymers exist, they are rarely encountered as homopolymers. Copolymers are however fairly common and may well present difficulty in identification from the infra-red spectrum alone. In any case of doubt it is most important to consider the n.m.r. spectrum also (see p. 527).

Condensation products of the polyalkylene oxides, in which a proportion of the terminal hydroxyl groups is condensed with either an alcohol or a phenol, are widely used as surface active agents. Of these, the more difficult to characterise from the spectrum are the alcohol condensates, since terminal hydroxyl groups are replaced by aliphatic ether groups which are difficult to observe in the presence of an excess of chain alkyl ether. Some decrease of hydroxyl content occurs but, since the concentration of hydroxyl varies with molecular weight, this is not a useful observation on completely unknown materials. However, these condensates are frequently terminated with long-chain aliphatic alcohols such as stearyl, oleyl, lauryl or cetyl, which show the long-chain methylene band at 13·9 μm (split if crystalline). Examples are the polyethylene oxide-cetyl alcohol condensate (Spectrum 10.3) and the polyethylene oxide-oleyl alcohol condensate (Spectrum 10.4).

Condensates with phenols on the other hand are much more easily recognised. The ether group introduced is now an aryl ether and the corresponding C—O band near 8 μm is evident in addition to the usual spectral features associated with the presence of an aromatic nucleus. Many of the phenols used are p-substituted, and this introduces a difficulty because there is a qualitative similarity between the spectrum of such a substance and that of a p-substituted aryl alkyl ether (e.g. an epoxy resin). However in the latter substances the alkyl ether band is usually at greater wavelength ($\sim$9·5 μm) and is relatively much weaker.

p-octyl phenol-terminated polyethylene oxide is representative of these materials (Spectrum 10.5). The ether bands at 8 μm and 9 μm are accompanied by the p-substitution band at 12·0 μm, sharp aromatic ring bands in the 5 μm to 6 μm region and a weak hydroxyl band at 2·9 μm arising from unsubstituted terminal hydroxyl groups. There are a number of these condensates which have rather similar infra-red spectra.

For the complete identification of these condensates with alcohols or phenols, the n.m.r. spectrum may be used to great advantage. Some examples are given on p. 530.

A similar series of substances is the ester-terminated polyethylene oxides. These also are chiefly used as surface active agents. The spectra are similar to that of unmodified polyethylene oxide, but an ester carbonyl band occurs near 5·8 μm. An example is polyethylene oxide stearate (Spectrum 10.6). The identification of the terminating acids is straightforward, since they are readily released by hydrolysis and can be converted into the barium salt for infra-red identification.

The reaction product of a glycol, such as ethylene glycol, with epichlorhydrin, i.e. an aliphatic epoxy resin, is effectively a dialkyl ether polymer, and has its strongest band near 9·0 μm, as with the other resins described above. However these materials have additional bands due to the epoxy ring at about 10·9 μm, 11·8 μm and 13·2 μm. These resins are discussed further on p. 525.

POLYFORMALDEHYDE

This resin has a highly distinctive spectrum (Spectrum 10.7) with extremely strong bands in the 8 μm to 11 μm region. These arise from vibrations of the C—O group, the complex C—O pattern being typical of a methylene diether.

Polyformaldehyde is inherently an unstable polymer and if unmodified decomposes rapidly on heating by an 'unzipping' reaction. This is prevented either by blocking the hydroxyl end groups, or by copolymerising the formaldehyde with other monomers which act as blocks along the chains. In the case of the end-blocked polymer, there is evidence of ester groups from a carbonyl band at about 5·8 μm. It is, on the other hand, difficult to observe the modification of the spectrum due to the presence of minor amounts of copolymerised ethers. If there is no evidence of ester from the infra-red spectrum, it will be better to proceed to the n.m.r. spectrum where the identification and determination of such minor constitutents is usually comparatively straightforward.

The infra-red spectrum of polyformaldehyde is sensitive to crystallinity, and the resin has different crystalline forms with different spectra[6]. These effects are, however, unlikely to cause difficulty in qualitative identification.

INFRA-RED SPECTRA OF DIARYL ETHER POLYMERS

The common example of a diaryl ether polymer is poly-2:6-dimethyl phenylene oxide, usually known as polyphenylene oxide (Spectrum 10.8). As in the case of the dialkyl ether polymers, the spectrum is dominated by the band of the C—O—C— group which, in the case of diaryl ethers, occurs at about 8·4 μm. Further evidence of the predominantly aromatic character of the resin is afforded by the pair of strong bands in the 6 μm to 7 μm region. Because the benzene ring is heavily substituted, the 'substitution pattern' band occurs at the comparatively short wavelength of 11·7 μm. The weak sharp band at 7·3 μm presumably arises from the methyl group.

INFRA-RED SPECTRA OF EPOXY RESINS

Epoxy resins contain one or more terminal epoxy groups,

in their molecules. By far the most commonly encountered epoxy resin is that prepared from epichlorhydrin and Bisphenol A, (p. 247) which has the general formula shown on p. 526.

The spectra of resins of this type are quite distinctive (Spectrum 10.9) but have some similarity to those of phenol-formaldehyde resins, and the spectral features which enable these two series to be distinguished have been discussed (see p. 496). The features which distinguish the phenol-based epoxy resins from phenol-terminated polyethylene oxide have also been mentioned (p. 524). The major bands arise from hydroxyl groups

(2·9 μm), aromatic ether (8·1 μm), aliphatic ether (9·6 μm) and the
p-substituted benzene nucleus (12·0 μm). The most obvious band of the
terminal epoxy group appears at 10·9 μm, with further less obvious
bands at about 11·8 μm and 13·2 μm. The identity of these bands is clear
if the spectra of a series of resins of different molecular weight are compared, e.g. Ref. 7 Figures 79 to 81; the bands reduce in intensity with
increasing molecular weight.

The 10·9 μm band can be used to estimate the epoxy content of epoxy
resins but, since good chemical methods are available for this[8], infra-red
quantitative measurements of epoxy content have found their major

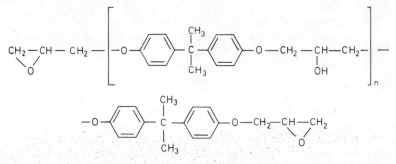

Epoxy resin from epichlorhydrin and Bisphenol A

application in estimating the degree of cure of epoxy resins. In the curing
process the terminal epoxy groups disappear, and their disappearance
may be conveniently followed by measurement of the intensity of the
10·9 μm band. A quantitative method is given by Burge and Geyer[8].

The —CH_2— group within the epoxy ring also has a characteristic
absorption band in the 2 μm region, and Goddu and Delker[9] have used
this to estimate epoxy contents of various materials.

The spectra of cured epoxy resins based on epichlorhydrin and Bisphenol A are not markedly different in their major features from that of
the uncured resin, other than in the disappearance of the 10·9 μm epoxy
band mentioned above, but additional bands may appear in the spectrum
due to the cross-linking agent (Nyquist[10] Figures 28, 29 and 32).

As an alternative to Bisphenol A, other polyhydric phenols may be
used, and of these the most important appears to be phenol-formaldehyde
resin. The spectrum of such a product is given by Nyquist[10] (Figure 30).
The spectrum has all the features of that of the epoxy resin based on
epichlorhydrin and Bisphenol A but in addition the substitution pattern
bands of phenol-formaldehyde resin appear at about 12·2 μm and 13·3 μm.
This product may be distinguished from a mixture of phenol-formaldehyde
resin and epoxy resin (Bisphenol A-epichlorhydrin) by the absence of a
strong phenolic hydroxyl band near 3·0 μm. In the spectrum of the amine-
cured product (Nyquist[10] Figure 31) the major difference, as expected,
is the virtual disappearance of the epoxy ring bands.

It is also possible to base epoxy resins on glycols rather than polyhydric phenols. The resulting material contains the dialkyl ether group (rather than the alkyl aryl ether as in the epoxy resins considered above) together with the epoxy group. Thus the spectra of these aliphatic epoxy resins (Hummel[1] Figure 1023 and Ref. 7 Figures 91 and 92) show an intense band near $9\cdot0\,\mu$m due to dialkyl ether, together with the epoxy ring bands mentioned above, namely at $10\cdot9\,\mu$m, $11\cdot8\,\mu$m and $13\cdot2\,\mu$m. Since these bands are not present in the infra-red spectra of other dialkyl ether polymers, this method of examination is a most satisfactory way of distinguishing the aliphatic polyether epoxy resins from the other dialkyl ether polymers.

Epoxy resins are frequently encountered in mixtures with other resins, either as stabilisers or as true resinous constituents of compositions, particularly adhesives. The epoxy resin stabilisers appear to be readily soluble, and the usual plasticiser-extraction methods will remove them; thus epoxy resins may be found mixed with conventional plasticisers. In such cases care must be taken to distinguish phenol-based epoxy resins from other phenolic stabilisers. The important feature of the infra-red spectrum of the epoxy resins appears to be the prominent $9\cdot6\,\mu$m alkyl ether band; very few phenolic stabilisers have a band of comparable width and intensity in this position. It is also necessary to distinguish between aliphatic epoxy resins and polyethylene oxide surface active agents. Here the bands in the spectrum due to the epoxy ring, as detailed above, are a satisfactory means of differentiation. This does however serve to emphasize the problems which may be encountered in the identification of additives in synthetic resin compositions.

The epoxy resins present in adhesive compositions are often cross-linked and, if a composition is to be examined for the presence of epoxy resin, it is sometimes possible to dissolve away other resinous constituents, leaving an insoluble residue which is substantially epoxy resin. Where this method is unsuccessful, treatment with solvents which will not dissolve the major resinous constituents (e.g. alcohol or ether) may remove epoxy resin which has escaped cross-linking. Infra-red examination of the residue, or solvent extract, as appropriate, will reveal the epoxy resin in such cases.

These extraction methods are in our opinion superior to methods based on the examination by infra-red (or other methods) of the total sample, although the series of spectra in Hummel[1] (Figures 1059 to 1062) may be found useful.

THE N.M.R. SPECTRA OF ALKYL ETHER POLYMERS

VINYL ETHER POLYMERS

A number of studies of vinyl ether homopolymers have been reported[11, 12] and many of these refer to the determination of the stereochemical arrangement of the polymer chain. The analyst may, however, be more interested in the use of the n.m.r. spectrum as a means of identification of the vinyl constituent of a vinyl ether copolymer, as has been outlined in Chapter 3.

POLYALKYLENE OXIDES

The n.m.r. spectrum is a particularly satisfactory method for the qualitative identification of polyalkylene oxides and for the qualititative and quantitative analysis of alkylene oxide copolymers. The values of the resonances of protons on the carbon atoms along the polymer backbone depend upon the proximity of the carbon atom to the ether oxygen atom and the values are given in Table 10.2.

Table 10.2. Resonances of protons on the carbon atoms along polyalkylene oxide and alkylene oxide copolymer backbones

'Backbone'	Group	τ value
$-O-CH_2-CH_2-O-$	$-CH_2-$	6·5
$-O-CH-CH_2-O-$ $\quad\quad\mid$ $\quad\quad CH_3$	$-CH-$	6·3
	$-CH_2-$	6·4
	$-CH_3$	8·8
$-O-CH_2-CH_2-CH_2-CH_2-O-$ $\quad\;(2)\quad\;(1)\quad\;(1)\quad\;(2)$	$-CH_2-$ (1) $-CH_2-$ (2)	8·4 6·4
$-O-CH_2-O-$	$-CH_2-$	5·1

The protons of polymethylene oxide and of polyethylene oxide are all equivalent, and thus each polymer shows a single resonance but that from polymethylene oxide is at much lower field. Polymers containing either propylene oxide or butylene oxide may be recognised by the presence of additional resonances in the spectrum at high field, as indicated in the table and having the correct area relationship to those near 6·4 τ.

The spectra of two copolyethers are given in Figures 10.1 and 10.2. The very intense single resonance in each case at 5·1 τ arises from copolymerised methylene oxide which is the principal component. In Figure 10.1 there is an additional sharp and weak resonance at 6·3 τ which must arise from copolymerised ethylene oxide units as no other structure appears other than the weak sharp resonance at about 5·2 τ which is reasonably attributed to the protons of the interchange unit (as shown in the figure). These additional resonances B and C are evidently of very similar strength, which would indicate that the $-O-CH_2-CH_2-O-$ groups are almost, if not entirely, all present as isolated units.

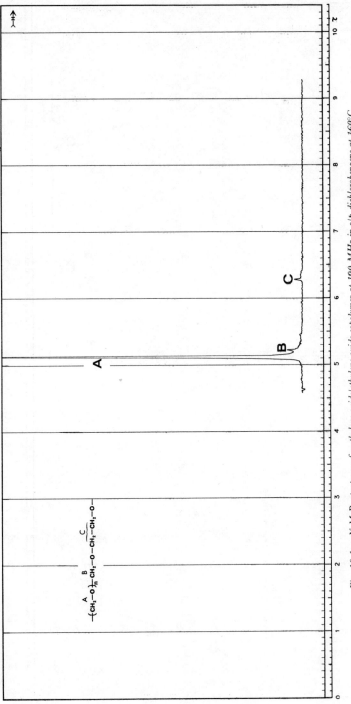

Fig. 10.1. *N.M.R. spectrum of methylene oxide/ethylene oxide copolymer at 100 MHz in o/p-dichlorobenzene at 160°C*

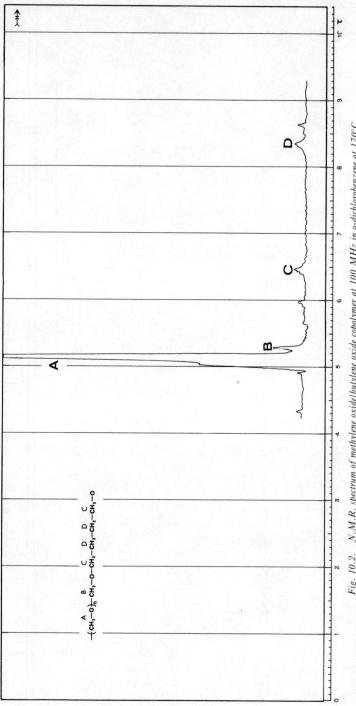

Fig. 10.2. N.M.R. spectrum of methylene oxide/butylene oxide copolymer at 100 MHz in o-dichlorobenzene at 170°C

In Figure 10.2 the major resonance at 5·1 τ is again accompanied by a weak resonance at about 5·2 τ which is also attributed to the interchange unit. However in this case we have 2 additional resonances at 6·4 τ and 8·4 τ of equal area, and evidently this is a copolymer with butylene oxide.

Other examples of the application of n.m.r. spectroscopy to the examination of polyalkylene oxide copolymers include the quantitative analysis of ethylene oxide/propylene oxide copolymers[13].

REACTION PRODUCTS OF POLYALKYLENE OXIDES

In addition to the polyalkylene oxide polymers so far described, which are presumably hydroxyl terminated, the polymer chemist will encounter reaction products of these substances. One important class of such compounds is the polyether urethane rubbers. The application of n.m.r. spectroscopy to these substances has been described (p. 462). A further series of compounds of this nature is the phenol-, ether-, or ester-terminated polyalkylene oxide surface active agents. The n.m.r. spectrum may be used to advantage in the examination of such materials, and particularly in the case of the phenol-terminated materials, of which an example is given in Figure 10.3.

The intense resonance at 6·5 τ is attributed to the protons of the polyethylene oxide chain. The resonances near 3 τ originate from protons on the aromatic nucleus, and the pattern clearly indicates *para*-substitution. The resonances in the 8 τ to 10 τ region relate to protons in the alkyl chain on the benzene ring, and are interpreted as shown in the figure. It is evident that a great deal of information may be obtained on substances of this nature from the n.m.r. spectrum.

THE INFRA-RED SPECTRA OF POLYSULPHONES

The —SO_2— group gives rise to a pair of very strong bands in the 7 μm to 9 μm region and these enable compounds containing this group to be recognised easily from their infra-red spectra. In particular the alkyl sulphones show these bands at 7·7 μm and 8·9 μm and in the diaryl sulphones they appear at 7·6 μm and 8·6 μm. Although the bands may be broad and irregular in shape and are often doublets in aryl compounds, they remain remarkably constant in position in the spectra of all sulphone resins.

The spectra of a large number of aliphatic sulphone resins appear in Hummel's collection[1] (Figures 1364 to 1380) and examination of these spectra shows both the considerable strength and the reliability of these bands of the sulphone group.

The diaryl sulphone polymers of current interest all contain the unit

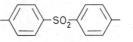

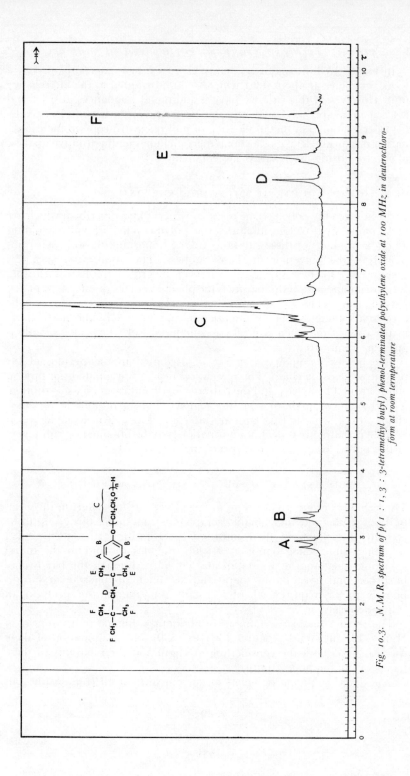

Fig. 10.3. N.M.R. spectrum of p(1 : 1, 3 : 3-tetramethyl butyl) phenol-terminated polyethylene oxide at 100 MHz in deuterochloro-form at room temperature

in the backbone. Thus the polysulphone made by the interaction of Bisphenol A and 4 : 4'-dihydroxydiphenyl sulphone, is evidently essentially

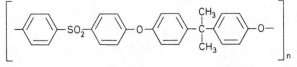

The infra-red spectrum of this resin (Spectrum 10.10) shows, in addition to the bands due to the sulphone group discussed above, a prominent band at 8·0 μm due to the diaryl ether group, and absorption near 12·0 μm indicative of *para*-substituted benzene nuclei. There is a general similarity between the spectrum of this material and that of poly-*p*-phenoxyphenyl sulphone

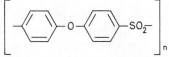

(Spectrum 10.11). The major absorption bands are almost coincident in the two spectra, but differences at 3·4 μm and 7·3 μm arise because methyl groups are present only in the resin based on Bisphenol A, and there are many other differences in the weaker structures in the spectra.

The spectrum of poly-*pp'*-diphenyl sulphone (Spectrum 10.12)

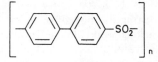

is again similar, but here the 8·0 μm band is absent because this resin does not contain the diaryl ether group.

Finally Spectrum 10.13 is that of poly(*pp'*-diphenylene ether—*p*-phenylene sulphone) and hence the polymer contains the last two units above. As expected, the spectrum contains the features of both Spectrum 10.11 and Spectrum 10.12.

N.M.R. SPECTRA OF POLYSULPHONES

The major interest in the n.m.r. spectra of polysulphones has been in the determination of 'faults' in the polymer structure, e.g. occasional *ortho*-substituted units

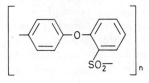

533

and trisubstituted benzene rings leading to chain branching of an otherwise entirely *para*-substituted polymer chain[14].

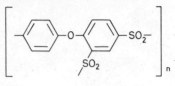

INFRA-RED SPECTRA OF CELLULOSE
AND DERIVATIVES

Cellulose contains ether and alcohol groups leading to strong absorption at 2·9 μm and between 9 μm and 10 μm (Spectrum 10.14). Various forms of cellulose, such as cotton, wood flour and regenerated cellulose in the form of fibre or transparent film, appear to show only minor spectral differences; such differences as may be noted in the spectra of cellulose films are often due to various additives (plasticisers and surface coatings) and largely disappear when these extraneous components are removed. Much fundamental work has been carried out on cellulose using infra-red measurements, such as measurement of the crystalline/amorphous ratio[15], but the analyst is more likely to be concerned with the origin of the material, for example distinguishing cotton from regenerated cellulose fibre, and identifying various wood flours. Infra-red measurements are in our experience of little value in such work and microscopic examination is probably the most satisfactory way to treat such samples.

In the many derivatives of cellulose, such as the ethers and esters, the distinctive hydroxyl absorption of the original cellulose is diminished or removed, depending upon the extent of modification, but the internal and external ether linkages are preserved. These groups are responsible for the rather broad band near 9·5 μm with irregular shoulders which appears in the spectra of all cellulose derivatives. It is, of course, somewhat obscured by the 9 μm ether band which appears in the cellulose alkyl ethers, but it is a distinctive feature of the cellulose esters, and appears to be the only reliable method of distinguishing the spectra of cellulose esters from the many ester resins described in Chapter 4.

There may be difficulty in the identification of cellulose derivatives from their infra-red spectra because the spectrum changes according to the extent of modification, for example the proportion of hydroxyl groups which are converted to ester or ether. Thus considerable variation may be noted in the structure in the 9 μm to 10 μm region because the bands at 9 μm, 9·8 μm and 10 μm due to hydroxyl groups all decrease in intensity as the hydroxyl groups are replaced. Furthermore, considerable differences may be noted in the intensity of the hydroxyl stretching band near 2·9 μm, on account of differences both in the hydroxyl content of the resin, and in the water content of the sample, which tends to decrease as the polar hydroxyl groups are replaced. It is important to bear these points in mind when comparing the spectra of unknown materials with the reference spectra of cellulose derivatives reproduced here and elsewhere in the literature.

CELLULOSE ESTERS

Cellulose may be acetylated to the mono-, di- or tri-acetate, and the infra-red spectrum shows differences according to the degree of acetylation. In the mono-acetate (Spectrum 10.15) the bands at 5·8 μm, 7·3 μm and 8·1 μm arise respectively from the C=O, methyl and C—O groups of acetate. Residual alcohol groups are evident from the 3 μm and 9 μm bands. With greater acetylation (Spectrum 10.16: cellulose triacetate) the 2·8 μm and 9 μm bands become very much weaker.

The bands of the acetate group dominate these spectra and care is required to differentiate, for example, cellulose acetate and polyvinyl acetate (Spectrum 3.2). However the 9·5 μm ether band is a feature only of the cellulose acetate spectrum and there are significant spectral differences between these resins beyond 10 μm wavelength. The hydroxyl band at 2·8 μm is not a satisfactory means of distinguishing the vinyl and cellulose esters, since vinyl esters are often partially hydrolysed and then contain alcoholic hydroxyl groups.

The C—O band of propionates and butyrates occurs at about 8·6 μm, hence the spectra of cellulose tripropionate (Hummel[1] Figures 240 and 241) and cellulose tributyrate (Hummel Figure 242) are very similar in their major features. The rather sharp band at 12·4 μm in the propionate spectrum would appear to be a useful distinguishing feature. Again there is considerable similarity between the spectra of cellulose tripropionate and polyvinyl propionate (Spectrum 3.4) and these esters also are distinguished by the presence of the ether band near 9·5 μm in the spectrum of the cellulose compound.

The mixed cellulose ester cellulose acetobutyrate (Spectrum 10.17) has two prominent bands at 8·2 μm and 8·6 μm arising from the acetate and butyrate respectively. It is therefore comparatively simple to distinguish this material from the simple esters by means of the infra-red spectra.

The other important cellulose ester is cellulose nitrate. This has a most useful spectrum for identification (Spectrum 10.18) with prominent bands at 6·1 μm and 11·9 μm. Somewhat surprisingly this resin may be confused, in its spectrum, with natural rubber and gutta percha since the trisubstituted double bond also has prominent bands at roughly these positions. Once again the reliable feature of the cellulose ester spectrum is the ether absorption near 9·5 μm.

CELLULOSE ESTERS

All cellulose esters show in their spectra bands of relatively high intensity in the 9 μm to 10 μm region, all other features of the spectrum being comparatively weak. The spectra of the more common ethers, methyl cellulose and ethyl cellulose (Spectra 10.19 and 10.20), show considerable similarity to those of the alkyl vinyl ethers (Spectra 3.8 to 3.11) but the cellulose ethers show a broader band in the 9 μm region with pronounced absorption at 9·5 μm. Effective differentiation of cellulose esters from materials such as vinyl ethers and the polyethers discussed earlier in this chapter therefore depends upon observing the spectra of samples which are sufficiently thin ($\sim$0·0003 in or 0·007 mm thickness) to enable the ester structure to be clearly seen.

Methyl and ethyl cellulose are themselves distinguished by bands in the 10 μm to 12 μm region. Benzyl cellulose (Hummel[1] Figure 257) is easily identified from the mono-substituted benzene ring pattern in the 13 μm to 15 μm region and the usual sharp 'aromatic bands' between 6 μm and 7 μm.

OTHER CELLULOSE DERIVATIVES

The spectra of a number of other cellulose derivatives such as hydroxyethyl cellulose, methyl hydroxyethyl cellulose and ethyl hydroxyethyl cellulose are given by Hummel[1] (Figures 253 to 255) but the spectra have broad features which are by no means ideal for qualitative identification. The same may be said, in general, of the spectra of similar compounds reported by Zhbankov[16].

The spectrum of sodium carboxymethyl cellulose is more interesting (Nyquist[10] Figure 25 and Hummel[1] Figure 256), having a strong band at 6·2 μm indicative of the presence of the carboxylate ion. However, liquid water has a strong band close to this position, and in case of doubt it is better to dry the sample thoroughly before drawing conclusions as to the presence of the carboxylate ion in hydrophilic substances such as cellulose and its derivatives.

SILICONES

The analyst in the plastics industry is likely to meet polymeric silicones as such or, indeed, articles that may have been treated with silicone polymers. This aspect of his work is rather specialised, but we have been fortunate in prevailing on Mr. J. C. Smith, formerly Chief Analyst of Midland Silicones Ltd, and Mr. J. F. Lees, Chief Analyst of the same company, to provide us with all the relevant material that forms the basis of this section.

In the first place it should be realised that Smith has provided useful reviews on the subject[17, 18].

When an analytical chemist is confronted with a material which he suspects to be a polymeric silicone there are certain simple preliminary tests for elements that can be performed that will help him in his decision.

PRELIMINARY QUALITATIVE TESTS FOR ELEMENTS

SILICON

It is important to realise that siliceous material of any description will give a positive result in the test for silicon and contaminating siliceous material should be removed before applying the test for silicon to the suspected silicone.

The test is carried out by fusion of a small amount of the sample with sodium peroxide and fusion mixture in a platinum spoon or loop. The fused mass is dissolved in water and the solution is tested for silicate by the molybdate-benzidine-hydrochloric acid procedure of Gilman, Ingham and Garsich[19]. Alternatively, 2 to 3 drops of the sample are mixed with 2 ml of fuming sulphuric acid and the acid removed by evaporation. The residue should consist of silica and this should be confirmed by treatment with hydrofluoric and sulphuric acids.

FLUORINE, CHLORINE AND NITROGEN

The presence or absence of these elements should be confirmed either by sodium fusion tests or by the Oxygen Flask Combustion procedure of Haslam, Hamilton and Squirrell[20] when this is applicable. The presence of fluorine may point to a methyl trifluoropropyl siloxane, of nitrogen to a methyl nitrile siloxane and of chlorine to a chlorophenyl siloxane.

The majority of the silicones that are commercially available are linear and cross-linked methyl and phenyl polysiloxanes or copolymerisation products of siloxanes with alkyds etc.

QUANTITATIVE TESTS

CARBON AND HYDROGEN

A variety of methods has been described in the literature for the determination of carbon and hydrogen in organo-silicon compounds and various procedures have been suggested in order to avoid the formation of silicon carbide[21-27]. A recent investigation[28] has shown that good results can be obtained for a variety of siloxanes when a combustion tube containing magnesium oxide as catalytic filler at 850°C is used. This last paper also gives a useful review of the subject in general.

The great majority of silicones encountered for identification are materials containing carbon, hydrogen, oxygen and silicon only. Hence the determinations of carbon and hydrogen are quite important. With further knowledge of the silicon content and calculation of the oxygen content by difference it is often possible to use these four figures, together with the results of functional group tests, to assign a probable structure to a silicone polymer.

SILICON

The best method of determination is by decomposition with sodium peroxide in a Parr bomb[29].

In certain favourable cases the acid digestion process will yield satisfactory results although in many cases the results may tend to be low due to loss of silicon. The initial breakdown is carried out carefully at low temperature as follows: 0·3 to 1 g sample is weighed into a weighed platinum dish. After cooling in dry ice, 2 to 4 ml of fuming sulphuric acid are added and the mixture is allowed to warm up to room temperature. The digestion is completed by gentle ignition over a Meker burner followed by final ignition at 850°C in a muffle furnace. The ignition is repeated until constant weight of the silica is obtained.

CHLORINE

This determination is best carried out by fusion of the sample with sodium peroxide in a Parr bomb and subsequent electrometric titration of the ionised halide with standard silver nitrate solution.

ADDITIONAL TESTS

REACTION OF THE SILOXANE WITH A SOLUTION OF HYDROFLUORIC ACID IN PROPANOL

This method, which removes siloxanes quantitatively, works very well with a variety of siloxane systems, providing evidence of their presence and amount. Moreover, after its application, other additives or components are often left behind in a pure condition. It should be noted however that any silica present in the siloxane system will also be removed by this test. The test is carried out by taking a small amount of the test material (0·3 g), adding 25 ml *n*-propanol and 2·5 ml of 40% w/w aqueous hydrofluoric acid, and then evaporating to dryness in a platinum dish. With siloxanes, initially non-volatile, the loss in weight is equivalent to the weight of the siloxane, providing silica was not present in the original material.

Alternatively, the reaction may be carried out in an all-copper system and the volatile product may be collected in dilute potassium hydroxide solution. Subsequent acidification and extraction should yield an extract useful in identification work.

TESTS FOR FUNCTIONAL GROUPS

The determination of functional groups may yield very useful information about the type of siloxane that may be present.

1. $\equiv$SiH groups*

Siloxanes containing $\equiv$SiH groups are used widely as textile water repellents and for paper treatment. The gasometric determination of $\equiv$SiH is accomplished[30] by reaction of the sample with an alcoholic solution of sodium butoxide in n-butanol.

The reaction proceeds as follows:

$$\equiv SiH + HO \cdot CH_2 \cdot CH_2 \cdot CH_2 \cdot CH_3 + BuONa \rightarrow \equiv SiOBu + H_2$$

This can be used as a simple qualitative test or made quantitative by measurement of the hydrogen evolved.

Alternatively, $\equiv$SiH groups may be determined by reaction of the sample with excess of mercuric acetate in methanol. Provided there is no olefinic unsaturation present one $\equiv$SiH bond yields two equivalents of acetic acid, as follows:

$$\equiv SiH + 2Hg(OOC \cdot CH_3)_2 + CH_3OH \rightarrow \equiv Si(OCH_3) + $$
$$Hg_2(OOC \cdot CH_3)_2 + 2\ CH_3 \cdot COOH$$

The procedure is very similar to that used by Martin[31] for the determination of olefinic unsaturation.

2. Silanol ($\equiv$SiOH) groups

The determination of silanol groups is important since it provides a means of assessing whether an active siloxane is present. The linear siloxanes very rarely contain any appreciable amounts of silanol groups and in most

* In this section on silicone chemistry, the symbol $\equiv$Si denotes three groups bonded to silicon, rather than a triple bond.

cases the importance of silanol determinations is in connection with silicone resins. In most resins that are commercially available the —OH content will lie between 0·1% and 6% w/v.

The following methods have been used with success.

(a) Measurement of the volume of methane produced on reaction of hydroxyl-containing compounds with methyl magnesium iodide. The most accurate procedure so far reported is that of Guenther[32] who uses methane as an inert gas. If other inert gases are used, corrections must be applied for the solubility of the methane generated in the test in the solvent used.

$$\equiv SiOH + CH_3MgI \rightarrow \equiv SiOMgI + CH_4$$

Note: The method is not specific for SiOH groups. Active hydrogen from other sources reponds to the test.

Two methods that are specific for SiOH groups are: (b) titration with Karl Fischer reagent; and (c) condensation of silanol groups with the aid of an acid or alkaline catalyst.

(b) The Karl Fischer method[33] has been reported as being satisfactory for the determination of hydroxyl groups in silanols. Silanol groups react with the methanol in the Fischer reagent, and one molecule of water is produced per SiOH group.

$$\equiv SiOH + CH_3OH \rightarrow \equiv Si(OCH_3) + H_2O$$

The water formed reacts with the Fischer reagent and the excess reagent is then determined as usual. The method is accurate in a semi-quantitative way but cannot be relied on to give accurate results in the presence of siloxanes because of the insolubility of many siloxanes in the methanol-pyridine solvent. Also, Smith and Kellum[34] have shown that hexamethyl cyclotrisiloxane interferes with the determination, giving results higher than theoretical.

(c) The condensation of silanol groups with the aid of an acid or alkaline catalyst and determination of the water produced is usually carried out in a Dean and Stark apparatus. The sample is dissolved in toluene and any free water is distilled off by azeotropic distillation. The mixture is then cooled and a trace amount of acid or alkali is added. The water produced is distilled off by azeotropic distillation and determined by titration with Karl Fischer reagent.

Acid Condensation

$$\equiv SiOH + H^+ \rightarrow \equiv SiOH_2{}^+$$

$$\equiv SiOH + \equiv SiOH_2{}^+ \rightarrow \equiv Si-O-Si \equiv + H_2O + H^+$$

Alkaline Condensation

$$\equiv\!SiOH + OH^- \rightarrow \equiv\!SiO^- + H_2O$$

$$\equiv\!SiOH + \equiv\!SiO^- \rightarrow \equiv\!Si\!-\!O\!-\!Si\!\equiv + OH^-$$

Condensation reactions are discussed by Eaborn[35].

3. Methyl silyl groupings

Most commercial silicones contain $Si\!-\!CH_3$ groupings and it is necessary to know the relative amounts of $-Si(CH_3)_3$, $=Si(CH_3)_2$ and $\equiv\!SiCH_3$ groups present. In the procedure of Heylmun and Pikula[36] the organosiloxane is heated with boron trifluoride in diethyl ether to convert the methyl silyl groupings into their corresponding fluorosilanes, e.g.

$$2CH_3SiO_{3/2} + 2BF_3\!\cdot\!Et_2O \rightarrow 2CH_3SiF_3 + B_2O_3 + 2Et_2O$$

$$3(CH_3)_2SiO + 2BF_3\!\cdot\!Et_2O \rightarrow 3(CH_3)_2SiF_2 + B_2O_3 + 2Et_2O$$

$$6(CH_3)_3SiO_{1/2} + 2BF_3\!\cdot\!Et_2O \rightarrow 6(CH_3)_3SiF + B_2O_3 + 2Et_2O$$

The mixed fluorosilanes are then resolved by a gas chromatographic method and from the areas of the fluorosilane peaks the relative concentrations of the various methyl silyl groups can be calculated.

However, when $(CH_3)HSiO$ is present, direct reaction produces a mixture of methyl fluorosilanes and methyl hydrogen fluorosilanes. To avoid this the siloxane should first be alkoxylated by reacting with ethanol in the presence of sodium methoxide.

$$(CH_3)HSiO + EtOH \xrightarrow{\ \ CH_3ONa\ \ } CH_3(EtO)SiO + H_2$$

The resulting solution is then reacted with boron trifluoride as described above. This converts the ethoxylated siloxanes into methyl trifluorosilane, e.g.

$$CH_3(EtO)SiO \xrightarrow{\ \ BF_3\!\cdot\!Et_2O\ \ } CH_3SiF_3$$

If the original siloxane contained $CH_3Si\!\equiv$ it is necessary to determine the SiH content of the siloxane in order to calculate how much of the final CH_3SiF_3 is due to $(CH_3)HSiO$ and how much to $CH_3SiO_{3/2}$ in the original siloxane.

The shorthand nomenclature used above is based on the suggestions made by Eaborn[35] in which

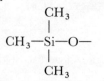

is represented by $(CH_3)_3SiO_{1/2}$ or M (Monofunctional) and

$$CH_3$$
$$|$$
$$-O-Si-O-$$
$$|$$
$$CH_3$$

is represented by $(CH_3)_2SiO_{2/2}$
or $(CH_3)_2SiO$ or D(Difunctional)
and so on.

T (Trifunctional) is $(CH_3)SiO_{3/2}$ and a compound of the molecular

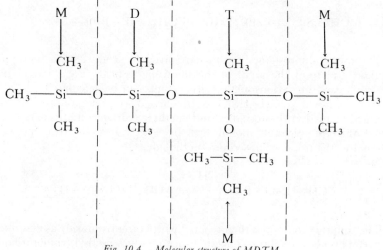

Fig. 10.4.   *Molecular structure of MDTM₂*

structure given in Figure 10.4 can be conveniently written as MDTM$_2$.

PYROLYSIS/GAS CHROMATOGRAPHY OF SILICONE RESINS

A distinctive feature of the silicone rubbers is their high thermal stability, a temperature of 750°C being required to produce a complete pyrolysis in 15 s. Pyrolysis at 550°C does however produce the same pattern but with the peaks much reduced in size. Retention times relative to styrene of the main peaks are 0·65 and 2·3. Phenyl silicone resin gives peaks at $R_T = 0·25$ and 0·75.

The physical properties of a siloxane can give some indication of the type of system present. The appearance of a siloxane after drying in air and for 1 h at 100°C should indicate the presence of cross-linked material:

Fluid appearance—in most cases the material would be a linear polymer.

Resinous appearance—contains cross-linked material.

Brittle residue—highly cross-linked.

Apart from this, other physical tests are of limited value unless the class of silicone is known.

For the linear siloxanes (silicone fluids) there is a great deal of information available on physical properties such as Refractive Index and Specific Gravity, and this can be fully employed in characterisation[37–39]. There is little comparative information on the silicone resins.

INFRA-RED EXAMINATION OF SILICONES

The spectra of silicone (siloxane) polymers, whether oils, rubbers or rigid resins have, as their most prominent feature, a broad band of extremely high intensity in the 9 μm to 10 μm region, often with two or more distinct maxima, which arises from the Si—O—Si group.

Strong broad bands in this region also occur in aliphatic ether polymers, such as the vinyl ethers discussed in Chapter 3, and the linear polyethers and cellulose and its derivatives, which are described elsewhere in this chapter. The silicones can in most cases be distinguished from the polyethers by the presence in the silicone spectra of absorption bands characteristic of alkyl or aryl groups directly attached to silicon. The prominent absorption bands of these structures are given on p. 545. The group most frequently encountered is the Si—Me group, which has a highly characteristic sharp band at 7·9 μm. This band is particularly easy to observe, even in the presence of other materials absorbing in this region of the spectrum.

In many cases the 8 μm Si—Me absorption band will enable the presence of a silicone to be established in mixtures with other resins including polyethers; silicones containing the Si—H group are also easily recognised in mixtures, since few other substances have significant absorption where the strong Si—H band appears (4·5 μm to 4·7 μm).

Silicones containing neither methyl nor hydrogen as substituents on silicon are more difficult to recognise, and there is, for example, a strong similarity between the spectrum of an all-phenyl silicone[40] and that of benzyl cellulose (Hummel[1] Figure 257), although the 'substitution pattern' bands near 14 μm are different in these two materials. The phenol-terminated polyethylene oxides also have spectra superficially similar to those of the silicones, although the appearance of the spectrum in the 11 μm to 15 μm region is usually sufficiently different to act as a distinguishing feature.

Many inorganic compounds of silicon, including silica and amorphous silicates, show broad and featureless bands in the 9 μm to 10 μm region. Examples of such spectra appear in Ref. 7, Figures 442 to 454. Thus the infra-red spectra of plastics compositions containing such compounds as fillers may appear similar to those of silicone polymers. In a soluble composition this presents no problem, as the absence of bands due to alkyl or aryl groups in the spectrum of the separated filler will show immediately its inorganic nature. In insoluble compositions the recognition of bands associated with alkyl or aryl groups bonded to silicon is the most satisfactory method of ascertaining whether a silicone is present.

Many common organic groups when linked to silicon give rise to well-defined and easily recognised bands which remain remarkably constant in position as the other substituents about the silicon atom are changed, and reliable conclusions may be drawn from the spectra of silicones regarding the substituent groups attached to silicon. This is considered in detail by Smith[41] but the data on those groups likely to be encountered in the synthetic resin field are summarised here.

IMPORTANT GROUP ABSORPTION BANDS IN THE SPECTRA OF SILICONES

Silicon-methyl

This is probably the grouping most frequently encountered in commercial

Table 10.3. Bands of silicon-methyl groups

Group	8 μm *region*	12 μm *region*
$>$Si—CH$_3$	7·9 μm (sharp, single)	12·8 μm—13·1 μm (one band)
Si with two CH$_3$	7·95 μm (sharp, single)	$\sim$ 11·7 μm and 12·4 μm—12·5 μm (11·7 μm band weaker)
—Si—CH$_3$ with three CH$_3$	7·9 μm and 8·0 μm (sharp doublet)	$\sim$ 11·9 μm and 13·1 μm—13·3 μm (11·9 μm band stronger)

silicones. It gives rise to one or more sharp bands between 7·9 μm and 8·0 μm, and between 11·5 μm and 13·5 μm, as shown in Table 10.3.

Silicon-alkyl

A band corresponding to the 8·0 μm band of Si—Me occurs in the 8·2 μm to 8·5 μm region, the wavelength increasing with increasing alkyl chain length[41].

Silicon-vinyl

The vinyl group attached to silicon shows a normal C=C stretching band at 6·2 μm. The bands occurring in vinyl olefines at 10·1 μm and 11·0 μm (Table 6.5) appear as doublets at 9·8 μm to 10 μm and 10·2 μm to 10·5 μm.

Silicon-phenyl

The silicon-phenyl linkage has a pattern of sharp bands at 7·0 μm, 8·9 μm and 10·0 μm, and a group of two or three bands, corresponding to the benzene ring 'substitution pattern', between 13·5 μm and 14·5 μm. In the silicones the 8·9 μm and 10·0 μm bands appear as sharp well-defined peaks superimposed on the broad Si—O—Si absorption band.

This complex band pattern also appears in the spectra of compounds containing phenyl groups attached to other Group IV B metals, except for the 7·0 μm band, which apparently is a characteristic band for silicon-phenyl.

Silicon-hydrogen

This group is easily identified from the Si—H stretching band occurring between 4·5 μm and 4·7 μm. This fairly strong band falls in a region where very few absorption bands appear and it is thus particularly suitable for identification purposes.

Silicon-hydroxyl

Silicon-hydroxyl is not easily identified from the infra-red spectrum. The O—H stretching band occurs near 3·0 μm; this cannot be distinguished from other types of hydroxyl group and may be confused with absorption due to water. The Si—O stretching band of this group occurs as a broad band between 11·2 μm and 12·0 μm, but may be obscured by the Si—Me bands.

Silicon-alkoxy

The group Si—O—CH$_3$ has a useful characteristic band at 3·55 μm and this, together with a band at 8·4 μm, identifies the group exclusively. Si—O—Et shows useful bands at ∼8·5 μm and ∼10·5 μm. A pair of sharp doublets at 6·8 μm to 6·9 μm and 7·2 μm to 7·3 μm with the absence of absorption at 10 μm and 11 μm, serve to identify the isopropoxy group attached to silicon. Other alkoxyl groups show less useful patterns.

Si—O—Si *group*

The character of the intense Si—O—Si band between 9 μm and 10 μm is of some interest. In short-chain compounds a single band appears between 9·2 μm and 9·8 μm, but with increasing chain length the band splits into two broad peaks centered on 9·2 μm and 9·8 μm. This prominent structure is one of the most easily recognised features of the infra-red spectra of silicone polymers, although the presence of bands in this region in the spectra of other compounds containing Si—O groups has been noted above.

THE INFRA-RED SPECTRA OF SILICONE POLYMERS

The methyl silicones appear to be the most common of the commercially available silicones. The simplest spectra are those of methyl silicone oil (Spectrum 10.21) and methyl silicone rubber (Spectra 10.22). In addition to the intense 9 μm to 10 μm structure, bands due to the CH$_3$—Si—CH$_3$ group appear at 3·4 μm, 7·1 μm, 7·9 μm, and 12·5 μm. Methyl silicone oils are often chain terminated with —Si(CH$_3$)$_3$ groups but in most cases there is no evidence of this group from the spectrum. Methyl hydrogen silicone oil (Spectrum 10.23) also has a simple spectrum, in which the bands due to ≡SiCH$_3$ are seen at 7·9 μm and 13·0 μm, and the strong Si—H bands appear at 4·6 μm. The group of bands between 11 μm and 12 μm are also due to the Si—H group, but are of lesser value than the 4·6 μm band for characterising this group as they are less constant in position, and may be obscured by bands due to

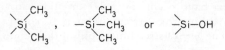

546

The so-called silicone resins differ from the silicone oils in that they have a chain-branched rather than a linear structure. Thus the all-methyl resin (Spectrum 10.24) is significantly different from the methyl oil in the 12 μm to 13 μm region, due to the presence of $\equiv$SiCH$_3$ and —Si(CH$_3$)$_3$ in addition to the $=$Si(CH$_3$)$_2$ group. As would be expected, the spectra do not differ markedly elsewhere in the 2·5 μm to 15 μm region.

Another silicone resin which may be encountered is that which contains both methyl and phenyl substituents, and here the spectrum is dependent upon the way in which the phenyl groups are distributed (Spectra 10.25 and 10.26). When phenyl groups are introduced in pairs on one silicon atom (Ph—Si—Ph group) three clear bands appear in the benzene 'substitution pattern' region, and the structure near 12 μm is indicative of substantially $=$Si(CH$_3$)$_2$ (Spectrum 10.25). Where phenyl groups are introduced singly, as in CH$_3$—Si—Ph (Spectrum 10.26), the 'substitution pattern' has only two clear bands, and the methyl structure near 12 μm becomes more complex. This resin also contains some Si—OH groups, as indicated by the absorption band near 3 μm and the broad band in the 11·4 μm region.

Quantitative determination of the ratio of phenyl to methyl groups in these mixed resins has been investigated by several workers, the determination being based on the measurement of the relative intensity of bands which are as far as possible structurally independent. Thus Lady *et al.*[42] use the 7 μm and 8 μm bands for phenyl and methyl respectively. Their method uses cast films, but with soluble oils and resins solutions may be preferred, as this gives much easier control of sample thickness.

The alternative method of Frischl and Young[43] is based upon the bands at 3·29 μm (Si—Ph) and 3·39 μm (Si—CH$_3$). This method appears to give more consistent results, possibly because the 8 μm band of the Si—CH$_3$ group is sensitive to the number of methyl groups attached to silicon, giving band broadening and a peak absorbance not strictly proportional to methyl content. The alternative method is however subject to serious interference from hydrocarbon impurities.

It is clear that the well-defined bands arising from Si—CH$_3$, Si—Ph, Si—H and Si—OH will enable the majority of silicone resins to be characterised from their spectra.

ULTRA-VIOLET ABSORPTION SPECTRA

The ultra-violet region of the spectrum may be used very effectively for the determination of the phenyl content of silicones. The absorption peak at 264 nm is frequently used[44]. It is suggested that the preliminary qualitative tests, the quantitative, additional and functional group tests, as well as the information on infra-red and ultra-violet absorption spectra, should help the analyst in the plastics industry seeking to identify an unknown silicone resin.

Analytical chemists have other duties to perform and one of them may concern the identification of silicones on fabrics etc. The following notes have been prepared to assist in this identification.

SILICONES ON FABRICS AND OTHER ARTICLES

The presence of silicones on fabrics and other articles is best established by first carrying out a solvent extraction. In some cases it is possible merely to rub a piece of cotton wool (saturated with solvent) over the surface and then to rub a sodium chloride plate with the moist cotton wool. Infra-red examination of the smear will often give a clear indication of the presence of a silicone.

The safest procedure is to carry out a solvent extraction using a Soxhlet apparatus. The solvent should be removed from the extract and the residue dried and tested as in the preliminary tests. If a silicone is indicated, a weighed portion of the extract should be subjected to a hydrofluoric acid-propanol treatment; this should give some indication of the weight of siloxane present in the extract. Infra-red examination before and after treatment may also be useful in assessing the type of siloxane present, and provide information about other materials present in the extract.

The choice of solvents is governed by the type of silicone treatment that has been applied. Siloxanes containing active groups such as $\equiv$Si—H or $\equiv$Si—OH can be extracted provided they have not been cross-linked by curing. Toluene, benzene, carbon tetrachloride, propanol and tetrahydrofuran can all be used satisfactorily.

If the siloxane has been cured only a small proportion of the siloxane will be removed by the above solvents, but even this non-quantitative extract may provide valuable information if its infra-red spectrum is determined.

CROSS-LINKED SILOXANES

In these cases it is necessary to use diethylamine or piperidine to extract the siloxane. Evaporation of the extract should yield the siloxane which may be identified by its infra-red spectrum. In the case of water-repellent textile treatments, siloxanes containing Si—H are frequently used, and basic solvents will bring about cleavage of the Si—H bond. In these cases, therefore, it is not possible to rely on the extract obtained by the use of basic solvents for analysis; the only way to assess whether any such material is present is to carry out a solvent extraction with a neutral solvent and to make an infra-red examination of the extract.

A quantitative determination of the Si—H may be carried out on the shredded sample using the normal procedure. A micro-apparatus may sometimes be necessary since the Si—H content may be small.

These two tests, together with an examination of a diethylamine or piperidine extract[45-46], should provide the information required to reach some useful conclusions as to the type of silicone coating present.

REFERENCES

1. HUMMEL, D. O., *Infra-Red Analysis of Polymers, Resins and Additives, An Atlas*, **1,** Pt II, Wiley Interscience (1969)
2. HUMMEL, D. O., *Kunststoff-, Lack- und Gummi-Analyse*, Carl Hanser, Munich (1958)
3. KLINE, G. M., (Ed.), *The Analytical Chemistry of Polymers*, Interscience, New York, FERRI, C.M., Pt I, 221 (1959); BRAUER, G. M., and NEUMAN, S. B., Pt III, 141 (1962)
4. RUBENS, E., *Z. ges. Shiess-u. Sprengstoffw.*, **28,** 172 (1933)
5. MILLER, R. G. J., and WILLIS, H. A., *J. appl. Chem., Lond.*, **6,** 385 (1956)
6. ZAMBONI, V., and ZERBI, G., *Jnl. Polym. Sci., Pt. C*, **1,** 153 (1964)
7. *Infra-Red Spectroscopy. Its use in the Coatings Industry*, Federation of Societies for Paint Technology, Philadelphia (1969)
8. BURGE, R. E., and GEYER, B. P., (Ed. KLINE, G. M.), *The Analytical Chemistry of Polymers*, Interscience, New York, Pt I, 123 (1959)
9. GODDU, R. F., and DELKER, D. A., *Analyt. Chem.*, **30,** 2013 (1958)
10. NYQUIST, R., *Infra-Red Spectra of Polymers and Resins*, The Dow Chemical Co., Midland, Mich., 2nd edn (1961)
11. RAMEY, K., *Jnl Polym. Sci., Pt. A*, **3,** 2885 (1965)
12. RAMEY, K., *Jnl Polym. Sci., Pt. B*, **2,** No. 9, 865 (1964)
13. MATHIAS, A., and MELLOR, N., *Analyt. Chem.*, **38,** 472 (1966)
14. CUDBY, M. E. A., FEASEY, R. G., JENNINGS, B. E., JONES, M. E. B., and ROSE, J. B., *Polymer,* **6,** 589 (1965)
15. MANN, J., and MARRINAN, H. J., *Trans. Faraday Soc.*, **52,** 487 (1956)
16. ZHBANKOV, R. G., (Transl. DENSHAM, A. B.), *Infra-Red Spectra of Cellulose and its Derivatives*, Consultants Bureau, New York (1966)
17. SMITH, J. C. B., *Analyst, Lond.*, **85,** 465 (1960)
18. SMITH, J. C. B., *Analysis of Organo-Silicon Compounds*, Midland Silicone Sales Publications, S.P. 363 (1965)
19. GILMAN, H., INGHAM, R. K., and GARSICH, R. D., *J. Am. Chem. Soc.*, **76,** 918 (1954)
20. HASLAM, J., HAMILTON, J. B., and SQUIRRELL, D. C. M., *Analyst, Lond.*, **86,** 239 (1961)
21. BELCHER, R., and SPOONER, C. E., *J. chem. Soc.*, 313 (1943)
22. KLIMOVA, V. A., KORSHUN, M. O., and BEREZNITSKAYA, E. G., *Dokl. Akad. Nauk SSSR*, **84,** 1175 (1952); **96,** 81 (1954); KLIMOVA, V. A., and BEREZNITSKAYA, E. G., *Zh. anilit. Khim.*, **11,** 292 (1956)
23. KAUTSKY, H., FRITZ, G., SIEBEL, H. P., and SIEBEL, D., *Z. analyt. Chem.*, **147,** 327 (1955)
24. BRADLEY, H. B., *Analyt. Chem.*, **27,** 2021 (1955)
25. LAMMER, H., *Chem. Tech., Berl.*, **4,** 491 (1952)
26. KORBL, J., and KOMERS, R., *Chemické Listy*, **50,** 1120 (1956)
27. BALIS, E. W., *Ohio St. Univ. Engng Exp. Sta. News*, **25,** No. 5, 58 (1953)
28. GAWARGIOUS, Y. A., and MACDONALD, A. M. G., *Analytica chim. Acta*, **27,** 300 (1962)
29. MCHARD, J. A., (Ed. KLINE, G. M.), *The Analytical Chemistry of Polymers*, Interscience, New York, Pt I, 365 (1959)
30. MCHARD, J. A., (Ed. KLINE, G. M.), *The Analytical Chemistry of Polymers*, Interscience, New York, Pt I, 374 (1959)
31. MARTIN, R. M., *Analyt. Chem.*, **21,** 921 (1949)
32. GUENTHER, F. O., *Analyt. Chem.*, **30,** 1118 (1958)
33. GRUBB, M. T., *J. Am. Chem. Soc.*, **76,** 3408 (1954)

34. SMITH, R. C., and KELLUM, G. E., *Analyt. Chem.*, **38,** 647 (1966)
35. EABORN, C., *Organosilicon Compounds*, Butterworths (1960)
36. HEYLMUN, G. W., and PIKULA, J. E., *Jnl Gas Chromatog.*, **3,** 266, Aug. (1965)
37. ROCHOW, E. G., *Chemistry of Silicones*, Wiley, New York (1951)
38. *Silicone Fluids*, Midland Silicones, Technical Data Sheet No. G.I., 6th edn
39. *Silicones, Products, Properties and Applications*, Imperial Chemical Industries (Nobel Division), Silicones Dept., 5th edn, 40/14/364 (1964)
40. HUNT, J. M., WISHERD, M. P., and LAWRENCE, C. M., *Analyt. Chem.*, **22,** 1478 (1950)
41. SMITH, A. L., *Spectrochim. Acta*, **16,** 87 (1960)
42. LADY, J. H., BOWER, G. M., ADAMS, R. E., and BYRNE, F. P., *Analyt. Chem.*, **31,** 1100 (1959)
43. FRISCHL, W., and YOUNG, I. G., *Appl. Spectrosc.*, **10,** 213 (1956)
44. URIN, T., *J. chem. Soc. Japan*, **62,** 1421 (1959)
45. STANLEY, E. S., *R.A.E. Tech. Memo. Chem.*, 95 (1953)
46. GRANTHAM, R. L., and HASTINGS, A. G., *J. Polym. Sci.*, **27,** 560 (1958)

11

PLASTICISERS, FILLERS AND SOLVENTS

This chapter is mainly concerned with the examination of plasticisers. However, in plastics compositions, these plasticisers are often accompanied by such substances as fillers and pigments. Moreover, certain plastics preparations may contain solvents in addition to plasticisers.

For convenience, therefore, observations are also made in this chapter on the examination of fillers, pigments and solvents.

In the first place attention is drawn to some preliminary tests that may be carried out on plasticisers. The chapter then deals with the various chromatographic methods of separation of plasticiser mixtures and the identification of the separated components.

Information is provided on the quantitative tests that may be carried out on plasticisers and on the recovery of alcohols produced by the hydrolysis of ester type plasticisers and on their subsequent gas chromatographic examination.

Useful methods of examination of individual plasticisers are detailed and some hints are given on the direct gas chromatographic examination of plasticiser mixtures.

The infra-red section follows and this deals almost entirely with the classification of plasticisers from the features of their infra-red spectra.

Some information is provided on the knowledge that may be gained from the n.m.r. examination of plasticisers.

The section on solvents is mainly concerned with the methods of isolation of solvents from polymer preparations and with the gas chromatographic, infra-red and chemical examination of solvent mixtures.

The chapter concludes with a short description of the methods of examination of pigments and fillers.

PLASTICISERS

The number of plasticisers and extenders met with as process additives in the plastics industry is very great indeed. These substances may be isolated by the analyst during the examination of plastics materials by the methods outlined in previous chapters. It may be useful, in the first place, to list below some of the compounds that have been encountered on one occasion or another in order to give some indication of the complexity of the problems involved.

Dialkyl (7–9) phthalate (Mixed alcohols C_7—C_9)
Dibutyl phthalate
Dicapryl phthalate
Di-2-ethyl hexyl phthalate
Di-iso-octyl phthalate
Dinonyl phthalate
Dimethyl phthalate
Diethyl phthalate
'Cellosolve' phthalate
Butyl acetyl ricinoleate
Dialkyl (7–9) sebacate
Dialkyl (7–9) adipate
Dibutyl adipate
Dibutyl sebacate
Di-iso-octyl adipate
Di-iso-octyl sebacate
Dioctyl adipate
Dioctyl sebacate
1 : 3-Butanediol adipate
Trialkyl (7–9) phosphate
Tritolyl phosphate
Trixylenyl phosphate
Trioctyl phosphate
Polypropylene adipate
Polypropylene adipate/laurate
Polypropylene sebacate
'Iranolin' and other mineral oil extenders
'Cereclor' (chlorinated paraffin wax)
Mesamoll (aryl ester of alkyl sulphonic acid)
Toluene sulphonamides
Poly-α-methyl-styrene
Castor oil
Dialkyl tin compounds
Oleates, stearates, adipates and benzoates of higher glycols
Ethyl palmitate
Butyl lactate
Butyl phthalyl butyl glycollate
Methyl phthalyl methyl glycollate
Benzyl butyl phthalate

It should be realised that other additives may also be isolated from a composition by the particular extraction method used, so that the ether extract from a polyvinyl chloride composition, for example, may also contain a cumarone or epoxy resin and/or organo-tin stabiliser.

In 1951 Haslam, Soppet and Willis[1] described methods for the chemical and infra-red examination of plasticisers used in polyvinyl chloride compositions. Within the last few years however it has become evident that a great deal of speculation and difficulty in the examination of mixed plasticisers, or of plasticiser extracts contaminated with other additives, could be avoided if the fact that a mixture was present was recognised in preliminary tests. A separation of the constituents of the mixture could then be carried out before attempting final identification of the components.

Useful preliminary tests are as follows.

QUALITATIVE TESTS FOR ELEMENTS IN PLASTICISERS

Previously the sodium fusion test was used, but today the Oxygen Flask Combustion test (p. 82) would be used to yield evidence of the presence or otherwise of additional elements such as nitrogen, sulphur, chlorine or phosphorus.

These tests provide pointers to e.g.

1. The presence of sulphonamides by the detection of sulphur and nitrogen.
2. The presence of Cereclor by the detection of large proportions of chlorine.
3. Suspicions of the presence of hydrocarbons such as Iranolin by the detection of traces of sulphur.
4. The presence of alkyl and aryl phosphates by the detection of phosphorus.
5. The presence of Mesamoll type plasticisers by the detection of sulphur and small amounts of chlorine.

The Oxygen Flask Combustion method should always be applied in a semi-quantitative manner, in order to obtain the maximum amount of information from the test.

TEST FOR PHTHALATES

About 0·05 g of resorcinol and phenol are weighed into separate test tubes ($6 \times \frac{3}{8}$ in or 150×12 mm). 3 drops of the plasticiser and 1 drop of conc. sulphuric acid are added to each tube and the tubes are then immersed in an oil bath at 160°C for 3 min. After cooling, 2 ml of water and 2 ml of 10% w/v sodium hydroxide solution are added to each tube and the mixture stirred. In the presence of phthalates the solution obtained in the resorcinol test shows a pronounced green fluorescence, whilst a red colour due to the production of phenolphthalein is obtained in the phenol test.

Both sebacates and ricinoleates give faint positive fluorescent reactions in the resorcinol test but they do not, of course, respond to the phenol test.

In certain cases this excellent test may fail e.g. in the presence of castor oil due to the carbonisation of the castor oil by the hot sulphuric acid.

However, it is well to be aware of a very good polarographic test for phthalates worked out by Whitnack and Gantz[2]. In favourable circumstances this test will detect as little as 2% of an ester such as dibutyl phthalate in a plasticiser. 0·1 g of the plasticiser and 2·5 ml of 0·1 M tetramethylammonium iodide are diluted to 10 ml with methyl alcohol. After removal of oxygen the solution is polarographed over the range −1·0 to −2·0 V. In the presence of dibutyl phthalate the reduction occurs at about −1·45 V. Such substances as tritolyl phosphate, butyl acetyl ricinoleate, tributyl citrate, triethyl citrate, dibutyl sebacate, Cereclor, Mesamoll and castor oil do not interfere under the conditions of the polarographic test. Of course ester phthalates other than dibutyl phthalate, i.e. such phthalates as dimethyl phthalate, dihexyl phthalate, dioctyl phthalate and dinonyl phthalate, are also reduced at about −1·5 V.

Incidentally, once the nature of the phthalate is known, the test may be applied on a quantitative basis.

FLUORESCENCE

A drop of the plasticiser is placed on a filter paper and examined under the U.V. lamp. As a rule ester plasticisers do not fluoresce, but substances such as Iranolin show distinct fluorescence.

2 : 6-DIBROMOQUINONE-CHLOROIMIDE TEST FOR PHENOLIC AND CRESYLIC PLASTICISERS

Reagents

0·5 N alcoholic potash.
N hydrochloric acid.
Sodium borate buffer solution.
23·4 g sodium borate ($Na_2B_4O_7 \cdot 10H_2O$) is dissolved in 900 ml of warm water, 3·27 g sodium hydroxide is added in the form of a 20% solution, and the solution is diluted to one litre.
2 : 6-dibromoquinone-chloroimide solution.

0·1 g 2:6-dibromoquinone-chloroimide is dissolved in 25 ml of 95% ethyl alcohol. The solution should be freshly prepared.

Procedure

10 mg of the plasticiser are dissolved in 5 ml 0·5 N alcoholic potash. The beaker is immersed in a rapidly boiling water bath for 10 min, and 2 ml of water are added to the evaporated residue followed by 2·5 ml of N hydrochloric acid. 1 ml of the resulting neutralised solution is transferred to a clean test tube and to it are added 2 ml of borate buffer solution and 5 drops of the 2:6-dibromoquinone-chloroimide solution. In the presence of the phenolic or cresylic type of plasticiser a blue indophenol colour develops at once.

The test is extremely sensitive. A blank test carried out on the reagents should give no colour.

TEST FOR EPOXY RESINS IN PLASTICISERS

4 drops of a saturated aqueous solution of glucose are added to 1 drop of the plasticiser. 6 drops of conc. sulphuric acid are added and the tube is swirled gently. A violet coloration is obtained in the presence of epoxy resins.

These tests, taken in conjunction with the infra-red spectrum of the sample, will usually provide sufficient evidence to indicate whether the sample or extract is a mixture or a single plasticiser.

SEPARATION OF PLASTICISERS

Having ascertained the general nature of the sample under test, the subsequent method of analysis may take several forms, but separation or partial separation of a mixture is a necessary prerequisite whatever the final method of identification adopted. Prior to the development of thin-layer chromatographic methods, the column chromatographic method described by Cachia, Southwart and Davison[3] was recommended. In its simplest form the procedure is well illustrated by the separation of equal parts of Cereclor, dioctyl phthalate and tritolyl phosphate.

A chromatographic column approximately 300 mm long and 10 mm in diameter is packed with a mixture of equal parts of silica gel 200/300 and Celite 545 by means of the wet sludge method with carbon tetrachloride— 30 ml of carbon tetrachloride are run through the column. 2 drops of the mixed plasticiser dissolved in 1 ml of carbon tetrachloride are transferred to the top of the column and then elution is carried out successively with 40 ml of each of the appropriate solvent mixtures as shown below.

1. Carbon tetrachloride.
2. Carbon tetrachloride+1·5% v/v Isopropyl ether (I.P.E.).
3. Carbon tetrachloride+2% v/v Isopropyl ether.
4. Carbon tetrachloride+3% v/v Isopropyl ether.
5. Carbon tetrachloride+4% v/v Isopropyl ether.

The eluates are collected in 10 ml fractions and the solvent is removed by evaporation, using a flow of compressed air over the tubes.

In those cases where a significant residue is obtained this residue is submitted for infra-red examination; the results obtained in a particular experiment are given below.

Eluate	Fraction No.	Infra-red Report
CCl_4	2, 3, 4	Cereclor
$CCl_4 + 1·5\%$ I.P.E.	7	Dioctyl phthalate
$CCl_4 + 2\%$ I.P.E.	11	Tritolyl phosphate

With polyvinyl chloride and related compositions containing both polymeric and monomeric plasticisers, the identification of the polymeric plasticiser is usually fairly easy. The reason for this is that in the preliminary extraction of the composition with ether an extract containing all the monomeric and some polymeric plasticiser is obtained. In a further extraction with hot methanol, however, the polymeric plasticiser is recovered in pure form.

It is the mixture of monomeric and polymeric plasticiser which causes greatest difficulty by conventional means, but our experiments have shown e.g. that the application of the Cachia and Southwart procedure[3] to a mixture of dioctyl phthalate and polypropylene adipate is attended with success. The polypropylene adipate remains on the column whilst the dioctyl phthalate is recovered from the carbon tetrachloride+1·5% isopropyl ether eluate.

Similarly, a mixture of Iranolin, dialkyl (7–9) phthalate and tritolyl phosphate is successfully resolved; the Iranolin is found in the CCl_4 eluate, the dialkyl phthalate in the $CCl_4 + 1·5\%$ isopropyl ether eluate, and the tritolyl phosphate in the $CCl_4 + 2\%$ isopropyl ether eluate.

The great versatility of the method is shown in a particular case involving a mixture of 50% hydrocarbon, 30% butyl benzyl phthalate and 20% propylene glycol dibenzoate. In this case the hydrocarbon is eluted first, the butyl benzyl phthalate second and the propylene glycol dibenzoate is retained on the column. It is subsequently removed from the column in an identifiable condition by means of ether.

A cumarone resin-dialkyl phthalate-tritolyl phosphate mixture is successfully resolved by the above procedure, the elution taking place in the above order.

The column chromatographic method is time consuming and produces many fractions, all of which have to be examined in order to identify the components that are separated. Thin-layer chromatographic methods, on the other hand, are carried out simply and rapidly and, in most cases, they give clear-cut separations that are visible on the chromatographic plate. Only a minimum number of fractions has to be washed from the support for identification.

Within recent years several papers have been published on the thin-layer examination of plasticisers. Those by Haase[4], Ligotti, Piacenti and Bonomi[5] and Braun[6] are particularly informative and all point out that R_F values give only a guide to the identification of the separated components; infra-red examination of the separated components provides the final method of identification and chemical breakdown of the plasticiser and examination of the products produced may also be necessary.

Our experience supports the conclusions of these authors and the procedures we find most useful in the examination of ether extracts are summarised below.

THIN-LAYER CHROMATOGRAPHY/INFRA-RED METHOD

Apparatus: unit development cell

This is shown diagrammatically in Figure 11.1 and consists of two essential parts.

1. A 200×200 mm glass plate, to three edges of which glass strips 10 mm wide $\times 3$ mm thick are fixed with Araldite adhesive.

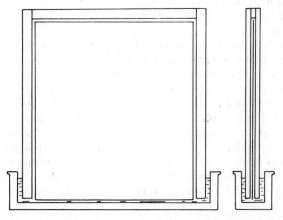

Fig. 11.1.   Thin-layer chromatography 'sandwich' development cell

2. A wooden block $12 \times 4 \times 1\frac{1}{2}$ in $(300 \times 100 \times 40$ mm) deep having a central well $9 \times \frac{3}{4} \times 1$ in $(230 \times 20 \times 25$ mm) deep. The well is lined with aluminium foil and serves as a container for the developing

solvent and also supports the development cell in a near-vertical position. Alternatively, this container may be made in stainless steel.

For the preparation of thin layer plates a leveller and coating spreader are also needed.

Reagents

Kieselgel G.

Solvent: benzene-ethyl acetate (95/5).

U.V. Indicator: Fluolite C 180.

Preparation of plates

In the preparation of three 1000 μm thick 200 mm $\times$ 200 mm plates, 60 g of Kieselgel G are weighed into a 250 ml conical flask. 100 ml water, in which 10 mg Fluolite C180 has been dissolved, are added and the mixture shaken for 90 s. The slurry is poured slowly into the spreader and the plates are coated without delay. The coated plates are allowed to set for 10 to 20 min, then removed carefully and placed in an oven at 110°C for 90 min. Thin-layer chromatography aluminium sheets, silica gel pre-coated, are also available commercially.

Sample addition and subsequent elution

Approximately 50 mg plasticiser are dissolved in about 0·5 ml ether. The solution is added to the plate, dropwise, in such a way as to form a continuous line approximately 20 mm from the edge of the plate, and is then allowed to dry.

10 mm wide strips of adsorbent are removed from the two edges of the prepared plate at right angles to the line of sample application and from the edge furthest from this line. This plate containing the sample is placed on the plate described under Apparatus 1 above so that the sample application line is adjacent and parallel to the open end of the cell that is assembled. The two plates are fixed together by means of spring clips.

The open end is immersed in the solvent which is contained in the well of the cell container so that the solvent covers the lower edge of the adsorbent but does not reach the line of application of the sample.

The cell is removed from the container when the solvent front has moved approximately 100 mm, then dismantled, and the plate dried with a hair dryer.

Identification of the components after separation

The dried plate is examined under U.V. light and the positions of the separated components, which appear as bands, are marked. With the aid of a spatula, each band is transferred carefully to a 100 ml beaker. Approximately 20 ml ether are added and the solution is filtered through a Whatman No. 1 filter paper. The filtrates are collected in small evaporating dishes and the ether is removed by evaporation on a water bath. Infra-red spectra of the residues are then obtained.

Notes

1. In our experience most of the samples submitted for test respond to the above treatment. All the possible combinations of the numerous plasticisers have, however, not been examined and it is unlikely that every mixture will separate readily.

   In cases where separation is not complete it is useful to carry out a double elution with the same solvent i.e. the procedure outlined above is repeated as far as the examination of the dried plate under U.V. light. The plate is then replaced in the solvent tank and again eluted. This procedure has been shown to be successful in the isolation of small amounts of Paraplex G 62 from phthalate esters.

   In certain cases it is advantageous to change the solvent. Benzene has been substituted for benzene-ethyl acetate to give a good separation of dioctyl phthalate from di-iso-octyl adipate.

2. The order of elution of the plasticisers is similar to that found in the column chromatographic method (p. 555) e.g. Cereclor moves with the solvent front, Paraplex G 62 remains at the starting point, and phthalate and phosphate esters are at intermediate positions.

QUANTITATIVE TESTS ON THE RECOVERED PLASTICISER

REFRACTIVE INDEX AND ABSORPTION

The refractive index of the plasticiser is determined at 20°C using an Abbé refractometer.

$E_{10\,mm}^{1\%}$ in ethyl alcohol. It is useful to have available tables giving the $E_{10\,mm}^{1\%}$ values in ethyl alcohol for all the commercial phthalates. For esters of *o*-phthalic acid 275 nm is a suitable wavelength to use, whilst for esters of terephthalic acid 284 nm is satisfactory.

If a substance under test proves to be a single phthalate these tables can yield useful confirmatory evidence.

ESTER VALUE

The ester value is extremely important in those cases where the preliminary tests have indicated that the plasticiser is a single substance of the ester type such as a phthalate. Hydrolysis is carried out by refluxing 2 g of the sample with 50 ml of 0·5 N alcoholic potash for 2 h. After cooling and diluting with water the excess of potassium hydroxide is titrated with 0·5 N hydrochloric acid using phenolphthalein as indicator. The ester value is calculated in the usual manner from the volume of potash consumed.

If mixtures of Cereclor and phthalate are being hydrolysed in the above manner the alcoholic potash will remove chlorine from the Cereclor in the course of the test. Allowance for this usage may be made by boiling down the neutralised solution from the ester value test to a small volume to remove alcohol, previous to potentiometric titration of an aliquot with 0·1 N silver nitrate solution.

DETERMINATION OF PHTHALATES

This test is carried out as follows: 2 g of the sample are refluxed for 2 h with 50 ml of N alcoholic potash prepared from absolute alcohol; the condenser is fitted with a calcium chloride trap. The precipitated di-potassium phthalate is filtered off on a sintered glass Gooch crucible (1 × G3) and the precipitate is washed with 50 ml boiling absolute alcohol. This operation is carried out rapidly to avoid contamination of the precipitate with potassium carbonate. The crucible and contents are dried for 4 h at 150°C and re-weighed. The proportion of phthalate in the plasticiser is calculated from the weight of dipotassium phthalate which is obtained.

This determination is quite satisfactory when carried out on individual phthalates and also on mixtures of phthalates and substances such as tritolyl phosphate, butyl acetyl ricinoleate and Mesamoll.

Cereclor interferes, as do sebacates, adipates and maleates.

The interference by Cereclor may be avoided by taking the impure precipitate of dipotassium phthalate and recovering the phthalic acid from it in the following manner.

The dry impure dipotassium phthalate is transferred to a beaker (100 ml) and the crucible washed well with water. The solution (total volume 50 ml) is treated with hydrochloric acid in an amount about 20% in excess of the potassium, on the assumption that the precipitate is pure dipotassium phthalate.

The acid solution is transferred to a continuous liquid/liquid extractor and extracted with ether for 3 h. The ether is removed from the extract and the residual phthalic acid dried to constant weight at 100°C. The above method is extremely useful in the examination of mixtures of phthalates and Cereclor. The influence of various substances on the dipotassium phthalate test and the subsequent test in which the phthalic acid is recovered as such may be seen from the figures given in Table 11.1.

As has been indicated, sebacates (and adipates) interfere because dipotassium sebacate is also precipitated along with the dipotassium phthalate.

Thus, a mixture of 86·54% dibutyl phthalate and 13·46% dibutyl sebacate, when subjected to the above test, gave a weight of precipitated

Table 11.1. Influence of various substances on the dipotassium phthalate test and the subsequent test in which the phthalic acid is recovered as such

Sample	Dipotassium phthalate		Phthalic acid		Acid value of recovered acid, mg KOH/g
	found %	*theory* %	*found* %	*theory* %	
Dibutyl phthalate (D.B.P.)	87·05	87·06	59·72	59·71	674
75·68% D.B.P. 24·32% Mesamoll	66·1	65·9	45·4	45·2	670
84·61% D.B.P. 15·39% Butyl acetylricinoleate	73·8	73·7	50·71	50·51	670
69·54% D.B.P. 30·46% Cereclor	73·9	60·47	41·8	41·5	671
72·55% D.B.P. 27·45% T.C.P.	62·92	63·16	43·2	43·31	672

potassium salt equivalent to 86·3% of the weight of the original mixture. On the assumption that both dipotassium phthalate and dipotassium sebacate were precipitated, the theoretical weight was 87·25%. The recovered acids amounted to 59·56% of the original sample (the theoretical figure for sebacic and phthalic acids present in the sample is 60·3%).

Within recent years polymeric plasticisers such as polypropylene adipate and polypropylene sebacate have been used increasingly and, if phthalate-adipate-sebacate combinations are subjected to the above phthalate determination, then the dipotassium salts of all the three acids are precipitated.

Of course, it has always been possible to get a measure of the phthalate content of such potassium salts by solution of the precipitate in 0·1 N hydrochloric acid solution and determination of the U.V. absorption of the solution at 276 nm.

Nevertheless, knowing as we have indicated above that the combined acids could readily be recovered from the dipotassium salts by acidification and continuous extraction with ether, it has seemed to us to be desirable to have available a method for the separation and determination of adipic and sebacic acids in adipic acid-sebacic acid-phthalic acid mixtures.

We have worked out a chromatographic method[7] for this test based on earlier work of Higuchi, Hill and Corcoran[8].

SEPARATION AND DETERMINATION OF ADIPIC AND SEBACIC ACIDS IN A MIXTURE WITH PHTHALIC ACID

The separation is effected on a silicic acid column buffered to pH 4·3. The sebacic acid is eluted from the column with 2% v/v butanol-chloroform and the adipic acid then eluted with 15% v/v butanol-chloroform. The phthalic acid is not eluted under the conditions of the test.

Apparatus

A chromatographic column 250 mm long and 15 mm in diameter, fitted at the bottom with a sintered-glass disk, grade 2 porosity.

Reagents

Mallinckrodt's silicic acid, 100 mesh.
Sodium citrate solution, 1 M.
Hydrochloric acid, N.
n-Butanol. Analar.
Chloroform. Analar.
m-Cresol purple indicator, 0·1% w/v in alcohol.
Tert-amyl alcohol.
Sodium hydroxide solution, 0·05 N alcoholic.

Procedure

Weigh 10 g of silicic acid into a beaker and add 10 ml of a citrate-buffered solution prepared by titrating 25 ml of N hydrochloric acid with 1 N sodium citrate to a pH of 4·30. Thoroughly mix the silicic acid with the buffer solution, then add 40 ml of 2% v/v *n*-butanol-chloroform solution and stir to form a slurry. Pour increments of the slurry into the chromatographic column and press each increment down by means of a flat-headed glass rod. When the column is filled, place a filter paper disk on top of the silicic acid and at no time let the column drain to dryness. Prepare a solution of the acid to be tested by dissolving 0·2 to 0·5 g of the acid in 10 ml of warm *tert*-amyl alcohol in a 50 ml standard flask and after cooling dilute to the mark with chloroform.

Transfer 5 ml of this solution by means of a pipette to the chromatographic column and wash through with two increments of 2 ml of 2% v/v *n*-butanol-chloroform solution and finally elute the column with 25 ml of the same solvent. Collect the eluate in a 150 ml conical flask and titrate with alcoholic 0·05 N sodium hydroxide using *m*-cresol purple as indicator. This titre gives a measure of the sebacic acid.

Now elute the column with 70 ml of 15% v/v *n*-butanol-chloroform solution, collect the eluate and again titrate with 0·05 N alcoholic sodium hydroxide solution. This titre gives a measure of the adipic acid.

Run a blank column alongside the test column and correct the titrations obtained in the test for that obtained in the blank.

Detecting the end-points with *m*-cresol purple will require some experience as the colour at the end-point tends to fade rapidly. The titration is best carried out fairly rapidly and the end-point is taken as that point where the change of colour from yellow to purple persists for more than 5 s.

Results

Application of this method to know mixtures of adipic, sebacic and phthalic acid gave the results shown in Table 11.2.

Table 11.2. Separation of mixtures of sebacic, adipic and phthalic acids

Mixture	Weight taken, g	Weight found, g
1. Sebacic acid	0·1624	0·1618
Adipic acid	0·1633	0·1604
Phthalic acid	0·1553	Not determined
2. Sebacic acid	0·1357	0·1367
Adipic acid	0·1581	0·1612
Phthalic acid	0·1270	Not determined

The method can be applied without modification to mixtures of adipic and sebacic acids, and application of the test to such mixtures has given the results shown in Table 11.3 (see p. 564).

The original paper should be consulted for observations on the behaviour of other dibasic acids under somewhat similar conditions of test.

CHLORINE

The chlorine content of plasticisers is determined by sodium peroxide fusion in a metal bomb as described in Chapter 3.

Alternatively, X-ray Fluorescence methods may be used. Cereclor-phthalate or Mesamoll-phthalate mixtures, for example, can be resolved by measurement of the chlorine or sulphur Kα intensities of the extract contained over a supporting window, after the mixture has been degassed in vacuo. The ratios of these intensities to those of instrument standards are calculated, and referred to calibration graphs prepared by the examination of known mixtures of dioctyl phthalate and Cereclor or Mesamoll. Should Cereclor and Mesamoll be present together then absorption effects occur and corrections must be applied.

Alternatively, suitable calibration standards are prepared so as to match closely the particular sample.

PHOSPHORUS

The phosphorus content of plasticisers such as tritolyl phosphate may be determined along the following lines: The sample (sufficient to contain 0·04 g P_2O_5) is weighed into a Kjeldahl flask (250 ml) and 10 ml of conc.

Table 11.3. Separation of mixtures of adipic and sebacic acids

Mixture	Weight taken, g	Weight found, g
1. Adipic acid	0·1034	0·1013
Sebacic acid	0·4038	0·4015
2. Adipic acid	0·2998	0·2996
Sebacic acid	0·2015	0·2035
3. Adipic acid	0·1605	0·1629
Sebacic acid	0·1572	0·1584

sulphuric acid and 5 ml of conc. nitric acid are added. The flask is heated until nitrous flumes cease to be evolved and then nitric acid is again added 0·5 ml at a time until a colourless or pale-yellow liquid is obtained. The flask is cooled, the solution diluted to 100 ml with water, and the phosphate determined by the ammonium phosphomolybdate volumetric procedure.

GENERAL SEPARATION OF THE ALCOHOLIC
CONSTITUENTS OF ESTER TYPE PLASTICISERS

In the examination of mixed plasticisers, e.g. mixtures of substances such as dinonyl phthalate, tricresyl phosphate and Cereclor, it is often desirable to establish with certainty the nature of the alcohol, e.g. nonyl alcohol, present as phthalate ester. Infra-red examination of the mixed plasticiser in its original condition is rarely satisfactory, but it is our experience that if the alcohol constituent is first isolated in as pure a condition as possible then the subsequent chemical and infra-red examinations are quite satisfactory and the alcohol is readily identified.

It should be pointed out that the alcohols used in the preparation of commercial phthalate plasticisers are usually branched-chain compounds, e.g. commercial hexyl, octyl and nonyl alcohols are usually 2-ethyl-butyl, 2-ethyl hexyl and trimethyl hexyl alcohols, the last containing a good deal of the 3 : 5 : 5-isomer. Nevertheless the isolated alcohols are reasonably constant in composition.

We have found the following method to be of very wide application in this connection: 2 ml of the plasticiser is refluxed for 1 h with 2 g of potassium hydroxide and 5 ml of ethylene glycol in a small pear-shaped flask (25 ml) fitted by a ground-glass joint to a small condenser. The ethylene glycol used is previously distilled to 190°C and the volatile fraction discarded. In the case of the alcohols of lower boiling point a certain amount of solid matter separates in the course of the hydrolysis. In the case of ethyl and methyl phthalates the hydrolysis takes place very soon after the liquid starts to boil and care must be taken to avoid loss by spurting. After hydrolysis the mixture is distilled until 1·5 to 2 ml of distillate have been collected in a stoppered separating funnel (5 ml capacity). The temperature at which the greater portion of the liquid distills over is noted. At no time should the temperature be allowed to rise above 175°C.

If the distillate separates into two layers the presence of butyl or higher alcohol is indicated. The lower aqueous layer is saturated with sodium

Table 11.4. Identification of the alcoholic constituents of ester type plasticisers

Plasticiser	Distillation temp. of alcohol, °C	n_D at 20°C	b.p. °C
Dibutyl phthalate	95–105	1·399	110
Dihexyl phthalate	115–125	1·420	122
Dioctyl phthalate	120–130	1·430	174
Dinonyl phthalate	140–155	1·432	179

chloride, separated, and reserved for possible examination for the presence of methyl or ethyl alcohols. The upper layer is washed with a further 1 ml of saturated salt solution, and the washed alcohol then pipetted into a 1 ml centrifuge tube containing $\sim\frac{1}{4}$ in (6 mm) layer of anhydrous potassium carbonate and allowed to stand overnight. The refractive index at 20°C and the boiling point by Siwoloboff's method are then determined on the dry alcohol. Results obtained on application of this method to commercial plasticisers are given in Table 11.4.

At this stage the alcohol is submitted to infra-red examination and valuable evidence in support of the chemical conclusions is thus obtained. This is particularly useful in the case of esters of hexyl alcohol because the hexyl alcohol obtained from commercial plasticisers is usually a mixture of several isomers.

If the distillate is present as one layer and the distillation temperature is below 105°C the presence of ethyl or methyl alcohol may be suspected. This distillate is treated with 20 ml of dichromate-sulphuric acid mixture (100 mg potassium dichromate dissolved in a mixture of 250 ml conc. sulphuric acid and 750 ml of water) and the mixture distilled until 1 to 2 ml of distillate has been collected. To 0·5 ml of this distillate contained in a 15 ml centrifuge tube is added 1 pellet of potassium hydroxide and the tube is placed in a beaker of boiling water. The presence of ethyl alcohol is indicated by the pellet first turning yellow after which the solution turns yellow and emits the disagreeable odour of aldehyde resin. A further 3 to 4 drops of the distillate are added to 2 ml of dilute sulphuric acid solution (100 ml of water and 150 ml of conc. sulphuric acid) contained in a test tube ($5 \times \frac{3}{8}$ in or 150×12 mm). A few crystals of chromotropic acid are added after which the tube is heated in a water bath at 60–70°C for 10 min. The appearance of a violet colour is indicative of the presence of methanol.

If the original distillate is present as one layer and the distillation temperature is above 105°C then the presence of a Cellosolve phthalate may be suspected. In this event a further 2 ml of the plasticiser is refluxed for 2 h with 5 g potassium hydroxide and 5 ml of water. The hydrolysis product is then distilled until 3 to 4 ml of distillate have been collected. This distillate is saturated with solid potassium periodate, and after standing for 30 min the upper layer of the liquid is decanted off and redistilled until 1 ml of distillate has been collected. Under these conditions a Cellosolve phthalate yields formaldehyde in the distillate which is confirmed by the application of the chromotropic acid test.

A method which has proved particularly useful when dealing with the alcohol from a very small amount of an individual phthalate plasticiser may be illustrated by the procedure which was adopted in the case of a sample of plasticiser recovered from a deaf-aid cord. The sample weighing 0·2 g was dissolved in 1 ml of ether and transferred to the bulb of the apparatus shown in Figure 11.2(a). The apparatus was constructed of Pyrex glass. After removal of the ether by placing the bulb in a water bath at 30–35°C and blowing a current of air over it, all remaining traces of ether were removed by heating the tube at 100°C for 10 min, 0·2 g of potassium hydroxide (2 pellets) and 0·5 ml of ethylene glycol (previously distilled to 190°C and the volatile fraction rejected) were added to the contents of the bulb, and the mixture refluxed on a small electric hotplate until hydrolysis was complete.

In the case of the sample in question the height to which the refluxing liquid rose in the tube was about 120 mm. If the sample is derived from

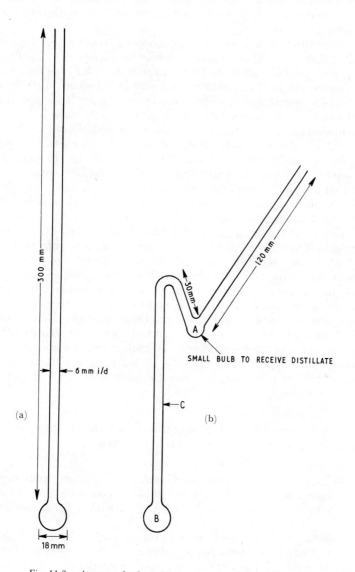

300 mm

6 mm i/d

18 mm

(a)

30 mm

120 mm

A

SMALL BULB TO RECEIVE DISTILLATE

C

B

(b)

Fig. 11.2.   Apparatus for the semi-micro separation of alcohols from plasticisers

an alcohol such as butyl alcohol then the refluxing height may be much greater than this, e.g. as high as 250 mm from the bulb. In this case it may be desirable to wrap a piece of wet cotton wool round the tube about 40 mm from the top.

The apparatus was then cooled and the upper part of the tube bent in a blow-pipe flame to yield an apparatus similar to the one shown in Figure 11.2 (b) with a small bulb at A. The contents of bulb B were then heated over a bunsen flame. The alcohol was collected in the small bulb A; it was easily possible to determine by observation when the alcohol had distilled over and only the glycol was refluxing to and fro in the tube C.

On completion of the distillation the alcohol in the bulb was withdrawn in a capillary dropper and transferred to a test tube $(80 \times 2 \cdot 5$ mm$)$ containing a small amount (about 20 mg) of sodium chloride. The separated alcohol was then withdrawn with a capillary dropper and transferred to a similar test tube containing a volume of saturated sodium chloride solution approximately equal to the volume of alcohol. In doing this the tip of the capillary was placed at the bottom of the test tube and the alcohol blown out in such a manner that it bubbled slowly through the sodium chloride solution. After separation the alcohol was withdrawn and placed in a third test tube $(30 \times 2 \cdot 5$ mm$)$ containing about 10 mg of potassium carbonate, and allowed to dry. The boiling point and refractive index of the dry alcohol were then determined, the refractive index being measured on the sample used for the boiling point determination. The following results were obtained in the case of the plasticiser derived from the deaf-aid cord: b.p. 170°C; n_D at 20°C 1·4306. The identity of the alcohol as octyl alcohol was confirmed by infra-red examination.

A comparable experiment on dibutyl phthalate gave a b.p. of 105°C and n_D at 20°C of 1·3985 for the alcohol isolated in the above manner.

GAS CHROMATOGRAPHIC SEPARATION AND IDENTIFICATION OF ALCOHOLS RECOVERED FROM PLASTICISERS

With the higher alcohols the distillate considered on p. 565 will separate into two layers. Both layers are examined. The chromatographic separation is effected on a gas chromatograph fitted with a flame ionisation detector.

The column, 2 m, ($\frac{1}{8}$ in or 3 mm o.d. stainless steel), 20% w/w nonyl-phenoxy-polyethylene-oxyethanol on silanised Celite or polytetrafluoroethylene, is maintained at 160°C with a nitrogen carrier gas inlet pressure of 10 lbf/in^2 (70 kN/m^2) and flow rate of 8 ml/min.

Figures 11.3 (a) and (b) indicate how it is possible to distinguish between the complex mixed alcohols obtained from dialkyl (7–9) phthalate and also mixed alcohols from di-iso-octyl phthalate from the type of chromatographic pattern obtained.

Problems of this kind cannot be solved by direct infra-red examination.

The behaviour on gas chromatographic examination of some of the alcohols used in plasticiser manufacture is shown in Figure 11.4.

The combination of gas chromatographic separation and infra-red examination of the separated components provides conclusive identification.

METHODS OF RECOVERY OF GLYCOLS FROM GLYCOL ESTERS

These methods are capable of adaptation to other problems.

The first method has been shown to be useful in the recovery of propylene glycol from propylene glycol sebacate.

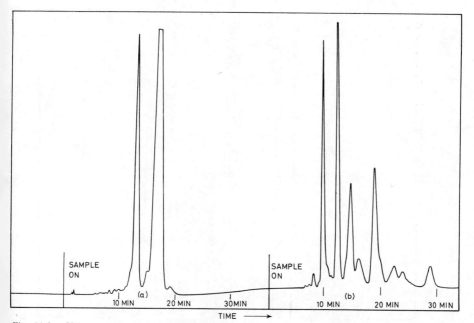

Fig. 11.3. Chromatograms of mixed alcohols, obtained from (a) di-iso-octyl phthalate and (b) dialkyl (7–9) phthalate

2 g of the ester are heated under reflux in a 100 ml flask with 3 g of KOH and 15 ml distilled water. The heating is continued for about 4 h. The solution is cooled and solid potassium carbonate is added in excess. The whole is transferred to a continuous liquid/liquid extractor and then extracted with ether for 6 h. On evaporation of the ether a residue of propylene glycol of high purity is obtained, quite suitable for infra-red identification.

A second method has been shown to be valuable in the recovery of 1 : 4-dihydroxymethyl cyclohexane from a polyester film and, in appropriate cases, should be applied to ester plasticisers. The film proved to contain a large proportion of the terephthalate of the above alcohol.

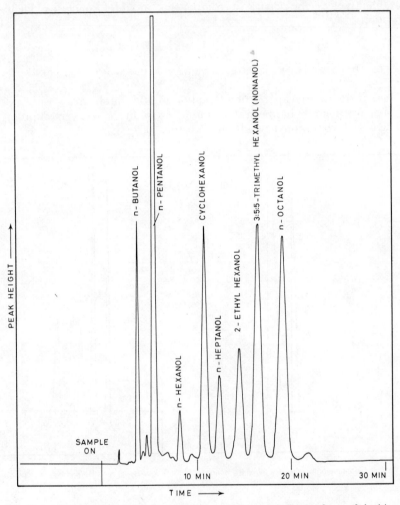

Fig. 11.4.   Chromatogram of synthetic mixture of some alcohols used in the manufacture of plasticisers

After hydrolysis with alcoholic potash and filtration from dipotassium terephthalate, the alkaline filtrate is neutralised with hydrochloric acid and then saturated with potassium carbonate. After standing, the mixture is filtered and the 1 : 4-dihydroxymethyl cyclohexane recovered in first-class condition by evaporation of the filtrate.

THE RECOVERY OF PHENOLS, CRESOLS AND XYLENOLS FROM
PLASTICISERS CONTAINING PHOSPHATE

This method may be used when it is desirable to obtain a sample of the
phenol, cresol or xylenol for further examination by chemical or infra-red
methods.

About 2 ml of the plasticiser is hydrolysed by the ethylene glycol-
potassium hydroxide method previously described (see p. 565). If only an
individual plasticiser of the phosphate type is present, the hydrolysis
products are transferred to a distillation flask (50 ml) and diluted with
25 ml of water. If however the phosphate plasticiser is mixed with an ester
type plasticiser, the alcohol is boiled out before dilution and the diluted
solution is again boiled until it no longer smells of alcohol. The solution
is then saturated with carbon dioxide and distilled into a separating
funnel (50 ml) until 20 ml have been collected. The distillate is saturated
with sodium chloride and extracted twice with 10 ml and 5 ml portions
of ether. The combined ether layers are washed twice with 5 ml portions
of saturated sodium chloride and filtered through a dry paper into an
extraction flask (100 ml). The ether is evaporated off on a water bath,
the last traces being removed by passing a stream of compressed air over
the residue. Care should be taken to avoid prolonged heating as this will
result in loss of the cresols.

BUTYL ACETYL RICINOLEATE

When this ester is present in a plasticiser the filtrate from the precipitation
of dipotassium phthalate (p. 560) will contain potassium ricinoleate and
potassium acetate in quantitative proportions. This filtrate is diluted to
50 ml with water, before evaporating the alcohol and rendering the filtrate
acid to methyl orange with hydrochloric acid. In the presence of ricinoleates a solid is precipitated and this turns to an oily liquid on standing.
The solution is then extracted three times with ether using 25, 20 and 10 ml
respectively, and the combined ether extract is washed three times with
10 ml of water on each occasion. The ether is removed by evaporation and
the recovered acid dried to constant weight at 100°C. The acid value of
the residue is determined as a check on the purity of the recovered
ricinoleic acid.

The acetyl content of butyl acetyl ricinoleate itself may be determined
by the following method, for the details of which we are indebted to
the late Mr. S. M. Whettem:

About 1 g of the sample is refluxed for 2 h with 50 ml of 0·5 N alcoholic
potash, after which the solution is diluted with water and evaporated free
from alcohol. The solution is then acidified with sulphuric acid and a
solution of 0·67 g of silver sulphate in about 100 ml of water is added with

continuous stirring. The precipitate is filtered off, washed well, and the filtrate acidified with 20 ml of phosphoric acid (Analar; specific gravity 1·75). The filtrate is then steam distilled through a spray trap until 250 ml has been collected. The receiver is then changed and the distillation continued. A stream of air is bubbled through the distillate for 20 min after which the distillate is titrated to phenolphthalein with 0·1 N sodium hydroxide. The steam distillation is continued until a 250 ml portion gives a titration of 0·2 ml or less.

A blank determination is carried out simultaneously.

When plasticisers such as tritolyl phosphate are present in addition to butyl acetyl ricinoleate, the ricinoleic acid recovered by the method described above will be quite impure and contaminated with cresols.

In such cases 0·5 g of the mixture, accurately weighed, is added to about 50 ml of water. Conc. ammonia solution is added dropwise until the solution is definitely alkaline to methyl red indicator. The solution is then boiled until free from ammonia, cooled, and the ricinoleic acid precipitated by the addition of 10% w/v calcium chloride solution. After standing for 10 min the precipitate is filtered off and washed with water. The calcium precipitate is washed off the filter paper into the original beaker with about 30 to 40 ml of water. 5 ml of conc. hydrochloric acid are added and the mixture brought to the boil, then cooled. The liberated ricinoleic acid is extracted by successive extractions with 50 ml, 30 ml and 20 ml of ether respectively. The ether extracts are combined and washed with water until free from acid. The ether solution is evaporated to dryness and any traces of water are removed by the addition of a small amount of acetone which is then evaporated off. The ricinoleic acid is finally dried to constant weight at 100°C.

NOTES ON OTHER PLASTICISERS

It is, of course, impossible to discuss all the different plasticisers but it may be useful to comment on three where points of analytical principle are concerned.

n-ETHYL *p*-TOLUENE SULPHONAMIDE

It is useful to have available a method for the determination of the nature of the alkyl radical in such alkyl aryl sulphonamides. In this test the sample is hydrolysed with acid, after which the solution is made alkaline, the alkyl amine distilled in steam, collected in acid, and then converted to an identifiable crystalline derivative.

In the method 2 g of the plasticiser are heated with 10 ml of 80% v/v sulphuric acid solution for 30 min in an oil bath at a temperature of 160°C. The mixture is then cooled and transferred to a Kjeldahl type ammonia distillation apparatus. The solution is brought to the boil, then neutralised with 30% w/v sodium hydroxide solution, after which a further

10 ml of the alkali solution are added. The alkaline solution is boiled for 20 min and the alkyl amine distilled off and collected in 10 ml of N hydrochloric acid solution. The solution in the receiver is then evaporated to dryness on a water bath and finally dried in an oven at 100°C for 30 min. The hydrochloride (0·5 g) is now boiled under reflux with phthalic anhydride (0·5 g), 1 g of sodium acetate, and 5 ml of glacial acetic acid for 30 min. The solution is then cooled, diluted with approximately 10 ml water and cooled in ice. The N-substituted phthalimide derivative is precipitated, filtered off, and recrystallised from alcohol, prior to m.p. determination. In this way the original alkyl aryl sulphonamide is characterised.

BUTYL PHTHALYL BUTYL GLYCOLLATE

This plasticiser has been encountered in transparent beer tubing and it provides a good example of the method of characterisation employed with such a complicated plasticiser. For example,

1. No elements other than C, H and O were detected.
2. A positive test for phthalate was obtained in the qualitative tests.
3. In the ethylene glycol potash hydrolysis butyl alcohol was obtained.
4. The R.I. at 20°C was 1·4908.
5. The ester value was 480 mg KOH/g.
6. The insoluble potassium salts amounted to 72·1 g/100 g sample.

After filtering off the insoluble potassium salts the filtrate was diluted with water and boiled down to a volume of 20 ml. Boiling was continued for 3 to 4 h, maintaining the volume by repeated addition of water.

The solution was now cooled and extracted twice with ether to remove any extractable matter.

The aqueous solution was now acidified with conc. hydrochloric acid and extracted with ether for 5 h in a liquid/liquid extractor.

The ether extract was dried over anhydrous sodium sulphate, then evaporated and dried. The dry residue proved to be glycollic acid on infra-red test.

All these figures, and the infra-red spectrum of the plasticiser, fit the behaviour of butyl phthalyl butyl glycollate.

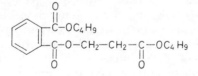

BENZYL BUTYL PHTHALATE

The main problem with this plasticiser is to obtain evidence of the benzyl and butyl alcohols. This is accomplished by hydrolysis with ethylene glycol potash and steam distillation of the products. The benzyl and butyl alcohols are collected, dried, fractionated, and identified.

In the original paper of Haslam, Soppet and Willis[1] it was made quite clear that, in the examination of plasticisers, chemical and infra-red tests are complementary.

All our experience in the interval supports this view.

GAS CHROMATOGRAPHIC ANALYSIS OF PLASTICISERS

In recent years gas chromatographic examination of plasticiser mixtures has been favoured by certain authors, although gas chromatographic data obtained from a single column are insufficient for positive identification.

Two or more columns are used or, alternatively, the separated fractions are trapped prior to infra-red and n.m.r. examination. Descriptions of useful procedures are given by Haase[9] and Guiochon and Henniker[10].

Haase uses a katharometer detector and employs three columns viz. (a) 20% silicone oil BR/SN on Sterchamol 0·2–0·3; (b) 10% silicone grease SE30 on Chromosorb 60/80, and (c) 10% silicone oil DC on Chromosorb 60/80. All relative retention times on each column are calculated against a reference sample of bis-β-methoxy-ethyl phthalate.

INFRA-RED SPECTRA OF PLASTICISERS

The usual method of removing plasticisers from resins is by ether extraction for monomeric plasticisers or by hot methanol extraction for polymeric plasticisers as previously described (p. 144). Since in these extractions substances other than plasticisers may be removed, the analyst must be prepared to encounter a number of materials such as heat stabilisers, lubricants and emulsifying agents in the examination of the extracts. In this discussion on the identification of plasticisers from their infra-red spectra some observations will therefore be included on other compounds which may be presented in mixtures with true plasticisers.

The present tendency, particularly in vinyl compositions, is in the direction of more-complex mixtures; this means that infra-red examination of the total plasticiser extract, without previous separation, has become much less satisfactory and may well be misleading. This is mainly due to large differences of the intensities of the absorption spectra of the various substances used in these mixtures. Phthalates, for example, absorb so strongly that components such as mineral oil or chlorinated paraffin wax, even in 30% to 40% concentration, cannot be seen qualitatively in the spectrum. The presence of aliphatic esters in mixtures with phthalates is often evident only from absorption bands near 8·6 μm which, while useful for the recognition of these constituents, do not enable them to be differentiated.

While the presence of phthalates obscures the spectra of other constituents, the presence of significant amounts ($>10\%$) of aromatic phosphates obscures the absorption bands in the 10 μm to 11 μm region which serve to differentiate many phthalate esters. It is evident, therefore, that infra-red examination of a total plasticiser, although a useful preliminary test, is of only limited value.

This situation has fortunately been greatly improved by the introduction of chromatographic separation methods which enable fractions, many of which are comparatively pure substances, to be recovered for identification. These methods introduce problems, however, because of the small size of the samples recovered from the chromatographic separation and some care in handling is required so that the sample is not lost. Fortunately plasticisers are relatively involatile, and we have found it satisfactory to dissolve the sample (which is often barely visible) by adding 2 or 3 drops of ethylene dichloride to the sample tube, and to transfer the solution with a small pipette on to the surface of a rock-salt plate of dimensions about 3×5 mm. The plate is warmed under a lamp bulb to evaporate the solvent, and is inserted at the sample focus of the spectrometer. Any part of the beam which does not pass through the sample is obscured. Alternative methods, such as examining the samples in carbon bisulphide solution in a liquid cell, are not fast enough to be practicable since, with the advent of this chromatographic separation, it is now necessary to examine up to 10 fractions, rather than the single unfractionated sample previously presented for examination.

One difficulty arising from the use of these chromatographic separation methods is that it is not possible to examine individual fractions for elements. Furthermore, there may be difficulties in the identification of a number of common plasticisers because they are mixtures which are partially separated by the chromatographic procedure. Thus the spectra of the fractions do not precisely match the published spectra of the original plasticiser. There is no doubt, however, that these disadvantages are greatly outweighed by the presentation of pure, or substantially pure, substances, or at worst simple mixtures, for spectroscopic examination, although it may mean that the identification of the fractions must be pursued almost entirely by the interpretation of the spectrum, with little assistance from other tests. Where the result of the examination is uncertain or ambiguous, enough information may be forthcoming to suggest what further tests may be conducted on the original sample to complete the identification, and in particular cases the G.L.C. examination of alcohols recovered by hydrolysis, and n.m.r. examination of the fractions, will be found very useful.

Owing to the very large number of known plasticisers, and the multitude of other additives which may be extracted from polymer compositions, it will be necessary in many cases to consider the elements detected

575

in the total plasticiser, and refer to the qualitative analysis section in Chapter 2. The following scheme, however, is based upon the positions of the strongest bands, and is intended to permit a classification of plasticisers entirely from the features of their infra-red spectra. It is not intended to enumerate all the common plasticisers or to reproduce their spectra here, but rather to present a method of examining the spectrum to ascertain the chemical nature of the material, using as examples the spectra of well known members of each chemical class. It is suggested that reference be made to a commercial collection of spectra of plasticisers, such as the *Sadtler Commercial Spectra Series* E^{11}, but with the provision that many commercial plasticisers are mixtures so that some care is required in matching spectra.

The compounds first considered are those without carbonyl groups, that is those without strong bands in the 5·5 μm to 6·5 μm region; the various carbonyl-containing compounds are then examined.

1. STRONGEST BANDS AT 3·4 μM AND 6·8 μM

Substances with their strongest bands in these positions are primarily of an aliphatic hydrocarbon nature. Iranolin (Spectrum 11.1), a hydrocarbon oil, shows additional weak bands in the 12 μm to 14 μm region due to minor aromatic constituents. Paraffin wax (Spectrum 11.2) can only be satisfactorily characterised if it is sufficiently pure to crystallise, when the doublet band at 13·8 μm appears. Rather similar spectra are given by aliphatic chlorinated hydrocarbon mixtures, but the appearance of the rather broad band near 8 μm is a useful distinguishing feature. Those commonly encountered in the plasticiser field are chlorinated paraffin wax (Spectrum 11.3) and chlorinated polythene (Spectrum 6.6).

Certain organo-metallic compounds cannot be distinguished from hydrocarbons by the general appearance of their spectra. An example is dibutyl tin dilauryl mercaptide (Spectrum 11.4). The presence of such compounds may be suspected if the infra-red spectrum does not match that of any common hydrocarbon plasticiser, particularly if metals are present (especially tin). The analysis of tin compounds of this nature is discussed in Chapter 3 (p. 162), and the spectra of other tin stabilisers are given in Spectra 11.41 and 11.44.

2. STRONGEST BANDS IN THE 10 μM REGION

Phosphates, which are common plasticiser constituents, have their strongest bands near 10 μm. The aliphatic phosphates (rather rare) usually have this band on the shortwave side of 10 μm (e.g. trioctyl phosphate, Spectrum 11.5, and trichloroethyl phosphate, Spectrum 11.6) while the aromatic phosphates absorb strongly at about 10·3 μm (tritolyl phosphate,

Spectrum 11.7; trixylyl phosphate, Spectrum 11.8, and triphenyl phosphate, Spectrum 11.9). The first two of these phosphates are mixed isomers, and some variation of relative intensity of bands in the 12 μm to 14 μm region may be expected with different samples. The spectra of a number of phosphate plasticisers, including some alkyl aryl phosphates, are given in the *Sadtler Commercial Spectra*[11].

3. STRONGEST BANDS BETWEEN 9 μM AND 9·5 μM

These materials are usually ethers, alcohols or alcohol ethers. If hydroxyl groups are present, the hydroxyl band near 3 μm should be seen. The spectra of some true ether plasticisers appear in the *Sadtler Commercial Spectra*[11], but in our experience the ethers which are most commonly encountered in solvent extracts from polymer compositions are the polyethylene oxide type surface active agents. These are described elsewhere (p.523).

The alcohols encountered are numerous and vary from simple long-chain alcohols such as stearyl alcohol (Spectrum 11.10) and lauryl alcohol (Spectrum 11.11) to polyhydric alcohols such as sorbitol (Spectrum 11.12) and sucrose (Spectrum 11.13). Again the spectra of most common materials of this nature are to be found in the *Sadtler Standard Spectra*[12] and the D.M.S. system[13].

Although the infra-red spectrum is a useful means of ascertaining that a fraction is an alcohol, it is rarely possible to complete the identification from the spectrum. The distinguishing features in the spectra of the alcohols are in most cases evident only in the crystalline state, and these chromatographic fractions are usually too impure to crystallise. In the case of crystallisable polyhydric alcohols such as the sugars we have occasionally succeeded in causing crystallisation to occur by thorough drying, and in this state the infra-red spectra are of very good quality and highly characteristic. In most cases, however, it will be better to use chemical or chromatographic methods for identification. Many of the alcohols used are mixtures and here the chromatographic method is particularly appropriate.

4. STRONGEST BANDS IN THE 12 μM TO 14 μM REGION

These are likely to be aromatic hydrocarbons or chlorinated aromatic hydrocarbons.

The aromatic hydrocarbons encountered as plasticisers are often resins, e.g. poly-α-methyl styrene (Spectrum 6.38) and butadiene/styrene copolymer (Spectrum 8.6). Although these may be of such a low molecular weight as to be viscous liquids, their spectra are qualitatively the same as those of the corresponding solid resins.

Polycumarone/indene is also a possibility. This material being a copolymer is variable in its spectrum (see Hummel[14] Figures 550 to 552) but an intense band at 13·3 μm is always present.

Of the chlorinated aromatic materials encountered as plasticisers, chlorinated biphenyl (Spectrum 11.14) and chlorinated polyphenyl (Hummel Figure 661) are the most common. Different samples of these materials have rather different spectra (e.g. see Hummel Figures 657 to 660) presumably as each of these substances is a complex mixture of isomeric forms.

5. OTHER CHEMICAL GROUPS GIVING RISE TO WELL CHARACTERISED ABSORPTION PATTERNS (OTHER THAN THE CARBONYL GROUP)

The sulphonamides are easily recognised with strong bands at 7·6 μm and 8·6 μm; p-toluene sulphonamide (Spectrum 11.15) is an example. The spectrum is rich in bands, in common with other compounds of this type; the various sulphonamides are distinguished mainly by the position and number of strong bands in the 'substitution pattern' between 12 μm and 15 μm. The N-substituted sulphonamides (e.g. N-ethyl p-toluene sulphonamide, Spectrum 11.16) contain the $=$NH rather than the $-$NH$_2$ group. This leads to a simplification of the structure in the 3 μm region, which serves as a useful distinguishing feature. The spectra of other sulphonamides appear in the *Sadtler Standard Spectra*[12].

Somewhat similar spectral features are shown by sulphonic acid esters, since these also contain the O$=$S$=$O group. Aryl esters of alkyl sulphonic acids (e.g. Mesamoll, Spectrum 11.17) show a pair of strong bands at 7·4 μm and 8·5 μm; the same pair of bands is shown by the alkyl esters of aryl sulphonic acids (e.g. ethyl p-toluene sulphonate, Sadtler Standard Spectrum No. 8956). Both these series of compounds have strong bands in the 12 μm to 15 μm region due to the aromatic nucleus present, and in this respect also their spectra are similar to the aryl sulphonamides, but it is possible to distinguish the sulphonamides and N-monosubstituted sulphonamides from the sulphonic acid esters by the presence of the sharp absorption bands in the 3 μm region due to NH groups.

Solvent extracts from resins may well contain antioxidants, and a number of aromatic amine antioxidants, such as p-phenylenediamine and diphenylamine derivatives, have highly characteristic spectra which are also somewhat similar to those of the sulphonamides, having in particular one or more sharp NH bands in the 3 μm region, and a strong band at about 7·7 μm. However the second strong band in the 8·6 μm region is not present in these compounds and this seems to be a reliable distinguishing feature. As these compounds are aromatic in nature, strong bands again occur in the 12 μm to 15 μm region. The spectra of many of these amines are given in the *Sadtler Standard Spectra*[12].

Extracts from latices and resins prepared from latices frequently contain sulphate or sulphonate surface active agents. These compounds show strong absorption in the 8 μm to 9 μm region. The infra-red spectra of compounds of this general nature are given by Hummel[15].

A strong absorption band near 3 μm together with one or more strong bands near 8 μm suggest a phenolic constituent. Phenols encountered in solvent extracts from polymer compositions are likely to be phenolic stabilisers, and the further identification of these additives is considered in the appropriate chapters according to the nature of the resin. Generally U.V. or n.m.r. identification will be preferred for materials of this nature.

One important group of substances having spectra resembling phenols are the epoxy resins (Spectrum 10.10) often encountered in extracts from resins. Although these resins have only terminal hydroxyl groups, they are often of low molecular weight, hence the 3 μm hydroxyl band appears at moderate intensity. The *p*-substitution band near 12 μm is prominent in the spectra of these resins, which are discussed more fully in Chapter 10.

FRACTIONS CONTAINING CARBONYL GROUPS

The differentiation of the various carbonyl-containing compounds is considered in Chapter 2. In the case of plasticisers, esters are by far the most common carbonyl-containing compounds, although ether extracts may also contain carboxylic acids, either in the free state or as acid salts.

CARBOXYLIC ACIDS

The carboxyl band at 5·9 μm, together with the broad hydroxyl band underlying the 3·4 μm C—H structure, denotes the presence of carboxylic acid. If the plasticiser appears to be a mixture containing a free acid there is little point in attempting identification of the acid in the presence of other constituents since the separation of an acid from a mixture by converting it to a water-soluble salt is usually straightforward.

The acids encountered in these compositions are usually long-chain aliphatic substances. If they are pure compounds and are of sufficiently high molecular weight they will be crystalline solids at room temperature: infra-red spectra measured on samples in this physical state are distinctive and characteristic for the individual acids, particularly in the weak but sharp band structure in the 7·8 μm to 8·5 μm region. Thus the spectrum of lauric acid (Spectrum A.4) can be clearly differentiated from that of stearic acid (Spectrum A.6). On the other hand, the lower fatty acids and unsaturated acids are liquid or so poorly crystalline at room temperature that their infra-red spectra are no longer useful for distinguishing the different acids, and it is necessary to convert the acids to their barium salts[16] which, being crystalline solids, again have useful spectra. Unfortunately these methods are not particularly useful for mixtures of acids such as are frequently encountered in commercial compositions, and it may well be necessary to use chromatographic separation if complete identification is required.

CARBOXYLIC ACID SALTS

These are identified by the pair of relatively strong bands at about 6·3 μm and 7·1 μm. The presence of the latter band serves to distinguish these from the *o*-hydroxy-benzophenone type stabilisers, which also show a strong and rather broad band at about 6·3 μm. The latter compounds also show evidence of their aromatic nature, but do not show the phenolic hydroxyl band at 3 μm.

Although there is some evidence that the wavelength and shape of the 6·3 μm band of carboxylic acid salts depends upon the nature of the anion, there is little point in attempting identification of the metal by this method as this is much more easily established by U.V. emission spectra. In many cases the conclusion that a carboxylic acid salt is present is sufficient, but if further identification is required it will usually be better to release the free acid for chromatographic or spectroscopic examination.

ortho-PHTHALATE PLASTICISERS

The *o*-phthalate pattern is very distinctive and easily recognised (see Table 4.4). If no additional bands of significant intensity appear in the 7 μm to 15 μm region, the material contains a simple dialkyl phthalate. The differentiation of the lower members of the series is possible from the infra-red spectrum, using the rather weak structure in the 10 μm to 12 μm region, provided there is no interference from other constituents. Examples are dimethyl, diethyl, di-*n*-butyl, di-iso-butyl, di(2-ethyl hexyl) and di-*sec*-octyl phthalate (Spectra 11.18 to 11.23). However, the phthalate esters of higher alcohols are almost impossible to distinguish from their spectra unless the alcohol has some unusual structural feature, as for example the geminal dimethyl group of the alcohol in di(3 : 5 : 5-trimethyl hexyl) phthalate (Spectrum 11.24) in which the split methyl band is seen at about 7·3 μm. Furthermore, the higher phthalate esters are in most cases based on complex mixtures of alcohols, and the spectra of these mixed esters are necessarily less distinctive. Thus, for example, the spectra of didecyl phthalate (Spectrum 11.25) and didodecyl phthalate (Spectrum 11.26) are almost indistinguishable, and many of the other higher phthalates have spectra which are very similar to these. The spectra of a number of mixed alkyl phthalates appear in the *Sadtler Commercial Spectra*[11] (e.g. butyl isohexyl phthalate, E 2821, and butyl isodecyl phthalate, E 2522). In these esters the absorption pattern would be expected to be similar to that of a mixture of the simple dialkyl esters derived from each of the alcohols present, and this is substantiated in the examples quoted above, in which the doublet band of the *n*-butyl ester ($\approx 10\cdot4$ μm and $10\cdot6$ μm) is supported on a broad band centred on $10\cdot5$ μm as is usually observed in the spectra of the esters of higher alcohols. Clearly, the identification from the spectrum of both the higher alkyl and mixed alkyl esters presents difficulty, and hydrolysis followed by the recovery of the alcohols for identification by gas chromatography will be necessary in many cases.

A common phthalate based on an unsaturated alcohol is diallyl phthalate (Spectrum 11.27). In addition to the distinctive structure in the 10 μm to 11 μm region, the sharp band at $6\cdot05$ μm indicates the presence of the C=C group. The only phthalate based on an alicyclic alcohol which we have encountered is dicyclohexyl phthalate; this has a useful spectrum (Spectrum 11.28) in which may be seen, in addition to the usual phthalate structure, a large number of sharp bands of medium intensity between 7 μm and 14 μm, often found in cyclohexyl compounds.

Phthalates based on aromatic alcohols and phenols are fairly easy to distinguish. The common representatives of this class are diphenyl phthalate (Spectrum 11.29) and dibenzyl phthalate (Spectrum 11.30). In both cases additional 'substitution pattern' bands appear in the 14 μm to 15 μm region and the C—H structure near $3\cdot4$ μm is extremely weak compared with that observed in the alkyl phthalates. The latter structure is particularly useful in distinguishing the aryl phthalates from alkyl aryl phthalates. Thus butyl benzyl phthalate (Spectrum 11.31) shows both the additional aromatic structure at $14\cdot3$ μm, and a C—H band of moderate intensity at $3\cdot4$ μm due to alkyl chains. The presence of the butyl group is also revealed by the similarity of the spectrum in the 10 μm to 11 μm region to that of dibutyl phthalate.

More complex esters, which show the basic phthalate pattern but with additional strong bands in the 7 μm to 10 μm region, include the alkyl phthalyl alkyl glycollates. An additional strong C—O alkyl band occurs in the spectra of this series at $8\cdot3$ μm; the short wavelength of the band distinguishes this group of esters from simple mixtures of phthalates with aliphatic esters, in which the corresponding C—O alkyl band is found at $8\cdot6$ μm to $8\cdot7$ μm. The common representatives of this group are methyl phthalyl methyl glycollate (Spectrum 11.32) and the ethyl-ethyl ester (Spectrum 11.33); several others appear in the *Sadtler Commercial Spectra*[11]. Once the chemical nature of these materials has been recognised from the presence of the $8\cdot3$ μm band in their spectra, further identification presents little difficulty.

An important variation of the alkyl phthalate series is the glycol phthalate group. In these, one hydroxyl group of the glycol is esterified with phthalic acid and the other condensed with an aliphatic alcohol. This introduces an aliphatic ether group, the evidence for which is a band of moderate to high intensity at $8\cdot9$ μm. The band is narrower and at shorter wavelength than that of a simple alkyl ether, thus enabling this class of compound to be distinguished from mixtures of phthalates with ethers such as polyethylene oxide derivatives. Di(methoxy ethyl) phthalate (Spectrum 11.34), and di(butoxy ethyl) phthalate (Spectrum 11.35) are representatives of this series; the spectra of many others are recorded in the *Sadtler Commercial Spectra*[11].

Even in those cases where the complete identification of a phthalate plasticiser is not possible from the infra-red spectrum, analysis of the spectrum, based on the observations above, may often suggest which further tests would be appropriate in particular cases. The infra-red examination of *o*-phthalate plasticisers may be summarised as follows.

1. Spectra showing the *o*-phthalate pattern only, with no other significant structure, are those of simple alkyl esters. The lower members, if presented in a clean state, may be unambiguously identified, but a complete identification is rarely possible in the case of the higher alkyl esters from the spectrum alone.

2. Spectra showing the *o*-phthalate pattern, but with additional sharp weak bands in the 9 μm to 12 μm region, are probably those of alicyclic esters.

3. Spectra showing additional strong bands in the 12 μm to 15 μm region are likely to be those of aromatic alcohol or phenyl esters, although the possibility of mixed aryl alkyl esters must be considered.

4. A strong additional band near 8·3 μm denotes esters of the alkyl phthalyl alkyl glycollate type, and these can be distinguished from mixtures of phthalates with simple alkyl esters.

5. Spectra showing an additional band near 8·9 μm are likely to be those of glycol ether phthalates, and these can be distinguished from mixtures of phthalates with alkyl ethers such as polyethylene oxide derivatives.

6. In some cases the *o*-phthalates extracted from compositions may be *o*-phthalate resins rather than plasticisers. Some evidence of this will be obtained if it is found to be impossible to remove all the phthalate from a composition by ether extraction; all the monomeric phthalates are, in our experience, ether soluble. Further evidence will be obtained by the examination of the hydrolysis products of the phthalate; if a glycol or other poly-functional alcohol is recovered, a phthalate resin rather than a plasticiser is indicated.

OTHER AROMATIC ESTER PLASTICISERS

These are recognisable by the usual features suggesting aromatic character. Among the most common of these esters are the benzoates, such as amyl benzoate (Spectrum 11.36), all of which have in their spectra a single strong band near 14 μm. Another series of aromatic esters which may be encountered in plasticiser extracts are the salicylates, particularly phenyl salicylate (Spectrum 11.37). These compounds, in common with others containing the phenolic hydroxyl group *ortho* to the carbonyl group, do not show the usual hydroxyl band at 3 μm.

ALIPHATIC ESTERS

Although the general features of the spectra of aliphatic esters enable them to be differentiated easily from other substances, the further identification of the ester from the infra-red spectrum presents considerable difficulty in most cases. Where esters have distinguishing features in their spectra these are noted in Table 4.4 but, except in these special cases, it is unwise to attempt identification of aliphatic esters on the basis of their infra-red spectra alone. Even when the spectrum of an unknown appears to be similar to a published spectrum, careful attention must be paid to the value of the spectral features used in the identification and, when doubt arises, hydrolysis of the plasticiser and subsequent identification of the recovered acid and alcohol or glycol, is the only satisfactory method of identification. In principle the n.m.r. spectrum can give information but it is important that the sample is pure otherwise the n.m.r. examination will be misleading.

All esters containing long aliphatic chains show a strong band in the infra-red spectrum near 13·8 μm, which appears as a doublet in the case of some long-chain esters in the crystalline state. Examples of such compounds are ethyl palmitate (Spectrum 11.38) and glyceryl distearate (Spectrum 11.39). The latter spectrum shows evidence of the presence of hydroxyl groups from the band at 3·0 μm, and the presence of this band in the spectrum of an ester will suggest that it is based on a polyhydric alcohol, some of the hydroxyl groups of which have not been reacted with acid in the preparation of the ester.

The acetate group is easily identified from the relatively strong band at 8·1 μm. This feature is seen in the spectra of the simple acetate plasticisers such as glyceryl triacetate (Ref. 17 Figure 654) and in the acetyl derivatives of esters of hydroxy acids, an example being butyl acetyl ricinoleate (Spectrum 11.40). The latter compound shows two C—O bands, at 8·1 μm and 8·6 μm, as two ester groups are present, and may be confused with esters such as adipates and sebacates, which show two C—O bands at about these wavelengths. However, the acetyl derivatives have a stronger band at 8·1 μm whereas the other esters mentioned show the stronger band at about 8·6 μm.

A feature of the spectra of 'epoxidised' esters is the appearance, in addition to the usual ester bands, of a doublet band of weak to medium intensity at about 12 μm and 12·2 μm due to the epoxy group. Examples here are in the *Sadtler Commercial Spectra*[11], E 2780, 2781, and 2784, all described as epoxy type plasticisers.

A series of tin stabilisers, the dialkyl tin dialkyl thioglycollates, contain aliphatic ester groups and their spectra are generally similar to those of other aliphatic esters. An example is dibutyl tin dinonyl thioglycollate (Spectrum 11.41). The prominent band at 7·7 μm appears distinctive but no doubt other dialkyl tin thioglycollate esters have similar spectra. The analysis of these compounds is considered in detail in Chapter 3, and the spectra of other tin stabilisers are given in Spectra 11.4 and 11.44.

Of the aliphatic esters based on dibasic acids, the alkyl adipates and sebacates appear to be most commonly encountered.

These are not very different in their spectra, and change in the nature of the alcohol, at least in the lower members of the series, seems to produce a larger effect than change in the nature of the acid.

Nevertheless the spectra of the lower esters seem to be reasonably satisfactory for identification purposes, and materials such as dibutyl adipate (Spectrum 11.42) and dibutyl sebacate (Spectrum 11.43) can be characterised from their infra-red spectra. Occasionally esters based on other dibasic acids are used as plasticisers and in particular azeleates and suberates may be used. While the infra-red spectrum is a useful and rapid test to determine the general nature of the material, if complete identification is required it will be necessary in most cases to hydrolyse the substance and isolate the acid and alcohol or glycol from the hydrolysis products for identification by infra-red spectroscopy or other test methods (e.g. n.m.r. or gas chromatography).

As discussed in the case of *o*-phthalate esters, ether alcohols rather than simple alcohols may be used to esterify a dibasic acid. The spectra of such products show, in addition to the aliphatic ester pattern, an additional band at about 8·9 μm due to the aliphatic ether group. An example of such a spectrum is that of dibutoxy ethyl adipate (*Sadtler Commercial Spectra*[11], E 2793).

Other plasticisers containing both aliphatic ester and ether groups are those based on di- or tri-ethylene glycol and monobasic acids. Here the ether absorption band appears as a broad band at about 9 μm, similar to that observed in polyethylene oxide derivatives (Spectra 10.3 to 10.6), to which these materials have an obvious chemical similarity. Examples of a plasticiser of this type are triethylene glycol dicaprylate and triethylene glycol dipelargonate (*Sadtler Commercial Spectra*[11], E 2527 and E 2528 respectively).

Another series of compounds based on dicarboxylic acids which may be found in plasticiser extracts are the half esters, half salts of dicarboxylic acids. Dibutyl tin dinonyl maleate is an example of such a compound (Spectrum 11.44). The features of both ester and carboxylic acid salt are evident in the spectra of these compounds and consequently they are difficult to distinguish from mixtures of esters and acid salts. Nevertheless, if such compounds contain tin, a tin stabiliser may be suspected and the method given in Chapter 3 may be applied. Spectra of other tin stabilisers are given in Spectra 11.4 and 11.41.

The dibasic acids are also used, in conjunction with diols, to produce polymeric (polyester) plasticisers. There is no pronounced spectral feature which will distinguish a monomeric ester based on a dibasic acid and an alcohol from a polyester based on a dibasic acid and a diol. However, since the majority of diols used are short-chain compounds, particularly ethylene and propylene glycol, the number of carbon atoms associated with the alcoholic constituent is usually far less in polymeric than in monomeric ester plasticisers. Thus a comparison of the spectra of poly-propylene glycol sebacate (Spectrum 11.45) and polypropylene glycol adipate (Spectrum 11.46) with those of dibutyl adipate (Spectrum 11.42) and dibutyl sebacate (Spectrum 11.43) shows that while the C—O structure near 8·5 μm is similar in all cases, the C—H band in the 3·4 μm region is relatively much weaker in the polymeric plasticisers. This test, while useful on reasonably pure substances, must be applied with caution to separated plasticiser fractions, as C—H-containing impurities may well be present. The best indications that an ester plasticiser is polymeric rather than monomeric are probably the relative insolubility of these materials in ether, and their relatively long retention in the chromatographic separation, as noted on p. 556.

N.M.R. EXAMINATION OF PLASTICISERS

N.M.R. spectroscopy is most effective in qualitative analysis when the samples examined are substantially pure compounds. This poses problems when dealing with plasticisers because most extracts from polymer compositions are mixtures and, if these are separated by thin-layer chromatography, the amount of the individual fractions is often too small for convenient examination by n.m.r. Furthermore, the original plasticisers themselves are often mixtures. Thus, tricresyl (tritolyl) phosphate is based on mixed cresols, while most of the higher phthalate esters are based on complex mixtures of alcohols.

However, in spite of these difficulties n.m.r. examination of plasticisers can often be rewarding. Thus, in the case of phthalate plasticisers, the lower alkyl phthalates are substantially pure compounds which give excellent n.m.r. spectra. Although it may not be possible to establish the precise chemical nature of the higher alkyl phthalates, the average chain length of the alkyl group can be calculated from the ratio of the areas of the resonances from aryl and alkyl protons. Similarly, the more complex phthalates (e.g. alkyl phthalyl alkyl glycollates) give useful spectra. The spectra of some common phthalates are given in Figures 11.5 to 11.10.

N.m.r. examination is particularly well suited to the examination of polyester plasticisers, and these may often be recovered from a composition in a comparatively clean state by hot methanol extraction (p. 144):

585

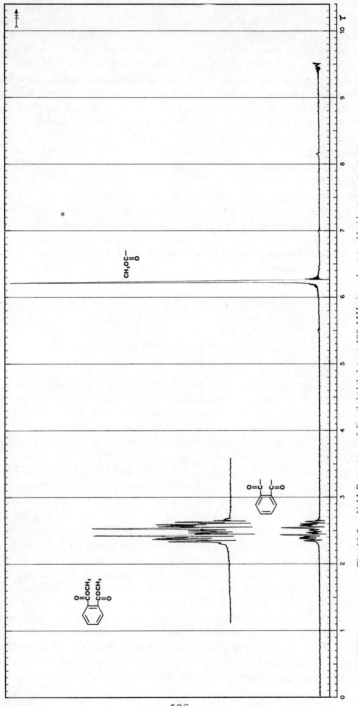

CH₃OC—
        ‖
        O

O=C   C=O

O=COCH₃
  COCH₃
     ‖
     O

Fig. 11.5.   N.M.R. spectrum of dimethyl phthalate at 100 MHz in carbon tetrachloride at room temperature

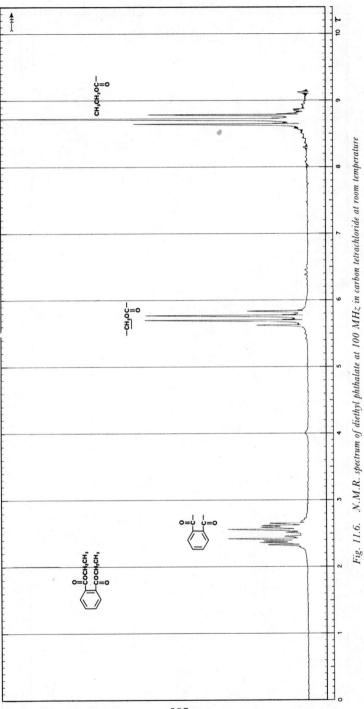

Fig. 11.6. N.M.R. spectrum of diethyl phthalate at 100 MHz in carbon tetrachloride at room temperature

587

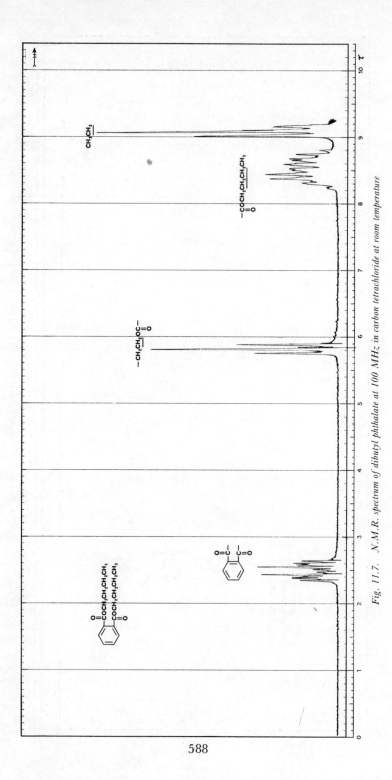

Fig. 11.7. *N.M.R. spectrum of dibutyl phthalate at 100 MHz in carbon tetrachloride at room temperature*

588

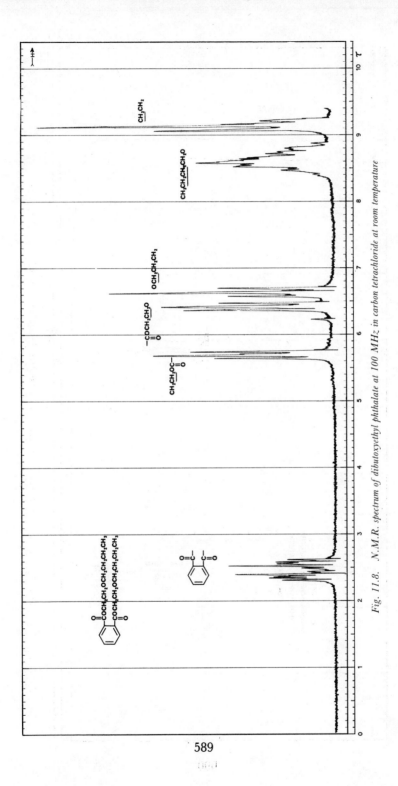

Fig. 11.8. N.M.R. spectrum of dibutoxyethyl phthalate at 100 MHz in carbon tetrachloride at room temperature

589

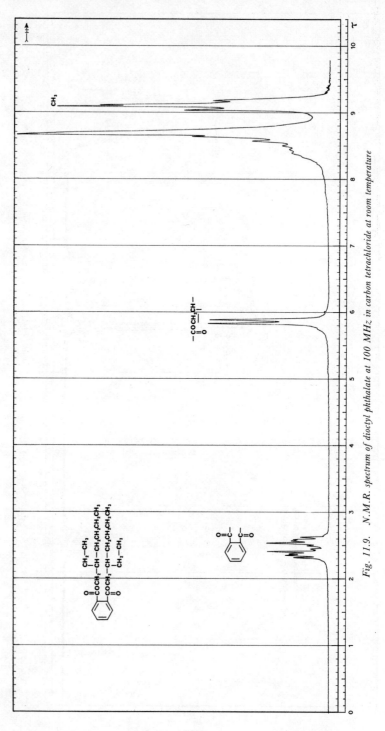

Fig. 11.9.   N.M.R. spectrum of dioctyl phthalate at 100 MHz in carbon tetrachloride at room temperature

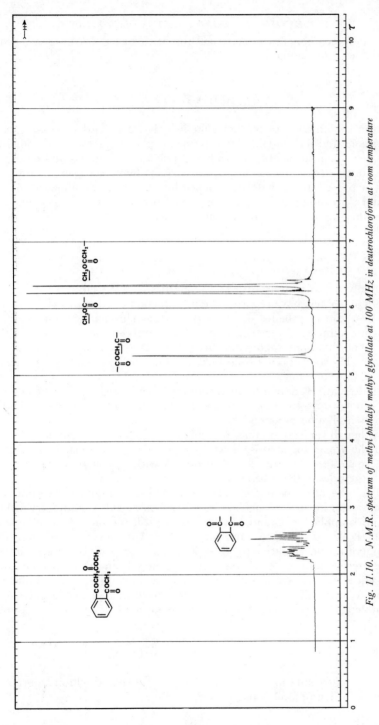

Fig. 11.10. N.M.R. spectrum of methyl phthalyl methyl glycollate at 100 MHz in deuterochloroform at room temperature

591

The identification of these materials from their spectra is considered (with examples) in Chapter 4.

SOLVENTS

In the examination of plastics adhesives, lacquers, surface coating preparations and inks suitable for printing on plastics materials, the analyst is called upon to identify not only the polymers, pigments and additives used but also the solvents present in what is often a complex mixture.

The procedure ultimately developed here involves a combination of gas liquid chromatographic separation of the solvent mixture[18] and infra-red identification of the separated components.

ISOLATION OF THE SOLVENT

The solvent is first isolated from the plastics adhesive etc.

Some workers have reported the introduction of adhesives etc. directly on to a pad of asbestos on the top of a gas chromatographic column. Viscous samples, however, are difficult to introduce and, moreover, the drops of the composition dry from the outside, trapping the solvent within. The solvent is not volatilised instantaneously and a true chromatogram is not obtained.

If alternatively a high-temperature vaporiser is used, there is a tendency with some compositions to obtain a chromatogram of solvent plus polymer depolymerisation products.

Although time consuming, preliminary isolation of solvent is preferred and the apparatus used is shown in Figure 11.11. Between 1 and 5 g of the preparation are transferred to tube A without contact of the sample with the sides of the tube at B.

The tube, after cooling, is sealed at B. Whilst maintaining the sample A in a refrigerant such as methanol-solid carbon dioxide mixture, a vacuum pump is applied at C and, with the pump still running, the capillary at point D is sealed with a hand torch. Tube A is now removed from the refrigerant and replaced by tube E. As the preparation warms up to room temperature, vacuum distillation of the solvent takes place and this solvent is recondensed in E. This process is assisted if necessary by placing tube A in a boiling water bath. When the distillation of the solvent is complete, the seal at D is broken and the bottom half of tube E is cut off.

PRELIMINARY EXAMINATION OF THE SOLVENT

A chromatogram on a general purpose gas liquid chromatographic column is first obtained under fixed conditions[18].

The column used is 6 ft (2 m) in length and of internal diameter $\frac{1}{4}$ in (6 mm) and is packed with 30% w/w dinonyl phthalate on 100–120 mesh Celite. This column is maintained at 100°C.

The general boiling range of the mixture is indicated by inspection of the chromatogram obtained. It may be desirable to repeat this preliminary

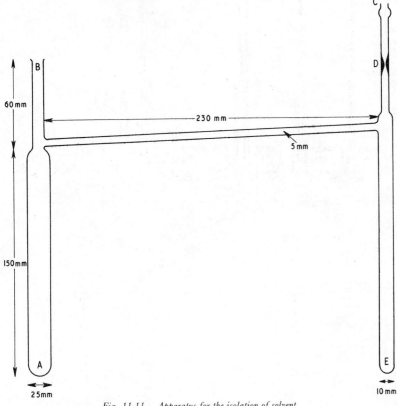

Fig. 11.11. Apparatus for the isolation of solvent

separation at a lower temperature (i.e. 50°C) if components are present that are rapidly eluted.

This preliminary test is particularly valuable for ascertaining the presence or absence of petroleum or naphtha fractions. These solvents produce chromatograms with standard 'patterns' that may be recognised immediately if standard chromatograms of comparable substances are available.

Two of these standard chromatograms are given in Figures 11.12 and 11.13.

Patterns of such substances as turpentine and commercial dipentene are again immediately recognisable.

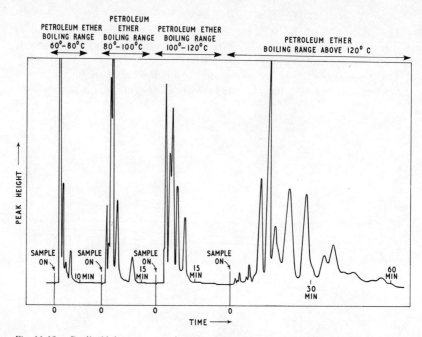

Fig. 11.12. *Gas liquid chromatograms of samples of petroleum ether on a 6 ft (2 m) column of 30% w/w dinonyl phthalate on Celite 545 at 100°C*

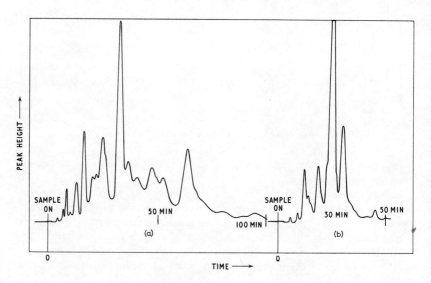

Fig. 11.13. *Gas liquid chromatograms of (a) white spirit and (b) solvent naphtha on a 6 ft (2 m) column of 30% w/w dinonyl phthalate on Celite 545 at 100°C*

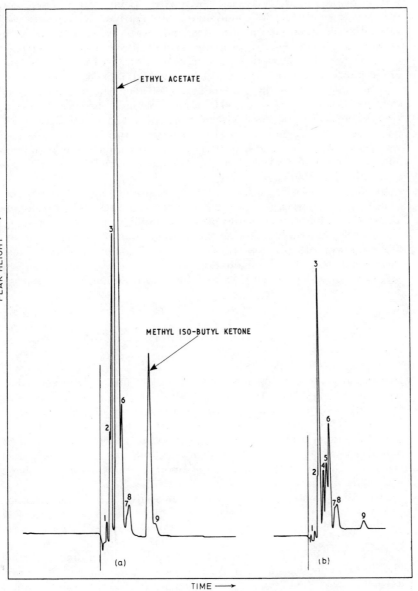

Fig. 11.14. *Chromatograms of (a) solvent from Neoprene adhesive and (b) petroleum ether (boiling range 60–100°C)*

The petroleum fractions and the mixed commercial xylenes show remarkable consistency in composition; naphtha fractions together with turpentine and commercial dipentene may vary considerably in composition according to differences in manufacturing processes and the country of origin. Aromatic hydrocarbon mixtures sold as paint solvents are usually characteristic of a particular method of manufacture.

Figure 11.14 (a) is the chromatogram of a solvent mixture isolated from an adhesive, and is a good example of those encountered in this work. Comparison of this chromatogram with that of 60°C to 100°C petroleum ether (Figure 11.14 (b)) shows that this is a principle constituent. Peaks 1 to 3 and 6 to 9 of the petroleum ether chromatogram are all visible, but peaks 4 and 5 are obscured by a large peak due to some other constituent.

In addition, there is evidence of the presence of a further constituent showing a strong peak near peak 9 of the petroleum ether chromatogram. Thus two solvents are evidently present in addition to petroleum ether. Infra-red examination, following the method outlined on p. 63, showed that these constituents were ethyl acetate and methyl isobutyl ketone. Measurement of the chromatogram shows that these constituents are present to the extent of 60% v/v and 10% v/v respectively.

IDENTIFICATION OF SOLVENTS OTHER THAN HYDROCARBON FRACTIONS

In their original work on this subject, Haslam and Jeffs attempted to separate at least 0·1 ml of the constituents other than petroleum ether in the above mixture in order to confirm the identity of these additional constituents from their infra-red spectra in the liquid state.

At that time relatively large infra-red cells were used and the mechanical handling losses were large. The system of collecting traps was also rather crude and the collection of such large amounts of separated components was a slow and laborious process.

Since 10 to 20 mg is a normal load for an analytical scale column using a katharometer detector, and overloading the column caused a vast loss of efficiency of separation, such a fractionation often took days to carry out.

At this stage we were looking forward for a less cumbersome method of identification. Dr. Rose and Mr. Ashmore, of the Government Laboratory, to whom many thanks are due, quickened our interest in a suggestion put forward originally by James and Martin[19], i.e. that identification of unknowns should be possible by comparison of relative retention times on columns of different polar character.

With Silocel firebrick as the support, paraffin wax and tritolyl phosphate were chosen as the stationary phases. Columns, 12 ft in length and $\frac{1}{4}$ in internal diameter were made up, enclosed in the same vapour jacket and maintained at 100°C.

Table 11.5 (taken from the *Analyst*[18]) gives the corrected relative retention times, i.e. relative to benzene, that were obtained by this procedure for a number of the more common solvents.

Table 11.5. Corrected relative retention times for various solvents on 12 ft columns of 33·3% paraffin wax on Silocel C22 and 33·3% tritolyl phosphate on Silocel C22

Solvent	Boiling point °C	Corrected relative retention time	Solvent	Boiling point °C	Corrected relative retention time
PARAFFIN WAX COLUMN			TRITOLYL PHOSPHATE COLUMN		
Acetaldehyde	20·2	0·09	n-Pentane	36·1	0·11
Methyl formate	31·5	0·11	Diethyl ether	34·6	0·17
Methanol	64·6	0·13	Acetaldehyde	20·2	0·19
Ethanol	78·3	0·17	Methyl formate	31·5	0·22
Acetone	56·0	0·18	Di-isopropyl ether	67·5	0·29
Methyl acetate	57·1	0·23	Methanol	64·6	0·36
Isopropyl alcohol	82·4	0·23	Iso-octane	99·2	0·37
Ethyl formate	54·2	0·23	Ethyl formate	54·2	0·38
Allyl alcohol	97·1	0·25	Methyl acetate	57·1	0·40
Diethyl ether	34·6	0·26	n-Heptane	98·4	0·42
Methylene dichloride	40·1	0·27	Cyclohexane	80·8	0·45
Tert-butyl alcohol	82·5	0·28	Acetone	56·0	0·46
n-Pentane	36·1	0·29	Methylene dichloride	40·1	0·47
Vinyl acetate	77·0	0·35	Ethanol	78·3	0·52
n-Propyl alcohol	97·2	0·36	*Tert*-butyl alcohol	82·5	0·53
Methyl ethyl ketone	80·0	0·40	Ethylidene dichloride	57·3	0·57
Ethyl acetate	77·2	0·43	Isopropyl alcohol	82·4	0·58
Ethylidene dichloride	57·3	0·44	Vinyl acetate	77·0	0·59
Sec-butyl alcohol	99·5	0·50	Methyl cyclohexane	100·8	0·66
Di-isopropyl ether	67·5	0·52	Ethyl acetate	77·2	0·70
Methyl propionate	79·9	0·54	Carbon tetrachloride	76·7	0·76
Isobutyl alcohol	108·1	0·55	Methyl propionate	79·9	0·77
Isopropyl acetate	88·9	0·58	Isopropyl acetate	88·9	0·79
Chloroform	61·2	0·60	n-Octane	125·6	0·85
Tetrahydrofuran	65·0	0·62	Tetrahydrofuran	65·0	0·86
Ethylene dichloride	83·5	0·72	Methyl ethyl ketone	80·0	0·89
Methyl n-propyl ketone	102·3	0·81	Chloroform	61·2	0·99
n-Butyl alcohol	118·0	0·83	*Sec*-butyl alcohol	99·5	1·14
n-Propyl acetate	101·6	0·93	n-Propyl alcohol	97·2	1·15
Ethyl propionate	99·1	0·94	Ethyl propionate	99·1	1·24
Carbon tetrachloride	76·7	1·06	Trichloroethylene	86·7	1·26
Methyl n-butyrate	102·3	1·09	Allyl alcohol	97·1	1·26
Dioxan	101·4	1·12	n-Propyl acetate	101·6	1·35
Cyclohexane	80·8	1·14	Ethylene dichloride	83·5	1·37
Propylene dichloride	96·4	1·17	Methyl n-butyrate	102·3	1·46
Isopropyl propionate	111·3	1·25	Isopropyl propionate	111·3	1·48
Isobutyl methyl ketone	116·8	1·28	Methyl n-propyl ketone	102·3	1·55
Trichloroethylene	86·7	1·35	Isobutyl alcohol	108·1	1·65
Iso-octane	99·2	1·37	Propylene dichloride	96·4	1·76
n-Heptane	98·4	1·40	Isobutyl acetate	117·2	1·93
Isobutyl acetate	117·2	1·51	Dioxan	101·4	1·95
Methyl cyclohexane	100·8	1·91	Toluene	110·8	2·10
n-Butyl acetate	126·2	2·13	Isobutyl methyl ketone	116·8	2·12
Toluene	110·8	2·32	n-Butyl alcohol	118·0	2·41
n-Octane	125·6	3·10	Di-n-butyl ether	142·4	2·41
Ethylbenzene	136·2	4·6	n-Butyl acetate	126·2	2·78
Di-n-butyl ether	142·4	4·8	Ethylbenzene	136·2	4·1
p-Xylene	138·4	5·2	*p*-Xylene	138·4	4·2
m-Xylene	139·3	5·3	*m*-Xylene	139·3	4·4
o-Xylene	144·0	6·1	*o*-Xylene	144·0	5·3

In the majority of cases this proved to be a valid method. If e.g. an unknown gave a figure of 0·87 on the tritolyl phosphate column then

n-octane	0·85
tetrahydrofuran	0·86
and methyl ethyl ketone	0·89

were all possibilities.

A figure of 0·63 for the unknown on the paraffin wax column definitely indicated tetrahydrofuran since n-octane gave a figure of 3·1 and methyl ethyl ketone a figure of 0·40 on this column.

However, a mixture of methyl acetate and acetone gave only one peak on each of the columns. At this level of retention time a difference of 0·08 unit would be necessary to show that two substances were present.

Other pairs of comparable performance on the two columns were

1. Acetaldehyde and methyl formate
2. m-Xylene and p-xylene
3. Methyl methacrylate and methyl n-butyrate.

Although identification of unknown solvents was possible in most cases, on a very small amount of sample there was always a slight uncertainty about the method.

Towards the end of 1958 a visit was paid to Dr. Anderson of the Edinburgh University and as a result of this visit interest in the infra-red identification of separated solvents was revived.

Anderson was obtaining the infra-red spectra of organic substances in the vapour state in an unheated gas cell, even of substances of comparatively high boiling point. This valuable work was published in the Analyst[20].

As a result of this visit Haslam, Jeffs and Willis re-investigated the whole problem of mixed solvent analysis.

They reached the conclusion that reliable procedures could be worked out based on

1. The preliminary isolation of the solvent or mixed solvent from plastics preparations by the H-tube method,
2. The gas chromatographic separation of the components of solvent mixtures,
3. The collection of the separated substances in cold traps, followed by
4. The transfer of those fractions boiling up to 100°C into a gas cell, and those boiling at higher temperatures into a micro-liquid cell, for infra-red examination.

The original paper of Haslam, Jeffs, and Willis[21] outlined the steps leading to these conclusions, and the apparatus used is described and illustrated in Chapter 1 (p. 63). Furthermore, the authors provided the infra-red spectra of 22 common solvents, examined in the gas phase, and these were published by Fisk[22].

The method described above was in use for several years, although some modifications were introduced, for example samples to be examined as liquids were collected directly in silver chloride cells held in a cold trap, because of the considerable transfer losses which ensued when transferring samples from cold traps in the liquid state. Eventually it was realised that it would be much easier to design apparatus specially so that all fractions could be delivered directly from the chromatographic column into a gas cell mounted permanently in the spectrometer. The cell is maintained at elevated temperature (about 150°C) so that condensation of fractions does not occur. Details of the construction of the apparatus, which has been found to be extremely satisfactory for the examination of fractions from solvent mixtures, are given in Chapter 1 (p. 68) together with an example of the identification of the constituents of a solvent.

The principles of this test method may be used in other circumstances, for example in the identification of alcohols obtained by potassium hydroxide hydrolysis of mixed phthalate plasticisers. The lower alcohols may be identified satisfactorily by direct measurement of their spectra in the gas phase, but for alcohols boiling above 250°C the examination of the fractions in the liquid state is preferred.

Infra-red spectra of alcohols and solvents in the liquid phase are given in Ref. 17.

FILLERS AND PIGMENTS

In the previous chapters mention has been made of the isolation and determination of specific fillers and pigments in various polymer compositions, e.g. carbon black in polythene (p. 332), inorganic additives in P.V.C. and fillers in urea-formaldehyde moulding powders.

General experience of the materials in common use in the plastics industry will usually assist the analyst in his search for fillers and in his quest for the correct method of isolation and determination.

Occasionally special attention has to be paid to the identification of fillers in particular compositions and in this connection the following general notes may be of value.

In the first place it is advisable to attempt the separation of the additive or filler from the composition in such a way that the recovered material is free from contaminants. Ashing methods should be avoided if at all possible.

In the case of compounds in which polymer and organic additives are soluble in a solvent at room temperature the insoluble filler may be isolated by filtration, then washed well with pure solvent and finally dried. Alternatively a centrifuging method may be adopted.

In the case of samples that are readily pyrolysed at low temperatures to yield volatile fragments, e.g. polythene and polypropylene, the principle of the method for the determination of carbon black in polythene (p. 332) may often be applied with advantage.

It should be borne in mind, however, that when such methods of isolation are employed other insoluble and non-volatile additives, not specifically added as fillers, may be recovered along with the filler residue. In such cases care must be taken in the subsequent identification.

The following series of general tests is useful in the examination of fillers:

1. The nature of the elements present is determined by emission and/or X-ray Fluorescence spectrography.
2. The filler residue is subjected to microscopic examination. The appearance of such substances as asbestos and glass fibres is very characteristic.
3. The behaviour of the filler on ashing is observed.
4. The material is tested for sulphate after decomposition by boiling with sodium carbonate solution.
5. The filler is tested for sulphide and also for carbonate.

Information provided by the above series of tests often points clearly to the presence or otherwise of such fillers as silica, titanium dioxide, barium sulphate, lead oxide, cadmium selenide, cadmium sulphide, basic lead sulphate, basic lead carbonate, calcium carbonate and the various clays used in the industry.

With this information in his possession, the analyst can proceed in any doubtful case to identification by either infra-red spectroscopy or X-ray diffraction methods.

IDENTIFICATION OF INORGANIC FILLERS BY INFRA-RED SPECTROSCOPY

Although infra-red spectroscopy has become widely accepted as a method of identification of organic substances, its use may be extended to the identification of inorganic materials.

Many of the fillers commonly used in polymer compositions are metallic salts in which the anion contains oxygen, for example carbonates, silicates and sulphates. Strong and characteristic absorption bands due to the C—O, Si—O and S—O vibrations within the ion occur in the 'fingerprint' region of the infra-red spectrum. These are listed in Tables 2.10 and 2.11 together with the characteristic bands of other anions which occur in this region. The presence of these bands in the infra-red spectrum of a filler will be good evidence of the presence of the appropriate anion, especially if these observations are supported by the results of a qualitative elementary analysis.

Although the infra-red spectrum gives no direct evidence on the nature of the cation, this will often be known from the elementary analysis. Furthermore, change of cation usually changes either the contour or the position of the absorption bands of the anion. Thus, as in the case of organic compounds, the major features of the infra-red spectrum (considered alongside the elementary analysis) indicate the general class of the compound, while the minor features of the spectrum will in many cases permit complete identification of an unknown by reference to standard spectra.

Other commonly used fillers are inorganic oxides. While oxides of light elements, e.g. silica and alumina, show useful spectra in the visible to 15 μm region, heavy metal oxides, for example lead oxide, show no significant bands in this region. The same applies generally to metallic sulphides, chlorides and bromides.

As with other methods of identification, the most satisfactory results are obtained by examining the filler in the purest possible condition. To obtain good spectra of inorganic materials it is important that the particle size is very small, but this is rarely a problem with fillers from plastics compositions because these are usually in a very finely divided form.

The infra-red spectra of fillers may be recorded either as alkali halide disks or in paraffin oil mulls. The former is usually easier and in the case of fillers rarely creates problems. There are several excellent collections of spectra of fillers[17, 23, 24], so that there is generally no difficulty in identification.

IDENTIFICATION OF INORGANIC FILLERS BY X-RAY
DIFFRACTION ANALYSIS

X-ray diffraction analysis is an extremely simple and reliable method for the qualitative identification of crystalline inorganic substances. Although it may be possible to identify the filler in a plastics composition by X-ray examination of the complete sample, this is to be avoided whenever possible since, as in almost all other forms of analysis, the most satisfactory results are achieved by examining the filler in the purest form in which it can be obtained. Thus it is most desirable to separate the filler by the methods suggested in Chapter 3.

Inorganic fillers in plastics compositions are usually in a very finely divided form and as such are ideal for X-ray diffraction study. The preferred sample size is a few milligrams, as it is easy to handle and gives a good pattern in 1 or 2 h. On the other hand, with prolonged exposure times it is possible to obtain satisfactory results from microgram quantities of sample.

The sample to be examined is placed in a capillary tube of low-absorption glass. The X-ray source generally used is copper Kα radiation (1·5418 Å). An exposure time of 1 h is used, and examination of the line intensities will indicate whether some other exposure time would have been preferable.

The powder pattern is identified by means of the intensities of the lines and their spacings (d values) by comparison with the substances recorded in the A.S.T.M. *Powder Diffraction File*[25]. Part of the file is classified according to the d value of the strongest line and the d values of the seven next strongest lines. By this means the majority of patterns may be matched and the unknown identified unambiguously.

There are however difficulties in that the patterns given by isomorphous substances are similar. Fortunately isomorphous compounds are frequently quite different in elementary composition. Thus it is extremely useful to have a knowledge of the elements present in a sample, especially as another section of the *Powder Diffraction File* is arranged in groups of compounds which contain particular chemical elements.

In this way the great majority of the common fillers used in plastics compositions may be identified by a comparatively simple test procedure, provided that they can be submitted in a reasonably pure form for examination.

REFERENCES

1. HASLAM, J., SOPPET, W. W., and WILLIS, H. A., *J. appl. Chem., Lond.*, **1**, 112 (1951)
2. WHITNACK, G. C., and GANTZ, E. ST. C., *Analyt. Chem.*, **25**, 553 (1953)
3. CACHIA, M., SOUTHWART, D. W., and DAVISON, W. H. T., *J. appl. Chem., Lond.*, **8**, 291 (1958)
4. HAASE, H., *Kautschuk Gummi Kunststoffe*, **21**, 9 (1968)
5. LIGOTTI, I., PIACENTI, R., and BONOMI, G., *Rass. chim.*, **1**, 3 (1964)
6. BRAUN, D., *Chimia*, **19**, 77 (1965)
7. CLASPER, M., and HASLAM, J., *J. appl. Chem., Lond.*, **7**, 328 (1957)
8. HIGUCHI, T., HILL, N. C., and CORCORAN, G. B., *Analyt. Chem.*, **24**, 491 (1952)
9. HAASE, H., *Kautschuk Gummi Kunststoffe*, **20**, 501 (1967)
10. GUIOCHON, G., and HENNIKER, J., *Industrie Plast. mod.*, **16**, 67 (1964)
11. *Sadtler Commercial Spectra Series E*, The Sadtler Research Laboratories, Philadelphia 2
12. *Sadtler Standard Spectra*, The Sadtler Research Laboratories, Philadelphia 2
13. *Documentation of Molecular Spectra*, Butterworths, London, and Verlag Chemie, Weinheim/Bergstrasse
14. HUMMEL, D. O., *Infra-Red Analysis of Polymers, Resins and Additives, An Atlas*, **1**, Part II, Wiley Interscience (1969)
15. HUMMEL, D. O., (Transl. WULKOW, E. M.), *Identification and Analysis of Surface-Active Agents by Infra-Red and Chemical Methods*, Interscience, New York (1964)
16. MEIKLEJOHN, R. A., MEYER, R. J., SANFORD, M., ARONOVIC, H. A., SCHUETTE, H. A., and MELOCH, V. W., *Analyt. Chem.*, **29**, 329 (1957)
17. *Infra-Red Spectroscopy. Its Use in the Coatings Industry*, Federation of Societies of Paint Technology, Philadelphia (1969)
18. HASLAM, J., and JEFFS, A. R., *Analyst, Lond.*, **83**, 455 (1958)
19. JAMES, A. T., and MARTIN, A. J. P., *Analyst, Lond.*, **77**, 915 (1952)
20. ANDERSON, D. M. W., *Analyst, Lond.*, **84**, 50 (1959)
21. HASLAM, J., JEFFS, A. R., and WILLIS, H. A., *Analyst, Lond.*, **86**, 44 (1961)
22. *Fisk's Paint Year Book and A-Z Buyers' Guide (1961-2)*, 169, compiled, edited and published by FISK, N. R., 58 College Road, Harrow, Middlesex
23. MILLER, F. A., and WILKINS, C. H., *Analyt. Chem.*, **24**, 1253 (1952)
24. HUNT, J. M., WISHERD, M. P., and BONHAM, L. C., *Analyt. Chem.*, **22**, 1478 (1950)
25. *Powder Diffraction File*, American Society for Testing and Materials, Philadelphia

INFRA-RED SPECTRA

INTRODUCTION

The infra-red absorption spectrum is a unique property of the molecule of a substance, and for this reason infra-red examination has proved to be among the more powerful methods of identification of organic materials. The art of infra-red spectroscopy has now advanced to a stage at which it is possible to obtain, on relatively cheap and commercially available equipment, a reliable infra-red spectrum of an unknown within a few minutes; the infra-red spectrum can therefore be considered alongside the more conventional tests used in characterisation, such as melting point, refractive index, and preparation of derivatives, as a general identification method. In the field of synthetic polymers, however, the more conventional tests are of restricted value and hence examination by the infra-red method is generally acknowledged to be the most important single test available for polymer identification.

The infra-red method is based both upon the assignment of absorption bands to particular structural groups, as outlined in Chapter 2, and upon matching the spectra of unknowns against those of well characterised standard substances prepared for examination in a definite manner. Thus for the proper use of the infra-red method a collection of such spectra is of the greatest importance.

Over many years we have been able to accumulate samples of a wide range of pure and well characterised synthetic polymers, and we present the spectra of these, together with the spectra of a number of associated substances, in this section of the book. Except where it is stated in the titles the spectra are those of samples prepared within the Research Department at I.C.I. Plastics Division, or they are authentic samples given to us by our many friends in the industry, including the Paints, Dyestuffs, Mond and Nobel Divisions of I.C.I., the Natural Rubber Producers' Research Association, and the Dunlop Rubber Company. We would like to thank all those who have helped us in this respect.

With the exception of Spectra 10.24, 10.25 and 10.26, which were kindly prepared for us by Midland Silicones Ltd., all the spectra were recorded in the Infra-Red Laboratory at I.C.I. Plastics Division. The majority were recorded on the Perkin-Elmer 137 (Infracord), particular attention being given to accuracy of wavelength calibration. In a few cases the original samples were no longer available, and the spectra in our files had been obtained on other instruments; such spectra have been redrawn on to Perkin-Elmer Infracord charts to preserve uniformity of presentation, every care being taken to avoid loss of detail in redrawing the spectra. Except in these few cases, the figures are reproduced directly from the records as produced by the instrument. We would like to express here our thanks to all the assistants in the Infra-Red Laboratory at I.C.I. Plastics Division, and in particular to Mr. B. J. Stay, for their careful work in producing these spectra. Our thanks are also due to Mr. Stay for help with the production of the diagrams throughout the text.

Mention is frequently made in the text of variations in the spectra of polymers which occur through changes in the physical state of the sample. Where such differences are significant we have usually included more than one spectrum of the sample, to indicate the nature of the differences which may occur, and in these cases it should be possible to obtain spectra similar to those given by following the preparation conditions stated.

Where the nature of the sample appeared to be appropriate, we have included A.T.R. (attenuated total reflection) spectra. This method is described in Chapter 1. All A.T.R. spectra were taken using a 90° isosceles thallium bromo-iodide prism, the sample being pressed into contact with the hypotenuse face, and the light entering and leaving the prism along the normal to the surface.

Bibliography of Published Infra-Red Spectra of Polymers and Associated Substances

The spectra presented here can be supplemented by a number of published collections. In our experience the following have proved valuable.

1. *Synthetic and natural polymers*

Three collections are particularly valuable here:
HUMMEL, D. O., *Infra-red Analysis of Polymers, Resins and Additives, An Atlas*, 1, Pt II, Wiley Interscience, New York and London (1969)
NYQUIST, R., *Infra-Red Spectra of Polymers and Resins*, The Dow Chemical Co., Midland, Michigan, 2nd edn (1961)
Sadtler Commercial Spectra Series D, The Sadtler Research Laboratories, Philadelphia 2, Pennsylvania (European sales, Heyden and Son Ltd, Spectrum House, Alderton Crescent, London NW4 3XX). This last is

a very large collection but the spectra are those of commercially available materials, many of which are mixtures, and most of the materials are described only by their proprietary names.

A smaller collection which is sometimes useful is *Infra-Red Spectroscopy. Its Use in the Coatings Industry*, Federation of Societies for Paint Technology, Philadelphia, Pennsylvania (1969).

Most of the spectra in the above publications are those of commercially available resins, while the spectra reproduced here are, with a few exceptions, those of research materials which we believe to be of high purity.

2. *Monomers*

The spectra of most monomers of commercial interest appear in *Sadtler Standard Spectra*, the Sadtler Research Laboratories, address above.

3. *Plasticisers*

Sadtler Commercial Spectra Series E, The Sadtler Research Laboratories, address above, and

Infra-Red Spectroscopy. Its Use in the Coatings Industry, referred to above, will usefully supplement the spectra given here. This book also contains a useful collection of spectra of solvents.

4. *Surface Active Agents*

A good collection of surface-active agents' spectra appears in

HUMMEL, D., (Transl. WULKOW, E. M.), *Identification and Analysis of Surface-Active Agents by Infra-Red and Chemical Methods*, Interscience, New York (1964)

5. *Hydrolysis products—acids, alcohols, glycols and bases*

Spectra of these are included in

Sadtler Standard Spectra, referred to above, and

Documentation of Molecular Spectra, Butterworths, London, and Verlag Chemie, Weinheim/Bergstrasse

NOTE FOR THE NEW EDITION

In revising the collection of infra-red spectra of polymers, we have tried to adhere to the principles by which the spectra were selected for the first edition. Thus the spectra relate only to pure and well characterised samples, and all the spectra and most of the samples have been prepared within this Research Department.

We have added a number of spectra to the collection, particularly those of polymers and copolymers which have achieved commercial importance in the past few years. Nevertheless, the collection has been kept small and representative rather than comprehensive, as it is presumed that the reader will if necessary refer to one of the large collections when he has ascertained the general nature of an unknown sample. The format of percentage transmission versus wavelength has been retained for uniformity, and this is in line with the largest collections of reference polymer spectra by Sadtler and Hummel.

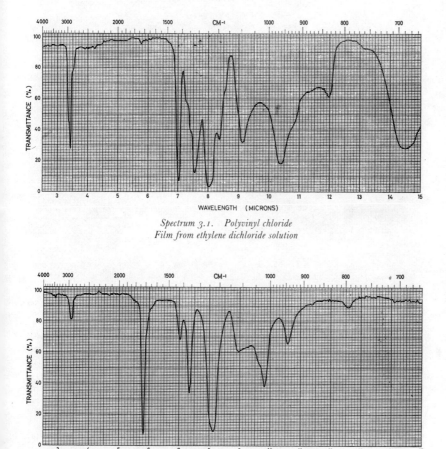

Spectrum 3.1. Polyvinyl chloride
Film from ethylene dichloride solution

Spectrum 3.2. Polyvinyl acetate
Film on NaCl plate from ethylene dichloride solution

609

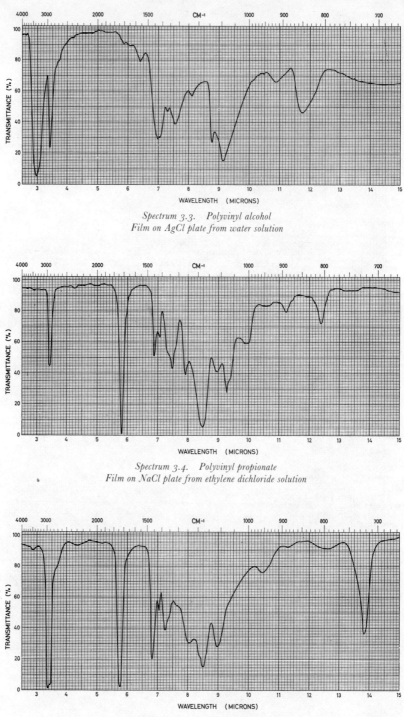

Spectrum 3.3. Polyvinyl alcohol
Film on AgCl plate from water solution

Spectrum 3.4. Polyvinyl propionate
Film on NaCl plate from ethylene dichloride solution

Spectrum 3.5. Polyvinyl laurate
Film on NaCl plate from ethylene dichloride solution

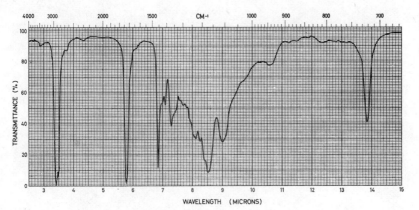

Spectrum 3.6. Polyvinyl stearate
Film on NaCl plate from ethylene dichloride solution

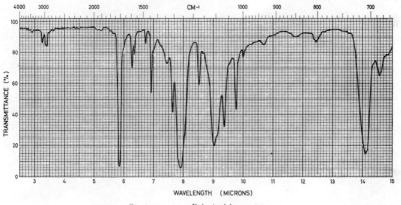

Spectrum 3.7. Polyvinyl benzoate
Film on NaCl plate from ethylene dichloride solution

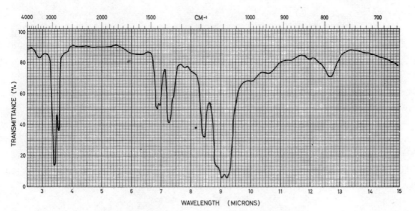

Spectrum 3.8. Polyvinyl methyl ether
Film on NaCl plate from ethyl acetate solution

611

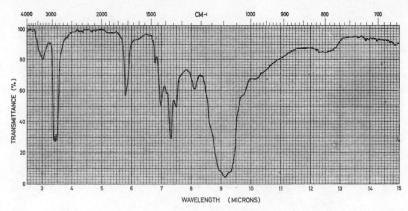

Spectrum 3.9.   Polyvinyl ethyl ether
Film on NaCl plate from ethylene dichloride solution

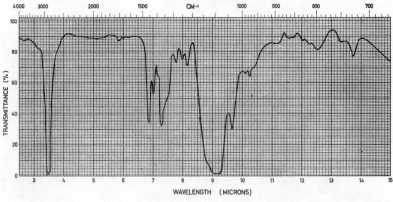

Spectrum 3.10.   Polyvinyl n-butyl ether
Film on NaCl plate from ethylene dichloride solution

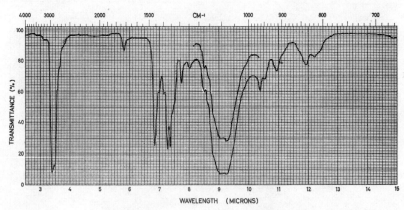

Spectrum 3.11.   Polyvinyl isobutyl ether
Film on NaCl plate from ethylene dichloride solution

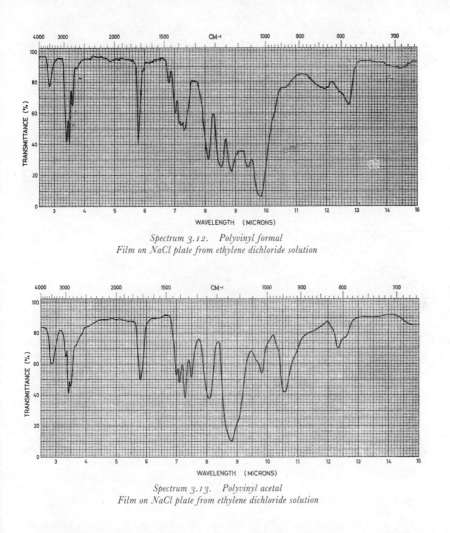

Spectrum 3.12. Polyvinyl formal
Film on NaCl plate from ethylene dichloride solution

Spectrum 3.13. Polyvinyl acetal
Film on NaCl plate from ethylene dichloride solution

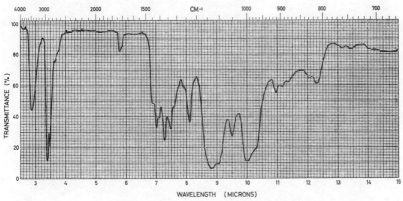

Spectrum 3.14. Polyvinyl butyral
Film on NaCl plate from ethylene dichloride solution

613

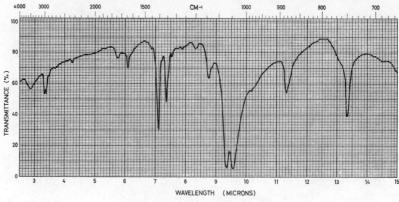

Spectrum 3.15. Polyvinylidene chloride
2% suspension in KBr disk

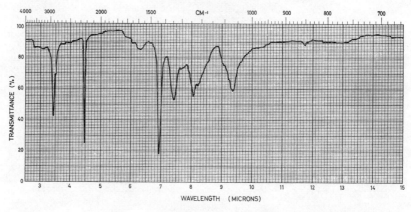

Spectrum 3.16. Polyacrylonitrile
Film from dimethyl formamide solution

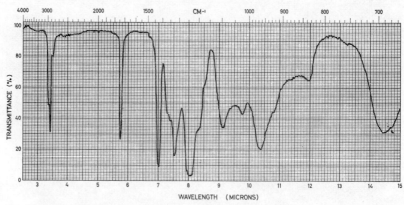

Spectrum 3.17. Vinyl chloride/vinyl acetate (98 : 2) copolymer
Film from ethylene dichloride solution

614

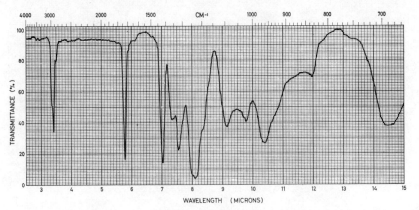

Spectrum 3.18. Vinyl chloride/vinyl acetate (95 : 5) copolymer
Film from ethylene dichloride solution

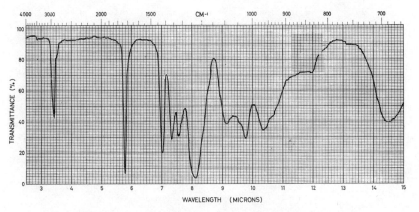

Spectrum 3.19. Vinyl chloride/vinyl acetate (89 : 11) copolymer
Film on NaCl plate from ethylene dichloride solution

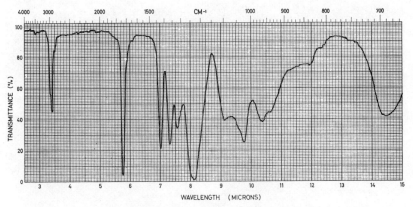

Spectrum 3.20. Vinyl chloride/vinyl acetate (84 : 16) copolymer
Film on NaCl plate from ethylene dichloride solution

615

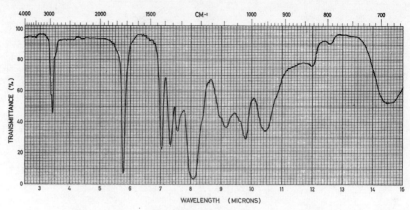

Spectrum 3.21. Polyvinyl chloride/polyvinyl acetate (84 : 16) blend
Film on NaCl plate from ethylene dichloride solution

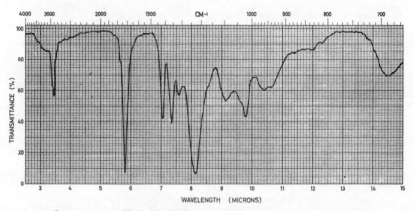

Spectrum 3.22. Vinyl chloride/vinyl acetate/maleic acid (81 : 16 : 3) terpolymer
Film on NaCl plate from ethylene dichloride solution

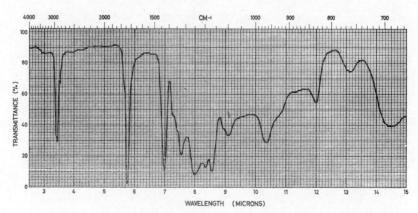

Spectrum 3.23. Vinyl chloride/methyl acrylate (87 : 13) copolymer
Film from ethylene dichloride solution

616

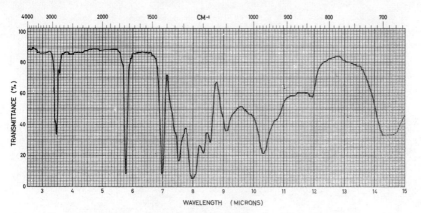

Spectrum 3.24.   *Vinyl chloride/dimethyl fumarate (maleate) (95 : 5) copolymer*
Film from ethylene dichloride solution

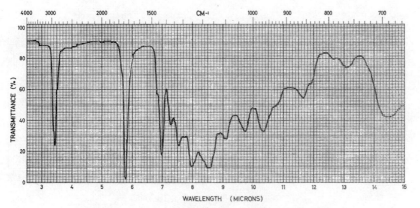

Spectrum 3.25.   *Vinyl chloride/ethyl acrylate (84 : 16) copolymer*
Film from ethylene dichloride solution

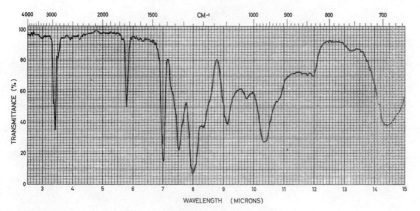

Spectrum 3.26.   *Vinyl chloride/diethyl fumarate (maleate) (95 : 5) copolymer*
Film from ethylene dichloride solution

617

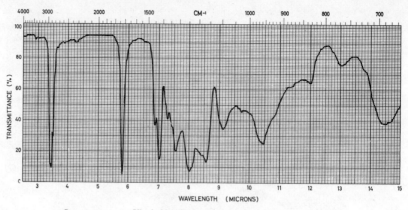

Spectrum 3.27. Vinyl chloride/2-ethyl hexyl acrylate (80 : 20) copolymer Film on NaCl plate from ethylene dichloride solution

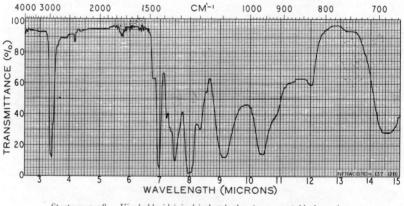

Spectrum 3.28. Vinyl chloride/vinyl isobutyl ether (~95 : 5) block copolymer Hot-pressed film

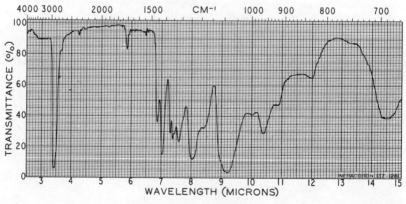

Spectrum 3.29. Vinyl chloride/vinyl isobutyl ether (87 : 13) copolymer Hot-pressed film

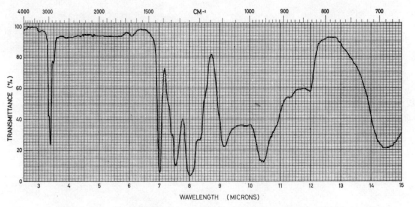

Spectrum 3.30.   Vinyl chloride/vinylidene chloride (96 : 4) copolymer
Film from ethylene dichloride solution

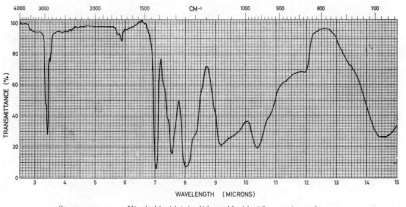

Spectrum 3.31.   Vinyl chloride/vinylidene chloride (80 : 20) copolymer
Film from ethylene dichloride solution

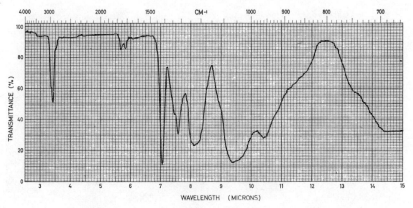

Spectrum 3.32.   Vinyl chloride/vinylidene chloride (60 : 40) copolymer
Film from ethylene dichloride solution

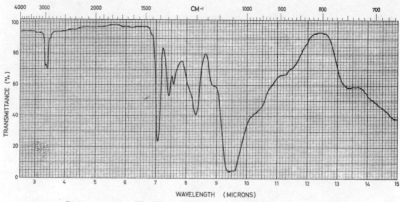

*Spectrum 3.33. Vinylidene chloride/vinyl chloride (74 : 26) copolymer
Hot-pressed film*

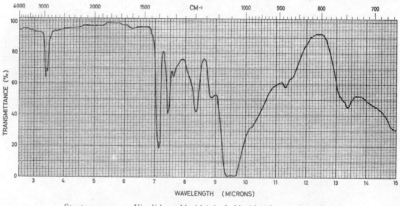

*Spectrum 3.34. Vinylidene chloride/vinyl chloride (87 : 13) copolymer
Hot-pressed film*

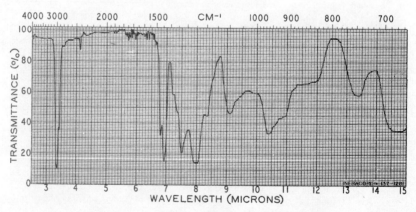

*Spectrum 3.35. Vinyl chloride/ethylene (90 : 10) copolymer
Film on NaCl plate from ethylene dichloride solution*

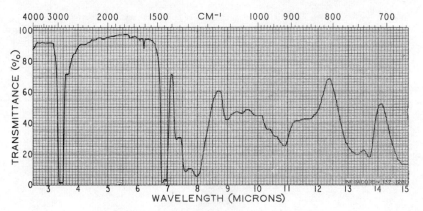

Spectrum 3.36. Vinyl chloride/ethylene (61 : 39) copolymer
Film on NaCl plate from ethylene dichloride solution

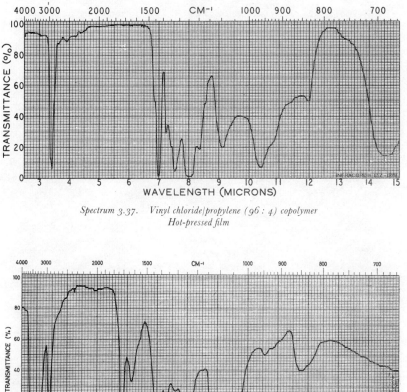

Spectrum 3.37. Vinyl chloride/propylene (96 : 4) copolymer
Hot-pressed film

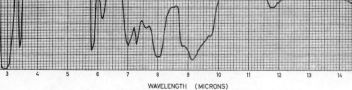

Spectrum 3.38. Partially hydrolysed polyvinyl acetate (12% acetate)
1% suspension in KBr disk

621

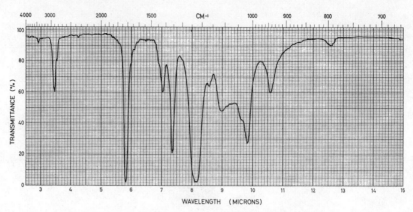

Spectrum 3.39.   Vinyl acetate/2-ethyl hexyl acrylate (85 : 15) copolymer
Film on NaCl plate from ethylene dichloride solution

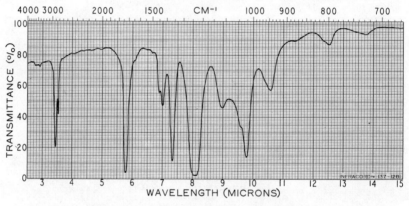

Spectrum 3.40.   Vinyl acetate/ethylene (85 : 15) copolymer
Hot-pressed between AgCl sheets

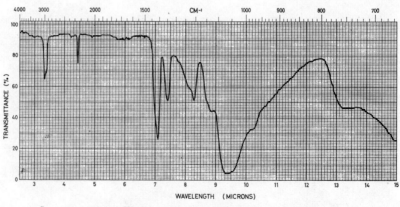

Spectrum 3.41(a).   Vinylidene chloride/acrylonitrile (91 : 9) copolymer. Amorphous
Hot-pressed film. Quenched

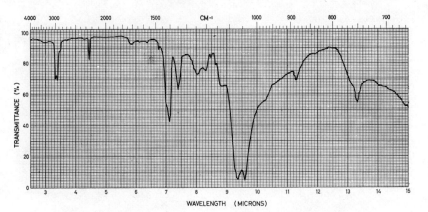

Spectrum 3.41(b). Vinylidene chloride/acrylonitrile (91 : 9) copolymer. Crystalline Film on NaCl plate from ethylene dichloride solution

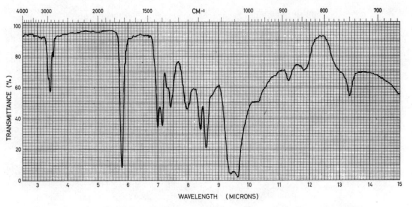

Spectrum 3.42. Vinylidene chloride/methyl acrylate (90 : 10) copolymer. Crystalline Film on NaCl plate from ethylene dichloride solution

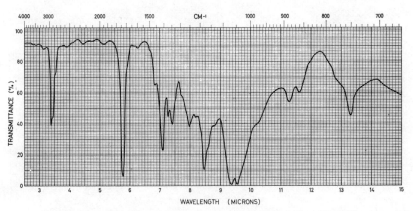

Spectrum 3.43. Vinylidene chloride/ethyl acrylate (90 : 10) copolymer. Crystalline Film on NaCl plate from ethylene dichloride solution

623

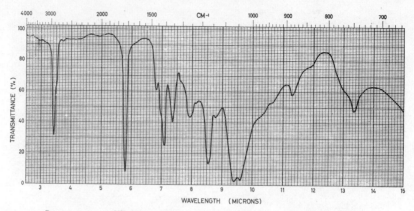

Spectrum 3.44. Vinylidene chloride/n-butyl acrylate (88 : 12) copolymer. Crystalline
Film on NaCl plate from ethylene dichloride solution

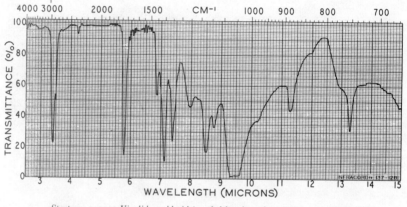

Spectrum 3.45. Vinylidene chloride/2-ethyl hexyl acrylate (90 : 10) copolymer
Hot-pressed film

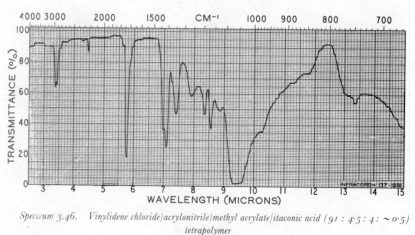

Spectrum 3.46. Vinylidene chloride/acrylonitrile/methyl acrylate/itaconic acid (91 : 4·5 : 4 : ~ 0·5)
tetrapolymer
Film cast from latex

624

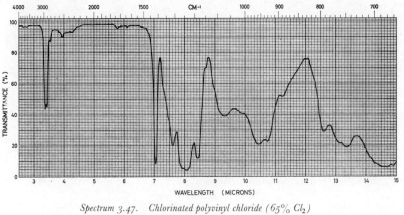

Spectrum 3.47.   Chlorinated polyvinyl chloride (65% Cl₂)
Hot-pressed film

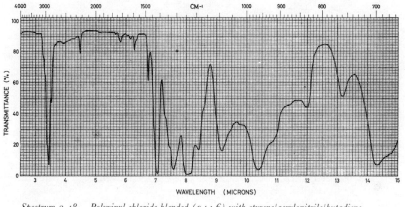

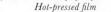

Spectrum 3.48.   Polyvinyl chloride blended (94 : 6) with styrene/acrylonitrile/butadiene
(54 : 31 : 15) terpolymer
Hot-pressed film

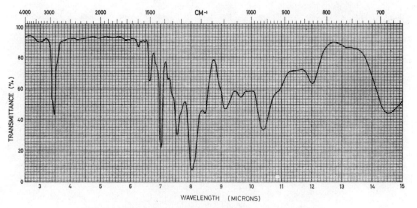

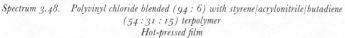

Spectrum 3.49.   Polyvinyl chloride/epoxy resin (97 : 3) blend
Film from ethylene dichloride solution

625

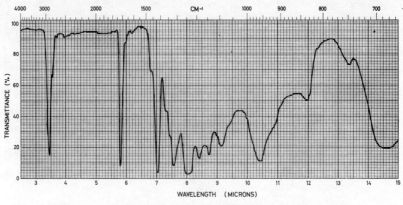

Spectrum 3.50.   Polyvinyl chloride/polymethyl methacrylate (95 : 5) blend
Hot-pressed film

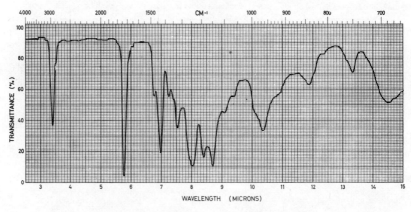

Spectrum 3.51.   Polyvinyl chloride/polymethyl methacrylate (80 : 20) blend
Film on NaCl plate from ethylene dichloride solution

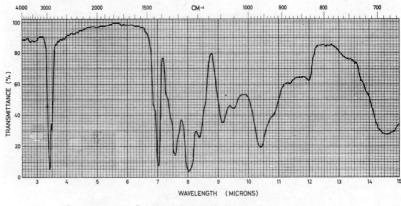

Spectrum 3.52.   Polyvinyl chloride/chlorinated polythene (85 : 15) blend
Hot-pressed film

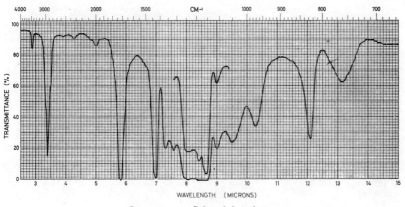

Spectrum 4.1.   Polymethyl acrylate
Film on NaCl plate from ethylene dichloride solution

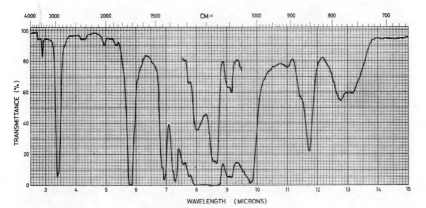

Spectrum 4.2.   Polyethyl acrylate
Film on NaCl plate from ethylene dichloride solution

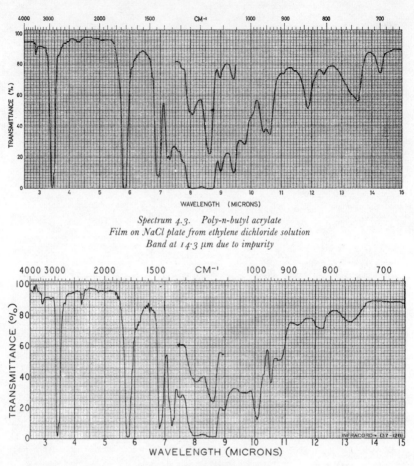

Spectrum 4.3. Poly-n-butyl acrylate
Film on NaCl plate from ethylene dichloride solution
Band at 14·3 μm due to impurity

Spectrum 4.4. Polyisobutyl acrylate
Film on NaCl plate from ethylene dichloride solution

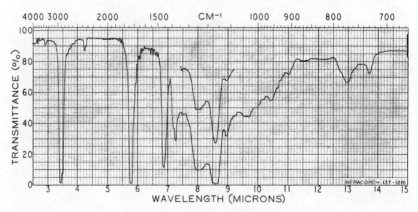

Spectrum 4.5. Poly-2-ethyl hexyl acrylate
Film on NaCl plate from ethylene dichloride solution

628

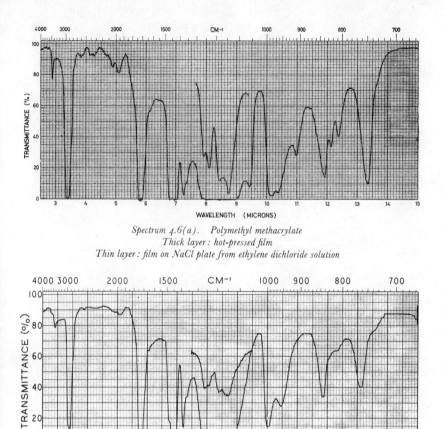

Spectrum 4.6(a).   Polymethyl methacrylate
Thick layer : hot-pressed film
Thin layer : film on NaCl plate from ethylene dichloride solution

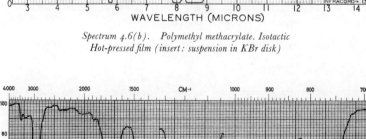

Spectrum 4.6(b).   Polymethyl methacrylate. Isotactic
Hot-pressed film (insert : suspension in KBr disk)

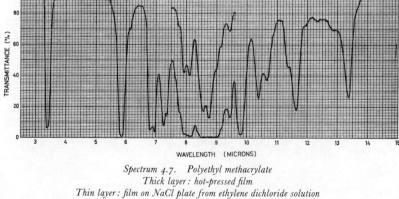

Spectrum 4.7.   Polyethyl methacrylate
Thick layer : hot-pressed film
Thin layer : film on NaCl plate from ethylene dichloride solution

629

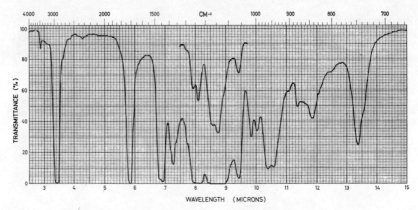

Spectrum 4.8.   Poly-n-butyl methacrylate
Film on NaCl plate from ethylene dichloride solution

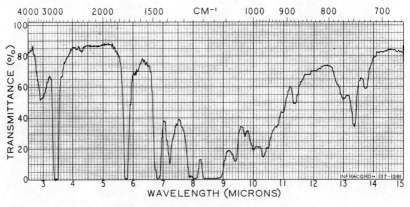

Spectrum 4.9.   Poly-2-ethyl hexyl methacrylate
Suspension in KBr disk

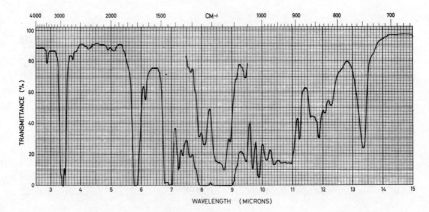

Spectrum 4.10.   Polycyclohexyl methacrylate
Thick layer: hot-pressed film
Thin layer: film on NaCl plate from ethylene dichloride solution

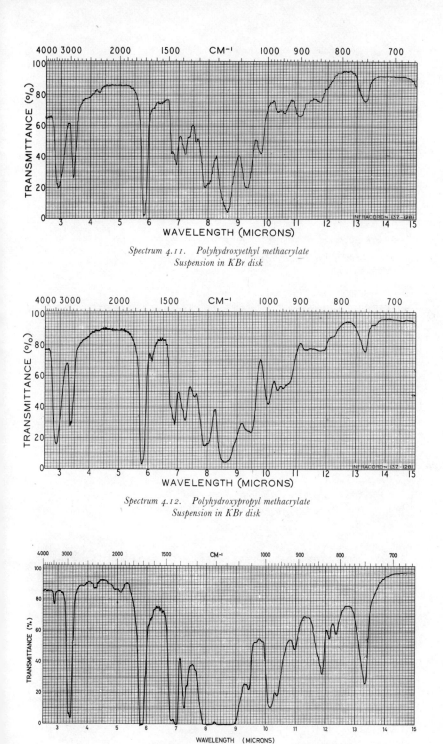

Spectrum 4.11.   Polyhydroxyethyl methacrylate
Suspension in KBr disk

Spectrum 4.12.   Polyhydroxypropyl methacrylate
Suspension in KBr disk

Spectrum 4.13.   Methyl methacrylate/ethyl acrylate (90 : 10) copolymer
Hot-pressed film

631

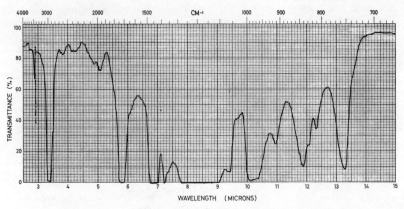

Spectrum 4.14.   Methyl methacrylate/methyl acrylate (90 : 10) copolymer
Hot-pressed film

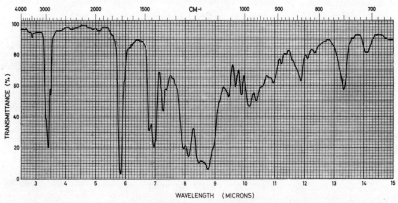

Spectrum 4.15.   Polymethyl methacrylate/polycyclohexyl methacrylate (50 : 50) blend
Film on NaCl plate from ethylene dichloride solution
Band at 14·1 μm due to residual solvent

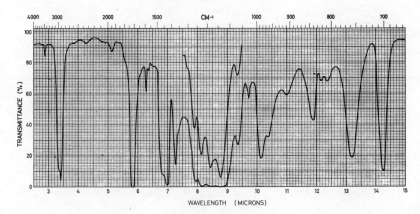

Spectrum 4.16.   Methyl methacrylate/styrene (80 : 20) copolymer
Thick layer : hot-pressed film
Thin layer : film on NaCl plate from ethylene dichloride solution

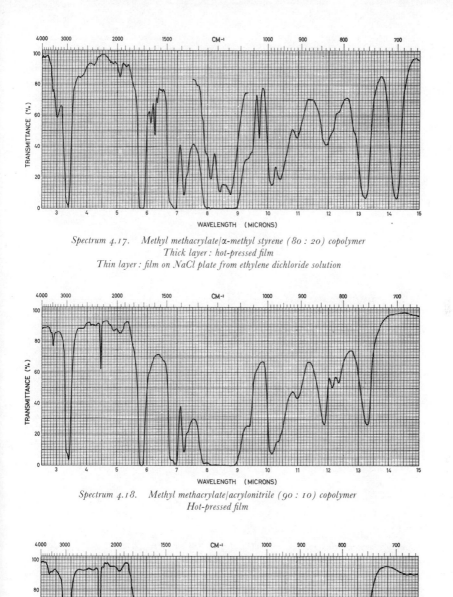

Spectrum 4.17.   Methyl methacrylate/α-methyl styrene (80 : 20) copolymer
Thick layer : hot-pressed film
Thin layer : film on NaCl plate from ethylene dichloride solution

Spectrum 4.18.   Methyl methacrylate/acrylonitrile (90 : 10) copolymer
Hot-pressed film

Spectrum 4.19.   Acrylonitrile/methyl methacrylate (55 : 45) copolymer
Hot-pressed film

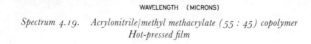

633

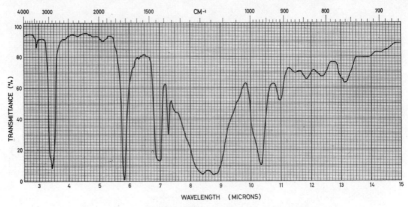

Spectrum 4.20. Methyl methacrylate/butadiene (60 : 40) copolymer
Film on AgCl plate cast from latex

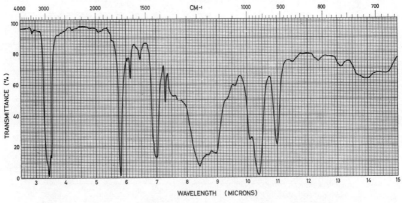

Spectrum 4.21. Butadiene/methyl methacrylate (70 : 30) copolymer
Film on AgCl plate cast from latex

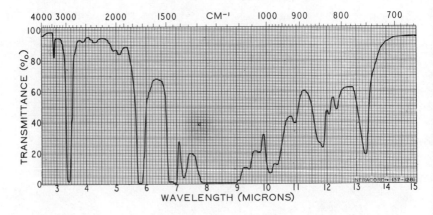

Spectrum 4.22. Polymethyl methacrylate/polyethyl acrylate (80 : 20) blend
Hot-pressed film

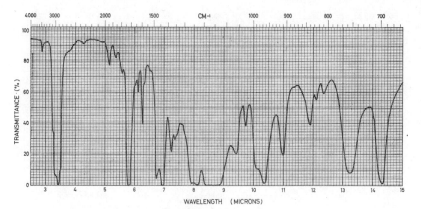

Spectrum 4.23. *Polymethyl methacrylate blended (50 : 50) with styrene/butadiene*
(60 : 40) copolymer
Hot-pressed film

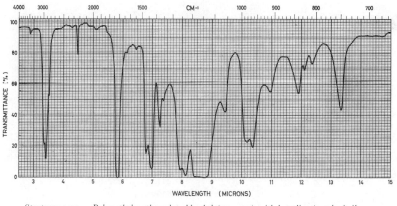

Spectrum 4.24. *Polymethyl methacrylate blended (70 : 30) with butadiene/acrylonitrile*
(70 : 30) copolymer
Hot-pressed film

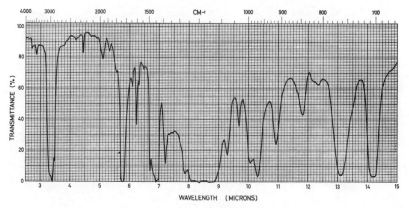

Spectrum 4.25. *ABS rubber/polymethyl methacrylate (60 : 40) blend*
Hot-pressed film

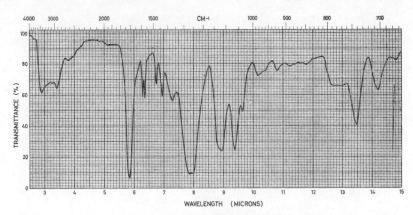

Spectrum 4.26.   Polyethylene o-phthalate
Film on NaCl plate from ethylene dichloride solution

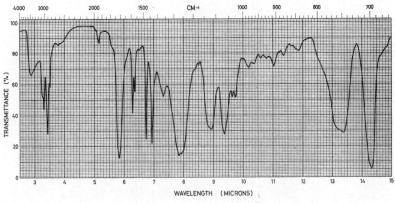

Spectrum 4.27.   Polyethylene o-phthalate/polystyrene blend
Film on NaCl plate from ethylene dichloride solution

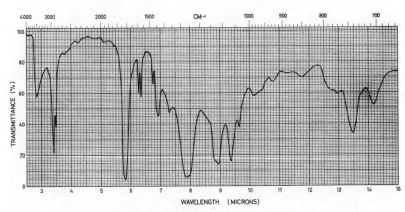

Spectrum 4.28.   Polyethylene o-phthalate. Oil modified
Film on NaCl plate from ethylene dichloride solution

636

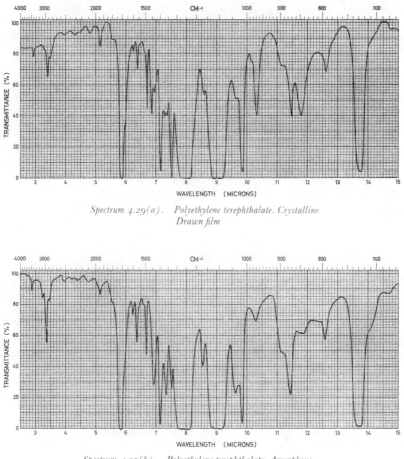

Spectrum 4.29(a).   Polyethylene terephthalate. Crystalline
Drawn film

Spectrum 4.29(b).   Polyethylene terephthalate. Amorphous
Quenched film

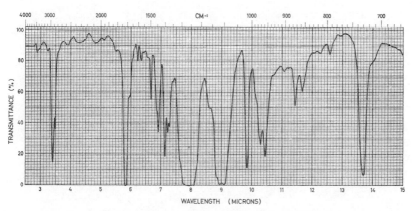

Spectrum 4.30.   Polybis-1 : 4-hydroxymethyl cyclohexane terephthalate
('Terafilm' ex Eastman Kodak Co.)

637

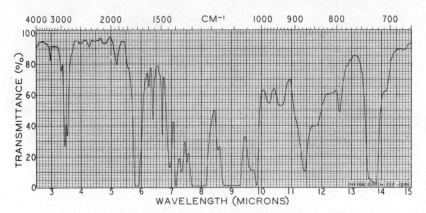

Spectrum 4.31. Terephthalic acid condensate with mixed (50 : 50 molar) ethylene
glycol and diethylene glycol
Hot-pressed film

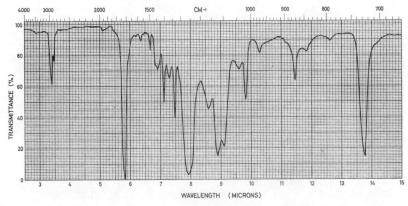

Spectrum 4.32. Ethylene terephthalate/sebacate (70 : 30) copolyester
Film on NaCl plate from ethylene dichloride solution

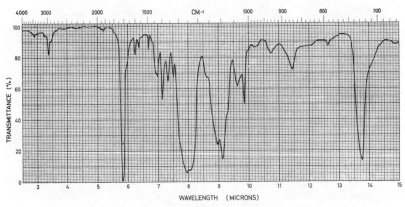

Spectrum 4.33. Ethylene terephthalate/isophthalate (60 : 40) copolyester
Film on NaCl plate from ethylene dichloride solution

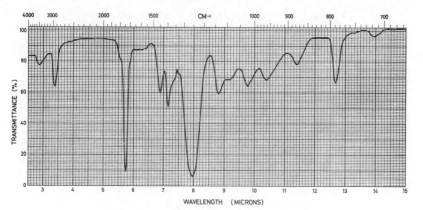

Spectrum 4.34.   Poly(diethylene glycol bisallyl) carbonate
Suspension in KBr disk

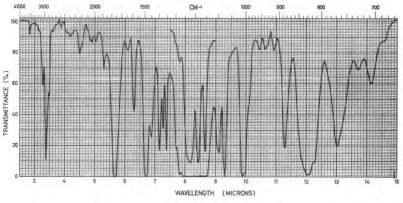

Spectrum 4.35.   Poly(diphenylol propane) carbonate
Films on NaCl plates from ethylene dichloride solution

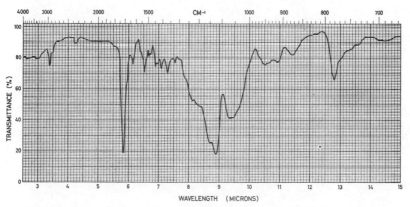

Spectrum 4.36(a).   Urethane rubber from polyethylene adipate and naphthalene di-isocyanate
A.T.R. spectrum

639

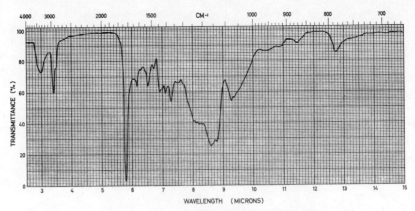

Spectrum 4.36(b). Urethane rubber from polyethylene adipate and naphthalene di-isocyanate
Suspension in KBr disk

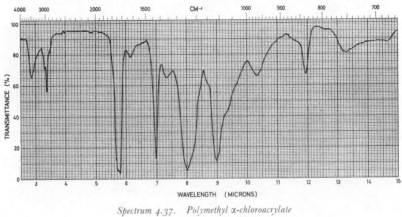

Spectrum 4.37. Polymethyl α-chloroacrylate
Suspension in KBr disk

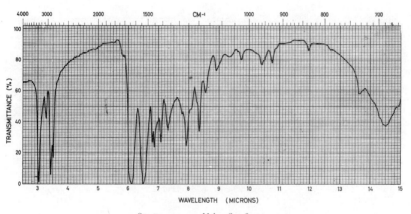

Spectrum 5.1.   Nylon 6 α-form
Film by Ziabicki's method (Ref. 16 chap. 5)

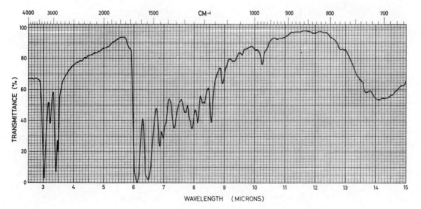

Spectrum 5.2   Nylon 6 δ-form
Film from formic acid solution

641

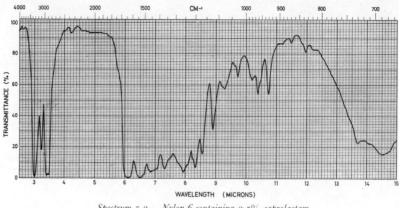

Spectrum 5.3.   Nylon 6 containing 3·5% caprolactam
Hot-pressed film

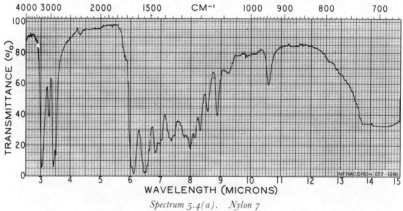

Spectrum 5.4(a).   Nylon 7
Film from formic acid solution. Annealed from 160°C to room temperature at 6°C/h

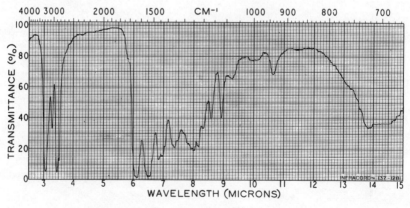

Spectrum 5.4(b).   Nylon 7
Film from formic acid solution. Quenched from 160°C in ice/water

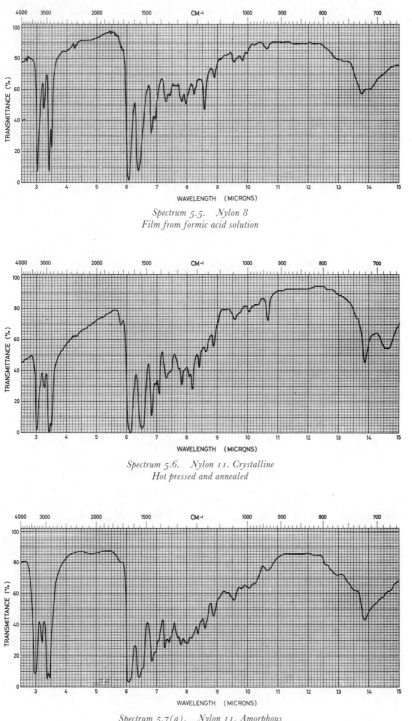

Spectrum 5.5. Nylon 8
Film from formic acid solution

Spectrum 5.6. Nylon 11. Crystalline
Hot pressed and annealed

Spectrum 5.7(a). Nylon 11. Amorphous
Film from formic acid solution

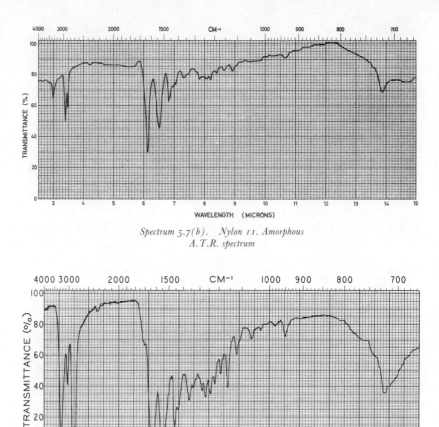

Spectrum 5.7(b). Nylon 11. Amorphous
A.T.R. spectrum

Spectrum 5.8(a). Nylon 12
Hot-pressed film. Annealed from 160°C to room temperature at 6°C/h

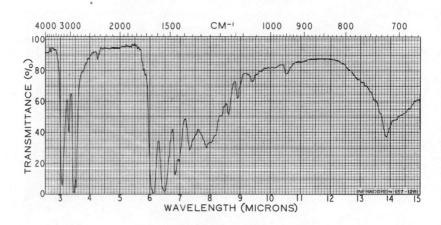

Spectrum 5.8(b). Nylon 12. Quenched

644

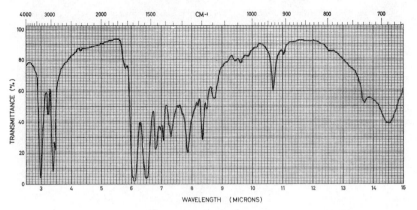

Spectrum 5.9.   Nylon 6 : 6
Film from formic acid solution

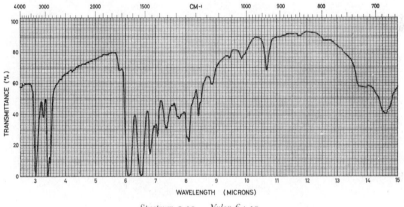

Spectrum 5.10.   Nylon 6 : 10
Film from formic acid solution

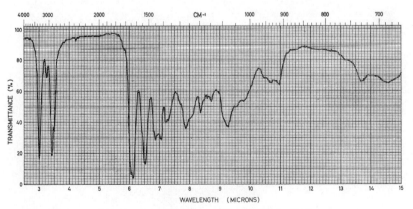

Spectrum 5.11.   Methoxy-methyl modified Nylon 6 : 6
Film from ethanol solution

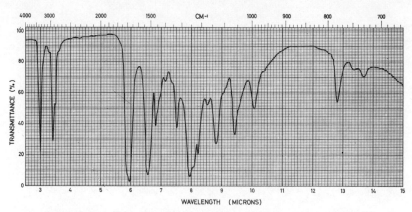

Spectrum 5.12. Polyurethane from hexamethylene di-isocyanate and 1 : 4-butane diol
Film from formic acid solution

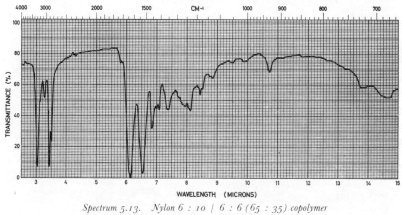

Spectrum 5.13. Nylon 6 : 10 / 6 : 6 (65 : 35) copolymer
Film from formic acid solution

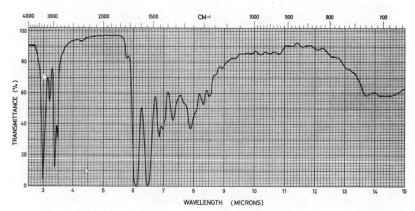

Spectrum 5.14. Nylon 6 : 6 / 6 (60 : 40) copolymer
Film on AgCl from methanol solution

646

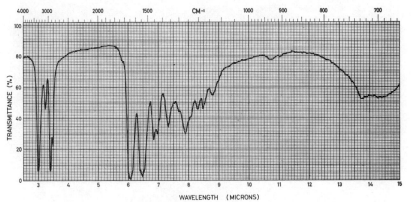

Spectrum 5.15. *Nylon 6 : 6 | 6 : 10 | 6 (40 : 30 : 30) terpolymer*
Film from formic acid solution

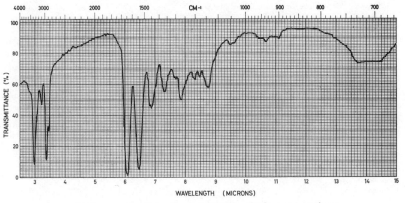

Spectrum 5.16. *Nylon 6 : 6 | PACM : 6 (60 : 40) copolymer*
Film on AgCl from formic acid solution

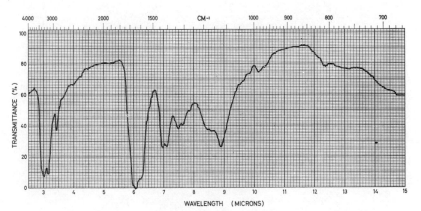

Spectrum 5.17. *Polyacrylamide*
Film on AgCl from formic acid solution

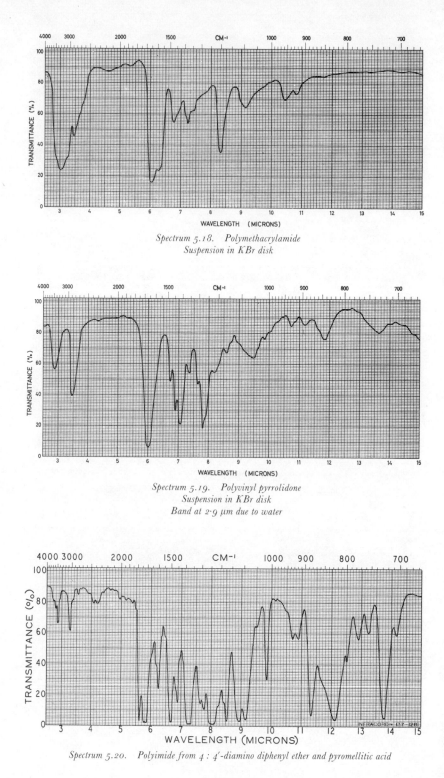

Spectrum 5.18. *Polymethacrylamide*
Suspension in KBr disk

Spectrum 5.19. *Polyvinyl pyrrolidone*
Suspension in KBr disk
Band at 2·9 μm due to water

Spectrum 5.20. *Polyimide from 4 : 4'-diamino diphenyl ether and pyromellitic acid*

648

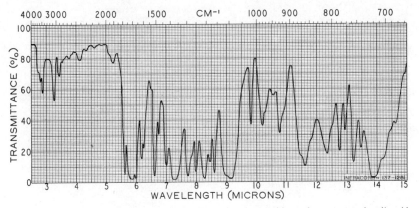

Spectrum 5.21.   Polyimide from 4 : 4'-diamino diphenyl methane and benzophenone tetracarboxylic acid

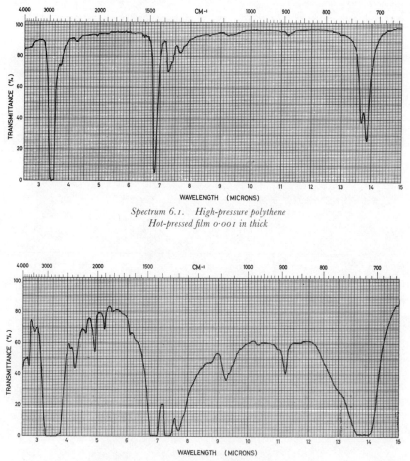

Spectrum 6.1. High-pressure polythene
Hot-pressed film 0·001 in thick

Spectrum 6.2. Polythene rich in pendent-methylene groups
Hot-pressed film of high-pressure polythene 0·02 in thick

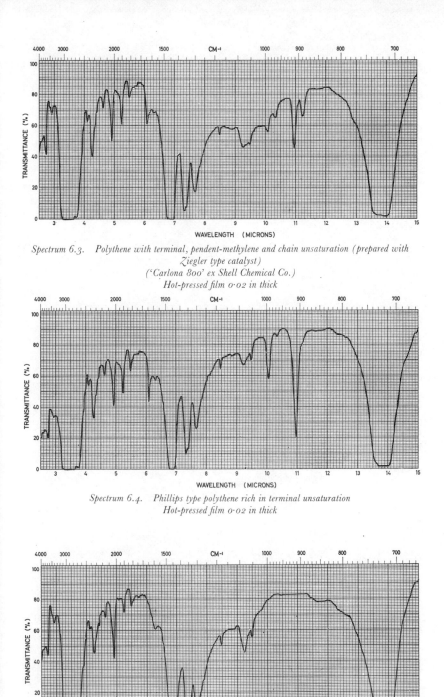

Spectrum 6.3.   Polythene with terminal, pendent-methylene and chain unsaturation (prepared with Ziegler type catalyst)
('Carlona 800' ex Shell Chemical Co.)
Hot-pressed film 0·02 in thick

Spectrum 6.4.   Phillips type polythene rich in terminal unsaturation
Hot-pressed film 0·02 in thick

Spectrum 6.5.   Polymethylene
Hot-pressed film 0·02 in thick

651

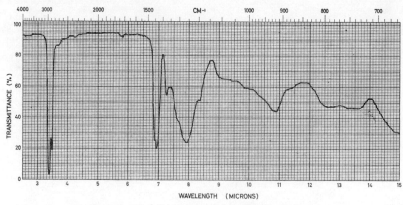

Spectrum 6.6. Chlorinated polythene. Random chlorination
Hot-pressed film

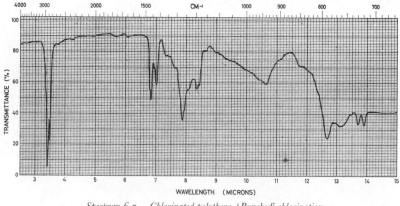

Spectrum 6.7. Chlorinated polythene. 'Bunched' chlorination
Hot-pressed film

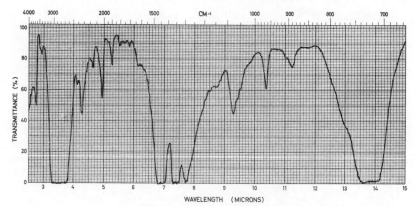

Spectrum 6.8. Irradiated high-pressure polythene. 20 megaroentgens dose
(Irradiated 'Alkathene WNC 18' ex Imperial Chemical Industries Ltd)
Hot-pressed film 0·03 in thick

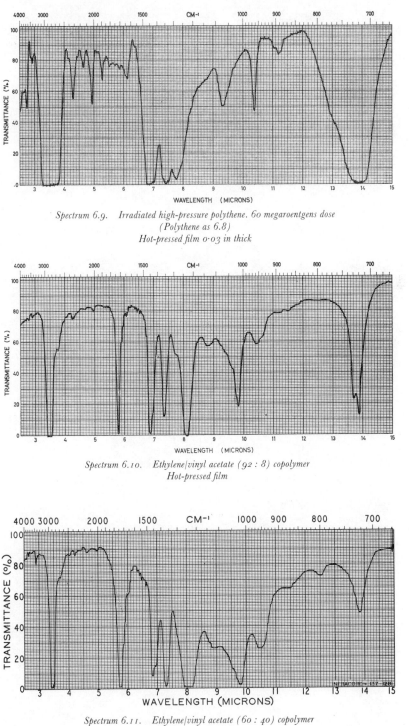

Spectrum 6.9. *Irradiated high-pressure polythene. 60 megaroentgens dose*
(Polythene as 6.8)
Hot-pressed film 0·03 in thick

Spectrum 6.10. *Ethylene/vinyl acetate (92 : 8) copolymer*
Hot-pressed film

Spectrum 6.11. *Ethylene/vinyl acetate (60 : 40) copolymer*
Hot-pressed film

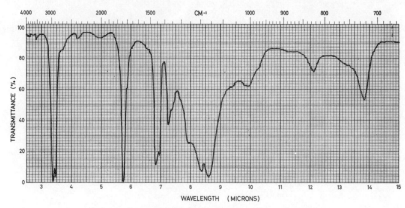

Spectrum 6.12.   Ethylene/methyl acrylate (87 : 13) copolymer
Hot-pressed film

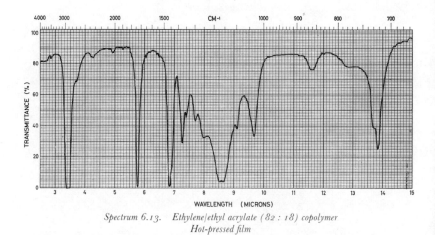

Spectrum 6.13.   Ethylene/ethyl acrylate (82 : 18) copolymer
Hot-pressed film

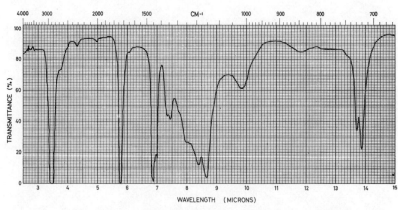

Spectrum 6.14.   Ethylene/dimethyl fumarate (maleate) (85 : 15) copolymer
Hot-pressed film

654

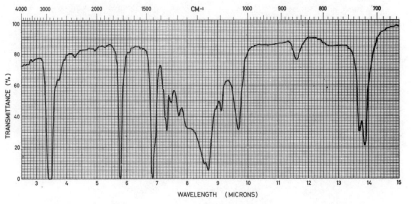

Spectrum 6.15. *Ethylene/diethyl fumarate (maleate) (80 . 20) copolymer*
Hot-pressed film

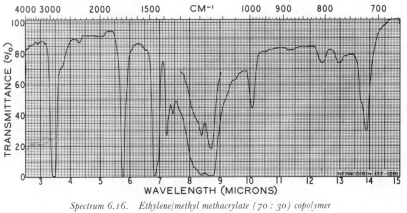

Spectrum 6.16. *Ethylene/methyl methacrylate (70 : 30) copolymer*
Hot-pressed films

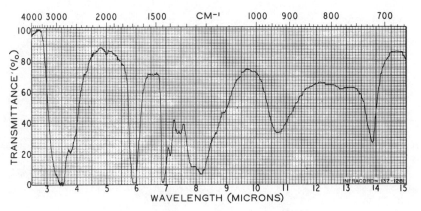

Spectrum 6.17. *Ethylene/acrylic acid (83 : 17) copolymer*
Hot-pressed film

655

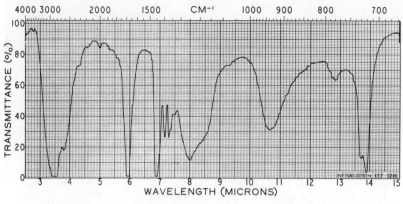

Spectrum 6.18.   Ethylene/methacrylic acid (85 : 15) copolymer
Hot-pressed film

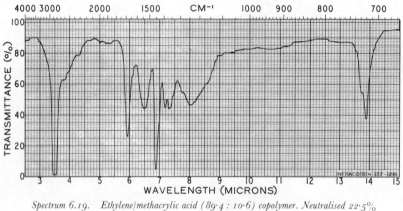

Spectrum 6.19.   Ethylene/methacrylic acid (89·4 : 10·6) copolymer. Neutralised 22·5%
Hot-pressed film

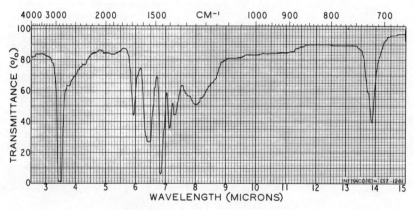

Spectrum 6.20.   Ethylene/methacrylic acid (89·4 : 10·6) copolymer. Neutralised 56%
Hot-pressed film

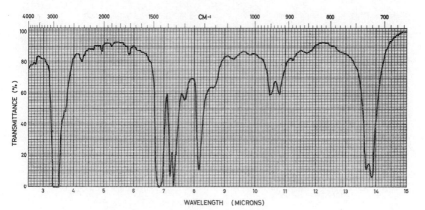

Spectrum 6.21.   Polythene/polyisobutene (75 : 25) blend
Hot-pressed film

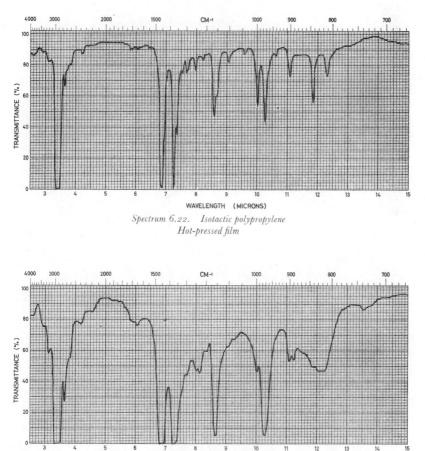

Spectrum 6.22.   Isotactic polypropylene
Hot-pressed film

Spectrum 6.23.   'Atactic' polypropylene
Layer between NaCl plates

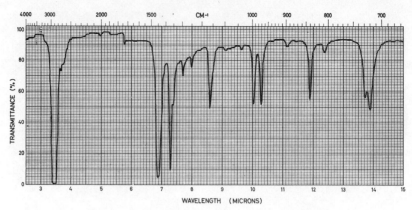

Spectrum 6.24.   Polypropylene/polythene (50 : 50) blend
Hot-pressed film

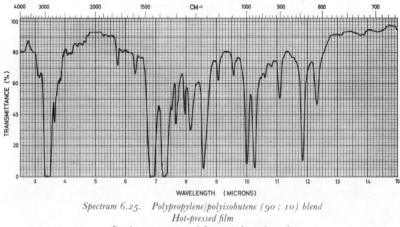

Spectrum 6.25.   Polypropylene/polyisobutene (90 : 10) blend
Hot-pressed film
Bands at 5·75 μm and 6·31 μm due to impurity

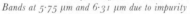

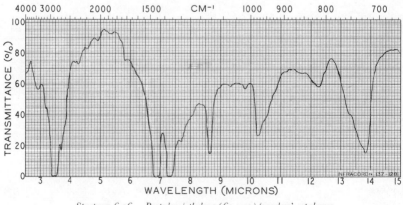

Spectrum 6.26.   Propylene/ethylene (60 : 40) 'random' copolymer
Hot-pressed film

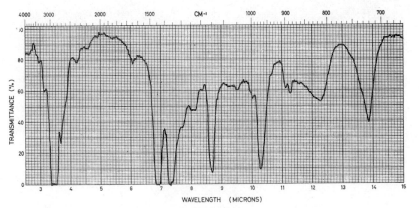

Spectrum 6.27. *Propylene/ethylene (80 : 20) copolymer*
Layer between NaCl plates

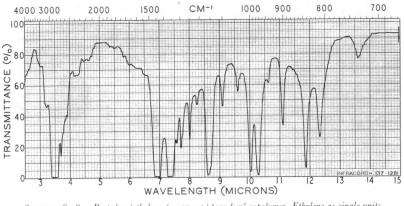

Spectrum 6.28. *Propylene/ethylene (~97 : 3) 'random' copolymer. Ethylene as single units*
Hot-pressed film

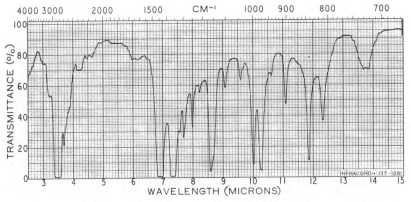

Spectrum 6.29. *Propylene/ethylene (~91 : 9) 'random' copolymer. Ethylene in short runs*
Hot-pressed film

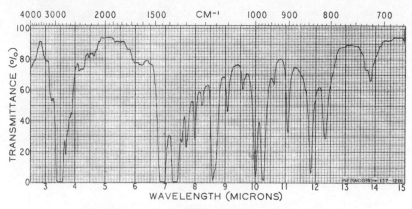

Spectrum 6.30.   Propylene/ethylene (~ 93 : 7) 'block' copolymer
Hot-pressed film

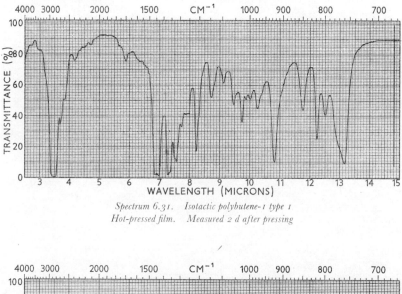

Spectrum 6.31.   Isotactic polybutene-1 type 1
Hot-pressed film.   Measured 2 d after pressing

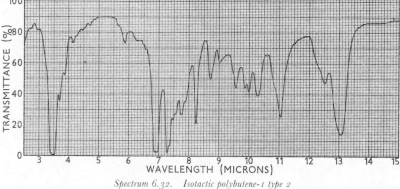

Spectrum 6.32.   Isotactic polybutene-1 type 2
Hot-pressed film. Measured within $\frac{1}{2}$ h of pressing

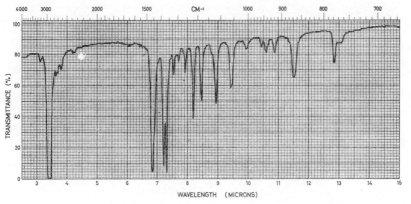

Spectrum 6.33.   *Isotactic poly-3-methyl butene-1*
Hot-pressed film

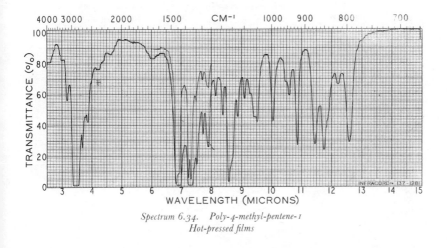

Spectrum 6.34.   *Poly-4-methyl-pentene-1*
Hot-pressed films

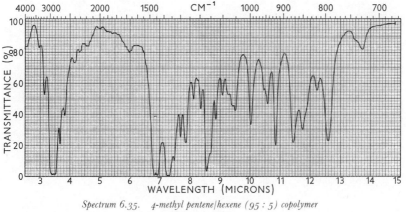

Spectrum 6.35.   *4-methyl pentene/hexene (95 : 5) copolymer*
Hot-pressed film

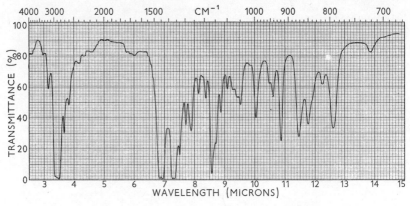

Spectrum 6.36.   4-methyl pentene/decene (96 : 4) copolymer
Hot-pressed film

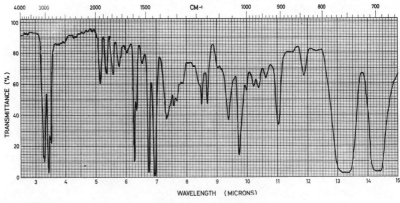

Spectrum 6.37.   Polystyrene
Film on NaCl plate from ethylene dichloride solution

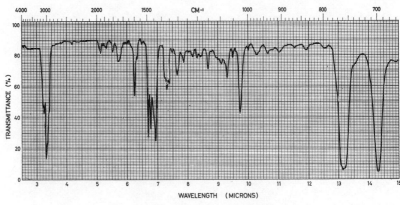

Spectrum 6.38.   Poly-α-methyl styrene
Layer between NaCl plates

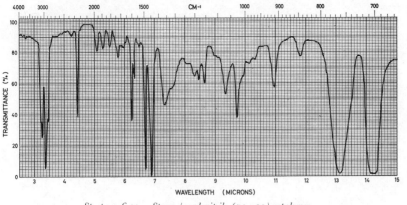

Spectrum 6.39. *Styrene/acrylonitrile (70 : 30) copolymer*
Hot-pressed film

FLUOROCARBON POLYMERS AND POLYMER EMULSIONS

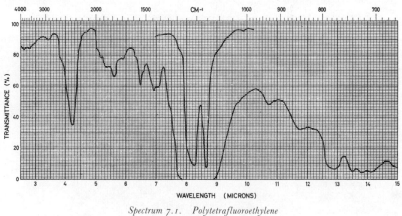

Spectrum 7.1. Polytetrafluoroethylene
Thick layer: microtomed layer
Thin layer: film on AgCl plate cast from aqueous dispersion

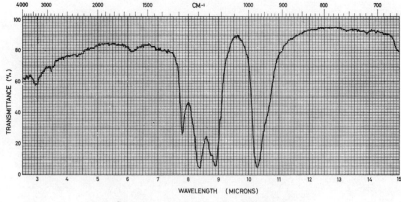

Spectrum 7.2(a). Polytrifluorochloroethylene
Suspension in KBr disk

664

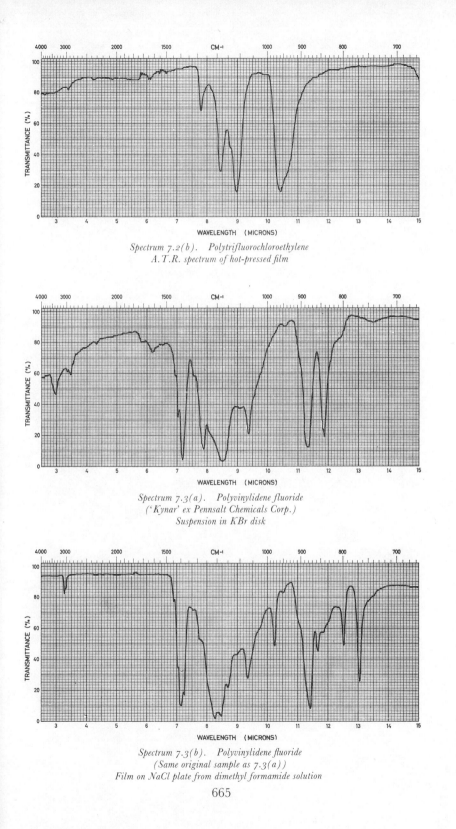

Spectrum 7.2(b). Polytrifluorochloroethylene
A.T.R. spectrum of hot-pressed film

Spectrum 7.3(a). Polyvinylidene fluoride
('Kynar' ex Pennsalt Chemicals Corp.)
Suspension in KBr disk

Spectrum 7.3(b). Polyvinylidene fluoride
(Same original sample as 7.3(a))
Film on NaCl plate from dimethyl formamide solution

665

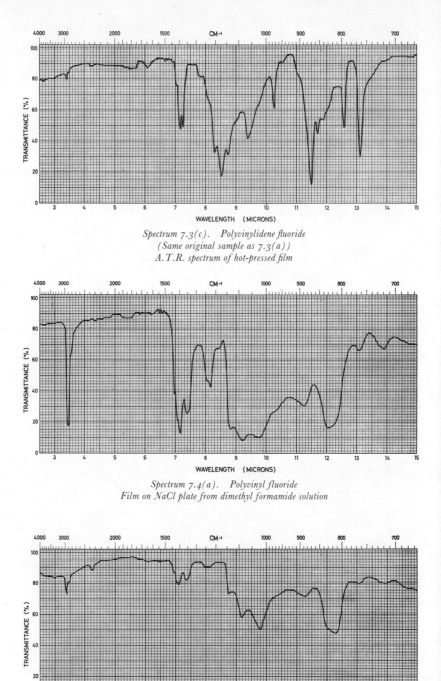

Spectrum 7.3(c).   Polyvinylidene fluoride
(Same original sample as 7.3(a))
A.T.R. spectrum of hot-pressed film

Spectrum 7.4(a).   Polyvinyl fluoride
Film on NaCl plate from dimethyl formamide solution

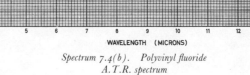

Spectrum 7.4(b).   Polyvinyl fluoride
A.T.R. spectrum

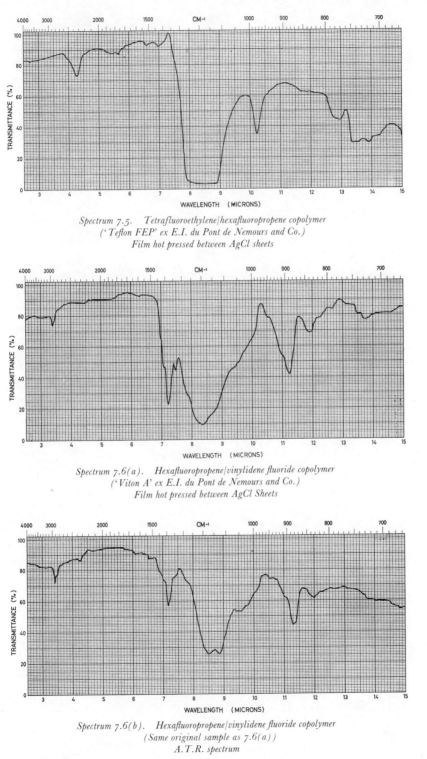

*Spectrum 7.5. Tetrafluoroethylene/hexafluoropropene copolymer
('Teflon FEP' ex E.I. du Pont de Nemours and Co.)
Film hot pressed between AgCl sheets*

*Spectrum 7.6(a). Hexafluoropropene/vinylidene fluoride copolymer
('Viton A' ex E.I. du Pont de Nemours and Co.)
Film hot pressed between AgCl Sheets*

*Spectrum 7.6(b). Hexafluoropropene/vinylidene fluoride copolymer
(Same original sample as 7.6(a))
A.T.R. spectrum*

667

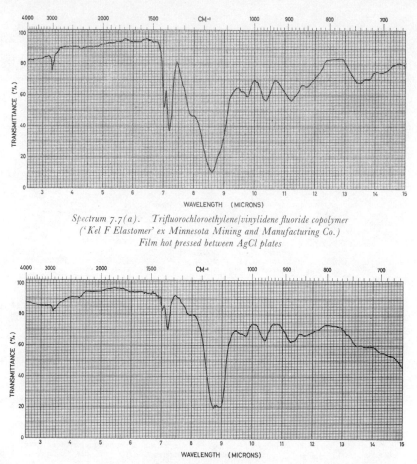

Spectrum 7.7(a). Trifluorochloroethylene/vinylidene fluoride copolymer ('Kel F Elastomer' ex Minnesota Mining and Manufacturing Co.) Film hot pressed between AgCl plates

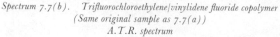

Spectrum 7.7(b). Trifluorochloroethylene/vinylidene fluoride copolymer (Same original sample as 7.7(a)) A.T.R. spectrum

668

RUBBER-LIKE RESINS

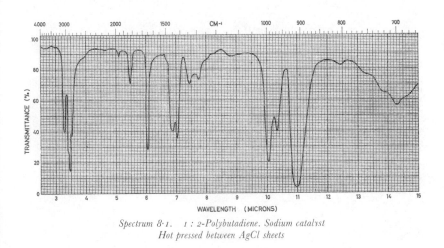

Spectrum 8·1.   1 : 2-Polybutadiene. Sodium catalyst
Hot pressed between AgCl sheets

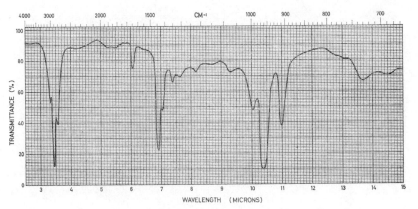

Spectrum 8.2.   1 : 4-Polybutadiene. Emulsion polymer with peroxide catalyst
Hot pressed between AgCl sheets

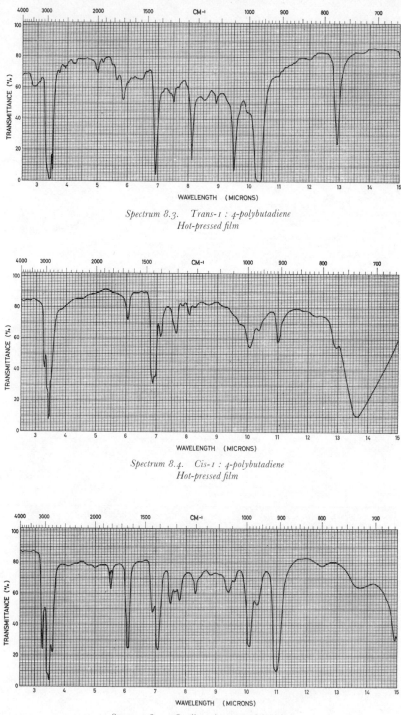

Spectrum 8.3.   Trans-1 : 4-polybutadiene
Hot-pressed film

Spectrum 8.4.   Cis-1 : 4-polybutadiene
Hot-pressed film

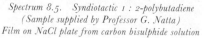
Spectrum 8.5.   Syndiotactic 1 : 2-polybutadiene
(Sample supplied by Professor G. Natta)
Film on NaCl plate from carbon bisulphide solution

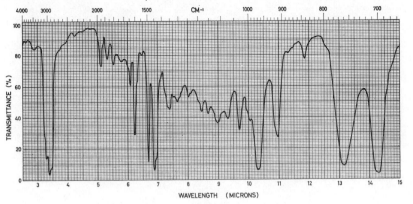

Spectrum 8.6. Styrene/butadiene (62 : 38) copolymer
Hot-pressed film

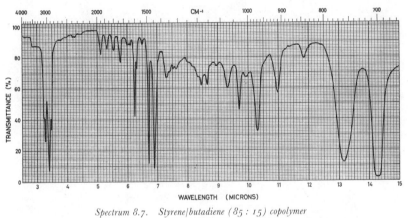

Spectrum 8.7. Styrene/butadiene (85 : 15) copolymer
Hot-pressed film

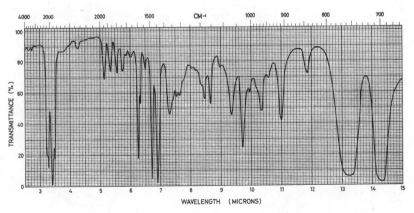

Spectrum 8.8. Styrene/butadiene (95 : 5) copolymer
Hot-pressed film

671

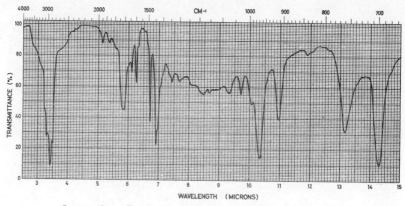

Spectrum 8.9.   Butadiene/styrene/carboxylic acid (50 : 45 : 5) terpolymer
Film on AgCl plate cast from latex

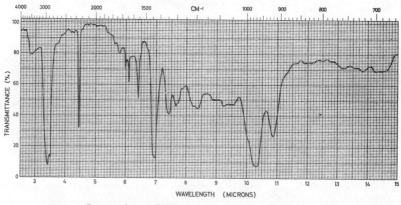

Spectrum 8.10.   Butadiene/acrylonitrile (65 : 35) copolymer
Film on AgCl plate cast from latex

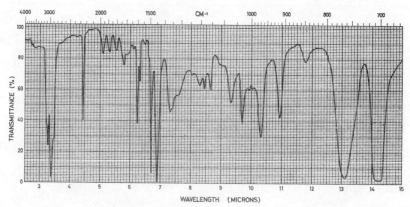

Spectrum 8.11.   Styrene/butadiene-butadiene/acrylonitrile blend
(Composition : styrene/acrylonitrile/butadiene (65 : 25 : 10))
Hot-pressed film

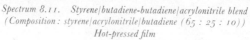

672

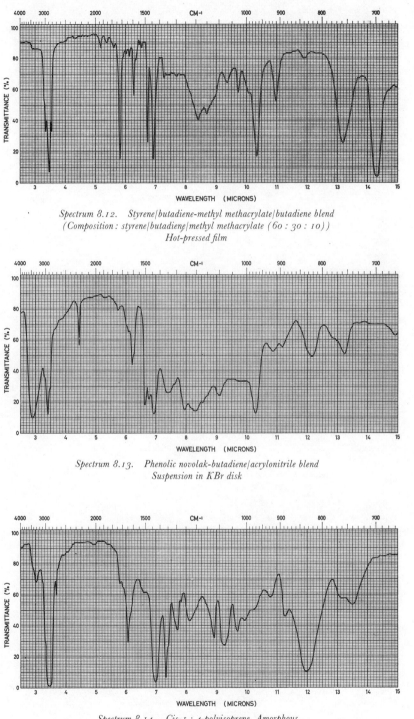

Spectrum 8.12. Styrene/butadiene-methyl methacrylate/butadiene blend
(Composition: styrene/butadiene/methyl methacrylate (60 : 30 : 10))
Hot-pressed film

Spectrum 8.13. Phenolic novolak-butadiene/acrylonitrile blend
Suspension in KBr disk

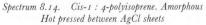

Spectrum 8.14. Cis-1 : 4-polyisoprene. Amorphous
Hot pressed between AgCl sheets

673

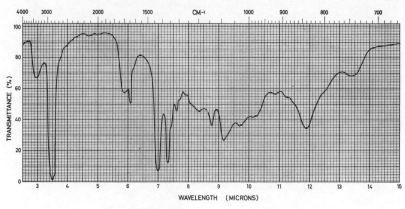

Spectrum 8.15. *Trans-1 : 4- polyisoprene. Amorphous*
Hot pressed between AgCl sheets and quenched

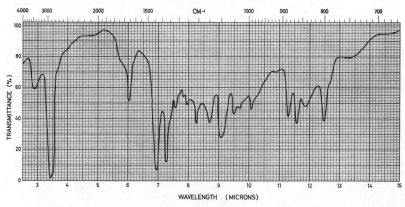

Spectrum 8.16. *Trans-1 : 4-polyisoprene. α-crystalline form*
Film on NaCl plate from benzene solution, solvent evaporated in vacuo at room temperature

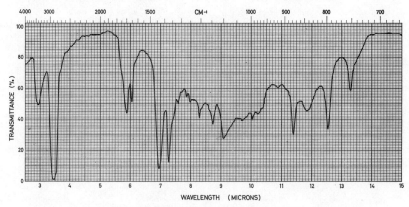

Spectrum 8.17. *Trans-1 : 4-polyisoprene. β-crystalline form*
Film as 8.16, heated to 70°C and cooled slowly

674

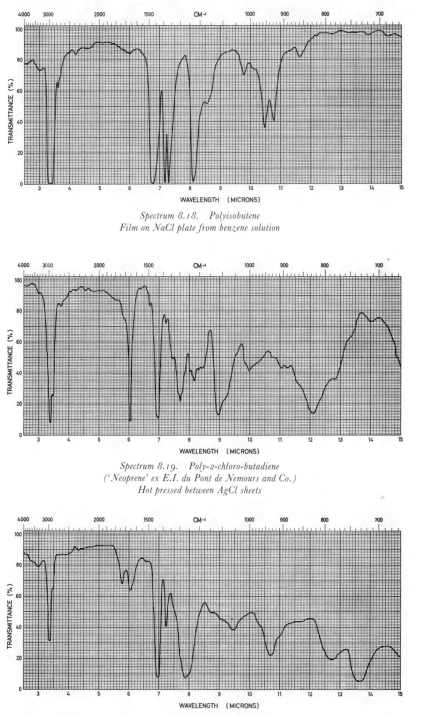

Spectrum 8.18.   Polyisobutene
Film on NaCl plate from benzene solution

Spectrum 8.19.   Poly-2-chloro-butadiene
('Neoprene' ex E.I. du Pont de Nemours and Co.)
Hot pressed between AgCl sheets

Spectrum 8.20.   Chlorinated rubber
Film on NaCl plate from benzene solution

675

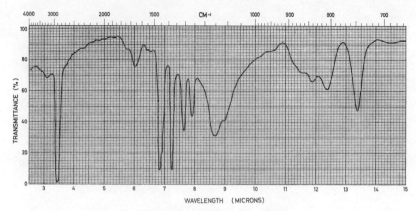

Spectrum 8.21.   Rubber hydrochloride
Film on NaCl plate from benzene solution

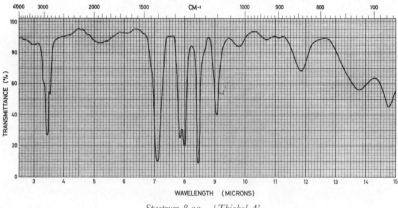

Spectrum 8.22.   'Thiokol A'
Hot pressed between AgCl sheets

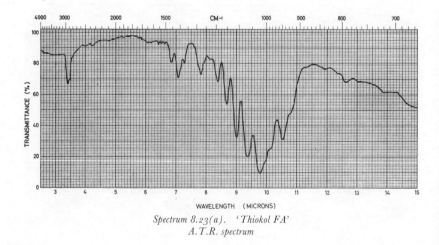

Spectrum 8.23(a).   'Thiokol FA'
A.T.R. spectrum

676

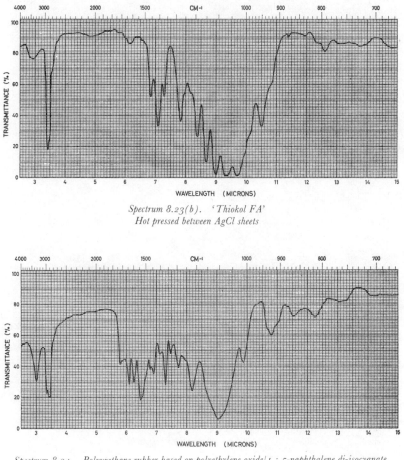

Spectrum 8.23(b). 'Thiokol FA'
Hot pressed between AgCl sheets

Spectrum 8.24. Polyurethane rubber based on polyethylene oxide/1 : 5-naphthalene di-isocyanate
Suspension in KBr disk prepared under liquid nitrogen

677

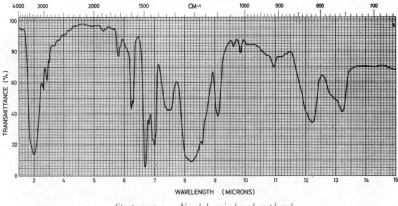

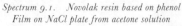

Spectrum 9.1.   Novolak resin based on phenol
Film on NaCl plate from acetone solution

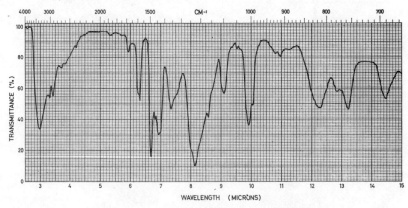

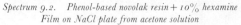

Spectrum 9.2.   Phenol-based novolak resin + 10% hexamine
Film on NaCl plate from acetone solution

678

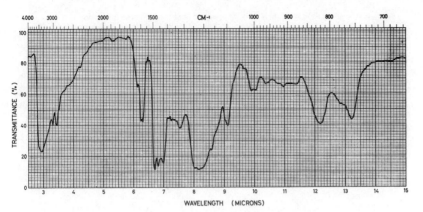

Spectrum 9.3.   Phenol-based novolak resin cross-linked with hexamine
Suspension in KBr disk

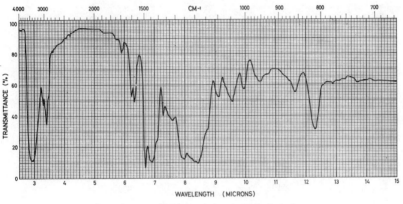

Spectrum 9.4.   Novolak resin based on mixed cresols
Film on NaCl plate from acetone solution

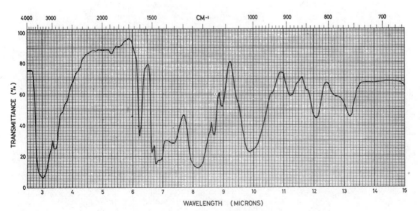

Spectrum 9.5.   Resole resin based on phenol
Film on AgCl plate from aqueous solution, solvent evaporated in vacuo at room temperature

679

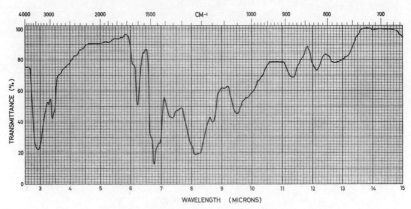

Spectrum 9.6.   Phenol-based resole resin, cross-linked by heating
Film as 9.5, heated 2 h at 150°C

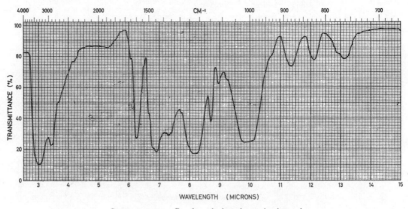

Spectrum 9.7.   Resole resin based on mixed cresols
Film on AgCl plate from aqueous solution, solvent evaporated in vacuo at room temperature

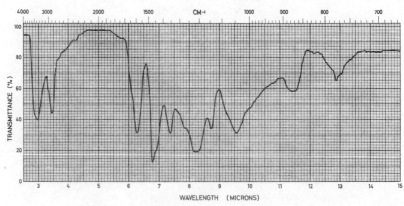

Spectrum 9.8.   Cresol-based resole resin, cross-linked by heating
Film as 9.7, heated 2 h at 150°C

680

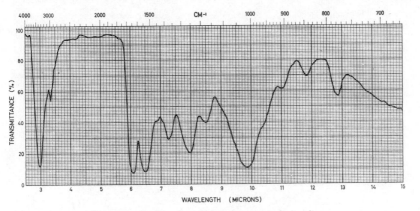

Spectrum 9.9.   Urea-formaldehyde (1 : 2·2, molar) resin
Film on NaCl plate from acetone solution, solvent evaporated at room temperature

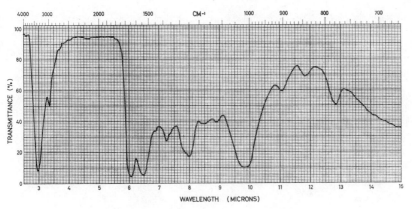

Spectrum 9.10.   Urea-formaldehyde (1 : 1·6, molar) resin
Film on NaCl plate from acetone solution, solvent evaporated at room temperature

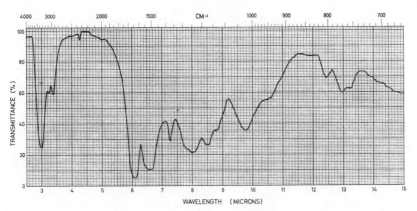

Spectrum 9.11.   Urea-formaldehyde resin, cross-linked by heating
Film as 9.10, heated 2 h at 160°C

681

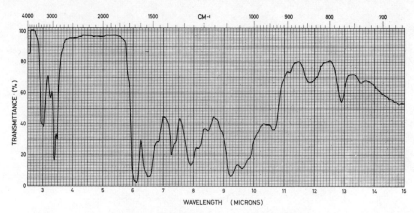

Spectrum 9.12. 'Butylated' urea-formaldehyde resin
Film on NaCl plate from solution as received

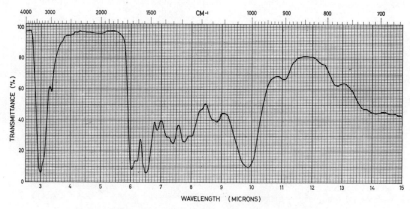

Spectrum 9.13. Urea-thiourea-formaldehyde (1 : 1 : 3, molar) resin
Film on NaCl plate from acetone solution, solvent evaporated at room temperature

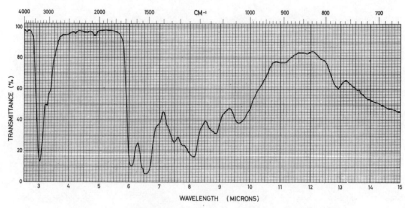

Spectrum 9.14. Urea-thiourea-formaldehyde resin cross-linked by heating
Film as 9.13, heated 2 h at 160°C

682

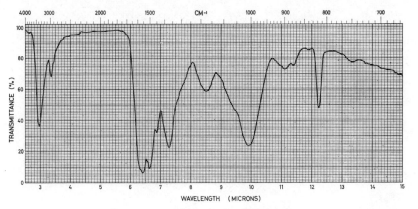

Spectrum 9.15.   Melamine-formaldehyde resin
Film on NaCl plate from acetone solution, solvent evaporated at room temperature

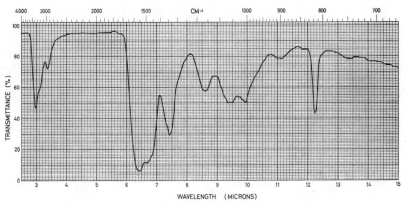

Spectrum 9.16.   Melamine-formaldehyde resin, cross-linked by heating
Film as 9.15, heated 1 h at 160°C

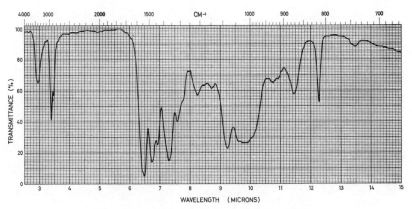

Spectrum 9.17.   'Butylated' melamine-formaldehyde resin
Film on NaCl plate from solution as received

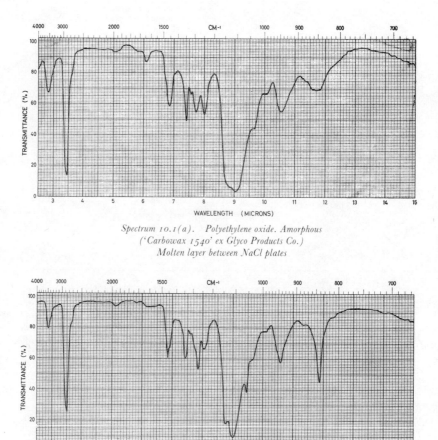

Spectrum 10.1(a). Polyethylene oxide. Amorphous
('Carbowax 1540' ex Glyco Products Co.)
Molten layer between NaCl plates

Spectrum 10.1(b). Polyethylene oxide. Crystalline
Sample as 10.1(a). Crystallised from melt

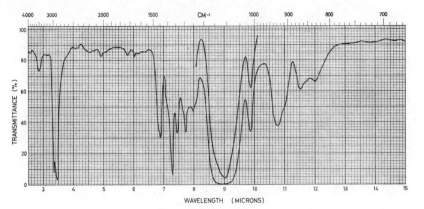

*Spectrum 10.2. Polypropylene oxide
Capillary film between NaCl plates*

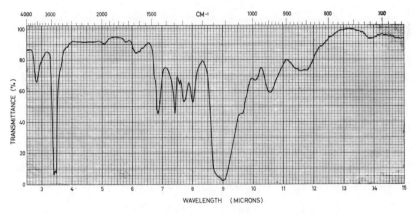

*Spectrum 10.3(a). Polyethylene oxide/cetyl alcohol condensate. Amorphous
Molten layer between NaCl plates*

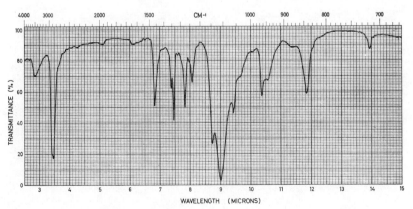

*Spectrum 10.3(b). Polyethylene oxide/cetyl alcohol condensate. Crystalline
Sample as 10.3(a). Crystallised from melt*

685

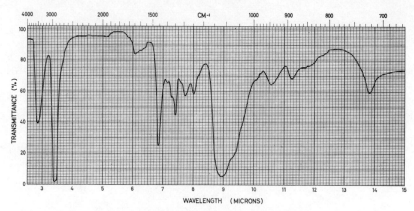

Spectrum 10.4.   Polyethylene oxide/oleyl alcohol condensate
Capillary film between NaCl plates

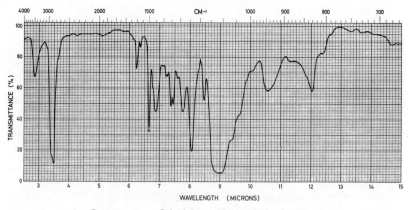

Spectrum 10.5.   Polyethylene oxide/p-octyl phenol condensate
Capillary film between NaCl plates

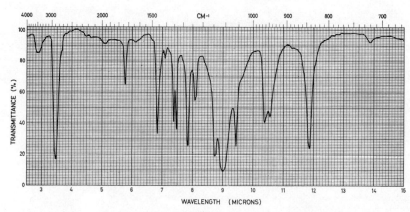

Spectrum 10.6.   Polyethylene oxide, stearate terminated
Capillary layer (crystalline) between NaCl plates

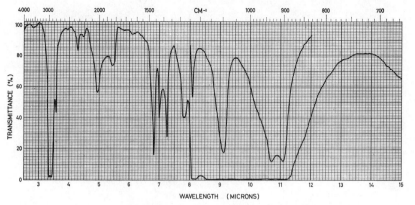

Spectrum 10.7.   Polyformaldehyde
Thick layer : cold pressed from powder in die used for making KBr disks
Thin layer : suspension in KBr disk

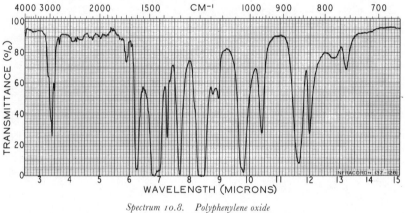

Spectrum 10.8.   Polyphenylene oxide
Film cast from chloroform solution

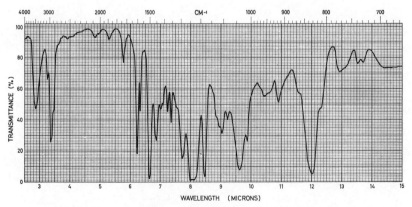

Spectrum 10.9.   Epoxy resin from epichlorhydrin and Bisphenol A
Film on NaCl plate from acetone solution

687

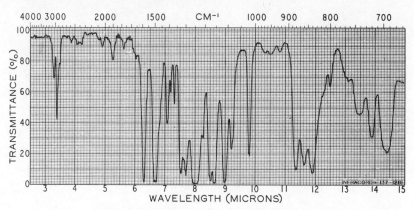

Spectrum 10.10. *Polysulphone from Bisphenol A and 4 : 4'-dihydroxydiphenol sulphone*
Film on NaCl plate from chloroform solution

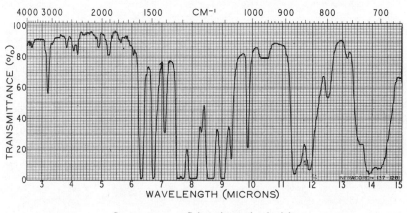

Spectrum 10.11. *Poly-p-phenoxyphenyl sulphone*
Hot-pressed film

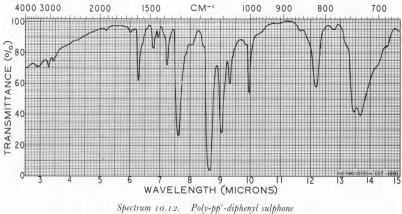

Spectrum 10.12. *Poly-pp'-diphenyl sulphone*
Suspension in KBr disk

688

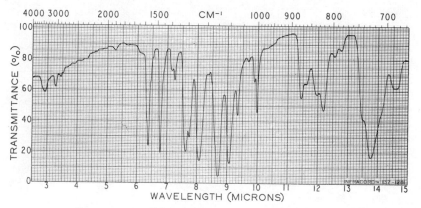

Spectrum 10.13. Poly(pp'-diphenylene ether-p-phenylene sulphone)
Suspension in KBr disk

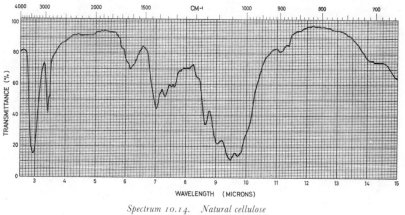

Spectrum 10.14. Natural cellulose
Suspension in KBr disk

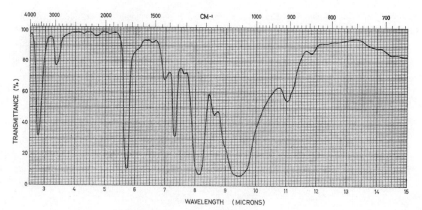

Spectrum 10.15. Cellulose monoacetate
Film from ethylene dichloride solution

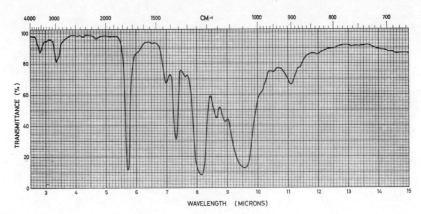

*Spectrum 10.16. Cellulose triacetate
Film from formic acid solution*

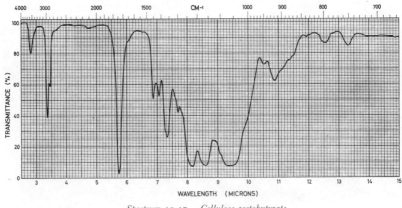

*Spectrum 10.17. Cellulose acetobutyrate
Film from ethylene dichloride solution*

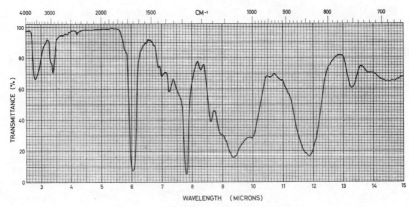

*Spectrum 10.18. Cellulose nitrate
Film from ethylene dichloride solution*

690

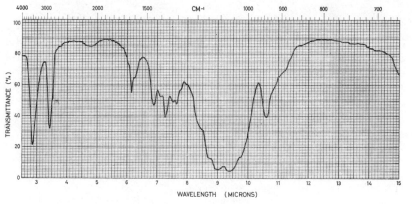

Spectrum 10.19.   Methyl cellulose
Suspension in KBr disk

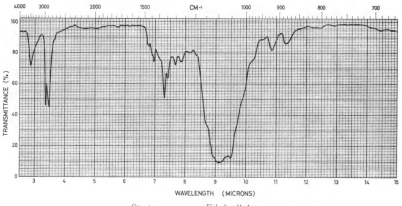

Spectrum 10.20.   Ethyl cellulose
Film on NaCl plate from ethylene dichloride solution

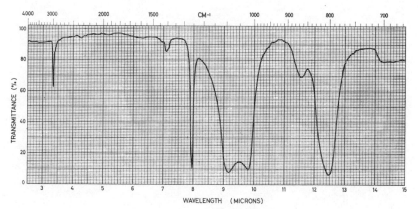

Spectrum 10.21.   Methyl silicone oil
Capillary layer between NaCl plates

691

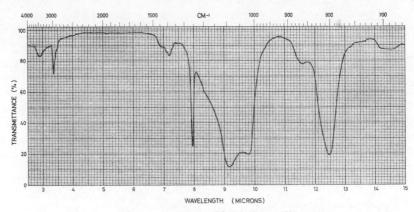

Spectrum 10.22(a).   Methyl silicone rubber
Suspension in KBr disk prepared under liquid nitrogen

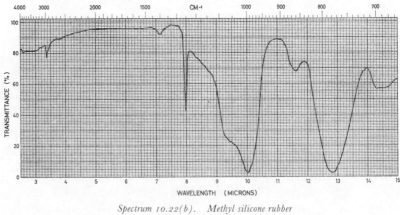

Spectrum 10.22(b).   Methyl silicone rubber
A.T.R. spectrum

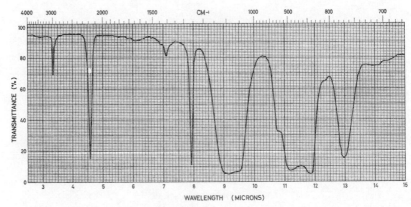

Spectrum 10.23.   Methyl hydrogen silicone oil
Capillary layer between NaCl plates

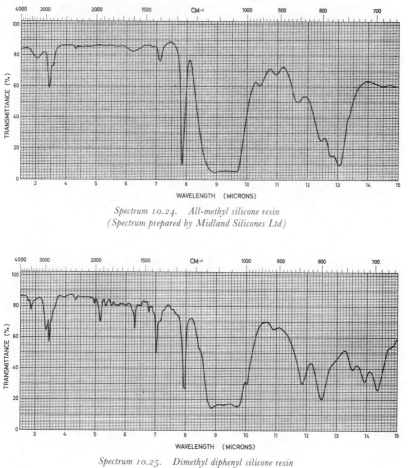

Spectrum 10.24.   All-methyl silicone resin
(Spectrum prepared by Midland Silicones Ltd)

Spectrum 10.25.   Dimethyl diphenyl silicone resin
(Spectrum prepared by Midland Silicones Ltd)

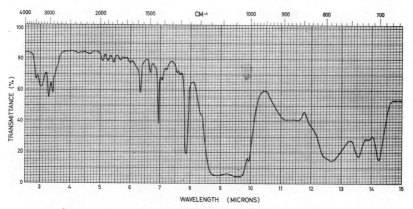

Spectrum 10.26.   Methyl phenyl silicone resin, containing Si-OH groups
(Spectrum prepared by Midland Silicones Ltd)

693

PLASTICISERS

Note *Except where otherwise indicated all plasticiser spectra were obtained on capillary layers between rock-salt plates.*

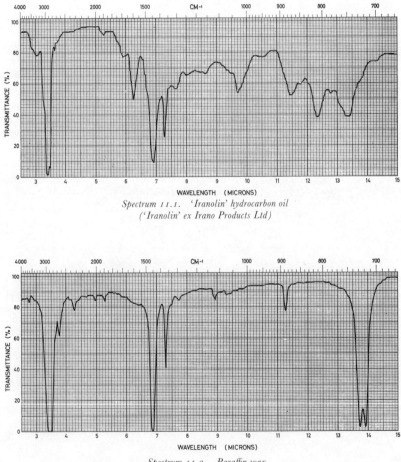

Spectrum 11.1. 'Iranolin' hydrocarbon oil
('Iranolin' ex Irano Products Ltd)

Spectrum 11.2. Paraffin wax
Crystallised from melt between NaCl plates

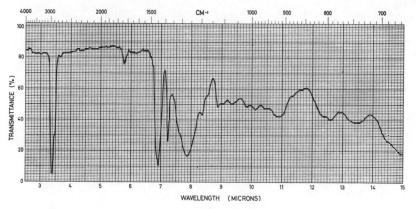

Spectrum 11.3. Chlorinated paraffin wax
('Cereclor 42' ex Imperial Chemical Industries Ltd)

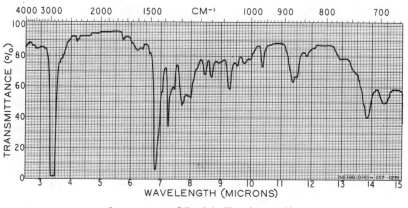

Spectrum 11.4. Dibutyl tin dilauryl mercaptide
(See also Spectra 11.41 and 11.44)

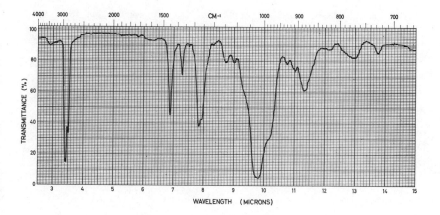

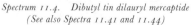

Spectrum 11.5. Trioctyl phosphate

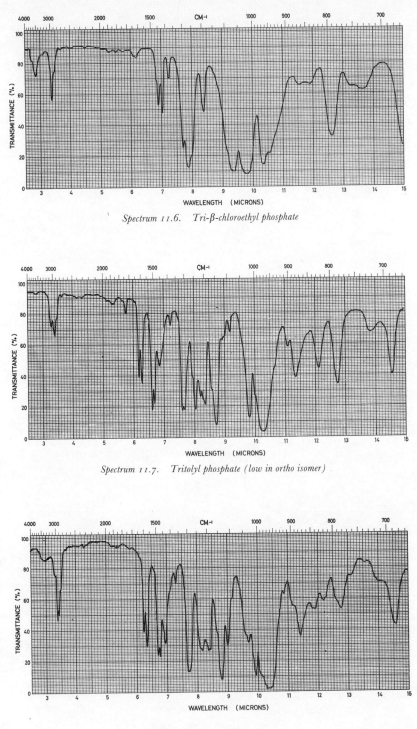

Spectrum 11.6.   Tri-β-chloroethyl phosphate

Spectrum 11.7.   Tritolyl phosphate (low in ortho isomer)

Spectrum 11.8.   Trixylyl phosphate

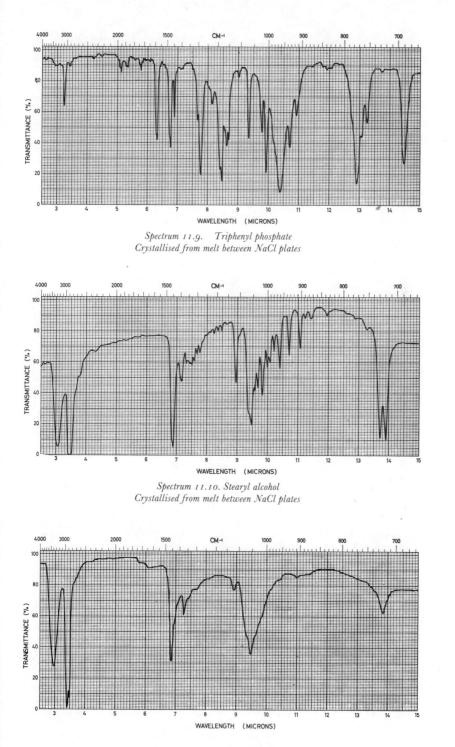

Spectrum 11.9.   Triphenyl phosphate
Crystallised from melt between NaCl plates

Spectrum 11.10. Stearyl alcohol
Crystallised from melt between NaCl plates

Spectrum 11.11.   Lauryl alcohol

697

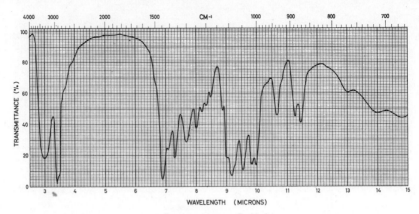

Spectrum 11.12.   Sorbitol
Suspension in liquid paraffin between NaCl plates

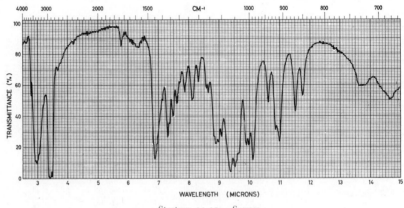

Spectrum 11.13.   Sucrose
Suspension in liquid paraffin between NaCl plates

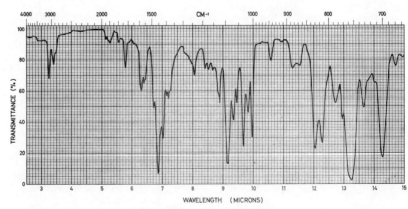

Spectrum 11.14.   Chlorinated biphenyl
('Aroclor 1232' ex Monsanto Chemical Co.)

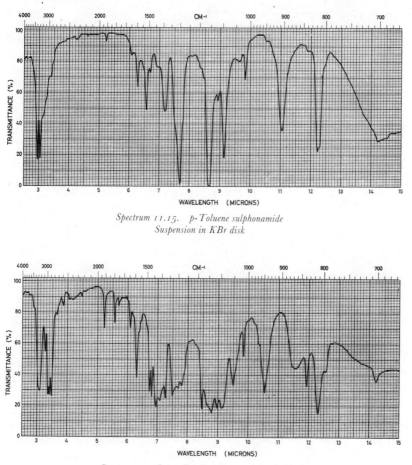

Spectrum 11.15. p-Toluene sulphonamide
Suspension in KBr disk

Spectrum 11.16. n-Ethyl p-toluene sulphonamide
Molten layer between NaCl plates

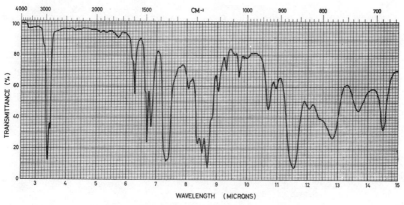

Spectrum 11.17. Mesamoll
('Mesamoll' ex Bayer A.G.)

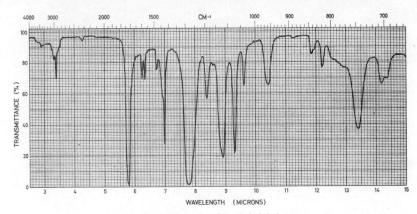

Spectrum 11.18.   Dimethyl phthalate

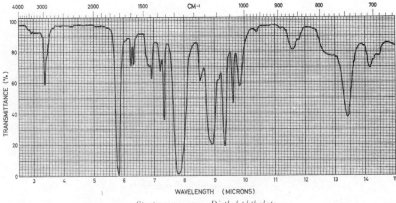

Spectrum 11.19.   Diethyl phthalate

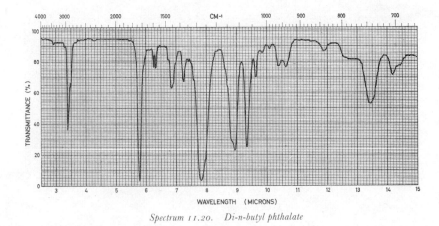

Spectrum 11.20.   Di-n-butyl phthalate

700

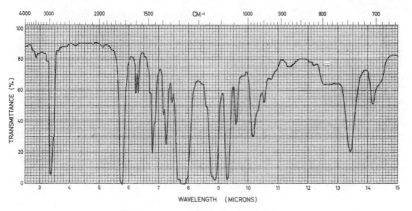

Spectrum *11.21.* *Di-iso-butyl phthalate*

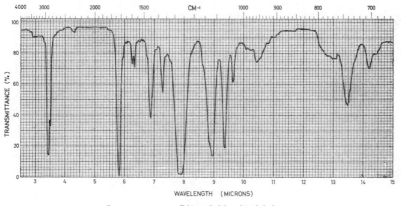

Spectrum *11.22.* *Di(2-ethyl hexyl) phthalate*

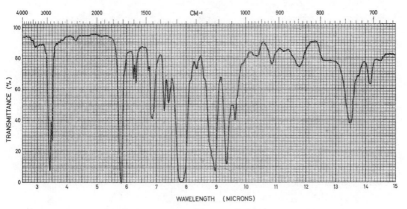

Spectrum *11.23.* *Di-sec-octyl phthalate*

701

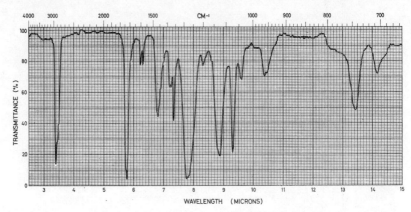

Spectrum 11.24.   Di(3 : 5 : 5-trimethyl hexyl) phthalate

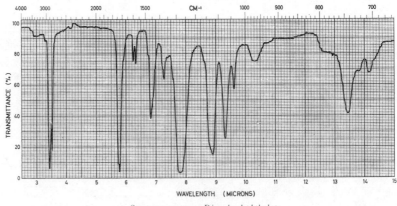

Spectrum 11.25.   Di-n-decyl phthalate

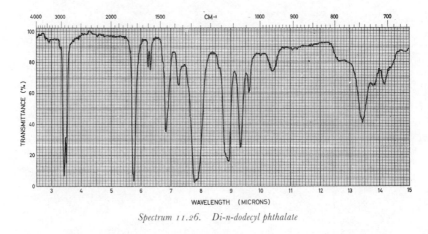

Spectrum 11.26.   Di-n-dodecyl phthalate

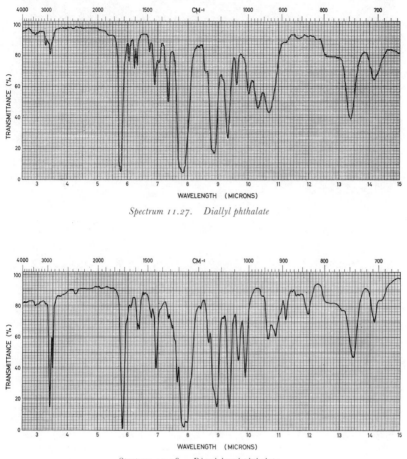

Spectrum *11.27.* *Diallyl phthalate*

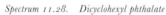

Spectrum *11.28.* *Dicyclohexyl phthalate*

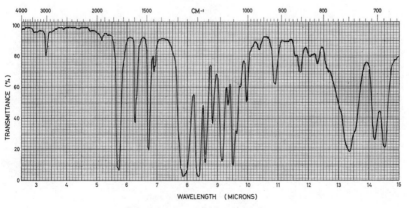

Spectrum *11.29.* *Diphenyl phthalate*

703

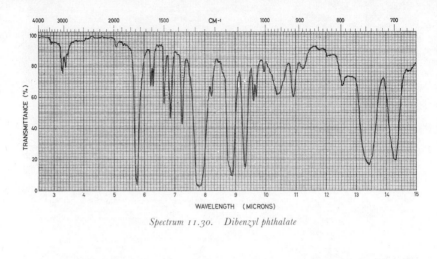

Spectrum 11.30. Dibenzyl phthalate

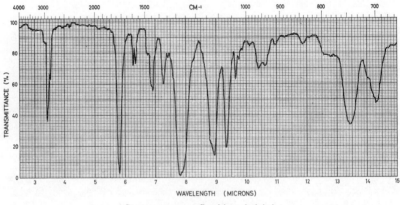

Spectrum 11.31. Butyl benzyl phthalate

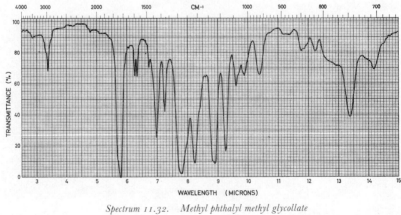

Spectrum 11.32. Methyl phthalyl methyl glycollate

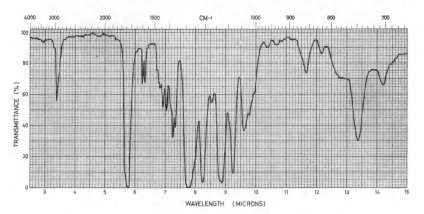

Spectrum 11.33. *Ethyl phthalyl ethyl glycollate*

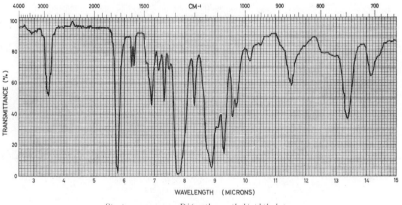

Spectrum 11.34. *Di(methoxy ethyl) phthalate*

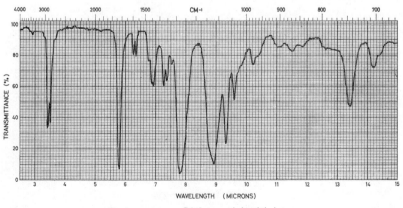

Spectrum 11.35. *Di(butoxy ethyl) phthalate*

705

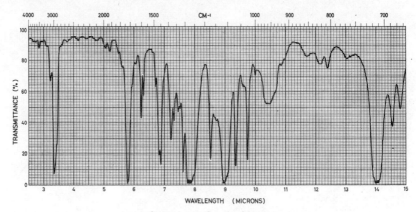

Spectrum 11.36. Amyl benzoate

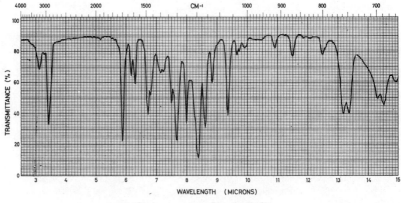

Spectrum 11.37. Phenyl salicylate
Suspension in liquid paraffin between NaCl plates

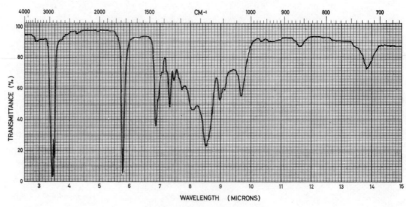

Spectrum 11.38. Ethyl palmitate

706

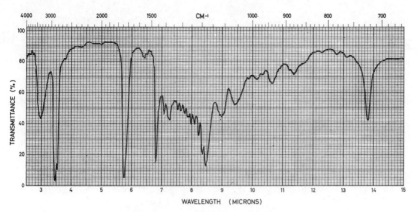

Spectrum 11.39.   Glyceryl distearate

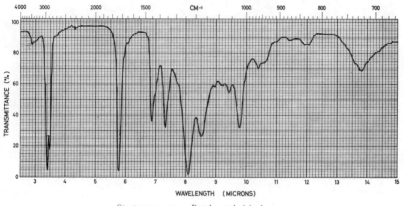

Spectrum 11.40.   Butyl acetyl ricinoleate

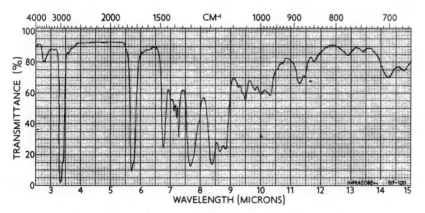

Spectrum 11.41.   Dibutyl tin dinonyl thioglycollate
(See also Spectra 11.4 and 11.44)

707

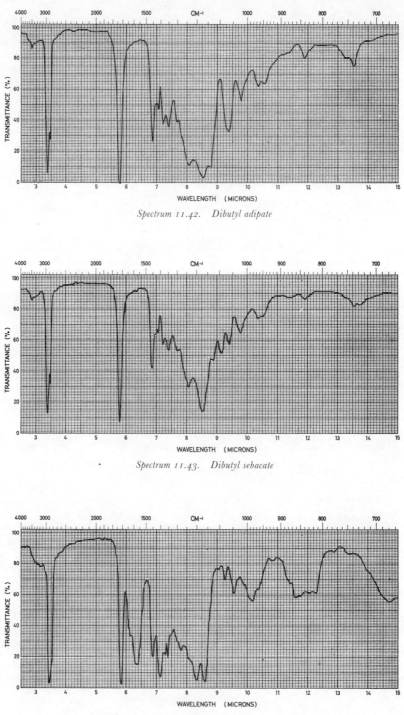

Spectrum 11.42.   Dibutyl adipate

Spectrum 11.43.   Dibutyl sebacate

Spectrum 11.44.   Dibutyl tin dinonyl maleate
(See also Spectra 11.4 and 11.41)

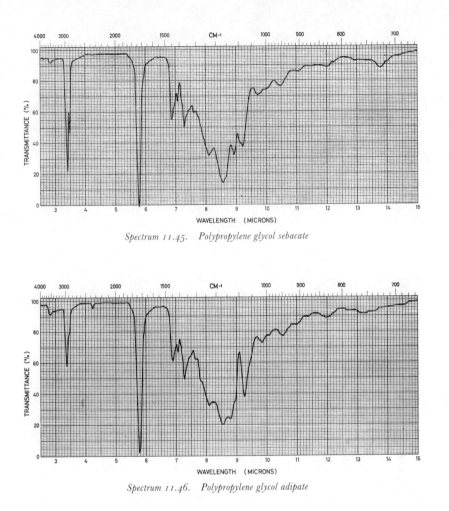

Spectrum 11.45. Polypropylene glycol sebacate

Spectrum 11.46. Polypropylene glycol adipate

709

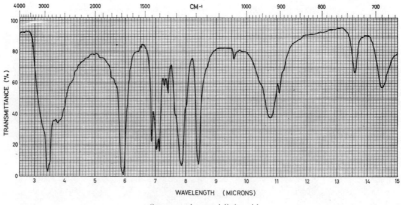

Spectrum A.1.   Adipic acid
Suspension in liquid paraffin between NaCl plates

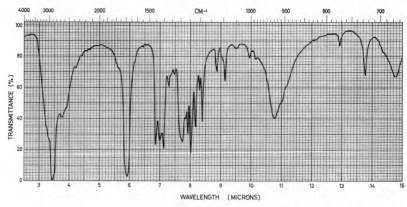

Spectrum A.2.   Azelaic acid
Suspension in liquid paraffin between NaCl plates

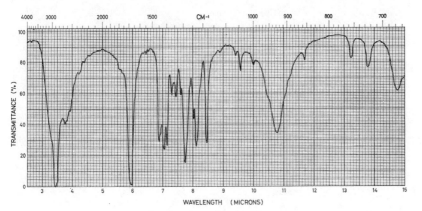

Spectrum A.3.   Sebacic acid
Suspension in liquid paraffin between NaCl plates

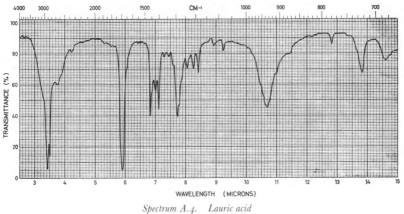

Spectrum A.4.   Lauric acid
Capillary layer crystallised from melt between NaCl plates

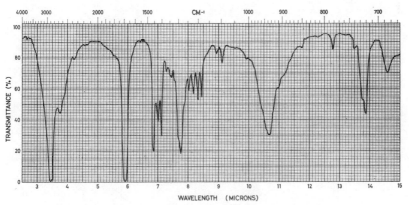

Spectrum A.5.   Palmitic acid
Capillary layer crystallised from melt between NaCl plates

711

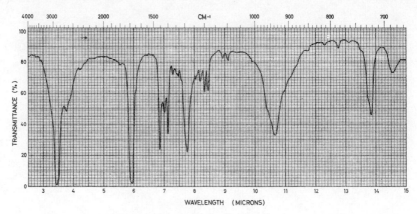

Spectrum A.6.   Stearic acid
Capillary layer crystallised from melt between NaCl plates

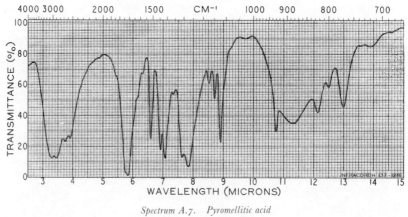

Spectrum A.7.   Pyromellitic acid
Suspension in KBr disk

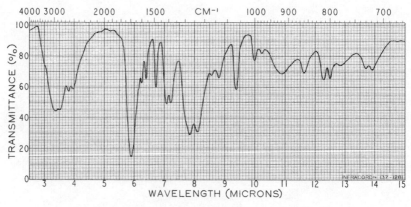

Spectrum A.8.   Benzophenone tetracarboxylic acid
Suspension in KBr disk

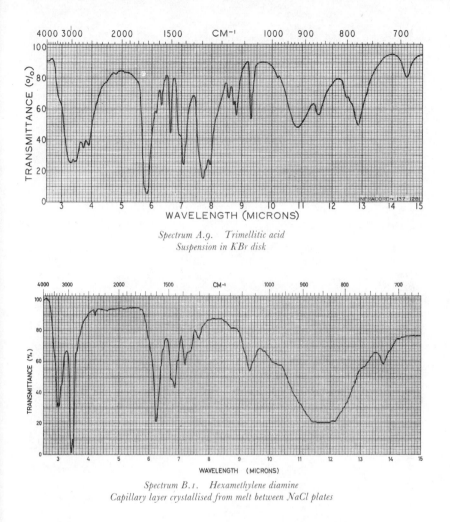

Spectrum A.9. Trimellitic acid
Suspension in KBr disk

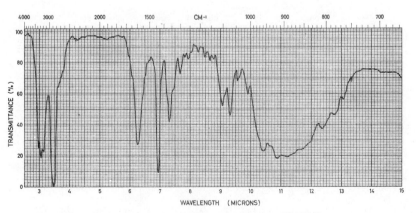

Spectrum B.1. Hexamethylene diamine
Capillary layer crystallised from melt between NaCl plates

Spectrum B.2. pp'-Diamino dicyclohexyl methane (PACM)

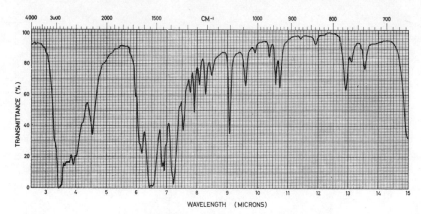

Spectrum B.3.   5-amino caproic acid
Suspension in liquid paraffin between NaCl plates

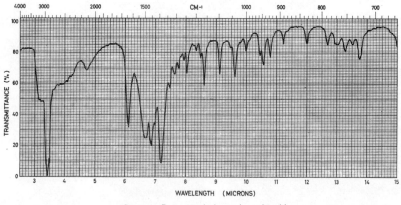

Spectrum B.4.   ω-Amino undecanoic acid
Suspension in liquid paraffin between NaCl plates

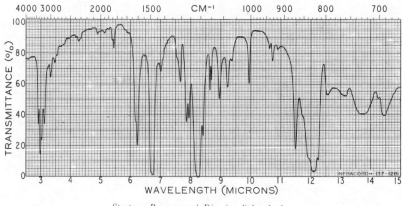

Spectrum B.5.   4 : 4'-Diamino diphenyl ether
1% suspension in KBr disk

714

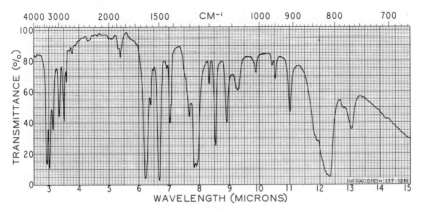

Spectrum B.6.   4 : 4'-Diamino diphenyl methane
1% Suspension in KBr disk

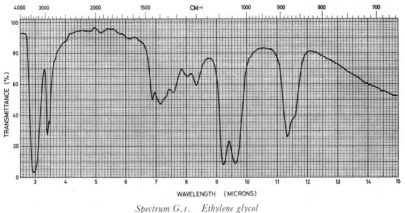

Spectrum G.1.   Ethylene glycol
Capillary layer between NaCl plates

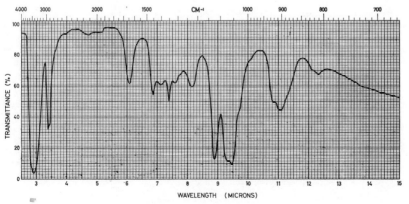

Spectrum G.2.   Diethylene glycol
Capillary layer between NaCl plates

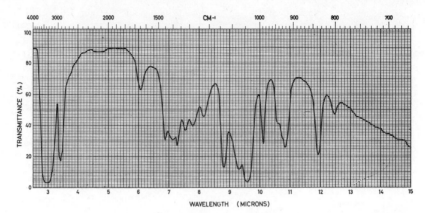

Spectrum G.3. *1 : 2-propane diol*
Capillary layer between NaCl plates

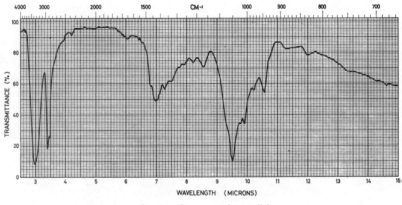

Spectrum G.4. *1 : 4-butane diol*
Capillary layer between NaCl plates

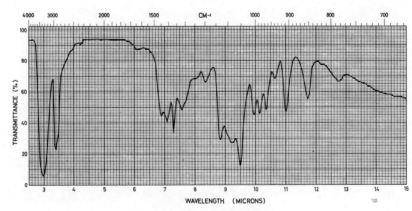

Spectrum G.5. *1 : 3-butane diol*
Capillary layer between NaCl plates

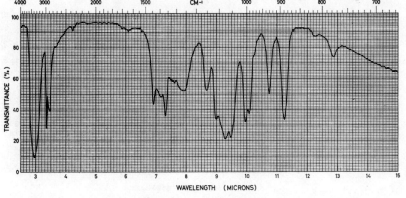

Spectrum G.6.   2 : 3-butane diol
Capillary layer between NaCl plates

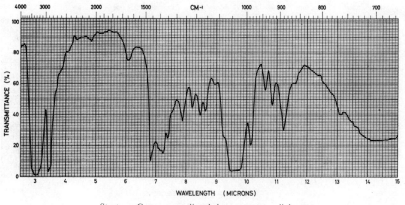

Spectrum G.7.   2 : 2-dimethyl 1 : 3-propane diol
Capillary layer between NaCl plates

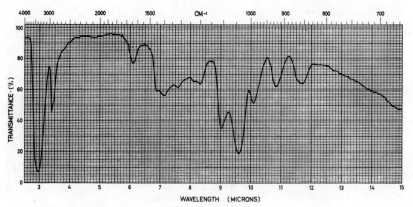

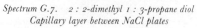

Spectrum G.8.   Glycerol
Capillary layer between NaCl plates

717

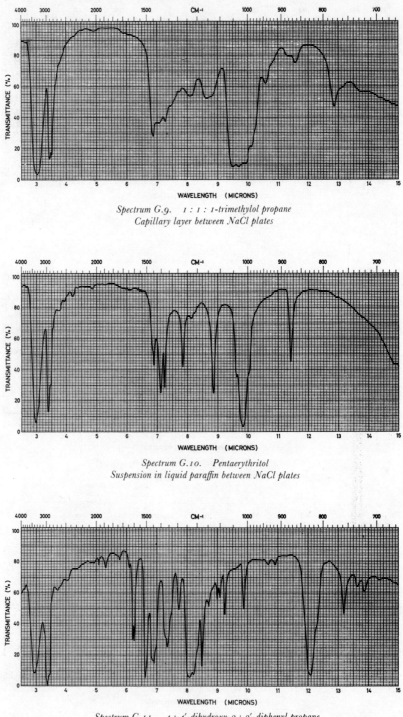

Spectrum G.9.   1 : 1 : 1-trimethylol propane
Capillary layer between NaCl plates

Spectrum G.10.   Pentaerythritol
Suspension in liquid paraffin between NaCl plates

Spectrum G.11.   4 : 4'-dihydroxy-2 : 2'-diphenyl propane
Suspension in liquid paraffin between NaCl plates

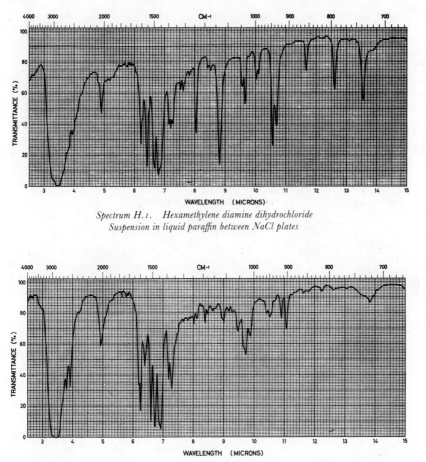

Spectrum H.1.   Hexamethylene diamine dihydrochloride
Suspension in liquid paraffin between NaCl plates

Spectrum H.2.   pp'-Diamino dicyclohexyl methane dihydrochloride (PACM dihydrochloride)
Suspension in liquid paraffin between NaCl plates

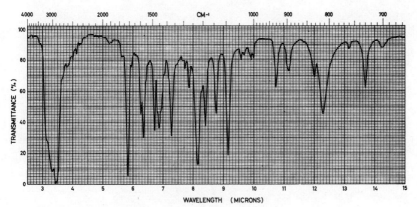

Spectrum H.3.   5-amino caproic acid hydrochloride
Suspension in liquid paraffin between NaCl plates

719

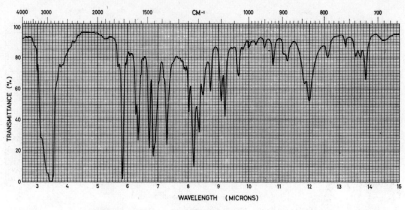

Spectrum H.4.   ω-amino undecanoic acid hydrochloride
Suspension in liquid paraffin between NaCl plates

INDEX TO INFRA-RED SPECTRA

721

726

727

INDEX TO N.M.R. SPECTRA

728

SUBJECT INDEX

730

735

747